Albatross
Biology and Conservation

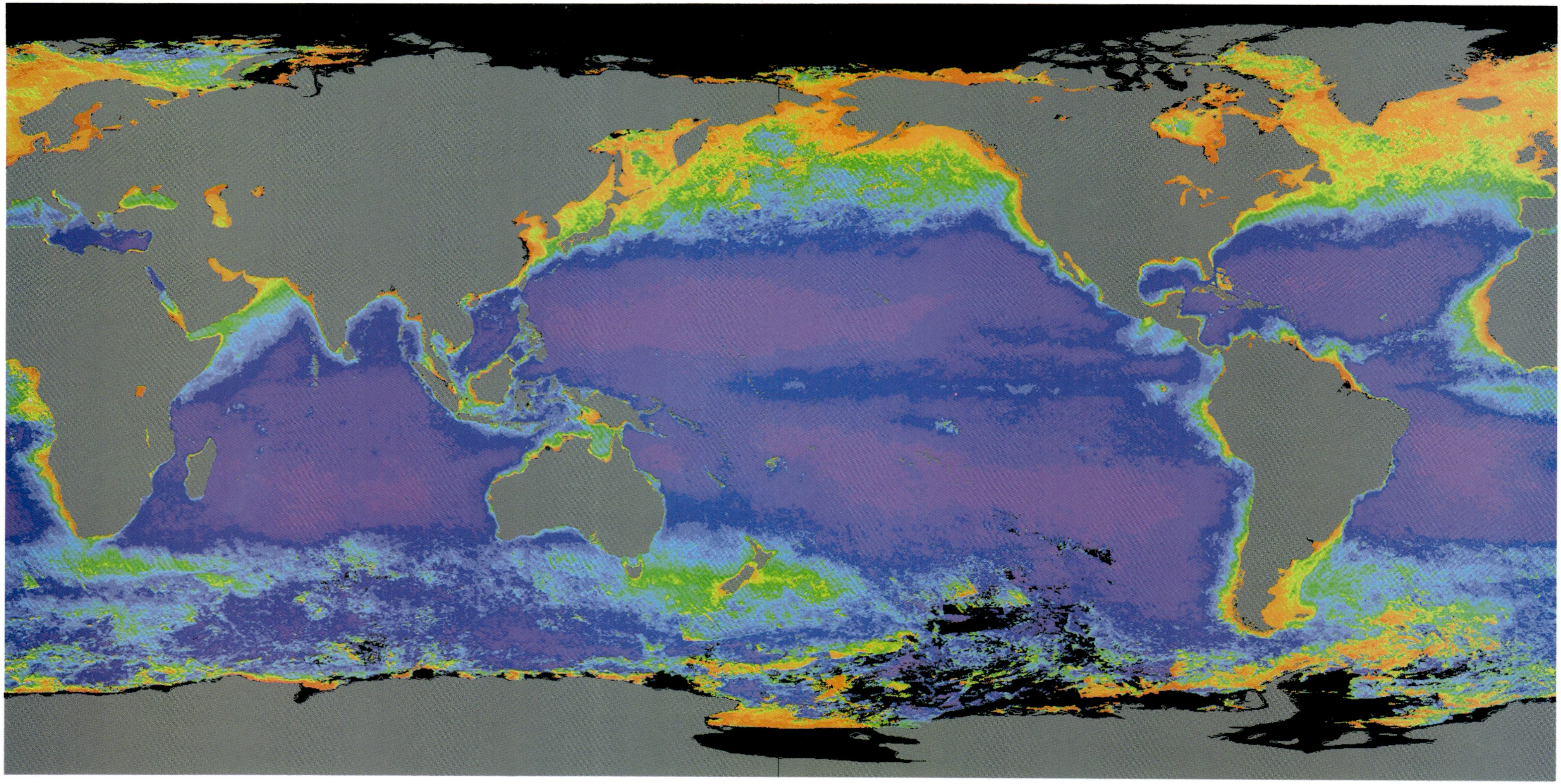

Satellite image of world oceans: zones of high primary productivity in oceans are reflected by the distribution of chlorophyll density. This is revealed by colour gradients, productivity being lowest (purple) in equatorial and mid-latitude regions and highest (yellow and red) in high latitudes and around continental shelves. Many chapters in this book describe albatross flights to areas of high productivity around continents, and to mixing zones of oceanic water masses. The mixing zones for sub-tropical and the cooler Arctic/Antarctic waters (transition between blue and purple) are frequented by albatrosses from islands in the Indian (Chapter 12) and North Pacific (Chapter 15) Oceans. Albatrosses from Japan traverse the North Pacific Ocean to forage in rich Alaskan waters (red and yellow) and albatrosses from the Galápagos Islands (Chapter 13) and New Zealand converge in rich upwelling regions off the Peruvian coast. South Georgia albatrosses (Chapter 11) forage extensively in Brazillian, Uruguayan and Argentine waters (Chapters 16, 17 and 18). These oceanic features of high productivity are also targeted by longline fisheries for tuna and various demersal fish species. The image was compiled from data collected by the Nimbus–7 satellite and was provided courtesy of the NASA Physical Oceanography Distribution Active Archive Centre at the Jet Propulsion Laboratory, California Institute of Technology, California, USA.

Albatross
Biology and Conservation

Edited by
Graham Robertson and Rosemary Gales

Published by
Surrey Beatty & Sons Pty Limited

The National Library of Australia
Cataloguing-in-Publication entry:

Albatross Biology and Conservation.

Includes index.
ISBN 0 949324 82 5.

1. Albatrosses. I. Robertson, Graham. II. Gales, Rosemary, 1960–

598.42

Published June 1998

PRINTED IN AUSTRALIA BY
SURREY BEATTY & SONS PTY LIMITED
43 Rickard Road, Chipping Norton, NSW 2170

This book is dedicated to the memory of
Peter Prince
whose energetic commitment to his work at South Georgia has contributed enormously to our understanding of the lives of albatrosses.

Preface

"The spirit who bideth by himself
In the land of mist and snow
He loved the bird that loved the man
Who shot him with his bow"

from *The Rime of the Ancient Mariner*
Samuel Taylor Coleridge (1798)

Albatrosses are a distinctive group of oceanic seabirds. They are the largest of the petrels, comprising 24 recognized taxa in four distinct genera, weigh 2–10 kg and have wingspans measuring 2–3 m. All species except the Waved Albatross, which breeds near the equator, inhabit oceanic islands in the cooler regions of both hemispheres between 30° and 55° latitude. These are the windy latitudes of the world and it is wind, perhaps more than any other factor, that defines the global distribution of albatrosses and facilitates their mobility. Their remarkable powers of gliding flight enable them to conduct foraging missions to waters vast distances from their breeding sites. Albatrosses spend much of their lives at sea, yet remain dependant upon land to which they must return to breed.

Due to their large size, colonial breeding habit and apparent docility when at the nest, albatrosses have been harvested in the past for their eggs, feathers and meat. Exploitation by people dates back to at least the 1700s, and it is likely that most albatross species have been affected at some time. Examples of exploitation abound, but few illustrate better the depredations by people, and the resilience of the albatrosses, than what took place at Midway Atoll in the early part of this century. An albatross stronghold, Midway Atoll comprises two small islands mid-way between the United States of America and Japan, and boasts an albatross nest every few metres or so. By the early 1900s plumage harvesters were estimated to have taken 500 000 albatrosses from Midway. The birds survived the feather hunters and lived in comparative peace until the advent of the Second World War when construction of a military base commenced, covering much of Midway's land area with roads, access points to underground installations, aircraft runways, buildings and communications facilities. Albatrosses were deemed to be a menace to aircraft, but attempts to remove them by clubbing, translocation, herding with bulldozers and flame throwers would all ultimately fail, the only long-term solution being the construction of runways which denied the birds access to soil and vegetation from which to construct nests and breeding colonies.

The tribulations at Midway were relatively short lived, and gave way, decades later, to perhaps the greatest danger albatrosses of all species have yet faced — their incidental capture and death in commercial fisheries. Between the 1950s and 1980s pelagic gillnets and driftnets ensnared tens-of-thousands of albatrosses, again causing population decreases. Public condemnation of the unacceptable levels of bycatch in these fisheries resulted in a global moratorium on driftnet fishing in 1993. However, the propensity for albatrosses to follow fishing vessels presented new threats with the global expansion of longline fisheries. Longline fisheries occur the world over, involve thousands of fishing vessels, and target a wide variety of fish species, ranging from those that occur near the sea bed to those closer to the ocean surface. Regardless of the species targeted, all longline vessels deploy, as the name implies, a long line — some reaching 130 km in length — from which are suspended baited hooks on branch lines of various lengths. Albatrosses forage in waters favoured by fishing vessels, they commute great distances to foraging grounds (which increases the chances of contact with fishing vessels), and are habitual ship followers and scavengers. These factors combine to make albatrosses vulnerable to baited hooks that remain in surface waters for any length of time when longlines are being set and retrieved. Longline fisheries have been implicated in the decreases measured over 20 years in some albatross populations, and it is likely that all albatross species are affected by this fishing practice.

It was the incidental capture and death of albatrosses in longline fisheries, and concern for the survival of albatross populations, that was the driving force behind the Inaugural Albatross Conference, which was held in Hobart in September 1995. The Conference attracted 120 seabird and fisheries scientists, fisheries managers, fishing industry representatives and conservationists from 11 nations. Being the first of its kind the Conference was able to attract the most current information on albatross biology and conservation, which was presented in 37 oral papers and 36 posters by Conference delegates. This volume is the direct result of the Conference.

Of the 23 chapters in this book 13 have been drawn from papers presented at the Conference and 10 have been solicited to broaden the context and to make the book as comprehensive as possible. The chapters are presented in six interlocking sections. The first section *(Systematics and Status)* deals with albatross taxonomy and the status of the various albatross species. A recent and major development in albatross taxonomy, presented in chapter two of the book, has been the reclassification of the 14 albatross species to 24 species, and the implications this has for the status and conservation of the various taxa. In all the texts we have included the new nomenclature in parenthesis after the old in each chapter. This approach will accommodate the existing taxonomy while ensuring the nomenclature used in the book maintains relevance in the future when the new taxonomy is finalized.

The second section *(Population Studies)* deals with studies at the population level and gives insights into the breeding biology and demography of selected albatross species, long-term trends in breeding population sizes and reproductive successes of the most closely monitored populations. These studies lead naturally to the third section of the book *(Foraging and Fisheries Relationships)* which describes the life of albatrosses at sea, in particular their dietary preferences and the extent of their spatial overlap with fisheries. This sets the scene for the fourth section *(Mortality due to fisheries and other factors)* which examines a range of issues considered to affect the survival of albatrosses, in particular longline fishing. The fifth section *(Perspectives)* includes a non-government organization perspective on albatross conservation issues and an assessment of the legal and political options available to minimize albatross mortalities in longline fisheries. The book ends with a single-chapter assessment of some fertile areas of endeavour in albatross research and conservation.

We are grateful to the authors for their contributions and their patience with the production of this volume. As part of the editorial process of the book all chapters were examined by two independent referees who are specialists in their respective fields of endeavour. The referees' contributions improved the final quality of the draft chapters and we thank them for their efforts.

We are particularly grateful to Eric Woehler, who critically read previous versions of the majority of chapters and added numerous improvements, and to Jenny Cole for her secretarial assistance. Barbara Wienecke organized the photo layout in the book, and we thank her for her creative efforts. We are grateful to Henri Weimerskirch, Gary Nunn, Hiroshi Hasegawa, Tui De Roy and Simon Pickering for providing us with the pick of their albatross photographs for use in the book.

In both the staging of the Albatross Conference and the production of this book we appreciate the support of the Australian Antarctic Division and the Tasmanian Parks and Wildlife Service. We are grateful to the Antarctic Foundation, Sirtrack Limited, CLS Argos and Gyrocast Pty Ltd for sponsoring the conference.

Given the focus in this book on albatross conservation we feel it is appropriate to acknowledge individuals that stand out to date in the international effort to understand and protect albatrosses: Nigel Brothers for his role in minimising albatross mortality in longline fisheries, Hiroshi Hasegawa for his efforts in saving the Short-tailed Albatross from extinction, John Croxall for the breadth of his scientific contributions and his leadership in global seabird conservation issues, and Henri Weimerskirsch and Peter Prince for their abundant contributions to the albatross sciences upon which the conservation effort depends.

Finally, our heartfelt thanks to our respective families for enduring our detachment from home life that ensued during the many months it took to edit this book.

Graham Robertson and Rosemary Gales

List of colour photographs

Contents

Systematics and Status

It is appropriate that this book commences with an up-to-date assessment of the status of albatrosses. To do this, of course, depends on knowledge of the number of species we are dealing with. It may seem surprising that a group of birds as large and conspicuous as albatrosses could come under review only recently to determine their phylogenetic relationships. This has been the case. Modern approaches to the taxonomy of albatrosses can be more illuminating if contrasted with those that existed in the past (Chapter 1) when birds were classified using purely descriptive criteria. Current techniques include the use of morphometric measurements and mitochondrial DNA analyses (Chapter 2) which are far more definitive than ways of old. Recent analyses based on these techniques have provided substantial evidence for the recognition of 24 albatross species, rather than the previously recognized 14 species. The updated taxonomy has important implications for the population and conservation status of the various taxa and the threats they face (Chapter 3). Such an assessment is not a simple task, and must be made using the best available knowledge of the population sizes of taxa, the number of breeding sites that exists and the trends in albatross numbers with time (Chapter 4).

The markings of the head and bill of albatrosses, as in this Buller's Albatross from New Zealand's Chatham Islands, may be used by taxonomists to identify the different albatross species.
GARY NUNN

Black-footed Albatrosses engaged in a territorial dispute, Midway Atoll, Hawaiian Leeward Islands.
GRAHAM ROBERTSON

CHAPTER ONE

Type specimens of albatrosses collected on Cook's second voyage

DAVID G. MEDWAY[1]

The first acceptable descriptions of three albatross species are based, wholly or in part, on specimens obtained at sea during James Cook's second circumnavigation (1772–75). They are the Yellow-nosed Albatross *Diomedea chlororhynchos* Gmelin, 1789, the Grey-headed Albatross *D. chrysostoma* Forster, 1785, and the Light-Mantled Sooty Albatross *Phoebetria palpebrata* (Forster, 1785). The type locality of *D. chlororhynchos* is at 35°13′S, 6°3′W, a position at sea in the South Atlantic Ocean a few hundred kilometres north-east of the Tristan da Cunha group; that of *D. chrysostoma* is at 50°15′S, 96°1′W, a position at sea in the south-east Pacific Ocean; and that of *P. palpebrata* is at 64°18′S, 38°48′E, a position at sea in the South Indian Ocean. No original specimen of these taxa is still in existence, but the bird depicted in George Forster folio 100, in The Natural History Museum in London, is designated as the lectotype of *D. chlororhynchos.*

Key words: Yellow-nosed Albatross *Diomedea chlororhynchos*, Grey-headed Albatross *D. chrysostoma*, Light-Mantled Sooty Albatross *Phoebetria palpebrata*, James Cook, Reinhold Forster, George Forster, Nominate Forms, Type Localities.

INTRODUCTION

FOR a long time there has been confusion about the actual type material and type localities of the Yellow-nosed and the Grey-headed Albatrosses. A comprehensive analysis of the best available evidence enables those matters to be determined for the two species, and also for the Light-mantled Sooty Albatross, about which there has been less confusion. The locations at sea where the specimens were obtained are determined from the co-ordinates given by Cook (in Beaglehole 1961) who recorded the date on which co-ordinates were obtained in ship time, with the day running from noon to noon. This chapter considers the descriptions and locations given two Centuries ago for the nominate forms of the Yellow-nosed, Grey-headed and Light-mantled Sooty Albatrosses. The conclusions reached for the nominate forms of these species should enable future consideration of albatross taxonomy (see Robertson and Nunn 1997) to proceed with a degree of certainty.

Yellow-nosed Albatross *Diomedea chlororhynchos* Gmelin, 1789

The first albatrosses obtained on James Cook's second circumnavigation (1772–75) were taken on Monday, 12 October 1772 at 35°13′S, 6°3′W, a position at sea in the South Atlantic Ocean a few hundred kilometres north-east of the Tristan da Cunha islands. Cook (in Beaglehole 1961:42) recorded:

> At 8 AM we brought too to wait for the Adventure and having but little Wind, we hoisted out a Boat and tryed the current, but found none, Mr Forster who was in the Boat shott an Albatross, and some sheer waters, of the Petrels or Mother Caries tribe.

George Forster was on the voyage as natural history draughtsman and assistant to his father Reinhold, who sailed as official naturalist. George (1772) gave a description of the first albatrosses he and his father were able to examine in the field:

> In the Afternoon we shot two Albatrosses, (Diomedea); they have a greyish white body and wings, but the quill and tail feathers are black with yellow or white rachis. The bills are black with a yellow streak on top, and a bright orange *rictus oris*; Under Jaws may be dilated. feet bluish white. Breadth from tip to tip about 6 foot. Length from head to tail about 2 foot. Size something less than a goose. These Albatrosses we were told were very small ones.

[1]PO Box 476, New Plymouth, New Zealand.
Pages 3–12 *in* ALBATROSS BIOLOGY AND CONSERVATION, ed by G. Robertson and R. Gales. Surrey Beatty & Sons, Chipping Norton. 1997.

Reinhold Forster (in Hoare 1982) included in his *Journal* a more detailed description of the albatrosses taken on 12 October 1772:

> In the afternoon we went out again & shot two *Albatrosses*. They have a dirty white body, & rump, black back, quill & tail feathers. The bill black with a yellow stroke all along its upperside. The under mandible is next to the feathers encompassed with a narrow fine orange yellow membrane, running down the *rictus oris* & a good way beyond it in a curved line downward. The expanded wings were 6 feet 8 inches. From the tip of the bill to the extremity of the middle toe 2 feet 9 inches. The shafts of the Quill feathers of a brownish yellow, of the tail feathers white, The Eyes black & round them a faint grey or blackish tinge towards the bill. The feet of a pale blueish hue. The Under Jaws may be dilated. Size less than that of a Goose. Their sternum is very short. They skim always on the Surface of the Sea & are never seen to rise high; they are found in the Southern Hemisphere only beyond the tropics & seem to be a new species. When skinned they afford a good palatable food. *(Diomedea chrysostoma).*

It is clear from these descriptions that the albatrosses taken on this occasion were of the taxon which was to become known as the Yellow-nosed Albatross *Diomedea chlororhynchos* Gmelin, 1789. There is no evidence that either of these specimens were retained. Indeed, Reinhold Forster's comment that "When skinned they afford a good palatable food" would tend to confirm that these birds were not preserved.

More albatrosses were obtained and examined eleven days later, on Friday, 23 October 1772 at 36°39′S, 11°6′E, as the *Resolution* approached the Cape of Good Hope from the west. On that day Cooper (1771–74) noted they "Hoisted the Boat out which went with Messrs Forsters shooting . . . Messrs Forsters &c shot sevl albatrosses & other Birds". It seems clear from what Cook (in Beaglehole 1961) said that all of the birds taken on this day were also eaten:

> Little Wind with some Calms, in the am hoisted out a Boat and shott some Albatrosses and other Birds on which we feasted the next day and found them exceeding good.

Reinhold Forster (in Hoare 1982) also gave a detailed account of the birds taken on 23 October 1772:

> . . . hoisted the boat out & Mr Charles Clerck, I and my Son went out a shooting: at some distance from the Ship we shot a (*Procellaria aequinoctialis* Linn) . . . We likewise killed some Albatrosses of the kind described before: one of them had a bill more yellow, than the rest. *(Diomedea chrysostoma).* And besides another kind of Albatross which is Linnaei *(Diomedea exulans)* . . . This bird afforded us, having been skinned, a good dish.

A detailed description of the albatross called "*Diomedea chrysostoma*" was included by Reinhold Forster (1772–75) in his *Descriptiones Animalium*. The description, along with those of several other new taxa, was apparently made when the Forsters were at the Cape of Good Hope between the end of October and the end of November 1772. The description of "*Diomedea chrysostoma*" is dated 12 October 1772 and was undoubtedly based on the two specimens taken on that date, although brief reference is also made to one of the specimens taken on 23 October 1772. It is apparent, however, that various amendments and additions were subsequently made to the description as originally entered in the manuscript. The original entry is shown in Appendix 1a. It, like the previously-quoted entry in Reinhold Forster's *Journal* on 12 October 1772, is clearly referable to *D. chlororhynchos*.

The albatross collected on 23 October 1772, which had a more yellow bill than the others, was referred to by Forster as having "rostro toto superiore flavo". It was almost certainly a *chlororhynchos* more mature than the other specimens taken. Gould (1840–48, 1865) noted that one of the characters by which younger members of that species might be easily distinguished from adults, especially when flying, was by the culmen of the bill being less distinctly marked with yellow. The colour pattern of the bill of adult *chlororhynchos* is well shown by the paintings in Gould (1840–48) and Salvadori (1914), and in the photograph published by Löfgren (1984:143). The distinctive yellow culminicorn stripe of *chlororhynchos* develops slowly, and the time taken to develop adult bill colour is unknown (Marchant and Higgins 1990).

George Forster's folio 100 is an almost completed painting of *D. chlororhynchos* (Fig. 1). It was probably based on the specimens taken on 12 October 1772 and is published here apparently for the first time. Lysaght (1959) considered the painting to be a representation of a bird that was not quite mature. The extent of the yellow culminicorn stripe and the dark maxillary unguis of the specimen depicted by Forster indicates an individual which has not reached maturity. The painting bears the name "*Diomedea chrysostoma*" and the following notations (Lysaght 1959):

> Irides brown. Under Eyelid White. Head dark pearly grey, gradually & very softly

Fig. 1. George Forster folio 100 of the Yellow-nosed Albatross. The bird depicted is herein designated as the lectotype of *Diomedea chlororhynchos* Gmelin, 1789. Reproduced by permission of The Natural History Museum, London.

> vanishing into a fine clear white on the neck. A more dark grey Spot over the Eye. The middle of the Back black gradually turning into pearl grey & then into White towards the Neck — But abruptly bounded by white on the Urypygium end. Feet pale greyish white.

Over the pencil outline of the foot is written "This Leg is something too large."

No species of small albatross other than the Yellow-nosed Albatross is known to have been obtained by the Forsters before they reached Cape Town at the end of October 1772. While there, Reinhold Forster communicated to Linnaeus some of his views about albatrosses based on observations made on the voyage to the Cape. When speaking of *D. exulans*, he included the following description (from the original Latin) which is clearly referable to *D. chlororhynchos*:

> . . . another kind of *diomedea* which I would prefer to call *chrysostoma* because of its black bill marked with a golden stripe above, and because of the golden inner margin of the gape on the lower mandible. It is about the size of *Anas anser*, or a little larger, the back rectrices and remiges are dusky-black, the rest with the rump smudgily white, a spot from the bill to the eyes is bluish-ash, the feet are bluish.

The amendments and additions to the October 1772 description of *"Diomedea chrysostoma"* in *Descriptiones Animalium* must therefore have been made at some time during the voyage after the Forsters had left the Cape of Good Hope at the end of November 1772. They could only have made those alterations after a close examination of further material which they considered referable to the same species. It so happens that the only other specimens of "yellow billed" albatrosses known to have been obtained on the voyage and examined by them were taken in the south-east Pacific at 50°15′S, 96°1′W on 12 February 1774. This position is in an area where *D. chlororhynchos* has never been recorded, but where the Grey-headed Albatross *D. chrysostoma* occurs (Marchant and Higgins 1990; Rogers 1980; Clark 1986; Tickell 1995). On 12 February 1774 Cook (in Beaglehole 1961) recorded:

> PM had Light airs next to a Calm, put a Boat in the Water in which some of the officers went and shott several Birds, Albatroses and Sheer-waters, and one of those we call Port Egmont Hens.

On the same day Clerke (1771–75) recorded: "This afternoon hoisted out the small Boat as near calm with which some of the officers went a shooting for a fresh dish — among other Game we kill'd a Port Egmont Hen"; Mitchel (1772–75) noted: "Hoisted the boat out, to shoott Albatross etc etc . . Shot many Albatrosses & some sea gulls"; and Cooper (1771–74) observed: "The Boat came on board having kill'd 10 Albatrosses".

Reinhold Forster (in Hoare 1982) included the following in his *Journal* entry for 12 February 1774:

> Some of our Officers went out in a boat, it being calm: They shot several Albatrosses & Shearwaters. The Albatrosses proved to be of the kind, I had named *Diomedea chrysostoma* in the Atlantic, before we reached the Cape. They had got one

> common Albatross *Diomedea Albatrus* & the Shearwaters were of the N. Holland or whiterumped kind. They likewise brought a *Port Egmont Hen* . . .

There is no evidence that any of these specimens were retained. Indeed, Cook (in Beaglehole 1961) recorded that they feasted on them the next day. It was undoubtedly on the basis of one or more of the several smaller albatross specimens taken on 12 February 1774 that the Forsters (Forster 1772–75) altered their manuscript description of "*Diomedea chrysostoma*". The altered version is shown in Appendix 1b.

The principal alterations made to the 1772 description included the lower mandible now being described as having "lineam flavam ante apicem terminatam" (a yellow line ending before the tip); and the bird now having "sub oculo a cantho postico ultra medium oculi, macula alba" (a white spot under the eye from the rear corner beyond the centre of the eye). These characteristics are among those diagnostic of *D. chrysostoma*. There is no doubt that the Forsters considered all "yellow billed" albatrosses examined by them in the course of the voyage to be of the same taxon. This is not particularly surprising considering that Hutton (1865), many years later, was of opinion that the Yellow-nosed, Black-browed and Grey-headed Albatrosses were the same species. The amendments and additions which the Forsters made in 1774 to their 1772 manuscript description of what was *D. chlororhynchos* effectively converted it in all material respects into a description of *D. chrysostoma*. It was this description in final form which was published by Lichtenstein (1844) who seems, however, to have inadvertently omitted the last sentence. Coues (1866), Gray (1869–71), Salvin (1896), Godman (1907–10) and Mathews (1910–28) all considered *chrysostoma* of Forster, 1844 to be referable to *chlororhynchos* of Gmelin, 1789. It was not until 1959 that Lysaght (1959) correctly stated that Forster's 1844 description refers "unmistakeably" to *chrysostoma* of Forster, 1785. She was also correct in saying it seemed clear Forster obtained both species but did not realize that they differed — "hence the placing of the description of *chrysostoma* under a locality and a date that in all probability pertained to *chlororhynchos*".

Latham (1781–85) described and illustrated his Yellow-nosed Albatross in 1785. He apparently saw a specimen in the British Museum which he referred to as a Yellow-nosed Albatross. Both the bulk of his description of the taxon, and his illustration of it, seem to have been based on that specimen. Latham's account was the sole basis for the *D. chlororhynchos* of Gmelin (1788–93):

> Br. Mus.
>
> LENGTH three feet: breadth seven. The bill four inches long, hooked at the end, but not very stout; the colour of it is black, except the upper ridge, which is yellow the whole length, quite to the tip, where it is hooked; the base of the under mandible is also yellow: irides brown: the head is grey: between the bill and eyes is an obscure black spot; just over the eye a dusky one: the hind part of the neck dusky, the lower part white: back, scapulars, and wings, dusky blue black: rump, and under part of the body, white: the tail dusky: the legs are pale yellowish white; the fore part of them, and the webs, dusky.
>
> This species is met with in the southern hemisphere, from 30 to 60 degrees, all round the pole. The specimen from whence the above description was drawn up, was taken off the *Cape of Good Hope*. Inhabits the *South Seas* without the *tropics*. Fly about five or six feet above the surface of the water.

Nothing further is known of Latham's specimen of the Yellow-nosed Albatross which he apparently saw in the British Museum in or shortly before 1785. Gray (1844) recorded that a specimen, apparently of *chlororhynchos* and said to be from the "South Sea", was in the collection of the British Museum. There is no way of knowing the true origins of that bird. It certainly cannot be identified with the specimen which Latham apparently saw in the collection over half a century earlier. In any event, no such specimen was mentioned by Salvin (1896), and Godman (1907–10), a few years later, stated it was no longer in existence.

Latham, like the Forsters, may have regarded all "yellow-nosed" albatross material as representative of the one species. In the absence of the specimen itself, there can be no guarantee that the bird apparently seen by Latham in the British Museum was in fact an individual of *D. chlororhynchos*. Furthermore, the hand coloured engraving of the Yellow-nosed Albatross as published by Latham in 1785, even assuming it was based on a specimen of *D. chlororhynchos*, cannot be relied upon, as Brooke *et al.* (1980) did in part, for the purpose of determining the nominate race of that species. For example, the colour of the bill as depicted is not an accurate representation of that feature as described by Latham who said "the colour of it is black, except the upper ridge, which is yellow the whole length, quite to the tip . . .". In the hand-coloured engraving as published only part of the culminicorn and the tip of the maxillary unguis is coloured yellow. The original of Latham's Plate 94 is folio 936 in the Latham collection of drawings of birds in The Natural History Museum, London. There is a print of Latham's drawing, coloured by his own hand, in a copy (CR 4671) of the second edition of Pennant's

Genera of Birds (1781) in the National Library of Wales. The bird depicted also has the bill coloured as in the published engraving.

George Forster's folios 100 and 101 of *"Diomedea chrysostoma"* were among the voyage drawings sold by Reinhold Forster to Sir Joseph Banks in August 1776. Two contemporary manuscripts exist which contain notes which are a useful supplement to the data written on many of the voyage drawings. They are the Catalogues B and C of Whitehead (1978). Latham not only had access to, and used, the Forster drawings; he also had access to, and used relevant data from, those manuscript catalogues. All of this material was then in the possession of Banks. In addition, Latham had access to other bird drawings, which were still in Reinhold Forster's possession, shortly before Forster left England in July, 1780. This is confirmed in Forster's letter of 11 April 1790 from Halle to his son (in Leuschner *et al.* 1982:402):

> When I was upon my departure from England a Surgeon in Kent one Mr Latham came to me *and* desired to see my drawings of birds, he saw those I had *and* after much entreaty got leave to copy them, *and* promissed fairly to give me a copy of his book in proportion as it would be published. He was then full of acknowledgments *and* kindness *and* has sent me the two first Volumes *and* then left off. He got acquainted, with Sir Joseph Banks, got a sight of our drawings, has made use of them, *and* then became proud, *and* haughty *and* spoke in his letters a language, I did not expect from a Man, whom I had so much countenanced: at last he became again kind *and* promissed to send me the following Volumes, but I never received them.

It presumably must have been at the time of this contact with Forster that Latham was also given access to, and made extracts from, the manuscript of *Descriptiones Animalium*. It is clear he had such access because, *inter alia*, some of his descriptions of avian species from Cook's second voyage were based almost exclusively on accounts in *Descriptiones Animalium* and could only have been obtained from it. This manuscript remained in Forster's possession until after his death in 1798, when it passed to the Royal Library in Berlin, but was not published until Lichtenstein did so many years later (Hoare 1982).

Thus, when compiling his 1785 description of the Yellow-nosed Albatross, Latham could have obtained information relative to his supposed species, not only from the specimen which was apparently in the British Museum, but also from each of the *Descriptiones Animalium* manuscript, the Forster drawings, and the catalogue manuscripts relating to them. It is apparent that he did use information from those Forster sources in his description. There is no reason to doubt Latham saw folios 100 and 101 and took them into account when describing the Yellow-nosed Albatross. Indeed, Catalogue B has the entry, which relates to Forster folios 100 and 101: *"Diomedea chrysostoma*. 2 Drawings. 30°–60°S all round the Pole. diminishd — about 7 feet". Latham's statement "This species is met with in the southern hemisphere, from 30 to 60 degrees, all round the pole" could only have come from the catalogue entry. His statement "Inhabits the *South Seas* without the *tropics*" is obviously from *Descriptiones Animalium* where it says "Habitat in Oceano australi extra Tropicum". And Latham's statement that the species flies "about five or six feet above the surface of the water" could not have come from other than *Descriptiones Animalium* where Forster said "Nunquam aethera petunt, sed pedes circiter 5 vel 6 supra undas volitant". The brown colour of the irides as given by Latham could not have come from a prepared specimen. It could, however, have been obtained from the notations on folio 100 where the irides are given as brown. Similarly, Latham's comment that his bird had a dusky spot just over the eye may also have come from the notations on folio 100 where it is recorded that there was "A more dark grey Spot over the Eye". It seems probable that the notations on folio 100 were written after the Forsters had examined the specimens of *"Diomedea chrysostoma"* taken on 12 February 1774.

Despite the reservation expressed earlier, it is not unreasonable on the basis of available information to regard all of the principal material used by Latham in his description of the Yellow-nosed Albatross as pertaining to *D. chlororhynchos*. Thus, the bird depicted in Forster folio 100, and the specimen apparently once in the British Museum, can be accepted as syntypes of *D. chlororhynchos* as defined in Article 73(b) of the *International Code of Zoological Nomenclature* (Ride *et al.* 1985).

Type locality of *D. chlororhynchos*

Article 73(b)(iii) of the *Code* provides that if syntypes of a nominal species-group taxon originated from two or more localities, the type locality is all of the places of origin. Latham said the specimen from which his description was made was taken off the Cape of Good Hope. However, in the absence of evidence, it is uncertain that this is the locality from which the apparent specimen in the British Museum originated. In these circumstances it is not appropriate to give "off the Cape of Good Hope" as a type locality of *D. chlororhynchos*. On the other hand, the place of origin of the first specimens of *chlororhynchos* taken by the Forsters, on which they based their original descriptions and folio 100, is known with certainty: 35°13′S, 6°3′W, in the South Atlantic a few hundred kilometres north-east of Tristan da Cunha where the nominate race of the species breeds (Rowan 1951). This can therefore be accepted as the correct type locality of the taxon. However, to put the matter beyond doubt, the syntype illustrated in George Forster folio 100 is hereby designated, per Article 74(a) of the *Code,* as the lectotype of *D. chlororhynchos*. The place of origin of that specimen is thereby confirmed, per Article 74(a)(iii) of the *Code,* as the type locality of *D. chlororhynchos.*

Grey-headed Albatross
D. chrysostoma Forster, 1785

According to Hoare (1976), Reinhold Forster became a corresponding member of the Academie des Sciences of Paris in 1776 and during a visit to that city in the same year promised to submit his *Memoire sur les Albatros* to the Academy. The description of *D. chrysostoma* which Forster (1785) completed for his *Memoire* must have been based on the description of that species in his *Descriptiones Animalium* manuscript. It is given here, translated from the original French.

> The Golden-billed Albatross *(Diomedea chrysostoma)* is as big as a goose, and of approximately the same shape as the common Albatross. Its wingspan is 6ft. 8in. and from the bill to the feet it is 2ft. 9in. long. It is white, grey-headed, the region above the eyes being a little blackish. The back, wings and tail, which is rounded, are black. The long feathers are brownish black. The scapes of the primaries are yellowish, those of the secondaries are white. The eyes are hazel nut-coloured and on the posterior angle, under each eyelid, a white spot can be seen. The bill is black, with an upper longitudinal yellow stripe, not on all its length, and the edges of the maxillae are also golden. The feet are bluish-grey. As for the rest, this species is entirely similar in manners and habits to the other Albatrosses and can be found in the same regions as the common species. Nevertheless we found that there were only very few of them near the Antarctic Polar Circle and in the Pacific Ocean.

Mathews (1910–28) noted that study of Forster's *Memoire*, "brought to light" by Sherborn (1902), enabled him to state that *D. chrysostoma* Forster, 1785, was applicable to the bird Gould (1843) renamed as *D. culminata*. It can be observed in passing that the existence of Forster's *Memoire* had in fact been noted in the English literature by Jonas Dryander (1796–1800) as early as 1796, but was subsequently overlooked for more than a century. Since Mathews, no doubt has been expressed about the identity of *chrysostoma* of Forster, 1785 with the taxon known as the Grey-headed Albatross.

In his *A Letter to the Right Honourable The Earl of Sandwich* . . . of 20 May 1778, George Forster (in Kahn *et al.* 1972) wrote his father had employed a "skillful painter" to finish paintings from the "rough draughts" George had done on the voyage. Reinhold Forster (in Scheibe 1978) also mentioned the paintings in a letter to Joseph Banks in mid-1777. Joppien (1976) was not able to identify the "skillful painter" employed to finish George Forster's drawings. He considered, however, that there is some evidence the artist may have been French. Likewise, Whitehead (1978) said that the artist had never been identified but he did name several talented natural history artists resident in London at the time who he thought would have been glad of such employment. There is in fact a reasonable amount of evidence, which cannot be traversed here, which suggests that John Frederick Miller, one of the persons named by Whitehead, was the "skillful painter" employed to finish the drawings intended for presentation to the King.

The subsequent history of these paintings has been given by Steiner and Baege (1971) and Joppien (1976). Twenty paintings depicting birds from the voyage were published by Steiner and Baege (1971) from the originals which are now in the Forschungsbibliothek, Gotha (see also Whitehead 1978). Three of them, of *D. exulans, D. chrysostoma* and *P. palpebrata,* formed the basis for the engravings, inferior in quality to the originals, which accompanied Reinhold Forster's *Memoire* of 1785.

The finished painting of *D. exulans* has no equivalent in the Forster collection of bird drawings in The Natural History Museum in London. Forster folio 101 in that collection, which is one of the two drawings identified as depicting *"Diomedea chrysostoma",* is only a pencil sketch (Lysaght 1959). It is similar in form to the bird depicted in the finished painting, which represents the Grey-headed Albatross. It seems unlikely, however, that an accurate painting which clearly depicts a bird different to the specimen of *D. chlororhynchos* featured in Forster folio 100, could have been satisfactorily completed from a pencil sketch alone, even if done under the supervision of the Forsters from their recollections of the bird being depicted. Nevertheless, it seems probable that both the finished paintings of *D. chrysostoma*, and *D. exulans*, were completed by the "skillful painter" from original Forster drawings which did not form part of the collection which went to Banks in 1776. Perhaps the originals were among the drawings of birds, earlier referred to, which Latham saw in

Reinhold Forster's possession shortly before Forster left England in 1780. The fate of those drawings is not known. The finished painting of *D. chrysostoma* no doubt depicts one of the specimens taken on 12 February 1774 and examined by the Forsters. It can therefore be accepted as an illustration of the type of *D. chrysostoma* Forster, 1785.

Type locality of *D. chrysostoma*

There has been considerable confusion over the correct type locality of *D. chrysostoma*. On the basis of what Forster said in 1785, it was given as "vicinity of Antarctic Circle and in Pacific Ocean" by, for example, Peters (1931), Jouanin and Mougin (1979), Brown *et al.* (1982) and Turbott (1990). It was given as "Cape Seas" by Lowe and Kinnear (1930), Mathews (1934) and Oliver (1955). Mathews (1937) designated it as "off Staten Land" giving as the reason that Forster in his *Memoire* said it was found "in the same place as the Common Albatros, in January 1775, off Staten Land" (Forster, in fact, said no such thing). Mathews and Hallstrom (1943) followed this incorrect designation, and the same locality given by McLachlan and Liversidge (1980) was probably also based on it. Murphy (1936) suggested South Georgia because Forster's statement that there were only very few *D. chrysostoma* near the Antarctic Polar Circle and in the Pacific Ocean "would seem to call" for an Atlantic locality, and because Forster met the species at South Georgia during the voyage of the *Resolution*. The species may very well have been among the many "yellow billed" albatrosses seen by Forster (in Lichtenstein 1844, in Hoare 1982) at sea in the vicinity of South Georgia in January 1775, but there is no evidence that a specimen was obtained there and examined by him. Murphy's designation was probably the basis for the statement by Johnson (1965) "Without type locality: South Georgia suggested". Most recently, Marchant and Higgins (1990) stated that "vicinity of Antarctic Circle and in Pacific Ocean" as given by Forster in 1785 " = South Georgia" (Murphy 1936) or "off Staten Island, South America" (Mathews 1943).

The correct type locality of the Grey-headed Albatross is now known to be none of these. It must surely be the place of capture of the only "yellow billed" albatrosses known to have been examined by the Forsters after the *Resolution* left the Cape of Good Hope at the end of November 1772. That was on 12 February 1774 at 50°15′S, 96°1′W in the south-east Pacific Ocean.

Light-mantled Sooty Albatross
P. palpebrata (Forster, 1785)

Determination of the type specimen and type locality of the Light-mantled Sooty Albatross is considerably less complicated than the same determinations for the preceding two taxa. On 5 December 1772, when at 47°10′S, 18°10′E, Reinhold Forster (in Hoare 1982:189) recorded that they saw "some *Quakerbirds* which we had never seen before". George Forster (1777:I:91) later wrote that "We likewise saw the two before mentioned species of albatrosses from time to time, together with a third, less than the others, which we named the *sooty*, and our sailors called the quaker bird, from its having a greyish-brown colour". However, no specimen of this "third" albatross was obtained until Tuesday 12 January 1773 when the *Resolution* was among loose ice in the South Indian Ocean at 64°18′S, 38°48′E. Cook (in Beaglehole 1961:76) recorded:

> At 6 o'Clock, having but little Wind, we brought to a mong some loose Ice, hoisted out the Boats and took up as much as filled all our empty Casks . . . while this was doing Mr Forster shott an Albatross whose plumage was of a Dark grey Colour, its head, uper sides of the Wings rather inclining to black with white Eye brows, we first saw of these Birds about the time of our first falling in with these Ice Islands and they have accompanied us ever sence. Some of the Seamen call them Quaker Birds, from their grave Colour . . . At 4 o'Clock in the PM hoisted in the Boats and made sail . . .

Reinhold Forster (in Hoare 1982) gave a more detailed description of this bird in his *Journal*:

> I went out to shoot a bird we saw swimming on the Sea. I winged it and it got some shot into the neck, so that we got the bird alive. It proved to be the *Quakerbird*, which I found to be a kind of Albatross, never described before, & I called it therefore *Diomedea palpebrata*. It is of a dusky grey, its head, wings & tail black, the bill of a jetty black, the feet of a dusky blue: the *(rachis)* shafts of the Quill and tail feathers white; the Eye which is of a hazel colour, is surrounded with a white except before: the Tail wedgeshaped.

There is no evidence that this bird was retained. No other specimen of the Light-mantled Sooty Albatross is known to have been examined by the Forsters during the voyage. They did shoot another *Quakerbird* on 23 January 1773 but, as Reinhold Forster (in Hoare 1982) noted, it "fell within 2 yards from the Ship into the Sea". George Forster's folio 102, which is a finished painting, bears the notations (Lysaght 1959): "G. Forster delin.1773.ad vivum in mari Antarctici. *Diomedea palpebrata*. die 20 Jan. Anni. 1773". It must have been based on the specimen taken on 12 January 1773. That specimen also formed the sole basis for the original description by

Reinhold Forster (1772–75) of *Diomedea palpebrata*, which is shown in Appendix 2. It was not published until Lichtenstein (1844) many years later.

Twelve years later, Reinhold Forster (1785) included a description of *D. palpebrata* in his *Memoire*. It was based on the one included in his *Descriptiones Animalium*. It is the first valid use of the name *D. palpebrata* but did not receive recognition until Forster's *Memoire* was "brought to light" by Sherborn in 1902. Forster's 1785 description was accompanied by an engraving which was based on the finished painting forming part of the collection, discussed earlier, which was intended by the Forsters for presentation to the King. That painting has been published by Steiner and Baege (1971). A very good contemporary copy of George Forster's folio 102 is included in a collection of second voyage bird drawings in the Mitchell Library in Sydney (Lysaght 1959). The first published description of *D. palpebrata*, translated from the original French, is as follows:

> The white eye-lidded Albatross *(D. palpebrata)* is as long as a goose. Its body appears lighter than those of the other Albatrosses. Its wingspan is 6 ft. 7 in. and from feet to bill it measures 2 ft. 7 in. The feathers are brownish-grey; the head is the colour of soot, as well as the wing and tail long feathers. The feathers in the middle of the tail are longer and have white scapes. The feathers covering the wings are blackish-brown. The bill is black and 4 in. long. The feet are dark grey, the eyes pale yellow and the upper eyelids and half the lower eyelids are white. This species is found from 47 degrees austral latitude to 71°10′ where before us no vessel had penetrated.

Reinhold Forster gave the range of *D. palpebrata* as from 47°S to 71°10′S. It has already been noted that the *Quakerbird* was first seen by him at 47°10′S on 5 December 1772. The *Resolution* reached its most southern point at 71°10′S in Longitude 106°54′W on 30 January 1774 where Cook (in Beaglehole 1961) saw "very few Birds of any kind". The Light-mantled Sooty Albatross must have been seasonally very common in parts of its range visited by the Forsters. Shortly after the Resolution left Breaksea Sound in New Zealand on 11 May 1773, Reinhold Forster (in Hoare 1982) noted that they saw "vast numbers of the *sooty Albatross (D. palpebrata)*" and, two days later, "a great many sooty Albatrosses" attended the ship. Latham (1781–5) also described the Sooty Albatross thus:

> Sooty, or brown Albatross, *Forst. Voy.*
> Albatross with a white eye-brow, *Cook's Voy.*
>
> SIZE of a *Goose*: length near three feet. Bill black: irides pale yellow: at each angle of the eye a nictitating membrane: general colour of the plumage brown: the head and tail inclining to black or *soot*-colour: for a small space above, behind, and beneath the eye, the feathers are white, but not on the fore part of it: quills and tail dark brown, nearly black; the shafts of both white; the last pointed in shape: legs pale brownish lead-colour: claws black. This species was first met with in lat 47. south, but was afterwards observed throughout the *Southern Ocean* within the *antarctic* circle. It was called by our sailors the *Quaker*, from its brown plumage.

At one time Mathews (1912:206) thought that the *Diomedea fuliginosa* of Gmelin (1788–93) was founded on George Forster's painting. Gmelin's account was based solely on the 1785 description by Latham of the Sooty Albatross. It is clear, however, that Latham's description was based almost entirely on the account in Reinhold Forster's manuscript *Descriptiones Animalium* to which Latham obviously had access before Forster left England in 1780. The specimen taken by Reinhold Forster on 12 January 1773 is therefore also the type of *Diomedea fuliginosa* Gmelin, 1789 which is a junior synonym of *P. palpebrata* (Forster, 1785). The correct type locality of *P. palpebrata* is 64°18′S, 38°48′E in the South Indian Ocean.

ACKNOWLEDGEMENTS

I am grateful to the Staatsbibliothek Preussischer Kulturbesitz in Berlin for providing a microfilm of Reinhold Forster's *Descriptiones Animalium*, and to The Natural History Museum in London for permission to reproduce George Forster's painting of *Diomedea chlororhynchos*. Sue Superville of the Hector Library at the Museum of New Zealand in Wellington efficiently provided me with photocopies of material required for this study. I am also grateful to Joy Chong of New Plymouth for the careful attention she has given to the typing of this paper, and to Lance Tickell and John Warham whose constructive comments have improved it.

APPENDIX 1a

Reinhold Forster's (Forster 1772–75) original (1772) description of the taxon that was to become known as the Yellow-nosed Albatross *Diomedea chlororhynchos.*

1772. Oceanus australis. Oct. XIImo Diomedea chrysostoma
20. Diomedea chrysostoma F. the yellowbilled Albatross.

Alis pennatis, alba, rectribus, dorso remigibus nigris, rostro supra linea, faucisque margine inferiore aureis.

Habitat in Oceano australi extra Tropicum.

Corpus magnitudine vix Anseris domestici. Alae expansae 6 pedum 8 unciarum. Longitudo a rostri apice in medii digiti unguem 2 pedum, 9 unciarum.

Caput album cum colla, abdomine, pectore & uropygio a rostro versus oculum canities.

Dorsum, tectrices alae & caudae nigra.

Remiges & rectrices fusco nigra; rachis priorum flavescens, posteriorum alba.

Rostrum nigrum, supra linea flava: mandibula inferior a plumis divisa margine elevato, membranaceo angusto aureo, decurrente postice per fauces & ultra sub oculis.

Irides nigrae.

Pedes aequilibres pallide glauci, palmati sine digito & ungue postico.

Digitus interior phalangibus 5, medius 4, exterior 3, & hic praeterea membrana marginali instructus, ut in congeneribus.

Obs. Vidi & quosdam rostro toto superiore flavo. Caro sapida, postquam cutis detracta est. Avis laesa vomit & mordet vehementer. Sternum breve. Nunquam aethera petunt, sed pedes circiter 5 vel 6 supra undas volitant.

APPENDIX 1b

Reinhold Forster's (Forster 1772–75) amended (1774) description of the taxon that was to become known as the Grey-headed Albatross *Diomedea chrysostoma.*

1772. Oceanus australis. Oct. XIImo Diomedea chrysostoma
20. Diomedea chrysostoma F. the yellowbilled Albatross.

Alis pennatis, alba, rectribus, dorso remigibus nigris, rostro supra linea, faucisque margine inferiore aureis.

Habitat in Oceano australi extra Tropicum.

Corpus magnitudine vix Anseris domestici. Alae expansae 6 pedum 8 unciarum. Longitudo a rostri apice in medii digiti unguem 2 pedum, 9 unciarum.

Caput canescens. Collum cum abdomine, pectore & uropygio album a rostro versus & circa oculum canities intensior & nigricans.

Dorsum, tectrices alae & caudae nigra.

Remiges & rectrices fusco nigra; rachis priorum flavescens, posteriorum alba.

Rostrum nigrum supra linea flava: mandibula inferior a plumis divisa, margine elevato, membranaceo angusto aureo, decurrente postice per fauces & ultra sub oculis, infra desinens in lineam flavam ante apicem terminatam.

APPENDIX 1b — *continued*

Lingua brevis, carnosa, lanceolata, apice cuspidato; lateribus postice, villis seu serraturis retro-flexis.

Palatum tribus ordinibus serraturarum retroflexarum, sub lingua mandibula inferior lateribus serie unica utrimque serraturae retroflexae. Sub oculo a cantho postico ultra medium oculi, macula alba.

Irides avellaneae.

Pedes aequilibres pallide glauci, Palmati sine digito & ungue postico.

Digitus interior phalangibus 5, medius 4, exterior 3, & hic praeterea membrana marginali instructus, ut in congeneribus.

Obs. Vidi & quasdam rostro toto superiore flavo. Caro sapida, postquam cutis detracta est. Avis laesa vomit & mordet vehementer. Sternum breve. Nunquam aethera petunt, sed pedes circiter 5 vel 6 supra undas volitant. Vidi alias rostro toto nigro, quas foeminis esse credo.

APPENDIX 2

Reinhold Forster's (Forster 1772–75) original description of *Diomedea palpebrata,* which was later to become known as the Light-mantled Sooty Albatross *Phoebetria palpebrata.*

1772. Oceanus antarcticus Xbr 5-1773.
Jan 13 Aves Diomedea palpebrata
71. *Diomedea palpebrata* F, *the sooty or the brown Albatross.*

Diomedea fusca, capite remigibus rectricibus fuliginoso-nigris, palpebris niveis.

Habitat in Oceano antarctico, a gradu latitud. austr. 47 in Circulum antarcticum.

Corpus magnitudine circiter Anseris domestici, cinereo-fuscum.

Rostrum atrum, *Nares* cylindricae.

Caput fuliginoso, nigrum. *Oculi* Iride pallide lutea; ad utrumque angulum oculi, est membrana nictitans, & supra oculos in parte posteriore & partim infra oculos plumae albae cingunt oculum, sed non in parte anteriore:

Remiges fuliginoso-nigri, rachi alba.

Rectrices 12 caudae cuneatae, fuliginoso-nigri, rachi alba.

Dorsum, collum, pectus, Abdomen, femora cinereo-fusca.

Pedes aequilibres tridactyli, rubicundi vel etiam glauci, palmati sine digito & ungue postico, totidemque phalangibus in digitis ac membrana laterali ut in congeneribus.

A Rostro in caudam 2 ped. 10 poll. in unguem medium 2 ped. 7½ poll. Alae expansae 6 ped. 7 poll. Rostrum 4 poll. A Rostro in medium oculi 5½ poll. Pedes nudi 9 poll. Digitus med. 5 poll. Unguis medius ½ poll. Cauda ab Uropygio in mediam rectricum 11 poll.

REFERENCES

Beaglehole, J. C. (ed), 1961. The Journals of Captain James Cook on his voyages of discovery. The voyage of the *Resolution* and *Adventure* 1772–1775. Hakluyt Society: Cambridge.

Brooke, R. K., Sinclair, J. C. and Berruti, A., 1980. Geographical variation in *Diomedea chlororhynchos* (Aves: Diomedeidae). *Durban Mus. Novit.* **12**(15)**:** 171–80.

Brown, L. H. Urban, E. K. and Newman, K., 1982. The Birds of South Africa. Volume 1. Academic Press: London.

Clark, G. S., 1986. Seabirds observed in the Pacific Southern Ocean during autumn. *Aust. Seabird Group Newsl.* **23:** 1–35.

Clerke, C., 1771–75. Log kept on the *Resolution*. British Library Add.MSS 8951–3.

Cooper, R. P., 1771–4. A Journal of the Proceedings of His Majesty's Sloop *Resolution* on a voyage upon Discovery, towards the South Pole. Adm 55/104, 109.

Coues, E., 1866. Critical Review of the Family Procellariidae; — Part V; embracing the *Diomedeinae* and the *Halodrominae*. *Proc. Acad. Nat. Sci. Philad. for* **1866:** 172–97.

Dryander, J., 1796–1800. Catalogus Bibliothecae Historico-Naturalis Josephi Banks. London.

Forster, G., 1772. Observationes Historiam Naturalem Spectantes quas Navigationes ad Terras Australes instituere coepit Pp. 93–107 *in* R. L. Kahn *et al.* (1972). G. F. Mense Julio, Anno MDCCLXXII.

Forster, G., 1777. A Voyage round the World, in His Majesty's Sloop, *Resolution*, commanded by Capt. James Cook, during the Years 1772/73/74 and 75. London.

Forster, J. R., 1772. Incomplete letter Forster to Linnaeus, Cape Town, 19 November 1772. Linnaean Correspondence. IV:277–80, *Linnean Society*, London. The reference to *D. chrysostoma* is in Sol. Z8 (Solander Slips, Aves at ff.153–6) in The Natural History Museum: London.

Forster, J. R., 1772–75. Descriptiones Animalium, quae in Itinere ad Maris Australis terras suscepto, collegit, descripsit and delineavit. Ms.lat.qu.133–136 in Staatsbibliothek Preussischer Kulturbesitz, Berlin.

Forster, J. R., 1785. Memoire sur les Albatros. Memoires Math. Phys. Paris (Acad. Sci.) **10:** 563–572, pls. xiii–xv.

Gmelin, J. F., 1788–93. Systema Naturae per Regna Tria Naturae. Editio decima tertia. Lipsiae.

Godman, F. Du C., 1907–10. A monograph of the petrels. Witherby and Co: London.

Gould, J., 1840–8. The birds of Australia. Published by the author: London.

Gould, J., 1843. Description of *Diomedea culminata*. *Proc. Zool. Soc. Lond. for* **1843:** 107–08.

Gould, J., 1865. Handbook to the birds of Australia. Published by the author: London.

Gray, G. R., 1844. List of the specimens of birds in the collection of the British Museum. Part I. — Accipitres. British Museum: London.

Gray, G. R., 1869–71. Hand-list of Genera and Species of Birds, distinguishing those contained in the British Museum. British Museum: London.

Hoare, M. E., 1976. The Tactless Philosopher: Johann Reinhold Forster (1729–98). Hawthorn Press: Melbourne.

Hoare, M. E. (ed), 1982. The Resolution Journal of Johann Reinhold Forster 1772–1775. 4 vols. The Hakluyt Society: London.

Hutton, F. W., 1865. Notes on some of the Birds inhabiting the Southern Ocean. *Ibis* **1:** 276–98.

Johnson, A. W., 1965. The Birds of Chile and adjacent regions of Argentina, Bolivia and Peru. 2 vols. Platt Establecimientos Graficos S.A.: Buenos Aires.

Joppien, R., 1976. Drawings from Captain Cook's Voyages. An unrecorded collection of fourteen ethnographical and natural history drawings relating to the second and third voyages. Hartnoll and Eyre Ltd: London.

Jouanin, C. and Mougin, J.-L., 1979. Order Procellariiformes in Peter's Checklist of the Birds of the World. Vol. 1, 2nd ed. Cambridge: Mass.

Kahn, R. L, Steiner, G., Fiedler, H., Popp, K-G. and Scheibe, S. (eds), 1972. Georg Forsters Werke. Sämtliche Schriften, Tagebücher, *Briefe*. Vol. 4. Akademie-Verlag: Berlin.

Latham, J., 1781–85. A General Synopsis of Birds. London.

Leuschner, B., Sheibe, S., Fiedler, H., Popp, K-G. and Schneider, A. (eds), 1982. Georg Forsters Werke. Sämtliche Schriften, Tagebücher, *Briefe*. Vol. 18. Akademie-Verlag: Berlin.

Lichtenstein, H. (ed), 1844. Descriptiones Animalium quae in itinere ad maris Australis terras per Annos 1772 1773 et 1774 suscepto collegit observavit et delineavit Joannes Reinoldus Forster. Berlin.

Löfgren, L., 1984. Ocean Birds: Their breeding, biology and behaviour. Croom Helm: London and Canberra.

Lowe, P. R. and Kinnear, N. B., 1930. British Antarctic ("Terra Nova") Expedition, 1910 Natural History Report. *Zoology*. Vol. IV(5):103–193. British Museum: London.

Lysaght, A., 1959. Some eighteenth century bird paintings in the library of Sir Joseph Banks (1743–1820). *Bull. Br. Mus. Nat. Hist. (Hist. Ser.)* **1:** 251–371.

Marchant, S. and Higgins, P. J., 1990. Handbook of Australian, New Zealand and Antarctic Birds, Vol. 1A. Oxford University Press: Melbourne.

Mathews, G. M., 1910–28. The Birds of Australia. London.

Mathews, G. M., 1912. A reference — list to the Birds of Australia. *Novit. Zool.* **18**(3)**:** 171–455.

Mathews, G. M., 1934. A Check-list of the Order Procellariiformes. *Novit. Zool.* **39:** 151–206.

Mathews, G. M., 1937. *Thalassogeron chrysostoma*. Grey-headed Mollymawk. *Ibis* **14 Ser.**(1)**:** 871.

Mathews, G. M. and Hallstrom, E. J. L., 1943. Notes on the Order Procellariiformes. Verity Hewitt Bookshop: Canberra.

McLachlan, G. R. and Liversidge, R., 1980. Roberts Birds of South Africa. 4th edition. John Voelcker Bird Book Fund: Cape Town.

Mitchel, B., 1772–75. Log kept on the *Resolution*. Adm 51/4555.

Murphy, R. C., 1936. Oceanic Birds of South America. American Museum of Natural History: New York.

Oliver, W. R. B., 1955. New Zealand Birds. 2nd edition. A. H. and A. W. Reed: Wellington.

Pennant, T., 1781. Genera of Birds: B. White: London.

Peters, J. L., 1931. Checklist of the Birds of the World. Vol. 1. Cambridge: Mass.

Ride, W. D. L., Sabrosky, C. W., Bernardi, G. and Melville, R. V. (eds), 1985. International Code of Zoological Nomenclature. Third Edition. International Trust for Zoological Nomenclature: London.

Robertson, C. J. R. and Nunn, G., 1997. Towards a new taxonomy for albatrosses. Pp. 13–19 *in* Albatross Biology and Conservation ed by G. Robertson and R. Gales. Surrey Beatty & Sons: Chipping Norton.

Rogers, A. E. F., 1980. Seabirds observed between Sydney and Buenos Aires. *Notornis* **27:** 69–78.

Rowan, M. K., 1951. The Yellow-nosed Albatross, *Diomedea chlororhynchos* Gmelin, at its breeding grounds in the Tristan da Cunha group. *Ostrich* **22:** 139–55.

Salvadori, T., 1914. Notes on some Species of the Genus Thalassogeron. *Ibis Ser.* **10**(2)**:** 503–06.

Salvin, O., 1896. Catalogue of the Tubinares in the Collection of the British Museum. Catalogue of the Birds: Volume 25. British Museum: London.

Scheibe, S. (ed), 1978. Georg Forsters Werke. Sämtliche Schriften, Tagebücher, Briefe. Vol. 13. Akademie-Verlag: Berlin.

Sherborn, C. D., 1902. Index Animalium. London.

Steiner, G. and Baege, L., 1971. Vögel der Südsee. 23 Gouachen und Aquarelle nach Zeichnungen Georg Forsters, entstanden während seiner Weltumsegelung 1772 bis 1775. Im Insel-Verlag: Leipzig.

Tickell, W. L. N., 1995. An Atlas of Southern Hemisphere Albatrosses. *Aust. Seabird Group Newsl. No.* **29:** 2–24.

Turbott, E. G. (conv.)., 1990. Checklist of the Birds of New Zealand and the Ross Dependency, Antarctica. Third. edition. Random Century: Auckland.

Whitehead, P. J. P., 1978. The Forster collection of zoological drawings in the British Museum (Natural History). *Bull. Br. Mus. Nat. Hist. (Hist. Ser.)* **6**(2)**:** 25–47.

CHAPTER TWO

Towards a new taxonomy for albatrosses

C. J. R. ROBERTSON[1] and G. B. NUNN[2]

Most previous revisions of albatross taxonomy have, like the original designations of taxa, been extrapolations based almost entirely on morphology from a poor specimen base. However, the use of mitochondrial DNA sequencing has demonstrated that both traditional biological and novel phylogenetic methods are largely supportive of each other. A review of the fossil albatrosses demonstrates a wider and more cosmopolitan distribution than those extant today. Of the at least 82 modern taxa recognized since the time of Linnaeus, a provisional present day total of 24 morphologically distinct terminal taxa (compared with the present 14 species) is recognized, within four genera (*Diomedea, Thalassarche, Phoebastria* and *Phoebetria*), rather than the present two. Final verification of albatross taxonomy will not be possible until molecular systematic analyses are available from all principal breeding localities.

Key words: Albatrosses, Taxonomy, Terminal Taxa, Morphology, Mitochondrial DNA.

INTRODUCTION

SINCE the birth of systematic taxonomy, knowledge of how to differentiate between taxa of albatrosses has been dominated by the work of museum specialists, with little reference until recently to any evidence obtained from field observations. Unfortunately for the development of taxonomic understanding the museum record is mostly old, often in poor condition, patchy in representation, rarely comprises adequate numbers of specimens and is dispersed world-wide. Most of the specimens have been collected at sea with no knowledge of location of natal island. No collection has an adequate number of specimens across a range of known-age birds for any albatross taxon. To our knowledge only three institutions world-wide have collections that contain representatives of all levels of the traditional polytypic species-level taxa. Specimens are often located at different parts of the world making cross comparisons almost impossible due to the cost of transporting material on loan. There is also reluctance by some institutions to hold collections of albatrosses, which are large and require considerable space to house. Adding to the problem is public distaste for taking live albatrosses for collections, the difficulty of collecting in remote locations and the cost of preparing, preserving and storing the material.

Other factors have also combined to limit understanding of albatross taxonomy. Many traditional taxa have no type specimen; for their identity we must rely on illustrations or a short and often generalized definition. Most revisions of albatross taxonomy have, like the original naming of taxa, been based on extrapolations, often with inadequate numbers of specimens from breeding localities. For example, in one current taxon, though suitable numbers of historic specimens were available — two museum collections at separate localities on either side of the Atlantic Ocean — it was nearly 80 years before specimens from the different breeding locations could be brought together for direct comparison. More recent activities by albatross taxonomists acted to counter the dispersed specimen base (e.g., see Robertson and Warham 1992). In this chapter we review albatross taxonomy at generic and specific levels with reliance upon new genetic data for higher level relationships.

The review does not presume to provide a complete and definitive revision of all "species" within the albatross family. However, in light of new genetic evidence and the application of a species concept pertinent to separating evolutionary pattern from underlying evolutionary processes which generate diversity, i.e., a phylogenetic species concept (Cracraft 1989), we

[1]Science and Research Division, Department of Conservation, PO Box 10420, Wellington, New Zealand.
[2]Department of Ornithology, American Museum of Natural History, Central Park West at 79th Street, New York, NY 10024 USA.
Present address: PE Applied Biosystems MS400, 850 Lincoln Centre Drive, Foster City, CA 94404 USA.
Pages 13–19 *in* ALBATROSS BIOLOGY AND CONSERVATION, ed by G. Robertson and R. Gales. Surrey Beatty & Sons, Chipping Norton. 1997.

endeavour to point out some significant areas where the present traditional biological species concept is too conservative. This conservatism restricts a growing view that islands host distinctive terminal taxa of albatross.

ALBATROSS TAXONOMY

There is a growing need for applying a pattern-based species concept to oceanic birds which spend a significant fraction of their life out sight of taxonomists. With current albatross taxonomy being based almost entirely on morphology from a relatively poor museum specimen base, the commencement of rigorous observational field studies has demonstrated a previously unsuspected diversity, as a widening range of taxonomic tools were applied to increasingly large samples of data.

Over the past 20 years there has been an increasing acceptance that life-history, behavioural and ecological data on albatrosses supported emerging conclusions from morphology-based taxonomy. The increasing ranges of field observational data, field techniques such as banding and, more recently, satellite tracking, are enabling us to "see" parts of the albatross life cycle previously only available from spot observations at sea. The recent development of molecular systematics as a taxonomic tool, in particular mitochondrial DNA sequencing, has enabled a reassessment of the relationship between traditional albatross taxonomy and their evolution. As the comb of techniques for explaining taxonomic differentiation gained more teeth, so it created a stable foundation upon which to conduct field investigations. Idiosyncratically among birds, the level of mitochondrial DNA sequence divergence between albatross taxa is relatively small compared to their diagnosable morphological and ecological character differences. Reassuringly, traditional taxonomic and novel molecular phylogenetic methods are largely supportive of each other.

THE FOSSIL RECORD

There is evidence of a range of fossil albatrosses with probably a wider and more cosmopolitan distribution than those extant today. Though it has been suggested that albatrosses developed during the Oligocene (Chattian), seemingly only one *(Diomedeoides minimus)* from near Leipzig in Germany (Fischer 1985) and one *(Plotornis delfortii)* from South Carolina, can be attributed to that period as the earliest records for the family. *Manu antiquus*, from the New Zealand early Oligocene, is reputedly earlier but, without additional material, has been disputed by Olson (1985a) as being indeterminate. Other than *D. minimus* there is little evidence until the Miocene and Pliocene of taxa with characteristics approaching those of the albatrosses of today.

The majority of the Miocene and Pliocene taxa (Olson 1985a) are from the northern hemisphere. Two taxa from California comprise one *(Diomedea milleri)* smaller than any extant taxa (but larger than the earlier *Plotornis*) and one *D. californica* that is similar in size or larger than the present *D. albatrus*. Howard (1982) and Chandler (1990) record the presence of further small but unnamed albatrosses from California. Fossil albatrosses are common in deposits in North Carolina, with possibly four species represented including one as large as *D. californica* (Olson 1985a). The lower Pleistocene *D. anglica*, from Suffolk, England, was smaller than the extant *D. exulans* but similar in size to *D. californica*. Fossil eggs of an albatross recently discovered in Bermuda are reputed to be similar to a moderately large albatross such as the extant *D. albatrus* of the Pacific. It seems clear that albatrosses were widespread in the north Atlantic into the late Tertiary, making their subsequent disappearance a Quaternary event (Olson 1985a).

It is probably not surprising that the now largely oceanic southern hemisphere has a poor avian fossil record (Olson 1985b,c). The oldest is a "small mollymawk", *D. thyridata* from Victoria, Australia, from the late Miocene (Wilkinson 1969). A large form (diagnosed from a single toe bone) and larger than *D. albatrus*, is recorded from Argentina from a similar period. These and a small species from the early Pliocene of South Africa are the oldest published records (Olson 1985c).

There are more recent fossil records from New Zealand. Those from sand deposits at Enderby Island (Auckland Islands) have been carbon-dated at about 2 000 years BP (McFadgen and Yaldwyn 1984) and have been identified as fitting *D. epomophora* (possibly breeding, as *in situ* egg shell is associated with complete skeletons). Though this species was exterminated on Enderby Island last century by whalers and sealers, recolonization is now occurring. At Pitt Island in the Chatham Islands there are also fossil deposits in consolidated sands (Falla 1960; Robertson 1991) carbon-dated as from 1 074–4 442 years BP. Previously thought to be a form of *D. exulans*, these now seem to be identifiable as *D. epomophora*, though some wing bones are about 10% longer than those of the extant species and larger than any other albatross alive today. Though Pitt Island is within 40 km of the present colonies of *D. e. sanfordi*, and the location has a demonstrated historic occupation by albatrosses, none breed there today. Human occupation of the Chatham Islands commenced between 300–600 years ago and albatross harvesting was actively pursued. We know of no other modern fossil record which may indicate a recent extirpation of albatross

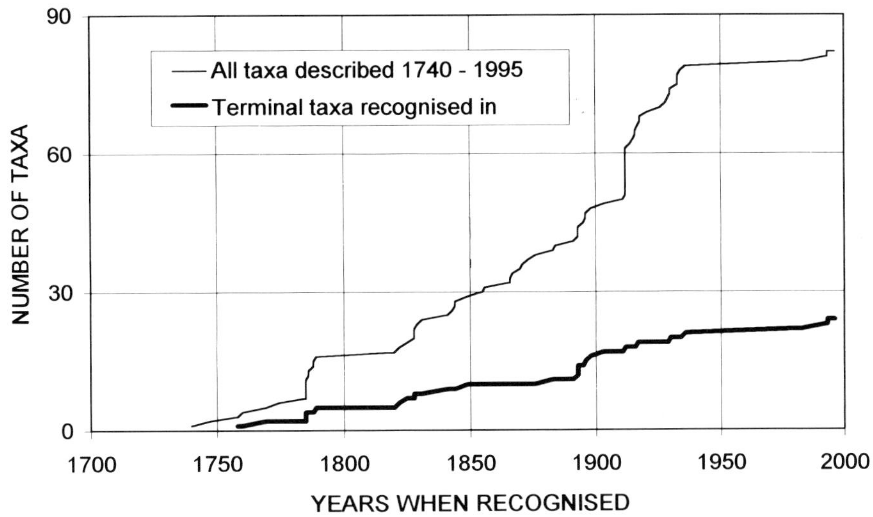

Fig. 1. The terminal taxa of albatrosses to 1995, showing the chronology of their original descriptions in comparison with the current species and subspecies recognized from 1740 to 1995.

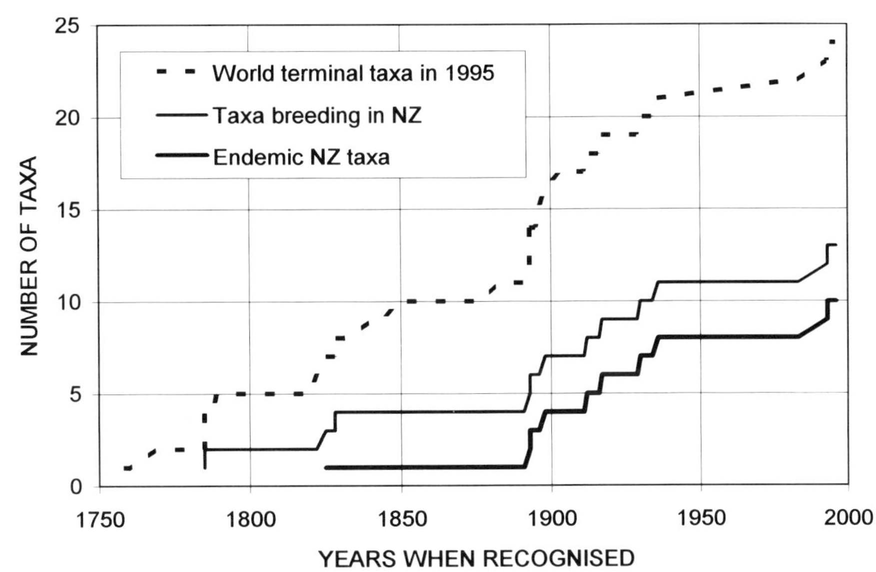

Fig. 2. Chronology of recognition for terminal albatross taxa (1995) breeding in, or endemic to, the New Zealand region, compared with the world taxa recognized in 1995.

colonies resulting from predation by humans, or by other predatory mammals, similar to or exceeding those which have been recorded, e.g., *D. albatrus* (Japan), *D. cauta cauta* (Australia), or *D. exulans* (Tristan da Cunha).

THE MODERN ALBATROSSES

Since the times of Linnaeus at least 82 different taxa of albatross have been recognized in the formal literature. However, with most specimens collected at sea, many taxa recognized and described were later found to be immature birds of the same species, and have subsequently been reduced to synonyms. Figure 1 shows the cumulative chronology of the description of albatross taxa covering species and subspecies.

Currently 14 species of albatross are recognized which comprise 24 terminal taxa by extension to subspecies within traditional polytypic species (Jouanin and Mougin 1979; Roux *et al.* 1983; Robertson and Warham 1992). New Zealand has the greatest diversity, and with the exception of *D. epomophora*, all endemic New Zealand albatross taxa have been recognized within the past 100 years (Fig. 2); this reflects the history of exploration of the area most remote from Europe. However, as the results from field studies and specific scientific expeditions have increased, so has the information for recognizing taxa more definitively.

DELIMITATION OF GENERA

Using biogeographic distribution, simple allometric relationship of wing and tail length, and characters of the divided plates making up the rhamphotheca of the bill, the extant albatross taxa fall into four broadly defined groups: (1) the North Pacific Albatrosses, (2) the Great Albatrosses, (3) the Southern Mollymawks, and (4) the Sooty Albatrosses (Warham 1990). Consideration of other simple morphological features has, from time to time, led to suggestions of splitting *Diomedea* as currently conceived into further generic groups such as *Thalassarche* and *Phoebastria* (Reichenbach 1852), *Thalassogeron* (Ridgway in Baird *et al.* 1884), *Diomedella* and *Nealbatrus* (Mathews 1912) and *Rhothonia* (Murphy 1917). None of these has gained acceptance and the number of albatross genera has remained obscure. In contrast, the "traditional" perception of relationships within the albatrosses distinguishes only two genera: the "primitive" Sooty Albatrosses *(Phoebetria),* and the remaining "White" Albatrosses *(Diomedea)* (Coues 1866; Peters 1931; Murphy 1936; Jouanin and Mougin 1979; Warham 1990).

Mitochondrial DNA

In an effort to resolve major lineages in the family, Nunn *et al.* (1996) used mitochondrial DNA sequences (the mitochondrial cytochrome-*b* gene) to determine phylogenetic relationships. In the majority of cases they made use of blood and tissue samples collected from breeding localities to' validate provenance. A small number of samples taken from adult birds at sea were used where land based samples were unavailable. Phylogenetic analysis, using both maximum parsimony and maximum likelihood techniques, identified a single best supported tree (Fig. 3). Further analyses of this data set using a greater number of outgroup taxa from the Procellariidae, Hydrobatidae and Pelecanoididae brought no change to the root location of the tree.

The results demonstrate an initial bifurcation in the Diomedeidae. These two main lineages each subdivide once more, giving rise to four phylogenetic groups of albatross — North Pacific Albatrosses, the Great Albatrosses, the Sooty Albatrosses and the Mollymawks, as traditionally conceived. However, the traditional genus *Diomedea* is shown to be paraphyletic. The southern hemisphere Mollymawks are in fact more closely related to the Sooty Albatrosses, and these together form a sister group to the remaining albatrosses. Consequently, to resolve the problem of paraphyly, Nunn *et al.* (1996) recommend re-instatement of the genera *Thalassarche* and *Phoebastria* (Reichenbach 1852), consistent with phylogenetic relationships. This results in *Thalassarche* comprising the Mollymawks (traditionally *Diomedea chlororhynchos, D. bulleri, D. cauta, D. chrysostoma* and *D. melanophrys*), *Phoebastria* containing the North Pacific Albatrosses (*D. albatrus, D. immutabilis, D. nigripes* and *D. irrorata*), and leaving *Diomedea* with only the Great Albatrosses (*D. epomophora, D. exulans* and *D. amsterdamensis*). The genus *Phoebetria* remains unchanged with *P. fusca* and *P. palpebrata.*

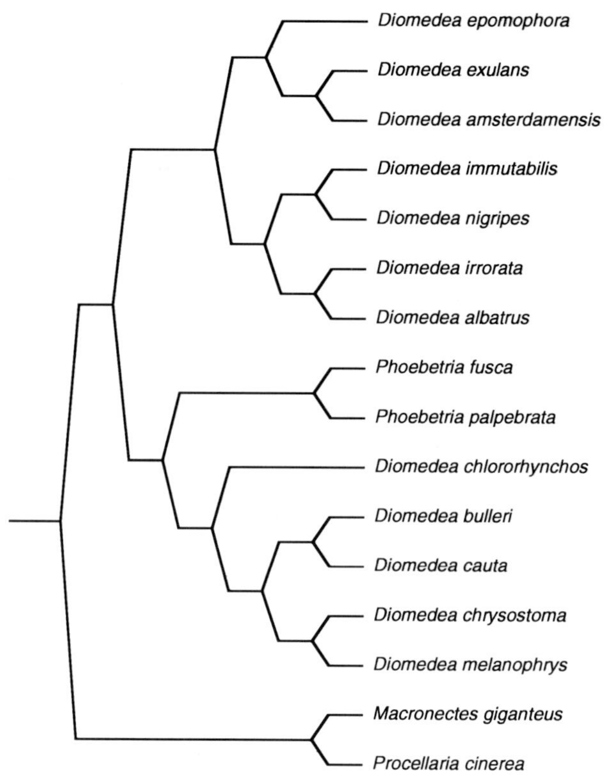

Fig. 3. Phylogenetic relationships among all traditional "species" of albatross. The tree is based on cladistic analysis of equally-weighted substitutional variation in the complete mitochrondrial cytochrome-*b* gene sequences of the 14 taxa (Nunn *et al.* 1996). The most parsimonious tree is 574 steps in length (consistency index, excluding uninformative characters = 0.591).

Using the higher level phylogenetic distinctions of genera by Nunn *et al.* (1996) it should now be possible to produce a generic key and updated species descriptions based on more than just the previous historic references to the bill and other simple allometric relationships.

THE TERMINAL TAXA, OR SPECIES

The clarification of generic lineages by phylogenetic analysis of mitochondrial DNA sequences has also stimulated a fresh insight into diagnosis of species of albatross. However, until specimens and taxa from all breeding locations are sampled and the series of specimens are enlarged, this differentiation of "terminal taxa" remains

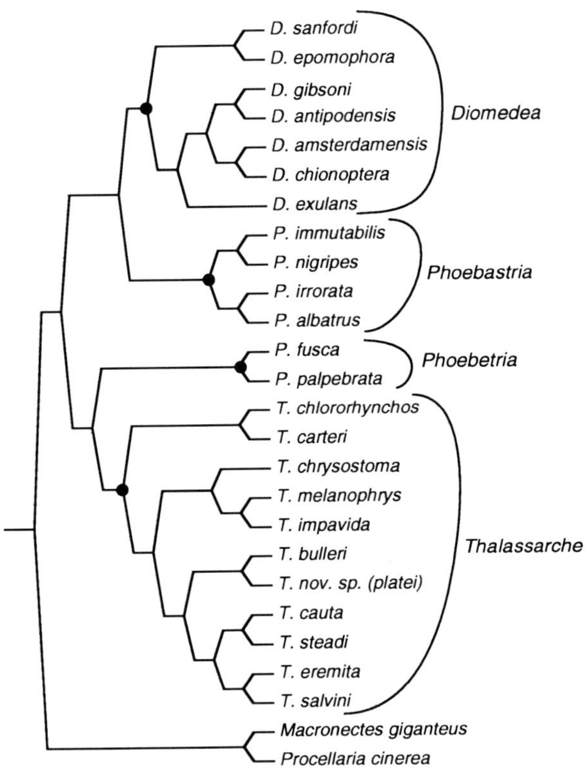

Fig. 4. Phylogenetic arrangement of the known morphologically diagnosable taxa of albatrosses. The tree is based on cladistic analysis of equally-weighted substitutional variation in the complete mitochondrial cytochrome-*b* gene sequences of the 24 taxa (Nunn *et al.* 1996; G. Nunn, unpubl. data). The most parsimonious tree is 601 steps in length (consistency index, excluding uninformative characters = 0.580). Solid circles represent the origin of lineage diversity in each higher-level group, i.e., genera (names are shown to the right).

provisional (Fig. 4), even though there are already clear phylogenetic and morphological differences for the separations shown.

North Pacific Albatrosses — *Phoebastria*

Geographically isolated from other albatross species, members of this genus can be found breeding at islands across the North Pacific Ocean. There are four described species, *P. albatrus, P. immutabilis, P. irrorata* and *P. nigripes* within the higher-level group with no traditional subspecies. Size variation is known between *P. nigripes* populations at Hawaii and islands off Japan, perhaps indicating a recent split between these archipelagoes. The acceptance of four terminal taxa is appropriate given current knowledge of diagnosis features within the genus.

Great Albatrosses — *Diomedea*

The largest and longest lived of all albatrosses, the Great Albatrosses *(Diomedea)* are complex in their own phylogeographic pattern. Currently *D. epomophora* is divided into two subspecies (*epomophora* and *sanfordi*) which breed only in New Zealand, but disperse along a southern circumpolar route (Robertson and Kinsky 1972; C. Robertson, unpubl. data). There are several key morphological distinctions between *epomophora* and *sanfordi* which support terminal taxa status.

The *D. exulans* complex grouped commonly under the vernacular name "Wandering Albatross", and including *D. amsterdamensis*, have proven difficult to analyse morphologically because of their complex plumage ontogeny; scoring systems have demonstrated changes occurring at 20+ years of age. Unlike *D. epomophora* with its circumpolar dispersal, particular "Wandering Albatross" taxa are much more restricted in oceanic distribution, and tend to differentiate into high latitude (lighter-plumaged) and low latitude (darker-plumaged) taxa. Again, a full selection of breeding locations have not been analysed, but five morphologically discernible terminal taxa can be presently recognized in the complex (Bourne 1989; Robertson and Warham 1992). Mitochondrial DNA sequences confirm differences between them, even though the diagnosis cannot be finalized until Crozet and Kerguelen material are included.

Analysis of mitochondrial cytochrome-*b* gene sequences demonstrate the very rare *D. amsterdamensis* is embedded phylogenetically among other "Wandering Albatross" taxa (G. Nunn, unpubl. data). As well as the Amsterdam Albatross *D. amsterdamensis*, the New Zealand "Wandering Albatrosses" are diagnosable morphologically and ecologically as two distinct taxa (*gibsoni* and *antipodensis*) each with a small breeding range similar to that of *D. epomophora* and *D. sanfordi*. There is no evidence that either of these New Zealand taxa overlap geographically with either the Gough Island *exulans* or *amsterdamensis*, nor of inter-breeding with the larger *chionoptera* taxon which migrates into their geographic feeding ranges.

The larger high latitude *chionoptera* taxon, commonly termed the "Snowy Wandering Albatross", breeds predominantly in the Atlantic and Indian Oceans. Inter-breeding of birds from this taxon of Crozet Island origin has been detected at Marion Island (J. Cooper, pers. comm.). Ecological and breeding behavioural differences suggest, however, that as in *Phoebetria* and the smaller "Wandering Albatross" taxa (i.e., *amsterdamensis, antipodensis* and *gibsoni*), a closer investigation is warranted of the populations of *chionoptera* in the Atlantic and Indian Oceans.

Mollymawks — *Thalassarche*

The most diverse group is the Mollymawks. These small southern albatrosses have become morphologically varied taxa since splitting from a common ancestor with Sooty Mollymawks *Phoebetria.* The Mollymawks have developed foraging patterns which, with the exception of *chrysostoma* (which breeds biennially), have enabled annual breeding cycles to evolve. Interestingly, phylogenetic sister-taxa of Mollymawk are generally divergent morphologically into lighter and darker head patterns. Similarly, the Mollymawk bill, with varying combinations of coloured plates and bare fleshy skin, has become a taxon-distinctive visual tool used in association with specific vocalizations during courtship and territorial display. At breeding islands around New Zealand where Mollymawk taxa mix an aversion is often observed when intertaxon display or vocalization is attempted.

The six terminal taxa within the distinctively New Zealand/Australasian groups of *"cauta"* and *"bulleri"* have diversified extensively within a small geographic area. The two groups of Mollymawk terminal taxa contain the best examples of extreme breeding isolation, where closely related but geographically disjunct taxa developed extensive gaps of several months between egg-laying periods. In the case of *T. bulleri* breeding is two months later at The Snares and Solander Islands than at the Chatham Islands *(T. platei)* and incubation stints are about three times the length. Though currently termed *T. platei*, the Chatham population is actually an undescribed taxon and *T. platei* should be reduced to a synonym, being just a juvenile plumage phase of *T. bulleri* (Murphy 1936; C. Robertson, pers. obs.).

Due to the isolation of their breeding sites the four terminal taxa in the *"cauta"* group are probably the most difficult to study and were, not surprisingly, the most recently described taxa. *T. cauta* and *T. steadi* can be differentiated by wing morphometrics, which do not overlap, though other differences seem less clear-cut. As with *T. bulleri* and its closest (unnamed) relative from Chatham Islands, there is a two month interval between the egg laying periods of *T. cauta* and *T. steadi.* It seems significant that populations of *T. steadi, T. salvini* and *T. eremita* are morphologically distinct, but breed within 1 000 km of one another, along with the phylogenetically close relatives *T. bulleri* and current *T. platei*, yet have evolved clearly unique morphological diagnosable features supporting their status as distinct terminal taxa.

Sooty Mollymawks — *Phoebetria*

Members of the genus *Phoebetria* arose from a common ancestor with Mollymawks *(Thalassarche)* and can be justifiably renamed Sooty Mollymawks as a common term. Only two terminal taxa are currently recognized in the genus — the circumpolar distributed *P. palpebrata* versus the more limited *P. fusca*. Analysis of further cytochrome-*b* DNA sequences (G. Nunn, unpubl. data) suggests differentiation within *P. fusca*. A closer examination of morphology and breeding biology of *P. fusca* in relation to geographic location is desirable in the future.

CONCLUSION

Both the fossil and the modern record for the Diomedeidae indicate a wide diversity of albatross taxa. The modern taxa are distinctive both in morphology and breeding ecology. In spite of being relatively well known oceanic seabirds, our knowledge of the relationships of breeding and feeding ecology for these taxa is limited, when compared with that for land birds. The research base, even in the most intensively studied albatross taxa, only covers 5–6 generations.

The modern tool of analysing DNA sequences has provided a refined insight into the phylogenetic relationships among albatross taxa and suggested a new hypothesis for their evolution. The combination of DNA cytochrome-*b* analysis and a better appreciation of morphological differences have produced a revision among albatrosses at the generic level (Nunn *et al.* 1996).

Also emerging from molecular evidence, coupled with a pattern-based species concept, is a less conservative approach than has been prevalent since Peters (1931) and Jouanin and Mougin (1979), towards the acceptance of morphologically distinct terminal taxa as species (see Appendix 1).

For those ornithologists who seek to unravel the complexities of an albatross lifestyle it is essential that there is a clear basis for recognizing and defining the animals on which we work. In an age when the protection of biodiversity is increasingly important, there is a demonstrable danger of the current taxonomy of 14 species subsuming the importance of the 24 morphologically diagnosable terminal taxa in arguments for their conservation. Clearly, a generous approach can be warranted for diagnosis of albatross taxa, which are, in general terms, decreasing. For instance management decisions relating to fisheries by-catch of "Wandering Albatross", become substantially more important if we are dealing with five morphologically and genetically distinct and localized taxa, than one all encompassing species called the "Wandering Albatross".

Until we have analysed a more complete range of molecular data and included all geographic representatives of the Diomedeidae, we can not be more positive about some of the gaps and possibilities suggested here. Similarly, there needs to be a concerted global effort among albatross researchers over the next 3–5 years to develop a comprehensive morphological database from field measurements and observations. This does not mean that specimen collecting should cease, but it needs to be concentrated in the areas of least knowledge, primarily in the recognition of known age sub-adult material and material from breeding locations. When this happens, there will have been significant progress towards the knowledge needed to proclaim definitive diagnosis of terminal taxonomy.

ACKNOWLEDGEMENTS

The authors wish to acknowledge those persons, agencies and museums world-wide who have given the required permission or ethical approvals to collect museum and other specimens, provided blood or tissue samples for DNA sequencing, or allowed access to their bird specimen collections. The DNA sequencing and laboratory work was funded by the Frank M. Chapman Memorial Fund, through a fellowship to G. B. Nunn, and carried out at the American Museum of Natural History, New York. The New Zealand Department of Conservation (and its predecessor the NZ Wildlife Service) funded participation in many field expeditions in New Zealand or Australia, and contributed towards museum research overseas by C. J. R. Robertson from 1967 to 1995. We are indebted to M. J. Imber, L. Clelland, D. Schodde and two anonymous referees that reviewed the manuscript and provided thoughtful comments.

APPENDIX 1

A revision of the family Diomedeidae giving genera, terminal taxa and recommended common names.

***Phoebastria* — Pacific Albatrosses**	
immutabilis 1893	Laysan Albatross
nigripes 1849	Black-footed Albatross
irrorata 1883	Waved Albatross
albatrus 1769	Short-tailed Albatross
***Diomedea* — Great Albatrosses**	
sanfordi 1917	Sanford's Albatross
epomophora 1825	Royal Albatross
gibsoni 1993	Gibson's Albatross
antipodensis 1993	Antipodean Albatross
amsterdamensis 1983	Amsterdam Albatross
chionoptera 1896	Snowy Albatross
exulans 1758	Wandering Albatross
***Phoebetria* — Sooty Mollymawks**	
fusca 1822	Sooty Mollymawk
palpebrata 1785	Light-mantled Mollymawk
***Thalassarche* — Mollymawks**	
chlororhynchos 1789	Yellow-nosed Mollymawk
carteri 1903	Carter's Mollymawk
chrysostoma 1785	Grey-headed Mollymawk
melanophrys 1828	Black-browed Mollymawk
impavida 1912	Campbell Mollymawk
bulleri 1893	Buller's Mollymawk
**nov. sp. *(platei)*	Pacific Mollymawk
cauta 1841	Tasmanian Mollymawk
steadi 1936	White-capped Mollymawk
eremita 1930	Chatham Mollymawk
salvini 1893	Salvin's Mollymawk

REFERENCES

Baird, S. F., Brewer, T. M. and Ridgway, R., 1884. The water birds of North America. Little, Brown and Co., Boston USA.

Bourne, W. R. P., 1989. The evolution, classification and nomenclature of the great albatrosses. *Gerfaut* **79:** 105–16.

Chandler, R. M., 1990. Fossil birds of the San Diego Formation, Late Pliocene, Blancan, San Diego County, California. *Ornithol. Monogr.* **44:** 73–161.

Coues, E., 1866. Critical review of the Family Procellariidae; — Part V; embracing the Diomedeinae and the Halodrominae. *Proc. Acad. Nat. Sci., Philadelphia* **18:** 172–97.

Cracraft, J., 1989. Speciation and its ontology: the empirical consequences of alternative species concepts for understanding patterns and processes of differentiation. Pp. 28–59 *in* Speciation and its consequences ed by D. Otte and J. A. Endler. Sinauer Associates: Sunderland, Massachusetts.

Falla, R. A., 1960. Notes on some bones collected by Dr Watters and Mr Lindsay at Chatham Islands. *Notornis* **8:** 226–27.

Fischer, K., 1985. Ein albatrosartiger Vogel (*Diomedeoides minimus* nov. gen., nov. sp., Diomedeoididae nov. fam., Procellariiformes) aus dem Mitteloligozn bei Leipzig (DDR). *Mitt. Zool. Mus. Berl.* **61.** *Suppl.: Ann. Orn.* **9:** 113–18.

Howard, H., 1982. Fossil birds from the Tertiary marine beds at Oceanside, San Diego County, California, with descriptions of two new species of the genera *Uria* and *Cepphus* (Aves: Alcidae). *Nat. Hist. Mus. LA County Contrib. Sci.* **341:** 1–15.

Jouanin, C. and Mougin, J.-L., 1979. Order Procellariiformes. Pp. 48–121 *in* Peter's Checklist of the Birds of the World ed by E. Mayr and G. W. Cottrell. Harvard University Press, Museum of Comparative Zoology: Cambridge, Massachusetts.

Mathews, G. M., 1912. Birds of Australia. Witherby: London.

McFadgen, B. G. and Yaldwyn, J. C., 1984. Holocene sand dunes on Enderby Island, Auckland Islands. *NZ J. Geol. Geophy.* **27:** 27–33.

Murphy, R. C., 1917. A new albatross from the west coast of South America. *Bull. Amer. Mus. Nat. Hist.* **37:** 861–64.

Murphy, R. C., 1936. Oceanic birds of South America. American Museum of Natural History: New York.

Nunn, G. B., Cooper, J., Jouventin, P., Robertson, C. J. R. and Robertson, G. G., 1996. Evolutionary relationships among extant albatrosses (Procellariiformes: Diomedeidae) established from complete cytochrome-*b* gene sequences. *Auk* **113:** 784–801.

Olson, S. L., 1985a. The Fossil Record of Birds. Avian Biology. Vol. III, 79–238. Academic Press: Orlando.

Olson, S. L., 1985b. An early Pliocene marine avifauna from Duinefontein, Cape Province, South Africa. *Ann. Sth Afr. Mus.* **95:** 147–64.

Olson, S. L., 1985c. Early Pliocene Procellariiformes (Aves) from Langebaanweg, Southwestern Cape Province, South Africa. *Ann. Sth Afri. Mus.* **95:** 123–45.

Peters, J. L., 1931. Check-list of birds of the world. Harvard University Press: Cambridge, Mass.

Reichenbach, L., 1852. Avium Systema Naturale. Leipzig: Germany.

Robertson, C. J. R., 1991. Questions on the harvesting of Toroa in the Chatham Islands. Science and Research Series No. 35. Department of Conservation: Wellington.

Robertson, C. J. R. and Kinsky, F. C., 1972. Dispersal movements of the Royal albatross *(Diomedea epomophora) Notornis* **19:** 289–301.

Robertson, C. J. R. and Warham J., 1992. Nomenclature of the New Zealand Wandering Albatrosses *Diomedea exulans. Bull. Brit. Ornithol. Club* **112:** 74–81.

Roux, J.-P., Jouventin, P., Mougin, J.-L., Stahl J.-C. and Weimerskirch, H., 1983. Un nouvel albatross *Diomedea amsterdamensis* n. sp. decouvert sur l'Ile Amsterdam (37°50'S, 77°35'E). *Oiseau* **53:** 1–11.

Warham, J., 1990. The Petrels. Their ecology and breeding systems. Academic Press: San Diego.

Wilkinson, H. E., 1969. Description of an Upper Miocene albatross from Beaumaris, Victoria, Australia, and a review of the fossil Diomedeidae. *Mem. Nat. Mus. Vict.* **29:** 41–51.

CHAPTER THREE

Albatross populations: status and threats

R. GALES[1]

The population and conservation status of the world's albatrosses are reviewed within the framework of recent taxonomic changes. The latest estimates of size of breeding populations (pairs) at all known localities of each of the 24 proposed species are presented; population trends are assessed where sufficient data are available. Despite increased efforts in population monitoring, the status (i.e., population trends) of two-thirds of the world's *ca.* 150 albatross populations remain unknown. For those that are known, almost half are decreasing. The threats currently facing each species are briefly reviewed. The best available evidence indicates that longline fishing is the most serious threat facing albatrosses today. Twenty-one of the 24 species are known to be killed on longline hooks, including rare and endangered species. Widespread implementation of appropriate mitigation measures is urgently required.

Key words: Albatrosses, Population Status, Threats, Conservation, Longline Fishing.

INTRODUCTION

THE status of albatross populations is increasingly a focus of attention for seabird biologists, conservation groups, fisheries scientists and managers, fishing practitioners and industry representatives. This is primarily due to realization of the high number of birds being killed during commercial fishing operations, resulting in the decrease of many albatross populations. Gales (1993) reviewed the status of albatross populations and the factors affecting them, and concluded that direct mortality associated with commercial fishing operations is the most serious threat facing albatross populations. The fishery most commonly reported to be associated with albatross bycatch is longlining, both demersal and pelagic.

Stimulated by increasing concern regarding the extent of these interactions, and the consequences for albatross populations, many new data have been collected since the 1993 review. These new data enable better understanding of the current status of albatross populations and the factors affecting them. Coincidentally, in a major revision of albatross systematics, Nunn *et al.* (1996) and Robertson and Nunn (1997) propose elevating the rank of several taxa, which would result in an increase from 14 to 24 species. The interaction between taxonomy, management and conservation raises many issues, particularly the problems of identifying the units of taxonomic concern (Groombridge 1993). This chapter incorporates the scientific nomenclature for the genera and species proposed by Robertson and Nunn, following the consensus reached at the 1995 Albatross Conference.

The purpose of this chapter is to update and revise the data and assessments presented in Gales (1993) in light of the new taxonomy. Status assessments inevitably focus at the level of taxonomically defined units, usually at the species level (see Croxall and Gales 1997). The extent of the recent taxonomic revision of albatrosses, and the fact that this may still undergo further development, makes it important also to review the status of albatrosses at the island population level. Consideration at the population level improves our understanding of the status of albatrosses and the threats they face, and is of particular importance in relation to management and conservation issues.

POPULATION NUMBERS AND DYNAMICS

Species profiles

The population size information is from the sources listed in Appendix 1. Species distributions and timing of breeding information are from the sources cited in Gales (1993). The status assessments were based on all available comparable data, not merely comparisons of the two most recent counts or surveys which may not

[1]Tasmanian Parks and Wildlife Service, Department of Environment and Land Management, 134 Macquarie Street, Hobart, Tasmania, Australia 7001.
Pages 20–45 *in* ALBATROSS BIOLOGY AND CONSERVATION, ed by G. Robertson and R. Gales. Surrey Beatty & Sons, Chipping Norton. 1997.

necessarily be comparable in methodology and accuracy. This is important given the annual fluctuations which may occur in albatross population size, particularly for the biennial-breeding species. The descriptions of the threats described in the species profiles are largely restricted to the threats currently affecting albatrosses. An historical account of all threats, past and present, is in Gales (1993).

Wandering Albatross *Diomedea exulans*

Previous name: Wandering Albatross *Diomedea exulans exulans.*

Breeding distribution and jurisdiction: South Georgia (United Kingdom); Crozet and Kerguelen Islands (France); Marion and Prince Edward Islands (South Africa); Macquarie Island (Australia).

Breeding frequency and season: Biennial breeder (when successful). Most eggs laid between December and February, the eggs are incubated for *ca.* 11 weeks, chicks hatch in March–April and fledge 40 weeks later between November and February.

Population size: Annual breeding population estimated at 8 500 pairs (see Appendix 1A), representing about 28 000 mature individuals, perhaps 55 000 birds in total. Reliability of survey data for this species is generally good, with time series population data for at least five island sites. Most Wandering Albatrosses breed in the South African sector, with the Marion Island and Prince Edward Island populations accounting for 36% of the global population.

Status: In 1995 the Macquarie Island population of Wandering Albatrosses was listed as an endangered species under the *Australian Endangered Species Protection Act 1992.*

The status of eight populations is known, and of these three are thought to be increasing, four are decreasing and one is stable. The tiny Macquarie Island population is currently stable after previous declines, currently with <10 pairs breeding annually.

The rates of population decrease vary according to population and the time frame over which changes are assessed. The Ile de la Possession (Crozet Islands) population has decreased by over 50% over the last two decades, compared to the 28% population decrease documented for the Bird Island (South Georgia) population since the early 1960s (Weimerskirch and Jouventin 1997; Croxall *et al.* 1997). The rate of decrease of the Crozet population has changed from –7% p.a. (1970–76), to –1.4% p.a. (1977–85) and has recently started to recover at 4% p.a. (1986–95) Weimerskirch *et al.* 1997). The Bird Island population, however, continues to decrease (Croxall *et al.* 1997).

Threats: The major factor affecting this species is mortality associated with commercial fishing operations, particularly longlining (Gales 1993), a view supported by assessments at the 1995 Albatross Conference and by Birdlife International (1995). Whilst other fishing operations also impact on the species (e.g., trawling and dropline fishing), the extent of mortality is much less than that associated with longline fishing.

Longline fishing does not affect Wandering Albatrosses equally across their range because different populations have different foraging distributions, and this affects the extent of overlap with fisheries. The South Georgia population may be most at risk from longline fishing operations throughout the southern sectors of the Atlantic, Indian and Pacific Oceans, whereas the Crozet population is more vulnerable to fishing operations within the Indian Ocean and Australian region. The vulnerability of Wandering Albatrosses to longline fishing may also differ within populations as a result of specific migration patterns during the breeding season and by birds of different sex, age and breeding status (and so influence extent of exposure to longline fishing: see Prince *et al.* 1997; Weimerskirch 1997).

The status of populations is dependent upon the spatial and temporal distribution of longline fishing effort. The recent signs of recovery, evidenced by increases in both adult and juvenile survival rates of the Crozet Island Wandering Albatross population, have been ascribed to the shift in longline fishing effort away from the birds foraging grounds (Weimerskirch *et al.* 1997).

Tristan Albatross *Diomedea dabbenena*

Previous name: Wandering Albatross *Diomedea exulans dabbenena.*

Breeding distribution and jurisdiction: Inaccessible Island (Tristan da Cunha group) and Gough Island (United Kingdom).

Breeding frequency and season: Biennial breeder (when successful). No published studies of breeding ecology, presumed most eggs laid between December and February and chicks fledge the following November to February.

Population size: The total breeding population is estimated at fewer than 2 000 pairs (see Appendix 1A). This suggests about 1 000 pairs breeding in any one year and perhaps 6 000–7 000 birds in the total global population. This species is virtually restricted to Gough Island with only two to three pairs breeding each year on Inaccessible Island in the Tristan da Cunha group (Ryan *et al.* 1990). The reliability of survey data for the population on Gough Island is poor, the only available data being a rough estimate in the 1980s (J. Cooper, pers. comm.).

Status: Endemic to territories of the United Kingdom.

As there are no time series data for the population on Gough Island, the status of this population is unknown. Given the trends of other albatross populations in the Indian Ocean it is likely that this population also has decreased since the 1980s. The population on Inaccessible Island was much larger in the past but the current population of 2–3 pairs appears to be stable (Ryan *et al.* 1990). Historically, Tristan Albatrosses also bred on the main island of the Tristan Group but were extirpated by humans at the turn of the Century (Watkins 1987).

Threats: Human persecution of Tristan Albatrosses at the breeding sites has been largely (if not totally) eliminated (J. Cooper, pers. comm.). Whilst small plastic particles have been collected from regurgitates of Tristan Albatrosses on Gough Island (J. Cooper, pers. comm.), the most likely threat to this population comes from longline fishing. Band returns confirm that Tristan Albatrosses from Gough Island are killed on longline hooks, and the foraging distribution of this species, which encompasses the South Atlantic Ocean and coastal regions of Southern Africa, place the species in contact with numerous longline fleets (J. Cooper, pers. comm.). Knowledge of the population status of this species and extent of fishing-related mortalities is urgently required.

Antipodean Albatross *Diomedea antipodensis*

Previous name: Wandering Albatross *Diomedea exulans antipodensis.*

Breeding distribution and jurisdiction: Antipodes Island and Campbell Island (New Zealand).

Breeding frequency and season: Biennial breeder (when successful). Egg laying starts in January (Antipodes Island) and February (Campbell Island) and chicks fledge between January and March the following year.

Population size: The annual breeding population is estimated at 5 150 pairs (see Appendix 1A), indicating a global population of 17 000 adults, or 33 000 individuals. The species is essentially restricted to Antipodes Island, although about six pairs also breed on Campbell Island each year.

Status: Endemic to New Zealand.

The current estimates of population size are reliable. A lack of comparable data prior to 1994, however, precludes any assessment of the status of the Antipodes Island population. Occasional surveys of the Campbell Island population of Antipodean Albatrosses since the 1960s suggests that this population has been stable at low numbers for at least three decades.

Threats: The only factor which has been identified as a threat to Antipodean Albatrosses is longline fishing. This species has been confirmed as being killed on longlines targeting tuna in New Zealand waters (Murray *et al.* 1993). Outside the breeding season, this species is known to migrate from New Zealand eastwards to Chile and the Patagonian shelf before returning to New Zealand sea mounts and the Tasman Sea (D. Nicholls, pers. comm.). The flights over the southern Pacific Ocean and Tasman Sea would put it in contact with oceanic longline fleets. The prolonged periods spent off the coast of Chile, an area where longline fishing effort is increasing, also presents a threat to individuals of this species.

Gibson's Albatross *Diomedea gibsoni*

Previous name: Wandering Albatross *Diomedea exulans gibsoni.*

Breeding distribution and jurisdiction: Auckland Island (New Zealand).

Breeding frequency and season: Biennial breeder (when successful). Most eggs laid between December and January and chicks fledge the following year between January and February.

Population size: The annual breeding population of Gibson's Albatrosses is estimated at about 6 200 pairs breeding each year (see Appendix 1A), perhaps 10 000 pairs in total, or 40 000 individuals. This species is restricted to breeding on three islands within the New Zealand sub-antarctic Auckland Island group: Adams Island (5 800 pairs; 95% population), Disappointment Island (250 pairs; 4% population) and Auckland Island (65 pairs; 1% population).

Status: Endemic to New Zealand.

The Gibson's Albatross population on Adams Island was estimated as 13 000 pairs in the 1970s but inconsistent survey techniques preclude valid comparisons with more recent information. This population has been studied annually since 1991 so knowledge of recent population trends should soon be forthcoming.

Threats: On Auckland Island the effects of introduced pests (cats and pigs) have the potential to limit the breeding success of Gibson's Albatrosses. The only other threat identified for the species is longline fishing. Gibson's Albatrosses were a significant bycatch species in the tuna longline fishery operating in New Zealand waters between 1988 and 1992 (Murray *et al.* 1993). This species is also killed on longline hooks deployed to catch tuna in the Australian region (Gales, Brothers and Reid, unpubl. data).

Satellite tracking studies have shown that Gibson's Albatrosses traverse areas over the Tasman Sea and eastwards into the Pacific

Ocean during the breeding season. Foraging areas used by males and females were mutually exclusive, with female birds frequenting the Tasman Sea in the vicinity of 40°S, whilst the males dispersed westwards at lower latitudes or, alternatively, travelled north-east towards the mid-Pacific Ocean (Elliot *et al.* 1995). These differences, and the sex-specific adult survival rates (males being higher than females; Walker and Elliot 1995) may not be coincidental. The continuing studies of this species should clarify the role of longline fishing in the population and conservation status of Gibson's Albatrosses.

Northern Royal Albatross *Diomedea sanfordi*

Previous name: Northern Royal Albatross *Diomedea epomophora sanfordi.*

Breeding distribution and jurisdiction: Chatham Islands (Big and Little Sister, and the Forty Fours), South Island (Taiaroa Head) of New Zealand. All New Zealand jurisdiction.

Breeding frequency and season: Biennial breeder (when successful). The full breeding cycle usually extends from November to September. Egg period lasts 79 days and nestling period about 240 days.

Population size: The annual breeding population is estimated at about 5 200 pairs, equivalent to a total breeding population of 8 500 pairs (see Appendix 1A), and perhaps 34 000 individuals in total. The Chatham Islands population accounts for >99% of the population, with <20 pairs breeding at Taiaroa Head each year. This small population includes five Southern Royal × Northern Royal Albatross hybrids, mixed pairs also occur at Enderby Island (C. J. Robertson, pers. comm.).

Status: Endemic to New Zealand.

The population of Northern Royal Albatrosses at the Chatham Islands is decreasing and this trend is expected to continue (see below). The population at Taiaroa Head, established in 1920, is slowly increasing being assisted by intensive human surveillance and management.

Threats: In the past, harvesting by humans has affected the Northern Royal Albatrosses on the Chatham Islands and though now illegal, small-scale harvesting of chicks is still thought to occur (C. J. Robertson, pers. comm.). The current population decrease of this population however, is thought to be mainly a result of nesting habitat degradation following severe storms in the 1980s (Robertson 1997). The effects of climatic changes and perturbations, which result in changes to the nesting habitat either through drying out or storm damage, are likely to have a significant effect on the status of this species for many years to come (Robertson 1997). The mainland colony of Northern Royal Albatrosses at Taiaroa Head (Dunedin, New Zealand) has increased, assisted by control of predators and human interference, and despite the incidence of flystrike which is responsible for some mortality of hatchlings (see Robertson 1997).

Fishing operations also affect Northern Royal Albatrosses. These birds are caught on longlines in the waters off southern Australia (Gales, Brothers and Reid, unpubl. data), and their extensive oceanic distribution exposes them to interactions with an array of longline operations. Fishing-related mortality at sea, whilst perhaps not the primary threat to this species, serves to hasten the decrease of the population.

Southern Royal Albatross *Diomedea epomophora*

Previous name: Southern Royal Albatross *Diomedea epomophora epomophora.*

Breeding distribution and jurisdiction: Campbell Island, Enderby Island, Adams Island and Auckland Island (New Zealand).

Breeding frequency and season: Biennial breeder (when successful). Most eggs laid in November–December, chicks hatch in February–March and fledge after eight months in October–November.

Population size: The annual breeding population is estimated as about 7 870 pairs, equivalent to a total breeding population of *ca.* 13 000 pairs (see Appendix 1A), and perhaps 50 000 individuals in total. There are four breeding sites but >99% of birds breed at Campbell Island, while the remaining three populations have fewer than 55 pairs breeding annually.

Status: The status of Southern Royal Albatrosses at Campbell Island is thought to be increasing but interpretation of counts is difficult due to inconsistent census efforts. The breeding population appears to have increased (at least until the 1980s), recovering from the effects of human predation during the sealing era, and the effects of burning and grazing during the farming era (until 1931). Fluctuations in the numbers of annual breeding population (23% between 1995 and 1996, P. Moore, pers. comm.) make recent trends difficult to interpret. The relict population at Enderby Island is currently increasing, following recolonization in 1940 after extirpation in the 1860s.

Threats: The impacts associated with the sealing and farming periods on Campbell Island ceased in the 1930s. At sea, trawling operations are known to have killed Southern Royal Albatrosses, but this appears to have been mitigated on vessels where the use of netsonde monitor cables were abolished (Bartle 1991). Today, longlining represents the major threat to Southern Royal Albatrosses. Southern Royal Albatrosses are known to have been caught on longlines in the

South Atlantic Ocean, the Indian Ocean and in the Australian Fishing Zone (Gales 1993 and references therein). All banded Southern Royal Albatrosses caught in the Australian Fishing Zone are from Campbell Island. The circumpolar dispersal of this species allows for extensive overlap with longline operations, and the nature and magnitude of the bycatch for most of these fleets remains to be quantified.

Amsterdam Albatross *Diomedea amsterdamensis*

Previous name: Amsterdam Albatross *Diomedea amsterdamensis.*

Breeding distribution and jurisdiction: Amsterdam Island (southern Indian Ocean), French jurisdiction.

Breeding frequency and season: Biennial breeder (when successful). Most eggs laid in February–March, chicks hatch in May and fledge in January–February of the following year.

Population size: Total breeding population estimated at 20 pairs (*ca.* 13 eggs laid each year), perhaps 90 birds in total (see Appendix 1A). One breeding site.

Status: Endemic to Amsterdam Island, an external territory of France. Classified as *Critically Endangered* (Collar *et al.* 1994). This species is listed by the Bonn Convention (Convention on the Conservation of Migratory Species of Wild Animals) as *Endangered* under Appendix 1.

Threats: Subfossil records show that population was historically much larger, restriction of current population being caused by fire and habitat degradation by cattle, and probably by deaths associated with fishing activities (Jouventin *et al.* 1989; Weimerskirch and Jouventin 1997). The impact of cattle is now restricted by exclusion fences, but predation by cats and rats remains a potential threat. The current population trend mirrors that of other Wandering Albatross populations in the Indian Ocean, and may have been similarly depleted by deaths on longline hooks set in proximity to the island during the 1970s and 1980s (Weimerskirch and Jouventin 1997). Whilst these local fishing activities have contracted, there are concerns regarding overlap between these birds and longliners operating in subtropical-tropical waters (H. Weimerskirch, pers. comm.). Amsterdam Albatrosses were recorded as bycatch on longliners operating south of Tasmania in 1992 (N. Brothers, pers. comm.). For this *Critically Endangered* species, any increase in mortality rates above natural levels would likely be catastrophic.

Short-tailed Albatross *Phoebastria albatrus*

Previous name: Short-tailed Albatross *Diomedea albatrus.*

Breeding distribution and jurisdiction: Torishima Island in Izu Islands, off southeastern Japan and Minami-kojima (Senkaku Islands) (Japan). Recent unsuccessful breeding attempt at Midway Atoll, Hawaii (USA).

Breeding frequency and season: Annual breeding. Eggs are laid in October–November, hatch in December–January, and chicks fledge in May–June.

Population size: While currently confined to two breeding locations, the Short-tailed Albatross was formerly an abundant species breeding on at least 11 Japanese islands. The toll taken by feather hunters and volcanic eruptions almost exterminated the species by the 1940s. In 1950, a relict population was discovered on the volcanic ash slopes of Torishima Island and this population is slowly recovering with 156 eggs laid in 1995. On Minami-kojima, the presence of a small population was confirmed in 1988, and in 1991 this population was estimated at 15 pairs (Hasegawa 1991). One Short-tailed Albatross egg was laid at Midway Atoll in 1993, but failed to hatch (Richardson 1994).

The total breeding population in 1995 was *ca.* 170 pairs (see Appendix 1B), perhaps 700–800 birds in total.

Status: Endemic to Japan.

Classified as *Endangered* (Collar *et al.* 1994). This species is listed by the Bonn Convention (Convention on the Conservation of Migratory Species of Wild Animals) as *Endangered* under Appendix 1.

Population status information is only available for the Torishima population which has been increasing at 7% p.a. (Hasegawa 1991).

Threats: The small size of the population of this species and restricted range (currently only two island breeding sites in Japan) combine to provide an unfavourable conservation status. Vulnerability is exacerbated by Torishima Island being an active volcanic island and the unstable nature of the ash slopes where the birds breed. Efforts to attract birds to nest on more stable slopes have been successful with one chick hatching in this new colony in 1996 (H. Hasegawa, pers. comm.).

Ingestion of plastics poses a potential risk to these birds (see Gales 1993), but the most serious risk to these birds is the mortality associated with fishing practices. Short-tailed Albatrosses are known to scavenge from longline vessels in the Pacific Ocean and Bering Sea/Gulf of Alaska region and birds have been seen at the colonies trailing fishing line from their beaks (H. Hasegawa, pers. comm.). To date there are three confirmed records of Short-tailed Albatrosses being killed on longline hooks set in the North Pacific (P. Gould, in litt.). All three birds were immatures although birds of all age classes are known to follow fishing vessels (Camp 1993). Given their primarily inshore distribution at sea, it is the short-range fisheries of Japan which potentially pose the greatest threat to this species (P. Gould, in litt.). However, the extent of fisheries-related mortality for this species in any fishery is unknown, and any efforts to document the extent of such mortality would be problematical given the extremely low population size and low level of observer coverage. For this *Endangered* species, however, the loss of any individuals to longline hooks represents a serious threat to the population.

Waved Albatross *Phoebastria irrorata*

Previous name: Waved Albatross *Diomedea irrorata*.

Breeding distribution and jurisdiction: Espanola Island (Galapagos Islands, Ecuador) and La Plata Island (Ecuador).

Breeding frequency and season: Annual breeding. Most eggs laid in May (April–June), chicks hatch in July and fledge in December.

Population size: Breeding population estimated at *ca.* 15 590 pairs (see Appendix 1B), perhaps 70 000–80 000 birds in total. Two breeding sites, with >99.9% (n = 15 580) eggs laid on Isla Espanola (Galapagos Islands, Ecuador), and only *ca.* 10 eggs laid each year on La Plata Island (Ecuador).

Status: Endemic to Ecuador. Considered *Near Threatened* (Collar *et al.* 1994).

Only two surveys have been conducted at Isla Espanola between 1970 ("at least 12 000 pairs", Harris 1973) and 1994 (two estimates from same survey: 15 581 pairs by Anderson 1995 and "approximately 17 000 pairs" by Douglas 1995); hence the population status cannot yet be confirmed. The status of the La Plata Island population is also unknown. Restriction of breeding range to only two sites greatly enhances the potential significance of any threat.

Threats: Mass abandonment of colonies and loss of eggs through egg rolling contribute to breeding failures (egg rolling behaviour occurs when no nesting material is available and is thought to result from thermal stress, Douglas 1995). The introduction of goats in the past may have had an indirect impact due to vegetation destruction and thermal exposure of the nest sites. Goats have since been removed from Isla Espanola but persist on La Plata Island. Flooding may also occasionally cause nest failure (Harris 1973).

It is not known if Waved Albatrosses have been caught on longline hooks as there have been no observations aboard fishing vessels in areas frequented by the birds. From satellite tracking studies it is clear that there is overlap between longline activities and the foraging areas of the birds but observations suggest that Waved Albatrosses do not readily follow boats (Anderson *et al.* 1997). Waved Albatrosses are, however, known to be scavengers of squid (Harris 1973) and Merlen (1996) warns of the potential of a learned response by the birds to taking squid baited hooks should longline fisheries become more concentrated in the vicinity of the breeding grounds. Both legal and illegal longline fishing efforts for tuna and other pelagic fish have increased in the waters surrounding the Galapagos Islands (Merlen 1996).

Examination of the presence of interactions with fishing practices in the foraging range of these birds remains a high priority. There are still no available data regarding seabird bycatch from the Peruvian region, although Greenpeace International have received reports of seabird bycatch in the area (M. Earle, pers. comm.).

Laysan Albatross *Phoebastria immutabilis*

Previous name: Laysan Albatross *Diomedea immutabilis*.

Breeding distribution and jurisdiction: Northwestern Hawaiian Islands, especially Midway Island and Laysan Island (USA), recently established on Mukojima (Bonin Islands, Japan); recent expansion to Guadalupe Island, Clarion Island and San Benedicto Island (Mexico).

Breeding frequency and season: Annual breeding. Most eggs are laid in November–December, hatch in January–February, and most chicks fledge in June–July.

Population size: The most numerous of the North Pacific Albatross species. The current population is approximately 607 000 breeding pairs (see Appendix 1B), representing perhaps two and a half to three million birds in total.

Status: Given the relative abundance of this species, compared to other species of albatross, its status is generally considered to be relatively secure at present (see Croxall and Gales 1997). However, of the 16 breeding sites documented for the species, two populations, representing 93% of the total breeding stock, are known to be decreasing. The population at Midway Atoll, which accounts for 70% of the global population, showed relatively rapid recovery following the cessation of human depredation at the turn of the Century until the 1960s, after which the rate of increase in numbers slowed (McDermond and Morgan 1993). Counts in 1991/92 showed a decrease of 15% (E. Flint, pers comm). The status of the Hawaiian Island colonies will become more clear with repetition of standardized counts. There is evidence of range expansion and re-occupation at some Hawaiian Island breeding sites, in addition to the recent colonizations in Mexico and Japan (see Appendix 1B).

Threats: A myriad of threats affect Laysan Albatrosses, but since the cessation of widespread harvesting of eggs and adults by Japanese 100 years ago, and the end of the intensive control programmes of the US military, mortality associated with fishing interactions represent the most recent and significant threat to the species.

Drift netting in the North Pacific Ocean since 1978 is estimated to have killed 212 900 Laysan Albatrosses (2.5% excess mortality/year: Gould and Hobbs 1992; Johnson *et al.* 1992). An estimated 17 548 were killed in 1990 alone, primarily by Japanese (54%) and Korean (27%) squid fisheries (Johnson *et al.* 1992). Whilst high seas drift netting was banned in 1992, similar rates of mortality are suspected for the North Pacific Ocean longline fisheries which have flourished since the cessation of drift netting (Ludwig *et al.* 1997). Whilst catch rates of albatrosses in North Pacific Ocean longline fisheries are not yet well known, indications suggest that they are similar to, or even in excess of, catch rates of Southern Ocean species. Japanese fishing masters have reported that the Pacific region is one of high bird catch rates (N. Brothers, pers. comm.). These reports are supported by observations on Hawaiian-based vessels targeting swordfish, where 85 Laysan Albatrosses were caught on one trip (Anon. 1996). Whilst the global Laysan Albatross population has tolerated pressure from fisheries bycatch in the past, it is not known at what level these anthropogenic impacts will be reflected in population trends. With the establishment of standardized and widespread population census efforts, and attempts to obtain catch rate information on fishing boats, this should become clearer in the future.

Black-footed Albatross *Phoebastria nigripes*

Previous name: Black-footed Albatross *Diomedea nigripes.*

Breeding distribution and jurisdiction: Hawaiian Island, (USA) and Izu Islands, Bonin and Senkaku Islands (Japan).

Breeding frequency and season: Annual breeding. Most eggs are laid in November, hatch in January, and most chicks fledge in June.

Population size: Current population approximately 58 500 breeding pairs (see Appendix 1B), perhaps 260 000–290 000 birds in total.

Status: By 1980 this species had largely recovered from the catastrophic population decreases resulting from indiscriminate harvesting by feather hunters at the turn of 20th Century. However, the recovery was retarded by habitat modification by rabbits and humans. The population recovery then halted as a consequence of bycatch associated with the North Pacific Ocean high seas squid and large mesh driftnet fisheries. These fisheries now illegal under a moratorium agreed in 1992.

The major populations are either decreasing or of unknown status. Of the 12 populations, only the three relatively small populations in Japan (which represent 4% of the global population) are increasing, the nine Hawaiian populations either decreasing (five populations representing 47% of the global population) or of unknown status. Clarification of rates of decrease are difficult due to variations in census techniques (McDermond and Morgan 1993) but current annual surveys now incorporate >70% of population (E. Flint, pers. comm.).

Threats: This species has faced an array of threats over the last 100 years (see Gales 1993), including direct human persecution up until the 1950s. Recently, the most important significant threat is bycatch associated with fishing operations. During the 1980s thousands of Black-footed Albatrosses were killed each year in drift nets, principally belonging to the Japanese and Korean squid fisheries (Northridge 1991). Whilst high seas drift netting has since been banned (due to the moratorium imposed as a result of devastating levels of bycatch), even higher rates of bycatch of Black-footed Albatrosses are now being detected in the rapidly expanding North Pacific Ocean longline fisheries that largely replaced drift netting (see Lugwig *et al.* 1997). Preliminary data from the Hawaiian-based longline fishery for swordfish alone indicates that more Black-footed Albatrosses are taken annually in longline interactions than were taken in the entire North Pacific Ocean high seas drift net fishery each year (Flint 1995).

Longline fishing fleets targeting tuna in the North Pacific Ocean and sablefish in Alaskan waters are both known to catch Black-footed Albatrosses (Flint 1995; Ogi 1995). Preliminary estimates of hooking rates indicate that the catch rates in the North Pacific exceed those that have caused Southern Ocean albatross populations to decline. An estimated 23 382 Black-footed Albatrosses were killed on longlines targeting swordfish and tuna between 1990 and 1994 (Anon. 1996). These bycatch rates constitute 91% of the human-caused mortalities, the remaining 9% being caused by chemical pollutant contaminants (Ludwig *et al.* 1997). Mortality rates of this magnitude which serve to exacerbate the mortality incurred from natural sources, primarily inundation of nests by waves, cannot be sustained by the population.

Black-browed Albatross *Thalassarche melanophrys*

Previous name: Black-browed Albatross *Diomedea melanophrys.*

Breeding distribution and jurisdiction: Falkland Islands, South Georgia (United Kingdom); Cape Horn (Chile); Crozet Islands, Kerguelen Islands, (France); Heard and McDonald Islands, Macquarie Island, Bishop and Clerk Islands (Australia); Antipodes Islands, Campbell Island and Snares Island (New Zealand).

Breeding frequency and season: Annual breeding. Timing of breeding varies with location, but generally the breeding season extends from September to April.

Population size: The most abundant of the southern albatross species (Appendix 1C). Current population is approximately 682 000 breeding pairs, perhaps 3 000 000 birds in total. This species is most numerous at the Falkland Islands (80% of global populations), particularly at Steeple Jason Island where about 250 000 pairs breed annually. The smallest Black-browed Albatross populations occur on the Australian and New Zealand sub-antarctic islands.

Status: Black-browed Albatrosses are the most widely distributed of all albatross species and their population status varies with respect to location of colony. The status of many of the smaller populations is not known, although current studies in both the Australian and New Zealand sectors should partially redress this situation. Whilst some colonies have shown signs of increases, low rates of adult survival and juvenile recruitment at the two most intensively studied populations (Bird Island and Kerguelen Island) are causing the populations to decline (see Croxall *et al.* 1997; Weimerskirch and Jouventin 1997).

Threats: The complex picture of the status of the various Black-browed Albatross populations is most likely a consequence of the differing degree of interactions between this species and various fishing operations. During the breeding season, Black-browed Albatrosses forage in continental shelf waters and birds from different colonies within island breeding sites may frequent different regions of these shelf waters (Prince *et al.* 1997; Weimerskirch 1997). The use of different foraging grounds by the different populations may explain disparate trends in status between colonies; birds from some colonies being afforded short-term gains by commuting to waters used by trawlers (and scavenge from discards), and birds from other colonies foraging in areas heavily fished by longliners (Weimerskirch 1997).

Given the distribution of fishing effort in relation to the colonies of this species, and the local nature of the birds foraging during the breeding season, Black-browed Albatrosses may face the greatest threats from fisheries of any albatross species (Weimerskirch and Jouventin 1997). Outside of the breeding season these birds travel to more distant shelf waters, and so interactions with fisheries depend on the concentration of fishing in each area. For example, the developing Hake longline fishery off the South African coast will most likely affect Black-browed Albatrosses from South Georgia, whereas Black-browed Albatrosses from the Falkland Islands most likely interact with longline operations off the Patagonian shelf (Prince *et al.* 1997; Schiavini *et al.* 1997; Neves and Olmos 1997).

Longline fishing off Australia by Japanese and Australian fleets is known to kill Black-browed Albatrosses from Kerguelen and Macquarie Islands (Gales, Brothers and Reid, unpubl. data). Black-browed Albatrosses are one of the commonest species in the seabird bycatch off Australia, but band returns from Black-browed Albatrosses killed on longlines off Australia are restricted to birds from Kerguelen Island (annual breeding population of *ca.* 3 000 pairs whose status is decreasing) and Macquarie Island (annual breeding population 38 pairs, with an unknown status).

Longlining thus constitutes the primary threat to Black-browed Albatrosses, with Australasian longlining representing the major threat to Indian Ocean and Australian populations. Longlining off Africa threatens the South Georgia population and longlining off South America threatens the Falkland and Chilean populations.

Campbell Albatross *Thalassarche impavida*

Previous name: Black-browed Albatross *Diomedea melanophrys impavida* (or New Zealand Black-browed Albatross).

Breeding distribution and jurisdiction: Campbell Island (New Zealand).

Breeding frequency and season: Annual breeding. Adults return to colonies in August, laying eggs in September-October. Successful breeders and chicks depart colonies in April–May.

Population size: The 26 000 pairs are restricted to Campbell Island (see Appendix 1C).

Status: Endemic to New Zealand.

Photographic evidence suggests decreases in the population between the 1960s and 1980s, one colony decreasing by 33% (Moore 1995). Current status during the 1990s not yet determined.

Threats: In the past human predation and sheep have affected the Campbell Albatross population. These threats no longer exist and predation of eggs and chicks by Skuas *Catharacta lonnbergi* and Northern Giant Petrels *Macronectes halli* (Moore and Moffat 1990) are the only identified threats to the species on land.

At sea, Campbell Albatrosses are mainly confined to the waters off southern Australia and New Zealand and the western Pacific Ocean (C. J. Robertson, pers. comm.). High capture rates of Campbell Albatrosses have been recorded from longliners operating off New Zealand and southern Australia (Murray *et al.* 1993; Gales, Brothers and Reid, unpubl. data). Bycatch of this species in New Zealand constitutes mainly new juveniles, but adults in breeding condition are also caught during summer off the Tasmanian coast. Of all returns recovered from banded albatrosses killed on longliners off Australia between 1987 and 1994, 13% were from Campbell Albatrosses (Gales, Brothers and Reid, unpubl. data). This evidence confirms that Australasian longlining is the major threat facing Campbell Albatrosses.

Buller's Albatross *Thalassarche bulleri*

Previous name: Buller's Albatross (or Southern Buller's Albatross) *Diomedea bulleri bulleri.*

Breeding distribution and jurisdiction: Snares and Solander Islands (Solander Island and Little Solander Island), (New Zealand).

Breeding frequency and season: The breeding biology of this species is poorly known. They are typically annual breeders, the adults returning to the colonies in December. Eggs are laid in January–February, their chicks hatch in March–April and fledge from late August to late October.

Population size: The breeding population estimated at about 11 000 pairs (see Appendix 1C), perhaps 50 000 to 55 000 birds in total. Most of these birds are from the Snares Islands population (77% of the population), with the remainder from Solander and Little Solander Islands.

Status: Endemic to New Zealand. Considered Near Threatened by Collar *et al.* (1994).

Given the paucity of reliable survey data for this species, determination of population status is not possible at present. Current research should soon clarify the current status of Buller's Albatrosses (P. Sagar, pers. comm.).

Threats: The only identified threat facing Buller's Albatrosses is mortality associated with fishing activities (Gales 1993). Many Buller's Albatrosses are known to have been killed by colliding with netsonde monitor cables on trawlers, but this source of mortality has been mitigated with the prohibition of this apparatus, at least in New Zealand waters (Bartle 1991). Buller's Albatrosses are also known to be hooked on longlines in waters off Australia and New Zealand (Murray *et al.* 1993; Gales, Brothers and Reid, unpubl. data). The autumn/winter breeding characteristic of Buller's Albatrosses focuses the risk from longlining as it is during the winter months that fishing effort is concentrated within the foraging range of breeding adults (Murray *et al.* 1993; Sagar and Weimerskirch, in press).

The current research into the demographic status and oceanic distribution of Buller's Albatrosses will assist understanding of impacts associated with longline fishing on this species.

Pacific Albatross *Thalassarche* nov. sp.

Previous name: Buller's Albatross (or Northern Buller's Albatross) *Diomedea bulleri platei.*

Breeding distribution and jurisdiction: Three Kings and Chatham Islands (Big and Little Sister Islands, and the Forty Fours) (New Zealand).

Breeding frequency and season: Annual breeding. Pacific Albatrosses lay eggs in November, hatch in January and chicks fledge in June.

Population size: The breeding population is estimated at about 18 000 pairs (see Appendix 1C), perhaps 80 000–90 000 birds. The population size estimates for the two large populations (Big Sister Island and the Forty Fours, with >95% of global population) are crude, assessed on the basis of area. The size of the Three Kings population (believed to be <1% of the total population) is also poorly known. Reliable population size data exists only for the Little Sister Island population (estimated at <4% of population).

Status: Endemic to New Zealand.

An evaluation of the status of this species is not possible given the lack of reliable population data.

Threats: Consistent with the lack of population data for this species, there is also little information regarding threats to the species. Distribution at sea is also poorly known as a result of confusion with Buller's Albatrosses. Despite this uncertainty, it has been suggested that the species is highly migratory during the non-breeding season, ranging eastward across the southern Pacific Ocean to the western South American coast (Lindsey 1986). During the breeding season the birds are known to forage in the area of the Challenger Rise, coincident with longline operations (C. J. Robertson, pers. comm.). To date there are no reports of deaths of these birds on longlines. However, in the areas prospected by these birds, both during the breeding and non-breeding season, longline operations proceed largely in the absence of any observer coverage.

Shy Albatross *Thalassarche cauta*

Previous name: Shy Albatross *Diomedea cauta cauta.*

Breeding distribution and jurisdiction: Albatross Island, Mewstone and Pedra Branca (Australia).

Breeding frequency and season: Annual breeding. Most eggs are laid in September, hatch in December and chicks fledge in April. Adults attend colonies in winter between breeding seasons.

Population size: The breeding population is estimated at about 12 200 breeding pairs (see Appendix 1C), approximately 55 000–60 000 individuals.

Status: Endemic to Australia.

The population estimates for this species are of moderate accuracy only, except for the Albatross Island colony in Bass Strait. This colony is showing signs of a slow recovery following the devastation executed by feather and egg collectors at the turn of the Century. The current population level on Albatross Island (5 000 pairs) constitutes about 25% of the estimated original island population size having increased from only 300 pairs (1.5% of initial population) in 1909. The trends of the Mewstone and Pedra Branca populations (estimated at 59% of the species population) is unknown.

Threats: A viral disease is known to reduce productivity rates during some years for the Albatross Island colony (see Gales 1993). There is, however, no evidence of this disease elsewhere in the species' range.

The major threat facing Shy Albatrosses is incidental mortality associated with fishing operations. In the Australian region, Shy Albatrosses constitute over 10% of the seabird bycatch on Japanese tuna longlines and most of these were adult birds (Gales, Brothers and Reid, unpubl. data). Based on assessments of foraging ranges as determined from satellite tracking (Brothers *et al.* 1977a) and fishing effort, it is likely that these birds are from the southern Tasmanian populations, whose status is unclear. Shy Albatrosses are also killed on longlines set by the Australian domestic fishing fleet (Brothers and Foster, in press). The spatial and temporal concentration of tuna fishing effort to the south and east coasts of Tasmania, and the recent increase in the Australian domestic longline fishery, is consequently of major concern with respect to this species. The three populations are differentially vulnerable as a result of their differences in foraging zones: the southern populations being vulnerable throughout their annual cycle. Band recoveries reveal that juveniles of the southern Tasmanian population migrate to waters off southern Africa and so are also placed at risk from longline fisheries in this region (Brothers *et al.* 1997b).

Recent proposals to increase the commercial squid *Nototodarus gouldi* fishery in Bass Strait, within the foraging ranges of the Albatross Island population, may also pose a threat. Direct competition for common prey may affect the birds if squid quotas increase with no regard to the food requirements of the albatrosses.

White-capped Albatross *Thalassarche steadi*

Previous name: Shy Albatross (or White-capped Albatross) *Diomedea cauta steadi.*

Breeding distribution and jurisdiction: Disappointment Island, Adams Island, Auckland Island (Auckland Island group) and Bollons Island (Antipodes Island group) (New Zealand).

Breeding frequency and season: Poorly known as there have been no detailed studies. Egg laying starts in mid-November, hatching in February and young fledging in mid-August (Robertson 1985). Likely to breed annually. No data available on breeding success, survival or recruitment for this species.

Population size: The breeding population is estimated at *ca.* 75 000 breeding pairs (see Appendix 1C), approximately 350 000–375 000 individuals.

Status: Endemic to New Zealand.

As there are no accurate estimates of population size for the species, there can be no reliable assessments of status or trends.

Threats: On Auckland Island habitat destruction by feral pigs has led to some reduction in nesting area, at least in the past.

At sea, the major threat facing White-capped Albatrosses is incidental mortality associated with fishing operations. In New Zealand waters White-capped Albatrosses constituted 85% of the bycatch associated with squid trawlers (Bartle 1991). This level of bycatch was not sustainable and has since been reduced by the prohibition on the use of netsonde monitor cables in New Zealand waters. White-capped Albatrosses are also known to be killed on bluefin tuna longlines in New Zealand waters (Murray *et al.* 1993), but due to difficulty in distinguishing this species from Shy Albatrosses, assessment of extent and magnitude of bycatch in other areas is problematical. This confusion also impedes understanding of the distribution of this species at sea. Knowledge of the status of this species, and its distribution at sea is urgently required.

Salvin's Albatross *Thalassarche salvini*

Previous name: Shy Albatross (or Salvin's Albatross) *Diomedea cauta salvini.*

Breeding distribution and jurisdiction: Snares and Bounty Islands (New Zealand). Penguin Island (Crozet Islands; France).

Breeding frequency and season: No detailed studies; assumed to be an annual breeder. Eggs hatch at Bounty Islands in mid-November, so laying presumed to occur in early October. Chicks fledge in late March to early April.

Population size: The breeding population is estimated at *ca.* 76 500 breeding pairs (see Appendix 1C), approximately 350 000–380 000 individuals. New Zealand populations account for >99% of the global population, <5 pairs nesting on Penguin Island (Crozet Group) in the Indian Ocean each year.

Status: Knowledge of the New Zealand populations is poor, the estimate of the major population on the Bounty Islands being derived from extent of breeding area. Given the lack of repeated counts for any of the populations of this species, the status and population trend of the species remains unknown.

Threats: There is no information regarding the threats faced by this species, either on land or at sea. The extensive marine distribution of this species (extending north to 5°S in the Humboldt Current, and also in the Indian Ocean and off the coasts of Australia and South Africa (Marchant and Higgins 1990), would place the species potentially at risk from both tropical and temperate longline operations. On the Pacific coast of South America, Salvin's Albatrosses concentrate over the continental slope region, their distribution coinciding with that of the developing longline fishery (Spear *et al.* 1995). It is likely that this fishery will impact on Salvin's Albatrosses.

Chatham Albatross *Thalassarche eremita*

Previous name: Shy Albatross (or Chatham Island Albatross) *Diomedea cauta eremita.*

Breeding distribution and jurisdiction: The Pyramid (Chatham Islands, New Zealand).

Breeding frequency and season: There have been no detailed studies of the breeding biology of this species. It is presumed to breed annually. Eggs are laid in August and September, and chicks are presumed to fledge in April (Robertson and van Tets 1982). Two Campbell Albatrosses are regularly seen amongst Shy Albatrosses on Albatross Island (Bass Strait, Tasmania) but there have been no observed attempts at breeding.

Population size: There have never been any ground counts of this species. Based on aerial photographs taken in 1972, the breeding population is estimated to be *ca.* 4 000 pairs (see Appendix 1C), perhaps 18 000 to 20 000 individuals.

Status: Endemic to New Zealand.

The accuracy of the single population estimate is low. The status of this species is therefore unknown.

Threats: A significant threat to this species on land is a reduction of the quality of nesting habitat as a result of storm damage and climatic change (C. J. Robertson, pers. comm.). These birds tend to be solitary at sea and their movement patterns are not clear. They are rarely seen in coastal regions. During the non-breeding (winter) season, they are reported to disperse towards the west coast of South America where they mostly frequent pelagic waters (Spear *et al.* 1995). In 1995 a banded juvenile Chatham Albatross was reported to have been killed on a longline targeting swordfish off Chile (C. J. Robertson, pers. comm.). Chatham Albatrosses also occur off the coast of Tasmania where they are known to interact with longline fishing operations (Reid and James, in press).

Atlantic Yellow-nosed Albatross *Thalassarche chlororhynchos*

Previous name: Yellow-nosed Albatross *Diomedea chlororhynchos chlororhynchos.*

Breeding distribution and jurisdiction: Tristan da Cunha group and Gough Island (United Kingdom).

Breeding frequency and season: Likely to be annual breeder. Most eggs laid in September–October, hatch in November–December and chicks fledge in April–May.

Population size: The breeding population is estimated at *ca.* 36 800 pairs (see Appendix 1C), corresponding to approximately 165 000–185 000 individuals in total. Population size estimates are crude as no population has ever been reliably surveyed.

Status: Endemic to territories of the United Kingdom.

The absence of any reliable information on population size precludes any assessment of population trends and status for the six populations. Unpublished information indicates that the Gough Island population has shown a significant decrease since the 1980s (J. Cooper, pers. comm.).

Threats: Since the recent cessation of harvesting by humans on Nightingale Island (Tristan da Cunha group), interactions with commercial fishing operations are the most serious threats faced by Atlantic Yellow-nosed Albatrosses. Little is known about the oceanic distribution of this species and confusion with Indian Yellow-nosed Albatrosses precludes quantitative assessment of bycatch rates in most areas. Yellow-nosed Albatrosses constituted a significant proportion of the seabird bycatch observed on tuna longlines off Brazil (Neves and Olmos 1997), and it is likely that these birds were Atlantic Yellow-nosed Albatrosses (T. Reid, pers. comm.).

Further information on the status of the population and the extent of bycatch of this species are urgently required.

Indian Yellow-nosed Albatross *Thalassarche carteri*

Previous name: Yellow-nosed Albatross *Diomedea chlororhynchos bassi.*

Breeding distribution and jurisdiction: Prince Edward Islands (South Africa), Kerguelen Islands, Crozet Island, Amsterdam and St Paul Islands (France).

Breeding frequency and season: Annual breeding. Most eggs laid in September–October, hatch in November–December and chicks fledge in March–April.

Population size: The annual breeding population is estimated at *ca.* 36 500 pairs (see Appendix 1C), corresponding to approximately 160 000–180 000 individuals in total. Population size estimates are generally poor for this species across its range, the only exception being the Amsterdam Island population which represents approximately 70% of the estimated global population (Weimerskirch and Jouventin 1997).

Status: The absence of reliable time series population data preclude any status assessment for all but the Amsterdam Island population. This population has decreased by over one-third since the early 1980s and is decreasing at 7% annually as a result of increased mortality of both adults and immatures (Weimerskirch and Jouventin 1997).

Threats: It appears that interactions with commercial fishing operations are the most serious threat faced by Indian Yellow-nosed Albatrosses. Previously this conclusion was largely inferential (Gales 1993), but the recently observed population decreases and the documented bycatch of these birds on tuna longlines confirms the connection. Although the Atlantic and Indian Ocean species were not distinguished, Yellow-nosed Albatrosses constituted 14% of the seabird bycatch observed on tuna longlines off Brazil (Neves and Olmos 1997), similar to the 13% that both species comprised in the bycatch in the Australian region (Gales and Brothers 1995). In the Australian sample, adults were primarily caught during winter whereas immatures were more often caught in the summer fishing season (Gales, Brothers and Reid, unpubl. data). The only other significant source of mortality identified for this species is a viral disease which in some years causes elevated chick mortality in some colonies (H. Weimerskirch, pers. comm.).

Grey-headed Albatross *Diomedea chrysostoma*

Previous name: Grey-headed Albatross *Diomedea chrysostoma.*

Breeding distribution and jurisdiction: South Georgia (United Kingdom); Diego Ramirez and Islands Ildefonso (Chile); Kerguelen and Crozet Islands (France); Marion and Prince Edward Islands (South Africa); Campbell Island (New Zealand); Macquarie Island (Australia).

Breeding frequency and season: Biennial breeder (when successful). Most eggs laid in October, hatch in December–January and chicks fledge in April–May.

Population size: The breeding population estimated at *ca.* 92 300 pairs each year (see Appendix 1D), corresponding to approximately 250 000 mature individuals, or 600 000 individuals in total. The population estimates for this species are reasonably reliable, with the notable exception of the two colonies within the French territories and the Chilean populations.

Status: Currently considered *Near Threatened* by Collar *et al.* (1994).

The status of the populations varies with breeding location. At South Georgia over the last two decades, the Bird Island population has decreased at an annual rate of 1.4–1.8%, mainly as a result of decreases in immature survival rate (from 35% to 5% recruitment) and reduced adult survival (from 95% to 93%) (Prince *et al.* 1994a; Croxall *et al.* 1997).

The population decrease of *ca.* 20% documented for Bird Island is alarming, particularly as the South Georgian population represents nearly two thirds (59%) of the world's population. Decreases of 79–85% since the 1940s have been observed for Campbell Island albatross colonies in which Grey-headed Albatrosses predominate (Moore 1995). The only population increase recorded for the species is a recent one (since 1992) for the Marion Island population (7% of global population) which had previously decreased at 0.7% p.a. since the 1970s (J. Cooper, pers. comm.).

Threats: Grey-headed Albatrosses are vulnerable to deaths associated with fishing practices, particularly longlining. Of the longline fisheries, it is the pelagic rather than the shelf-slope fisheries which probably pose the greatest threat to this species (Prince *et al.* 1997).

The decrease in the Grey-headed Albatross population at Bird Island is most probably attributable to deaths of immatures on longlines (Prince *et al.* 1994a). This is consistent with the predominance of immatures in the Grey-headed Albatross component of the seabird bycatch from Japanese tuna longlines operating off Australia (Gales, Brothers and Reid, unpubl. data) and New Zealand (Murray *et al.* 1993).

Grey-headed Albatrosses are also hooked on longlines set in Kerguelen waters (Cherel *et al.* 1996). The impacts of this mortality, however, is not known as the populations of this species in the French territories of Crozet and Kerguelen are not routinely monitored.

Of potential, rather than realized threat, is the development of commercial squid fisheries in the Southern Ocean (Prince *et al.* 1997). As Grey-headed Albatrosses direct their foraging efforts at the squid *Martialia hyadesi*, which is a target of increasing commercial fishing interests, expansion of the squid fishery should consider the requirements of these albatrosses so that the species is not subject to further adverse effects.

Sooty Albatross *Phoebetria fusca*

Previous name: Sooty Albatross *Phoebetria fusca.*

Breeding distribution and jurisdiction: Tristan da Cunha Islands and Gough Island (United Kingdom); Prince Edward Island and Marion Island (South Africa); Kerguelen, Crozet, Amsterdam and St Paul Islands (France).

Breeding frequency and season: Biennial breeder (when successful). Most eggs laid in October, hatch in December and chicks fledge in May.

Population size: The annual breeding population is estimated at *ca.* 15 655 pairs (see Appendix 1D), approximately 100 000 individuals in total. Six breeding assemblages exist, the major sites being Gough Island and the Tristan da Cunha group in the southern Atlantic Ocean. Of the 15 island sites, detailed monitoring only occurs at the small Possession Island (Crozet Islands, Indian Ocean) population (which represents about 2% of the estimated global population).

Status: Considered *Near Threatened* by Collar *et al.* (1994).

The status of Sooty Albatrosses is known only for Possession Island population. The population decrease documented for this species is the most extensive of the six albatross species studied within the French external territories of Crozet or Kerguelen. The population is currently decreasing at a rate of 3% p.a., previous rates of decrease were as high as 6.9% p.a. (between 1979–86). These rates translate to a total decrease in the population of 58% since 1980 (Weimerskirch and Jouventin 1997). Decreasing survival rates of both adults and immatures are responsible for the observed population trends.

The status of the remaining 14 populations (comprising approximately 98% of the estimated global population) is unknown.

Threats: The adult mortality rates of Sooty Albatrosses on Possession Island (Crozet group, Indian Ocean) are significantly related to the longline fishing effort that occurs in the oceanic sectors prospected by them (Weimerskirch and Jouventin 1997). Adult and immature Sooty Albatrosses are known to be killed on Japanese longlines set both inside and beyond the Australian Fishing Zone (Gales, Brothers and Reid, unpubl. data). Information detailing the composition of seabird bycatch within the foraging area of this oceanic species (e.g., high seas, especially Indian Ocean) exists but is currently confidential to fisheries managers. From the limited information which is available, however, it is clear that Sooty Albatrosses are caught in proportionately significantly higher numbers on the high seas than in Exclusive Economic Zone fisheries, reflecting the oceanic habit of this species (Gales, Brothers and Reid, unpubl. data). Without information to the contrary, and in the absence of identification of other anthropogenic factor affecting Sooty Albatrosses, the logical conclusion is that longline fishing is responsible for the observed population decreases and is the most serious threat facing this species.

Light-mantled Albatross *Phoebetria palpebrata*

Previous name: Light-mantled Sooty Albatross *Phoebetria palpebrata.*

Breeding distribution and jurisdiction: South Georgia (United Kingdom); Prince Edward and Marion Islands (South Africa); Kerguelen and Crozet Islands (France); Heard Island and Macquarie Island (Australia); Auckland, Campbell and Antipodes Islands (New Zealand).

Breeding frequency and season: Biennial breeder (when successful). Most eggs laid in October–November, hatch in December–January and chicks fledge in May–June.

Population size: The breeding population is estimated at *ca.* 21 600 pairs each year (see Appendix 1D), approximately 140 000 individuals in total, from the nine breeding sites distributed between southern Pacific, Atlantic and Indian Oceans. Accurate population estimates are available only for the Possession Island and Macquarie Island populations, each of which constitutes about 5% of the estimated global population.

Status: Information on population trends and status is restricted to the small population on Possession Island. This population has decreased by 13% since 1980, but the demographic parameters influencing this change are not yet clear (Weimerskirch and Jouventin 1997). Indirect evidence of the decrease of this species is evident in the decreased abundance of Light-mantled Albatrosses in their summer feeding grounds since 1980 (Woehler 1996).

Threats: Gales (1993) concluded that it was most likely that mortality associated with fishing activities constituted the major threat to Light-mantled Albatrosses. Since then, confirmation of the capture of Light-mantled Albatrosses on longline hooks set in the Australian Fishing

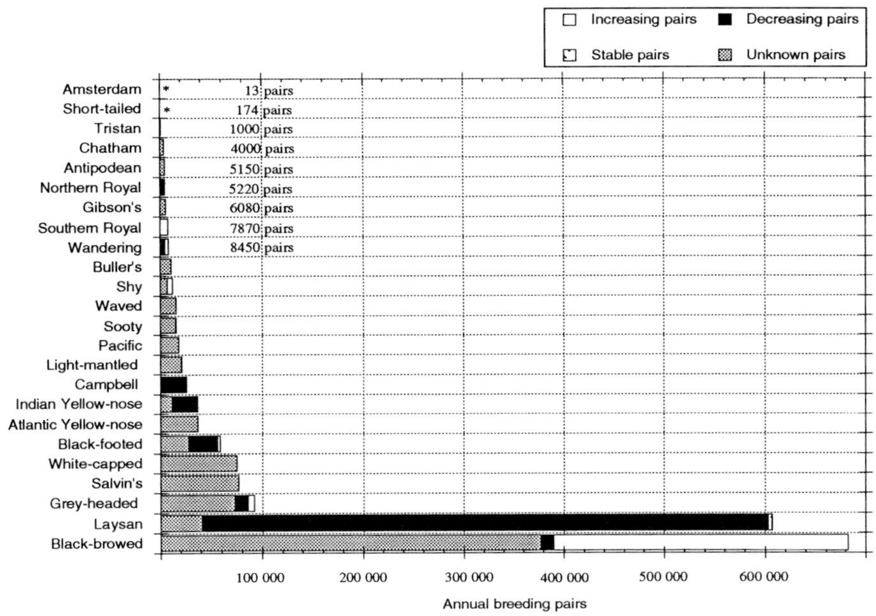

Fig. 1. Population size (annual breeding pairs) and trends of albatross species (species with less than 10 000 annual breeding pairs listed with population size for clarity).

Zone has been obtained (Gales and Brothers 1995). Satellite tracking studies indicate that interactions with longliners may be limited during the breeding season as a result of the southern distribution of the foraging grounds; however during the non-breeding periods, the birds move northwards and mix with pelagic fishing fleets (Weimerskirch 1997). The proficiency of the diving behaviour of *Phoebetria* albatrosses (Prince *et al.* 1994b) enhances their vulnerability to capture on longlines. Given this information and in the absence of other factors responsible for the decreases in the Light-mantled Albatross populations, interactions with longlining remains the most compelling factor responsible for their population decreases.

DISCUSSION

The status of albatrosses

Despite the recent increases in albatross population monitoring, the status of two-thirds of the *ca.* 150 populations of albatrosses worldwide remains unknown (Table 1). Of the 53 populations for which the current status is known, almost half (42%) are decreasing. The remaining populations are either currently increasing (43%) or stable (15%), albeit many after suffering declines in the past. These assessments of population status use the proportions of populations for which information is available rather than proportions based on the total population sizes. When population size is taken into account the situation becomes more serious as it is often the small/relict populations that are stable or showing signs of recent increase and recovery and the large populations which are either decreasing or of unknown status (Table 1). Based on the number of annual breeding pairs within populations of the different status categories, <1% are stable, 17% are increasing, and 37% are decreasing. Population trends for the remaining 48% are presently unknown.

When the trends are examined at the species level their relative abundance and status becomes more clear. Of the 24 albatross species in the proposed taxonomy, 21 species have populations that are decreasing or of unknown status for more than 50% of the entire/global population (Fig. 1). The evaluations of the conservation status of albatrosses under the IUCN guidelines (see Croxall and Gales 1997) testify to the threatened status of albatrosses worldwide.

The threats confronting albatrosses

The life history strategy of albatrosses is characterized by delayed maturity, low reproductive output, high rates of natural survival and longevity. Albatrosses are therefore particularly vulnerable to changes in fecundity and survival rates and breeding success. Of all the albatross demographic parameters, sensitivity analyses identified changes in adult survival as being the most immediately important factor influencing population trends, with a decrease of 1% p.a. being sufficient to cause a population decrease (Croxall and Rothery 1991; Weimerskirch *et al.* 1997). The next most influential demographic parameter affecting changes in albatross population trends is juvenile survival.

Table 1. Number, status and size of albatross populations (shading indicates those taxa for which ≥50% of the global population is decreasing or of unknown status).

Old taxonomy	New taxonomy	Estimated No. of annual breeding pairs	No. of populations	Status of population: Increasing	Stable	Decreasing	Unknown
Wandering Albatross	Wandering Albatross	8 448		43%	<1%	47%	10%
Diomedea exulans exulans	*Diomedea exulans*		9	3	1	4	1
	Tristan Albatross	1 003			<1%		>99%
Diomedea exulans dabbenena	*Diomedea dabbenena*		2		1		1
	Antipodean Albatross	5 154			<1%		>99%
Diomedea exulans antipodensis	*Diomedea antipodensis*		2		1	1	
	Gibson's Albatross	6 077					100%
Diomedea exulans gibsoni	*Diomedea gibsoni*		3				3
Amsterdam Albatross	Amsterdam Albatross	13		100%			
Diomedea amsterdamensis	*Diomedea amsterdamensis*		1	1			
Northern Royal Albatross	Northern Royal Albatross	5 218		<1%		>99%	
Diomedea epomophora sanfordi	*Diomedea sanfordi*		2	1		1	
Southern Royal Albatross	Southern Royal Albatross	7 872		99%	<1%		<1%
Diomedea epomophora epomophora	*Diomedea epomophora*		4	2	1		1
Short-tailed Albatross	Short-tailed Albatross	174		91%			8%
Diomedea albatrus	*Phoebastria albatrus*		2	1			1
Waved Albatross	Waved Albatross	15 591					100%
Diomedea irrorata	*Phoebastria irrorata*		2				2
Laysan Albatross	Laysan Albatross	607 059		<1%	<1%	93%	7%
Diomedea immutabilis	*Phoebastria immutabilis*		16	4	2	2	8
Black-footed Albatross	Black-footed Albatross	58 498		4%	2%	47%	47%
Diomedea nigripes	*Phoebastria nigripes*		12	3	1	5	3
Black-browed Albatross	Black-browed Albatross	682 316		43%	<1%	2%	55%
Diomedea melanophrys melanophrys	*Thalassarche melanophrys*		24	6	1	2	16
	Campbell Albatross	26 000				100%	
Diomedea melanophris impavida	*Thalassarche impavida*		1			1	
Buller's Albatross	Buller's Albatross	10 960					100%
Diomedea bulleri bulleri	*Thalassarche bulleri*		3				3
	Pacific Albatross	18 170					100%
Diomedea bulleri platei	*Thalassarche* nov. sp.		4				4
Shy Albatross	Shy Albatross	12 200		41%			59%
Diomedea cauta cauta	*Thalassarche cauta*		3	1			2
	White-capped Albatross	75 175					100%
Diomedea cauta steadi	*Thalassarche steadi*		4				4
	Salvin's Albatross	76 654					100%
Diomedea cauta salvini	*Thalassarche salvini*		3				3
	Chatham Albatross	4 000					100%
Diomedea cauta eremita	*Thalassarche eremita*		1				1
Yellow-nosed Albatross	Atlantic Yellow-nosed Albatross	36 750					100%
Diomedea chlororhynchos chlororhynchos	*Thalassarche chlororhynchos*		6				6
	Indian Yellow-nosed Albatross	36 492				69%	31%
Diomedea chlororhynchos bassi	*Thalassarche carteri*		6			1	5
Grey-headed Albatross	Grey-headed Albatross	92 275		7%		14%	79%
Diomedea chrysostoma	*Thalassarche chrysostoma*		10	1		2	7
Light-mantled Sooty Albatross	Light-mantled Albatross	21 567				5%	95%
Phoebetria palpebrata	*Phoebetria palpebrata*		14			1	13
Sooty Albatross	Sooty Albatross	15 655				2%	98%
Phoebetria fusca	*Phoebetria fusca*		15			1	14

Currently, with so many populations of albatrosses decreasing, the continued existence of some species is further compounded by the reduced size of many of the populations. One-third of all populations comprise fewer than 100 breeding pairs each year, and so are extremely vulnerable to stochastic events and/or catastrophes. Further, the high degree of philopatry of both juveniles and adults severely limits the ability of albatrosses to move away from sites facing adverse circumstances.

Against this background the precarious status of albatrosses worldwide is clearly evident. Affecting the status of albatross populations are a variety of factors (Table 2), some of which have always occurred, their effects being reflected in natural mortality rates. Currently, most factors

Table 2. Summary of documented threats which currently affect albatrosses. (Information from text and Gales (1993) and references therein.)

	Documented fisheries-related mortality			Other documented threats					
	Longlining	Trawling	Other	Plastic ingestion	Human predation/ disturbance	Other alien predators	Fire/floods volcanoes habitat disturbance	Oil/ chemicals	Avian pox
Wandering Albatross	x	x	x	x					
Tristan Albatross	x			x					
Antipodean Albatross	x								
Gibson's Albatross	x					x			
Northern Royal Albatross	x			x	x	x	x		
Southern Royal Albatross	x	x		x					
Amsterdam Albatross	x					?			
Short-tailed Albatross	x		?	x		?	x		
Waved Albatross	?				?	?	x		
Laysan Albatross	x		x	x	x	x	x	x	
Black-footed Albatross	x		x	x			x	x	x
Black-browed Albatross	x	x	x						x
Campbell Albatross	x								
Buller's Albatross	x	x							
Pacific Albatross	?			x	?		x		
Shy Albatross	x	x	x						x
White-capped Albatross	x	x				x			
Salvin's Albatross	?								
Chatham Albatross	x				x		x		
Atlantic Yellow-nosed Albatross	x			x					
Indian Yellow-nosed Albatross	x					x	x		x
Grey-headed Albatross	x	x		x		x			x
Sooty Albatross	x			x		x	x		
Light-mantled Albatross	x								

influencing the status of albatross populations however, involve human activities. Of these, interactions with longline fishing operations remains the single most important factor responsible for the current widespread decline of albatross populations.

Each year about 1.8 million pairs of albatrosses breed on remote island locations around the world. Each year over 100 million longline hooks are deployed to catch fish. It is these hooks which pose the greatest threat facing albatrosses today. Twenty-one of the 24 species are known to be killed on longline hooks and these deaths occur in every major oceanic sector. The evidence linking longline fishing to changes in the status of albatross populations across the world is compelling, and confirms the role of longline fishing in the observed changes in the status of many populations (see for example Weimerskirch *et al.* 1997). This conclusion has been reinforced in many of the chapters of this book.

The future for albatrosses

The future prospects for most albatross populations are clouded by uncertainty. On the one hand, if appropriate actions are not taken, the future is bleak, given that restoring albatross populations in some oceanic sectors will require substantial and urgent reduction in the catch rates of albatrosses on longlines (Croxall *et al.* 1997). Nevertheless, it appears that reductions in excess of 90% in catch rates can be achieved immediately with mitigation techniques and procedures which have already been identified (Brothers 1991; Alexander *et al.* 1997). The power of implementation ultimately lies with the longline fishers and so it is inevitably they who will dictate the future of these birds. It is, however, the responsibility of governments across the world to promote and expedite that process.

The responsibility for redressing the situation must lie in the global realm. The degree of national jurisdiction for albatross breeding grounds is highly focused, 16 of the 24 species being restricted to islands under the responsibility of single nations, 10 species breeding on New Zealand islands alone. Specifically, the breeding grounds of albatrosses occur on islands under the jurisdiction of 10 nations: Australia, Chile, Ecuador, France, Japan, Mexico, New Zealand, South Africa, United Kingdom, and USA (Table 3). Eight of these nations also operate longline fleets (and/or provide access to longline fleets of other nations) in waters where albatrosses are known to be killed on longlines. Most of these

Table 3. Estimated percentage of each albatross species for which each nation has jurisdiction. (Bold italics indicates nations which operate longline fleets (and or provide access to longline vessels) in waters where albatross-longline mortality is known to occur.)

Species	Breeding frequency	Annual breeding pairs	***Australia***	***Chile***	Equador	***France***	***Japan***	Mexico	***New Zealand***	***South Africa***	***UK***	***USA***
Wandering Albatross	Biennial	8 450	<1			38				36	26	
Tristan Albatross	Biennial	1 000									100	
Antipodean Albatross	Biennial	5 150							100			
Gibson's Albatross	Biennial	6 080							100			
Amsterdam Albatross	Biennial	13				100						
Northern Royal Albatross	Biennial	5 220							100			
Southern Royal Albatross	Biennial	7 870							100			
Short-tailed Albatross	Biennial	174					100					
Waved Albatross	Annual	15 590			100							
Laysan Albatross	Annual	607 100					<1	<1				>99
Black-footed Albatross	Annual	58 500					4					96
Black-browed Albatross	Annual	682 320	<1	5		<1			<1		95	
Campbell Albatross	Annual	26 000							100			
Buller's Albatross	Annual	10 960							100			
Pacific Albatross	Annual	18 170							100			
Shy Albatross	Annual	12 200	100									
White-capped Albatross	Annual	75 180							100			
Salvin's Albatross	Annual	76 650							100			
Chatham Albatross	Annual	4 000							100			
Atlantic Yellow-nosed Albatross	Annual	36 750									100	
Indian Yellow-nosed Albatross	Annual	36 490				81				19		
Grey-headed Albatross	Biennial	92 270	<1	11		15			7	8	59	
Light-mantled Albatross	Biennial	21 570	7			29			35	1	29	
Sooty Albatross	Biennial	15 660				17				18	65	

nations, and others such as Taiwan, Korea, Spain, Russia, Ukraine, Bulgaria and Argentina, operate longline fleets on the high seas. China has the potential to become a major longline fishing nation. Domestic fleets from Australia, South American and African countries also operate inshore fleets which catch seabirds, including albatrosses. The birds killed in these waters may have originated from any or all of the nations responsible for the breeding islands, such is the global nature of the oceanic habits of these birds. Whether nations are responsible for the albatrosses at their breeding islands or for the fishing operations that kill albatrosses, there can be no apportioning of blame but rather acceptance of the urgency of implementing appropriate solutions.

ACKNOWLEDGEMENTS

My sincere thanks to those who, many for the second time, provided their hard won field data to update the population estimates since the Gales (1993) review. In particular I thank Dave Anderson, Sandy Bartle, Nigel Brothers, John Cooper, John Croxall, Beth Flint, Hiroshi Hasegawa, Colin Miskelly, Peter Moore, Peter Prince, Paul Sagar, Ian Strange, Alan Tennyson, Chris Robertson, Kath Walker and Henri Weimerskirch. For improving earlier drafts I especially thank John Croxall, and also Peter Moore, Graham Robertson, Nigel Brothers, David Pemberton and April Hedd. This work was carried out whilst the author was a recipient of an Australian Research Council QEII Research Fellowship.

APPENDIX 1A

Annual population size and status of the Great Albatrosses (biennial breeding species).

Old taxonomy	New taxonomy	Location	Breeding pairs (annual)	Survey date	Survey reliability	Population status	Source*
Wandering Albatross	Wandering Albatross						
Diomedea exulans exulans	*Diomedea exulans*	**South Georgia**					
		Bird Island	1 314	1995–96	A	Decreasing	1
		Other islands	864	1984	B	?	1
		Crozet Islands					
		Ile de la Possession	349	1995	A	Increasing†	2
		Ile aux Cochons	1 060	1981	?	Decreasing	2
		Ile de l'Est	325	1982	?	Decreasing	2
		Kerguelen Islands	1 455	1992	B	Increasing†°	2
		Marion Island	1 794	1995	A	Increasing†	3
		Prince Edward Island	1 277	1984	B	Decreasing?	3,4
		Heard Island	0	1987–88	?	?	5
		Macquarie Island	10	1995	A	Stable†	6
	Tristan Albatross						
Diomedea exulans dabbenena	*Diomedea dabbenena*	Gough Island	*ca.* 1 000	1980s	C	?	3,4
		Inaccessible Island	2–3	1990	A	Stable†	7
	Antipodean Albatross						
Diomedea exulans antipodensis	*Diomedea antipodensis*	**New Zealand**					
		Antipodes Islands	5 148	1996	A	?	8
		Campbell Island	6	1995	A	Stable	9
	Gibson's Albatross						
Diomedea exulans gibsoni	*Diomedea gibsoni*	**New Zealand**					
		Auckland Island group					
		Adams Island	5 762	1995	A	?	8
		Disappointment Island	250	1993	B	?	8
		Auckland Island	65	1995	A	?	8
Northern Royal Albatross	Northern Royal Albatross						
Diomedea epomophora sanfordi	*Diomedea sanfordi*	**New Zealand**					
		Chatham Island	5 200	1995	B	Decreasing	10
		Taiaroa Head	18	1995	A	Increasing	10
Southern Royal Albatross	Southern Royal Albatross						
Diomedea epomophora epomophora	*Diomedea epomophora*	**New Zealand**					
		Campbell Island	7 800	1996	B	Increasing?	9
		Enderby Island	55	1995	A	Increasing	9
		Adams Island	15	1991	B	Stable?	8
		Auckland Island	2	1989	B	?	8
Amsterdam Albatross	Amsterdam Albatross						
Diomedea amsterdamensis	*Diomedea amsterdamensis*	Amsterdam Island	13	1995	A	Increasing†	2

*See Appendix 1E for sources of information.

APPENDIX 1B

Annual population size and status of Pacific Albatrosses (annual breeding species).

Old taxonomy	New taxonomy	Location	Breeding pairs (annual)	Survey date	Survey reliability	Population status	Source*
Short-tailed Albatross *Diomedea albatrus*	Short-tailed Albatross *Phoebastria albatrus*	**Japan**					
		Izu Islands (Torishima)	158	1995	A	Increasing	11
		Senkaku Islands (Minami-kojima)	*ca.* 15	1991	B	?	11
		Hawaii					
		Midway Atoll	1	1993	A	?	12
Waved Albatross *Diomedea irrorata*	Waved Albatross *Phoebastria irrorata*	**Galapagos Islands**					
		Isla Espanola	15 581	1995	B	?	13
		Ecuador					
		Isla de la Plata	10	1990	A	?	14
Laysan Albatross *Diomedea immutabilis*	Laysan Albatross *Phoebastria immutabilis*	**Hawaii**					
		Necker	500*	1995	B	Stable?	15
		French Frigate Schoals	2 871	1994	A	Stable	15
		Gardner Pinnacles	≤15	?	?	?	16
		Laysan	135 252	1994	B	Decreasing?	15
		Lisianski	26 500	1982	B	?	15
		Pearl and Hermes Reef	10 500	1979	B	?	15
		Midway Atoll	427 500	1991/92	B	Decreasing?	15
		Kure Atoll	3 500	1995	B	?	15
		Kauai	100	1994	A	Increasing	15
		Niihau	175	?	?	?	16
		Kaula	59	1993	B	?	15
		Oahu	10	1994	A	Increasing	15
		Moku Manu	0	1995	A	?	15
		Molokai	‡‡	?	?	?	16
		Japan					
		Bonin Islands (Mukojima)	20	1992	?	Increasing	11
		Mexico					
		Isla Guadalupe	*ca.* 50	1992	B	Increasing	17
		Isla Benedicto	≤5	1992	?	?	18
		Isla Clarion	≥2	1990	?	?	19
Black-footed Albatross *Diomedea nigripes*	Black-footed Albatross *Phoebastria nigripes*	**Hawaii**					
		Nihoa	31*	1994	B	Decreasing?	15
		Necker	112*	1995	B	Decreasing?	15
		French Frigate Schoals	2 742	1995	A	Decreasing	15
		Lysan	24 813	1995	B	?	15
		Lisianski	2 800*	1982	B	?	15
		Pearl and Hermes Reef	5 220	1996	B	Decreasing	15
		Midway Atoll	19 255	1995	A	Decreasing	15
		Kure Atoll	1 200*	1995	B	Stable?	15
		Lehua	‡‡	?	C	?	16
		Kaula	5	1993	B	?	15
		Japan					
		Senkaku Islands (Kita-kojima)	*ca.* 20	1991	C	Increasing	11
		Izu Islands (Torishima)	*ca.* 1 300	1995	B	Increasing	11
		Bonin Islands (Mukojima)	*ca.* 1 000	1993	C	Increasing	11

*See Appendix 1E for sources of information.

APPENDIX 1C

Population size and status of the annual breeding of Southern Albatrosses.

Old taxonomy	New taxonomy	Location	Breeding pairs	Survey date	Survey reliability	Population status	Source*
Black-browed Albatross	Black-browed Albatross						
Diomedea melanophrys melanophrys	*Thalassarche melanophrys*	**Falkland Islands**					
		Steeple Jason Island	250 826	1995/96	A	Increasing	20
		South Jason Island	350	1983/84	A	Stable	21
		Elephant Jason Island	600	1984/85	A	Increasing	21
		Beauchene Island	149 363	1995/96	A	?	22
		Bird Island	15 000–20 000	1995/96	C	?	23
		Grand Jason Island	50 000–100 000	1986/87	C	?	24
		West Point Island	15 400	1994/95	A	Increasing	25
		New Island	10 500	1995	A	Increasing	21
		North Island	14 625	1995	A	Increasing	21
		Saunders Island	12 505	1992/93	A	?	26
		Keppel Island	2 085	1987/88	A	?	26
		Grave Cove	170	1992/93	A	Increasing	27
		South Georgia					
		Bird Island	9 539	1995	A	Decreasing	28
		All other areas	86 713	1986	A	?	29
		Chile					
		Diego Ramirez					
		Isla Ildefonso	17 000	1985	C	?	30
		Isla Diego de Almagra	15 000	1985	B	?	30
		Crozet Islands	980	1981	A	?	31
		Kerguelen Islands	3 115	1995	A	Decreasing	32
		Heard Island	600–700	1987/88	B	?	33
		McDonald Island	82–89	1981	B	?	34
		Macquarie Island	38	1995	A	?	6
		Bishop & Clerk Island	141*	1993	B	?	35
		New Zealand					
		Antipodes Islands	*ca.* 100	1992	C	?	10
		Campbell Island	>30	1995	C	?	9
		Snares Island	1	1986	B	?	36
	Campbell Albatross						
Diomedea melanophrys impavida	*Thalassarche impavida*	Campbell Island	26 000	1992	B	Decreasing?	9
Buller's Albatross	Buller's Albatross						
Diomedea bulleri bulleri	*Thalassarche bulleri*	Snares Islands	8 460	1992	B	?	37
		Solander Island	4 000–5 000	1985	?	?	38
		Little Solander Island	300	1985	?	?	38
	Pacific Albatross						
Diomedea bulleri platei	*Thalassarche* nov. sp.	**Chatham Islands**					
		Big Sister Island	1 500	none	C	?	10
		Little Sister Island	650	1994–96	C	?	39
		Forty-Fours	*ca.* 16 000	none	C	?	10
		Three Kings	*ca.* 20	?	?	?	10
Shy Albatross	Shy Albatross						
Diomedea cauta cauta	*Thalassarche cauta*	**Tasmania**					
		Albatross Island	5 000	1995	B	Increasing†	35
		Mewstone	7 000	1995	B	?	35
		Pedra Branca	200	1995	B	?	35
	White-capped Albatross	**New Zealand**					
Diomedea cauta steadi	*Thalassarche steadi*	Disappointment Island	72 000	1993	B	?	10
		Adams Island	100	1993	B	?	8
		Auckland Island	3 000	?	C	?	10
		Antipodes Islands	50–100	1994	C	?	10
	Salvin's Albatross						
Diomedea cauta salvini	*Thalassarche salvini*	**New Zealand**					
		Bounty Islands	76 000	1978	C	?	40
		Snares Islands	≤650	1984	C	?	41
		Iles Crozet					
		Ile des Pingouins	4	1980s	A	?	42
	Chatham Albatross	**New Zealand**					
Diomedea cauta eremita	*Thalassarche eremita*	Chatham Islands	4 000	1992	C	?	10

*See Appendix 1E for sources of information.

Appendix 1C — *continued.*

Old taxonomy	New taxonomy	Location	Breeding pairs	Survey date	Survey reliability	Population status	Source*
Yellow-nosed Albatross	Atlantic Yellow-nosed Albatross						
Diomedea chlororhynchos chlororhynchos	*Thalassarche chlororhynchos*	**Tristan de Cunha Islands**					
		Tristan de Cunha	16 000–30 000	1972–74	C*	?	43
		Nightingale Island	4 500	1972–74	?	?	43
		Inaccessible Island	1 100	1982–83	C*	?	44
		Middle Island	100–200	1972–74	?	?	43
		Stolenhoff Island	500	1972–74	?	?	43
		Gough Island	*ca.* 5 000–10 000	1972–74	C*	?	3, 43
	Indian Yellow-nosed Albatross						
Diomedea chlororhnchos bassi	*Thalassarche carteri*	Prince Edward Island	7 000	1979	C	?	45
		Kerguelen Islands					
		Ile de Croy	50	1984–87	?	?	32
		Crozet Islands					
		Ile des Pingouins	3 200	1986	?	?	42
		Ile des Apotres	1 230	1981–82	?	?	46
		Amsterdam Island	*ca.* 25 000	1981–94	B	Decreasing	47
		St Paul Island	12	?	?	?	46

*See Appendix 1E for sources of information and explanatory notes.

APPENDIX 1D

Population size and status of biennial breeding Southern Albatrosses (see also Appendix 1A).

Current taxonomy	Proposed taxonomy	Location	Breeding pairs (annual)	Survey date	Survey reliability	Population status	Source*
Grey-headed Albatross	Grey-headed Albatross						
Diomedea chrysostoma	*Thalassarche chrysostoma*	**South Georgia**					
		Bird Island	6 500	1993–95	A	Decreasing	48
		All other areas	47 718	1985–86	A	?	29
		Chile					
		Diego Ramirez	10 000	?	?	?	49
		Isla Iledefonso	*ca.* 10	1984	B	?	50
		Kerguelen Islands	7 900	1984–1987	?	?	32
		Crozet Islands	5 946	1980–1982	?	?	31
		Marion Island	6 217	1995	A	Increasing*	3
		Prince Edward Island	1 500	1979	C	?	45
		Campbell Island	*ca.* 6 400	1995	B	Decreasing	9
		Macquarie Island	84	1995	A	?	6
Sooty Albatross	Sooty Albatross						
Phoebetria fusca	*Phoebetria fusca*	**Tristan da Cunha Group**					
		Tristan da Cunha	2 000–3 000	1972–74	C**	?	43
		Nightingale Island	100–200	1972–74	?	?	43
		Inaccessible Island	60+	1982–83	C**	?	44
		Stoltenhoff Island	25–50	1972–74	?	?	43
		Gough Island	5 000–10 000	1972–74	C**	?	43
		Prince Edward Island	700	?	?	?	51
		Marion Island	2 055	1987	B	?	51
		Kerguelen Island	≤5	1986–87	?	?	32
		Crozet Islands					
		Ile de la Possession	273	1995	A°	Decreasing	32
		Ile de l'Est	1 300	1981–82	?	?	46
		Ile aux Cochons	400–500	?	?	?	46
		Ile des Pingouins	250	1981–82	?	?	46
		Ile des Apotres	20–30	1981–82	?	?	46
		Amsterdam Island	300–400	1995	C	?	46
		St Paul Island	?	?	?	?	46
Light-mantled Sooty Albatross	Light-mantled Albatross						
Phoebetria palpebrata	*Phoebetria palpebrata*	South Georgia	5 000–7 500	?	?	?	1
		Prince Edward Island	40	1983–1990	C	?	51
		Marion Island	201	1987	B	?	51
		Kerguelen Islands	3 000–5 000	1984–87	?	?	32
		Crozet Islands					
		Ile de la Possession	996	1995	A	Decreasing	47
		Ile de l'Est	>900	1981–95	?	?	46
		Ile aux Cochons	50–100	1981–82	?	?	46
		Ile des Pingouins	30	1981–82	?	?	46
		Ile des Apotres	150	1981–82	?	?	46
		Heard Island	200–500	1954	C	?	52
		McDonald Islands	‡‡	?	C	?	53
		Macquarie Island	1 000–1 150	1994–95	A	?	6
		New Zealand					
		Auckland Islands	*ca.* 5 000	1972–73	C	?	54
		Campbell Island	>1 500	1995	C	?	9
		Antipodes Islands	<1 000	1969	C	?	55

*See Appendix 1E for sources of information and explanatory notes.

APPENDIX 1E

Sources of information for Appendix 1A–D.

1 J. P. Croxall and P. Prince (pers. comm.) and Croxall *et al.* (1990b).
2. P. Jouventin and H. Weimerskirch, pers. comm. The populations on the three Crozet Islands decreased by 4.9%, 2.6% and 6% (in listed order) between 1975 and 1982 (Weimerskirch and Jouventin 1987). Since then they have stabilized. Similarly severe declines were monitored at Kerguelen colonies between 1971 and 1985 (Weimerskirch *et al.* 1989) but have since stabilized (Weimerskirch *et al.*, in press).
3. J. Cooper, pers. comm. Marion Island population decreasing at *ca.* 0.7% per annum until 1992, since then increasing.
4. Watkins (1987).
5. Kirkwood *et al.* (1989). Only record is one pair in 1980 (Johnstone 1982).
6. Gales *et al.* (unpubl. data).
7. Ryan *et al.* (1990). Stables at this level for 50+ years (see text).
8. K. Walker (pers. comm.).
9. P. Moore (pers. comm.).
10. C. J. R. Robertson (pers. comm.).
11. H. Hasegawa (pers. comm.).
12. Richardson (1994).
13. Anderson (1995).
14. Oritz-Crespo and Agnew (1992).
15. E. Flint (pers. comm.) (USFWS, unpubl. data/State of Hawaii, unpubl. data).
16. Harrison (1990) and references therein.
17. Gallo-Raynoso and Figueroa-Carranza (1996).
18. Howell and Webb (1992) and references therein.
19. W. Everett (pers. comm.).
20. M. Riddy, unpubl. in litt. to J. P. Croxall. Earliest estimate of "30 000 young birds" in March 1962 (E. Goss to R. Napier to R. Woods). Possible increase from 200 000 to 230 000 in 1987.
21. I. J. Strange (pers. comm.) (New Island [South] Conservation Trust).
22. M. Riddy, unpubl. in litt. to J. P. Croxall. Estimate of 162 360 ± 10% (range 140 000–170 000) pairs in 1980/81 (Prince 1982) and 134 394 ± 10% pairs in 1991/92 (Thompson 1993).
23. S. and H. Poncet and M. Riddy, in litt. to J. P. Croxall. In 1939, 8 000 (out of probably 10 000) eggs were collected (R. Napier, pers. comm. to M. Riddy).
24. S. and J. Poncet, in litt. to J. P. Croxall.
25. Falklands Conservation (M. Bingham) in litt. to J. P. Croxall. Status based on comparison between estimates of 5 000–7 000 pairs in 1962/63 and 12 050 pairs (±10%) in 1989/90 (Thompson 1993).
26. Thompson (1993).
27. M. Riddy, unpubl. in litt. to J. P. Croxall. Increase from 120 pairs in 1987 (Falklands Conservation Seabird Monitoring Programme per M. Bingham).
28. Prince *et al.* (1994a) and Croxall *et al.* (1997). This population decreasing by 6.95% p.a. (since 1989/90).
29. Prince *et al.* (1994a).
30. Clark *et al.* (1992). Figure of 8 500 chicks on Isla Ildefonso translated to 17 000 pairs (J. P. Croxall, pers. comm.).
31. Weimerskirsch *et al.* (1986).
32. Weimerskirsch and Jouventin (1997).
33. Kirkwood and Mitchell (1992).
34. P. Keage (pers. comm.).
35. N. Brothers (unpubl. data).
36. A. Tennyson and C. Miskelly (pers. comm.).
37. Sagar *et al.* (1994).
38. Cooper *et al.* (1986).
39. Robertson (1991).
40. Robertson and van Tets (1982).
41. Miskelly (1984).
42. Jouventin (1990).
43. Richardson (1984).
44. Fraser *et al.* (1988).
45. Cooper and Brown (1990).
46. Jouventin *et al.* (1984) and H. Weimerskirch (pers. comm.).
47. H. Weimerskirsch (pers. comm.).
48. J. P. Croxall (pers. comm.). This population decreasing at *ca.* 1.42% p.a.
49. Schlatter (1984).
50. Clark *et al.* (1992).
51. Cooper and Brown (1990).
52. Downes *et al.* (1959).
53. Woehler (1991).
54. Bell (1975) in Marchant and Higgins (1990).
55. Warham and Bell (1979).

†Currently stable/increasing (whichever specified) at low levels after previous population declines.

°Extrapolated from partial survey.

*Indicates an extrapolation to total eggs from chicks counted later in the season.

**No complete population survey ever conducted.

‡‡Indicates breeding is suspected but not confirmed.

‡Indicates breeding is confirmed but no available information to quantify extent.

Status Reliability
A: High
B: Medium
C: Low

REFERENCES

Alexander, K., Robertson, G. and Gales, R., 1997. The incidental mortality of albatrosses in longline fisheries. Australian Antarctic Division. Tasmania. 44 pp.

Anderson, D., 1995. Census of the Waved Albatross, 1994. Unpublished Final Report to the Charles Darwin Research Station.

Anderson, D., Schwandt, A. J. and Douglas, H. D., 1997. Foraging ranges of Waved Albatrosses in the eastern Tropical Pacific Ocean. Pp. 180–85 *in* Albatross Biology and Conservation ed by G. Robertson and R. Gales. Surrey Beatty & Sons: Chipping Norton.

Anon., 1996. Hawaii longliners hook, kill thousands of albatrosses annually. *Environ. Hawaii, January* **1996:** 1–2.

Bartle, J. A., 1991. Incidental capture of seabirds in the New Zealand subantarctic squid trawl fishery, 1990. *Bird Conserv. Internat.* **1:** 351–59.

Birdlife International, 1995. Global impacts of fisheries on seabirds. A paper prepared by Birdlife International for The London Workshop on Environmental Science, Comprehensiveness and Consistency in Global Decisions on Ocean Issues. 30 November–2 December 1995.

Brothers, N. P., 1991. Albatross mortality and associated bait loss in the Japanese longline fishery in the Southern Ocean. *Biol. Conserv.* **55:** 255–68.

Brothers, N. and Foster, A., in press. Seabird catch rates: an assessment of causes and solutions in Australia's domestic tuna longline fishery. *Mar. Ornithol.*

Brothers, N., Gales, R., Hedd, A. and Robertson, G., 1997a. Foraging strategy of Australian Shy Albatrosses: Implications for interactions with longline fisheries. *Ibis.*

Brothers, N., Reid, T. and Gales, R., 1997b. Shy Albatross at-sea distribution derived from records of band recoveries and colour marked birds. *Emu.*

Camp, K., 1993. Observations of Short-tailed Albatrosses in the Bering Sea. *Colonial Waterbirds* **16:** 221–22.

Cherel, Y., Weimerskirch, H. and Duhamel, G., 1996. Interactions between longline vessels and seabirds in Kerguelen waters and a method to reduce seabird mortality. *Biol. Conserv.* **75:** 63–70.

Clark, G. S., Cowan, A., Harrison, P. and Bourne, W. R. P., 1992. Notes on the seabirds of the Cape Horn Islands. *Notornis* **39:** 133–44.

Collar, N. J., Crosby, M. J. and Stattersfield, A. J., 1994. Birds to Watch 2: The worlds list of threatened birds. Birdlife Conservation Series No. 4. Cambridge: Birdlife International.

Cooper, J. and Brown, C. R., 1990. Ornithological research on the sub-Antarctic Prince Edward Islands: a review of achievements. *Sth Afr. J. Antarc. Res.* **20:** 40–57.

Cooper, W. J., Miskelly, C. M., Morrison, K. and Peacock, R. J., 1986. Birds of the Solander Islands. *Notornis* **33:** 7–89.

Croxall, J. P., Rothery, P., Pickering, S. P. C. and Prince, P. A., 1990. Reproductive performance, recruitment and survival of Wandering Albatrosses *Diomedea exulans* at Bird Island, South Georgia. *J. Anim. Ecol.* **59:** 773–94.

Croxall, J. P. and Rothery, P., 1991. Population regulation of seabirds and implications of their demography for conservation. Pp. 272–96 *in* Bird Population Studies: their relevance to conservation and management ed by C. M. Perrins, J. D. Lebreton and G. M. Hirons. Oxford University Press: Oxford, U.K.

Croxall, J. P., Prince, P. A., Rothery, P. and Wood, A. G., 1997. Population changes in albatrosses at South Georgia. Pp. 69–83 *in* Albatross Biology and Conservation ed by G. Robertson and R. Gales. Surrey Beatty & Sons: Chipping Norton.

Croxall, J. P. and Gales, R., 1997. An assessment of the conservation status of albatrosses. Pp. 46–65 *in* Albatross Biology and Conservation ed by G. Robertson and R. Gales. Surrey Beatty & Sons: Chipping Norton.

Douglas, H., 1995. Population dynamics and ecology of the Waved Albatross *(Diomedea irrorata)*. Pp. 44 *in* Abstracts from First International Conference on the Biology and Conservation of Albatrosses, Hobart (Tasmania), 28 August–1 September, 1995.

Downes, M. C., Ealey, E. H. M., Gwynne, A. M. and Young, P. S., 1959. The birds of Heard Island. Australian National Antarctic Research Expedition Reports (Series B).

Elliot, G. P., Walker, K. J., Nicholls, D. G. and Murray, M. D., 1995. Foraging patterns of Wandering Albatross from the Auckland Islands. P. 19 *in* Abstracts from First International Conference on the Biology and Conservation of Albatrosses, Hobart (Tasmania), 28 August–1 September, 1995.

Flint, E. N., 1995. Albatross populations and conservation issues in the Northern Hemisphere. P. 9 *in* Abstracts from First International Conference on the Biology and Conservation of Albatrosses, Hobart (Tasmania), 28 August–1 September, 1995.

Fraser, M. W., Ryan, P. G. and Watkins, B. P., 1988. The seabirds of Inaccessible Island, South Atlantic Ocean. *Cormorant* **16:** 7–33.

Gales, R., 1993. Co-operative Mechanisms for the Conservation of Albatross. Division of Parks and Wildlife, Tasmania. Tasmanian Govt. Printer. 132 pp.

Gales, R. and Brothers, N., 1995. Characteristics of seabird bycatch in the Japanese tuna longline fishery in the Australian region, 1987–1994. CCSBT/ERS/95/

Gallo-Reynoso. J.-P. and Figueroa-Carranza, A.-L., 1996. The breeding colony of Laysan Albatrosses on Isla de Guadalupe, Mexico. *Western Birds* **27:** 70–76.

Gould, P. and Hobbs, R., 1992. Population dynamics of the Laysan and other Albatrosses in the North Pacific. *In* Proceedings of a Symposium on the Biology, Distribution and Stock Assessment of Species Caught in High seas Driftnet Fisheries in the North Pacific Ocean ed by W. Shaw, R. Burgher and J. Itô. International North Pacific Fisheries Commission, 4–6 November, Tokyo, Japan.

Groombridge, B., ed, 1993. IUCN Red List of Threatened Animals. IUCN: Gland, Switzerland and Cambridge, UK. lvi + 286 pp.

Harris, M. P., 1973. The biology of the Waved Albatross *Diomedea irrorata* of Hood Island, Galapagos. *Ibis* **115:** 483–510.

Harrison, C. S., 1990. Seabirds of Hawaii: natural history and conservation. Cornell University Press: New York.

Hasegawa, H., 1991. Red data bird: Short-tailed Albatross. *World Birdwatch* **13:** 10.

Howell, S. N. G. and Webb, S. W., 1992. Changing status of the Laysan Albatross in Mexico. *Amer. Birds* **46:** 220–23.

Johnson, D. H., Shaffer, T. L. and Gould, P. J., 1992. Incidental catch of marine birds in the North Pacific high seas driftnet fisheries in 1990. *In* Proceedings of a Symposium on the Biology, Distribution and Stock Assessment of Species Caught in High seas Driftnet Fisheries in the North Pacific Ocean ed by W. Shaw, R. Burgher and J. Itô. International North Pacific Fisheries Commission, 4–6 November, Tokyo, Japan.

Johnstone, G. W., 1982. Zoology. *In* Expedition to the Australian territory of Heard Island and McDonald Islands 1980. Canberra: Department of National Development and Energy, Division of National Mapping. *Technical Report* **31:** 33–34.

Jouventin, P., 1990. Shy Albatross *Diomedea cauta salvini* breeding on Penguin Island, Crozet Archipelago, Indian Ocean. *Ibis* **132:** 126.

Jouventin, P., Stahl, J.-C., Weimerskirch, H. and Mougin, J.-L., 1984. The seabirds of the French Subantarctic Islands and Adelie Land, their Status and Conservation. Pp. 609–25 *in* Status and Conservation of the World's Seabirds ed by J. P. Croxall. ICBP Technical Publication No. 2.

Jouventin, P., Martinez, J. and Roux, J.-P., 1989. Breeding biology and current status of the Amsterdam Island Albatross. *Ibis* **131:** 171–89.

Kirkwood, R. J. and Mitchell, P. J., 1992. The status of the Black-browed Albatross *Diomedea melanophrys* at Heard Island. *Emu* **92:** 111–14.

Kirkwood, R .J., Woehler, E. J. and Burton, H. R., 1989. Heard Island 1987–88 ANARE Report. Unpublished Report, Australian Antarctic Division: Hobart.

Lindsey, T. R., 1986. The seabirds of Australia. Angus and Robertson: Sydney.

Ludwig, J. P., Summer, C. L., Auman, H. J., Gauger, V., Bromley, D., Geisy, J. P. Rolland, R. and Colborn, T., 1997. The roles of organochlorine contaminants and fisheries bycatch in recent population changes of Black-footed and Laysan Albatrosses in the North Pacific Ocean. Pp. 225–38 *in* Albatross Biology and Conservation ed by G. Robertson and R. Gales. Surrey Beatty & Sons: Chipping Norton.

Marchant, S. and Higgins, P. J., 1990. Handbook of Australian, New Zealand and Antarctic Birds Oxford University Press: Melbourne.

McDermond, D. K. and Morgan, K. H., 1993. Status and Conservation of North Pacific Albatrosses. Pp. 70–81 *in* The Status, Ecology and Conservation of Marine Birds of the North Pacific ed by K. Vermeer, K. T. Briggs, K. H. Morgan and D. Siegel-Causey. Canadian Wildlife Service Special Publication: Ottawa.

Merlen, G., 1996. Scavenging waved albatross around the Galapagos Archipelago: at risk from long-liners. *The Seabird Group Newsl.* **74:** 2–5.

Miskelly, C. M., 1984. Birds of the Western Chain, Snares Islands 1983–84. *Notornis* **31:** 209–23.

Moore, P. J., 1995. Status and decline of mollymawks on Campbell Island. P. 7 *in* Abstracts from First International Conference on the Biology and Conservation of Albatrosses, Hobart (Tasmania), 28 August–1 September, 1995.

Moore, P. J. and Moffat, R. D., 1990. Mollymawks on Campbell Island. Science and Research Internal Report No. 59. Department of Conservation: Wellington, New Zealand.

Murray, T., Bartle, J. A., Kalish, S. R. and Taylor, P., 1993. Incidental capture of seabirds by Japanese southern bluefin tuna longline vessels in New Zealand waters 1988–1992. *Bird Conserv. Internat.* **3:** 181–210.

Neves, T. and Olmos, F., 1997. Albatross mortality in fisheries off the coast of Brazil. Pp. 214–19 *in* Albatross Biology and Conservation ed by G. Robertson and R. Gales. Surrey Beatty & Sons: Chipping Norton.

Northridge, S. P., 1991. Driftnet fisheries and their impacts on non-target species: a world-wide review. FAO Fisheries Technical Paper, No. 320. Rome, FAO. 115 pp.

Nunn, G. B., Cooper, J., Jouventin, P., Robertson, C. J. R. and Robertson, G. G., 1996. Evolutionary relationships among extant albatrosses (Procellariiformes: Diomedeidae) established from complete cytochrome-*b* gene sequences. *Auk* **113:** 784–801.

Ogi, H., 1995. Ingestion of plastic particles of black-footed and Laysan albatrosses. P. 12 *in* Abstracts from First International Conference on the Biology and Conservation of Albatrosses, Hobart (Tasmania), 28 August–1 September, 1995.

Ortiz-Crespo, F. and Agnew, P., 1992. The birds of la Plata Island, Ecuador. *Bull. B.O.C.* **112:** 66–73.

Prince, P. A., 1982. The black browed albatross *Diomedea melanophris* population at Beauchene Island, Falkland Islands. *Colloque sur les Ecosystemes Subantarctiques. 1981, Paimpoint. C.N.F.R.A.* **51:** 111–17.

Prince, P. A., Rothery, P., Croxall, J. P. and Wood, A. G., 1994a. Population dynamics of Black-browed and Grey-headed Albatrosses *Diomedea melanophrys* and *D. chrysostoma* at Bird Island, South Georgia. *Ibis* **136:** 50–71.

Prince, P. A., Huin, N. and Weimerskirch, H., 1994b. Diving depths of albatrosses. *Antarc. Sci.* **6:** 353–54.

Prince, P., Croxall, J. P., Trathan, P. N. and Wood, A. G., 1997. The pelagic distribution of South Georgia albatrosses and their relationships with fisheries. Pp. 137–67 *in* Albatross Biology and Conservation ed by G. Robertson and R. Gales. Surrey Beatty & Sons: Chipping Norton.

Reid, T. and James, D., in press. The Chatham Island mollymawk *Diomedea eremita* in Australia. *Notornis.*

Richardson, J., 1984. Aspects of the ornithology of the Tristan da Cunha group and Gough Island. *Cormorant* **12:** 122–201.

Richardson, S. A., 1994. Status of the Short-tailed Albatross on Midway Atoll. *'Elepaio* **54:** 35–37.

Robertson, C. J. R., ed, 1985. The Complete Book of New Zealand Birds. Reader's Digest Services: Sydney, Australia.

Robertson, C. J. R., 1991. Questions on the harvesting of Toroa in the Chatham Islands. Science and Research Series No. 35. Department of Conservation: Wellington, New Zealand.

Robertson, C. J. R., 1997. Factors influencing the breeding performance of the Northern Royal Albatross. Pp. 99–104 *in* Albatross Biology and Conservation ed by G. Robertson and R. Gales. Surrey Beatty & Sons: Chipping Norton.

Robertson, C. J. R. and Nunn, G. B., 1997. Towards a new taxonomy for albatrosses. Pp. 13–19 *in* Albatross Biology and Conservation ed by G. Robertson and R. Gales. Surrey Beatty & Sons: Chipping Norton.

Robertson, C. J. R. and van Tets, G. F., 1982. The status of birds at the Bounty Islands. *Notornis* **29:** 311–36.

Ryan, P. G., Dean, W. R. J., Moloney, C. L., Watkins, B. P. and Milton, S. J., 1990. New information on seabirds at Inaccessible Island and other islands in the Tristan da Cunha group. *Mar. Ornithol.* **18:** 43–54.

Sagar, P. M. and Weimerskirch, H., in press. Satellite tracking of Southern Buller's Albatrosses from the Snares, New Zealand. *Condor.*

Sagar, P. M., Molloy, J., Tennyson, A. J. D. and Butler, D., 1994. Numbers of Buller's mollymawks breeding at the Snares Islands. *Notornis* **41:** 85–92.

Schiavini, A., Frere, E., Gandini, P., Garcia, N. and Crespo, E., 1997. Albatross–fisheries interactions in Patagonian shelf waters. Pp. 208–13 *in* Albatross Biology and Conservation ed by G. Robertson and R. Gales. Surrey Beatty & Sons: Chipping Norton.

Schlatter, R. P., 1984. The status and conservation of seabirds in Chile. Pp. 261–69 *in* Status and Conservation of the World's Seabirds ed by J. P. Croxall, P. G. H. Evans and R. W. Schlatter. ICBP Technical Publication No. 2: Cambridge, U.K.

Spear, L., Ainley, D. G. and Webb, S. W., 1995. Distribution, abundance and behaviour of Buller's, Chatham Island, and Salvin's Albatrosses off Chile and Peru: Potential interaction with longliners. P. 25 *in* Abstracts from First International Conference on the Biology and Conservation of Albatrosses, Hobart (Tasmania), 28 August–1 September, 1995.

Thompson, K. R., 1993. Falkland Island seabird-monitoring programme: summary of results 1989/90 to 1992/93. Falklands Conservation Report SMP/3, Stanley (unpublished).

Walker, K. J. and Elliott, G. P., 1995. Albatross research on a shoe-string: Population biology of Auckland Island Wandering Albatross. P. 8 *in* Abstracts from First International Conference on the Biology and Conservation of Albatrosses, Hobart (Tasmania), 28 August–1 September, 1995.

Warham, J. and Bell, B. D., 1989. The birds of Antipodes Island, New Zealand. *Notornis* **26:** 121–69.

Watkins, B. P., 1987. Population sizes of King, Rockhopper and Macaroni Penguins and Wandering Albatrosses at the Prince Edward Islands and Gough Island, 1951–1986. *Sth Afr. J. Antarct. Res.* **17:** 155–62.

Weimerskirch, H., 1997. Foraging strategies of Indian Ocean albatrosses and their relationships with fisheries. Pp. 168–79 *in* Albatross Biology and Conservation ed by G. Robertson and R. Gales. Surrey Beatty & Sons: Chipping Norton.

Weimerskirch, H. and Jouventin, P., 1987. Population dynamics of the Wandering Albatross *(Diomedea exulans)* of the Crozet Islands: causes and consequences of the population decline. *Oikos* **49:** 315–22.

Weimerskirch, H. and Jouventin, J., 1997. Changes in population sizes and demographic parameters of six albatross species breeding on the French sub-Antarctic islands. Pp. 84–91 *in* Albatross Biology and Conservation ed by G. Robertson and R. Gales. Surrey Beatty & Sons: Chipping Norton.

Weimerskirch, H., Brothers, N. and Jouventin, P., 1997. Foraging dynamics of Wandering Albatross *Diomedea exulans* and Amsterdam Albatross *D. amsterdamensis* in the Indian Ocean and their relationships with long-line fisheries: conservation implications. *Biol. Conserv.* **79:** 257–70.

Weimerskirch, H., Jouventin, P. and Stahl, J.-C., 1986. Comparative ecology of six albatross species breeding on the Crozet Islands. *Ibis* **128:** 195–213.

Weimerskirch, H., Zotier, R. and Jouventin, P., 1989. The avifauna of the Kerguelen Islands. *Emu* **89:** 15–29.

Woehler, E. J., 1991. Status and conservation of seabirds of Heard Island and the McDonald Islands. Pp. 263–77 *in* Seabird Status and conservation. A Supplement ed by J. P. Croxall. Technical Publication 11. ICBP: Cambridge, England.

Woehler, E. J., 1996. Concurrent decreases in five species of Southern Ocean seabirds in Prydz Bay. *Polar Biol.* **16:** 379–82.

CHAPTER FOUR

An assessment of the conservation status of albatrosses

J. P. CROXALL[1] and R. GALES[2]

The conservation status of albatrosses has been re-evaluated using the recent suggested taxonomic revisions of albatrosses and the new IUCN criteria for identification of threatened taxa. The evaluations of threat status and categories were prepared by independent albatross specialists before being critically reviewed and endorsed by BirdLife International.

This evaluation reveals albatrosses to have the highest proportion of threatened species in any bird family that comprises more than a single species. Of the 24 albatross species recognized in the new taxonomy, two species (Amsterdam Albatross and Chatham Albatross) are Critically Endangered, two species (Northern Royal Albatross and Tristan Albatross) are Endangered, and 16 species are classed as Vulnerable. Currently only the two most abundant albatross species, the Black-browed (Near Threatened) and Laysan Albatrosses are classified as Lower Risk. For two species (Light-mantled Albatross and Atlantic Yellow-nosed Albatross) the present level of knowledge regarding population trend and status provides an inadequate basis for classification and these are assessed as Data Deficient.

Albatrosses are currently the most threatened and vulnerable group of all seabirds, and their protection both on land and at sea is an urgent requirement.

Key words: Conservation, Status, Albatrosses, IUCN Red List.

INTRODUCTION

THE identification of species whose existence is threatened is a key element in the process of defining global and regional priorities for the conservation of species and their habitats. For many years the IUCN (International Union for the Conservation of Nature) criteria for assigning threat status and category — as used in Red Data Books since the 1960s — have been increasingly unsatisfactory because of their very general definitions, substantial subjectivity in application and resulting unevenness of treatment, both within and between different taxonomic groups.

In the 1980s the IUCN started a process of replacing the existing criteria with an objective system using numerical criteria to derive a hierarchy of threatened categories of increasing risk of extinction, based on measured or estimated rates of decline, population levels and range size and distribution. The development of this system, described in Mace and Lande (1991), Mace *et al.* (1992) and Mace and Stuart (1994), culminated in the adoption by the IUCN Council of new Red List categories, the structure of which is summarized in Appendix 1, in December 1994 (IUCN 1994).

The 1994 IUCN Red List of threatened animals (Groombridge 1993) was, therefore, the last such list to be published using the old criteria. In 1995 the IUCN Specialist Groups were requested to evaluate species against the new criteria in preparation for contributing to the 1996 IUCN Red Data List (then under preparation) which will be the first to use the new criteria. For birds, however, BirdLife International (formerly the International Council for Bird Preservation), the body officially responsible for the bird Red Data List, had been closely involved in the development of these criteria. In its recent comprehensive and detailed review of the world's threatened birds (Collar *et al.* 1994), it used the latest version of the IUCN criteria then available (Mace and Stuart 1994), which differed only in a few details from that finally adopted by the IUCN.

When the First International Albatross Conference and Workshop were held in

[1]British Antarctic Survey, Natural Environment Research Council, High Cross, Madingley Road, Cambridge CB3 0ET, U.K.
[2]Tasmanian Parks and Wildlife Service, Department of Environment and Land Management, 134 Macquarie Street, Hobart, Tasmania, Australia 7001.
Pages 46–65 *in* ALBATROSS BIOLOGY AND CONSERVATION, ed by G. Robertson and R. Gales. Surrey Beatty & Sons, Chipping Norton. 1997.

Tasmania during August 1995, it was uncertain whether the identity and classification of threatened bird species would be further revised for the 1996 IUCN Red List or whether this would follow exactly the contents of Collar *et al.* (1994). In either case, however, the Workshop recognized that there was great importance in immediately undertaking a re-evaluation of the conservation status of albatrosses by applying the new IUCN criteria (Alexander *et al.* 1997).

The reasons for this included:

(a) The results of a major taxonomic revision of albatrosses (Nunn *et al.* 1996; Robertson and Nunn 1997), recommending the recognition of 10 new taxa at the species level. It was clear that several of these new species would be strong candidates for Threatened status;

(b) Substantial new data on the population status and trends of albatrosses and on the threats they face which had become available since BirdLife International undertook their evaluation. Generally these data suggested that several species, including most of those previously classified as Near Threatened, were likely to warrant threatened status;

(c) The presence at the Workshop of specialists with detailed knowledge of all albatross taxa, enabling a collective view to be taken of the new taxonomy and a start to be made on the work necessary for a critical re-evaluation of the conservation status of all species.

It was recognized that undertaking this task would need not only input from specialists on albatrosses but also independent review by experts in the application of threatened species criteria. Accordingly, the Workshop tasked J. P. Croxall and R. Gales with co-ordinating the initial assessments by albatross specialists, arranging independent reviews by the staff at BirdLife International responsible for Collar *et al.* (1994) and circulating the revised results of the exercise to participants.

In November 1995, 10 specialists were requested to prepare initial evaluations of threat status and categories of one or more of the 24 species recognized by the new taxonomy. To assist them in this task they were provided with: a) the booklet describing the new threatened status categories and their criteria (IUCN 1994). These are summarized in Appendix 2a (modified from Collar *et al.* 1994) and set out in full in Appendix 2b and c; b) the simplified explanation of the criteria by BirdLife International (Collar *et al.* 1994: pp. 14–21), together with an explanation of the changes between the version of the criteria used by BirdLife International and that finally adopted by IUCN (see Appendix 3); c) the texts for the three albatross species (Wandering Albatross *Diomedea exulans*, Amsterdam Albatross *D. amsterdamensis*, Short-tailed Albatross *D. albatrus*) listed as threatened in Collar *et al.* (1994), to indicate the nature of the information required; d) a note, prepared by A. J. Stattersfield of BirdLife International, on applying the IUCN criteria to seabirds (see Appendix 4).

It was expected that evaluators would consult widely with colleagues but it was recognized that this would not always be possible prior to producing the initial texts. A complete set of texts was finally available in March 1996. They were collated and edited for style by R. Gales and sent to BirdLife International in April 1996. The review of the application of criteria and classification into categories of threat was undertaken by A. J. Stattersfield and N. J. Collar. Their report was sent to all original compilers of evaluations in May 1996 with a request to undertake final revision by 30 June, 1996. All accounts were then circulated to participants at the Workshop and Conference for comments and resubmitted to BirdLife International for final evaluation and endorsement.

The next section comprises the species assessment texts as revised by compilers, and incorporating as far as possible all comments received from participants at the Workshop and Conference. At the end of this section there is a brief summary and discussion of: a) the changes in the conservation status of albatrosses compared with the evaluation by Collar *et al.* (1994); b) the reasons for these changes; and c) the consequences and implications for albatrosses.

The scientific nomenclature used follows Nunn *et al.* (1996) and Robertson and Nunn (1997), except that we have accepted Medway's (1993) arguments that *Diomedea exulans* Linnaeus is based on material of the widespread southernmost taxon of the Wandering Albatross group (sometimes called *D. (e). chionoptera* Snowy Albatross). The representative of the Wandering Albatross group at Tristan da Cunha and Gough Islands is therefore correctly known as *D. dabbenena*. The vernacular nomenclature is based on the same sources above but has been modified following the majority of the views expressed by participants in the Workshop.

SPECIES ASSESSMENTS

Wandering Albatross *Diomedea exulans*

Status: Vulnerable (A1a, d).

Rationale: On the basis of population reductions in excess of 20% over most of its range within the last three generations (based on direct observations) and potential levels of exploitation, this taxon qualifies as Vulnerable.

Text: Breeds at six main island groups (South Georgia, Crozet, Kerguelen, Prince Edward, Heard, Macquarie) in the Southern Ocean with a total annual breeding population of no more than 8 500 pairs (equivalent to a total mature population of about 28 000 individuals). With more than one count from all breeding sites and extensive time series from four sites, this is probably the best documented of all albatross species.

Data from the best documented sites at South Georgia (Bird Island) and Crozet (Possession Island) (together representing 30% of the world population) indicate steady (1% p.a. over 20 years) and steady-to-steep (1–7% p.a.) population declines respectively during the 1970s and 1980s (Weimerskirch and Jouventin 1987; Croxall *et al.* 1990; Weimerskirch *et al.* 1997). In total, breeding populations at Possession Island and Bird Island have been reduced by 54% (to 1986) and 28% (to 1996), respectively. Substantial population declines were also reported for Kerguelen, Marion, Prince Edward and Macquarie Islands (Watkins 1987; de la Mare and Kerry 1994; Weimerskirch *et al.* 1997). Populations are now stable or slightly increasing at Kerguelen, Crozet and Marion Islands (Weimerskirch *et al.* 1997; Gales 1993; J. Cooper, pers. comm.), stable (at very low numbers) at Macquarie Island (de la Mare and Kerry 1994) and still decreasing at South Georgia (Croxall *et al.* 1997). Longline fisheries are suspected to be the main causes of change in all these populations (Weimerskirch and Jouventin 1987; Croxall and Prince 1990; de la Mare and Kerry 1994; Croxall *et al.* 1997; Weimerskirch *et al.* 1997).

Prepared by: J. P. Croxall, British Antarctic Survey, Madingley Road, High Cross, Cambridge CB3 OET, United Kingdom and H. Weimerskirch, CEBC-CNRS, 79360, Beauvoir, France.

Tristan Albatross *Diomedea dabbenena*

Status: Endangered (B1 + B2e).

Rationale: Given an area of occupancy of <500 km^2 (Gough Island is 65 km^2), fewer than five breeding locations (colonies) and inferred and/or projected declines, this species qualifies as Endangered.

Text: Essentially confined to Gough Island, with 2–3 pairs also breeding at Inaccessible Island (Tristan da Cunha group) in 1990 (Ryan *et al.* 1990). In the 1980s the Gough Island population was estimated to contain about 1 000 pairs breeding annually (Watkins 1987) equivalent to a total breeding population of about 3 300 individuals for this biennially breeding species.

Given historical declines at Inaccessible Island (and probably also Gough Island) and the population trends of adjacent Wandering Albatross taxa (e.g., at South Georgia and Indian Ocean islands) it is hardly conceivable that there has not been further decrease since the 1980s. Limited evidence for longline mortality exists, based on recoveries of two birds banded at Gough Island (Cooper 1994). Furthermore, prospects for recovery/increase of a population likely to forage widely in the South Atlantic Ocean are not encouraging (see Prince *et al.* 1997).

Prepared by: J. Cooper, Percy FitzPatrick Institute, University of Cape Town, Rondebosch 7700, South Africa and J. P. Croxall, British Antarctic Survey, Madingley Road, High Cross, Cambridge CB3 OET, United Kingdom.

Antipodean Albatross *Diomedea antipodensis*

Status: Vulnerable (D2).

Rationale: The restricted range (fewer than five breeding locations) automatically qualifies this species as Vulnerable.

Text: Breeds at only two sites, the sub-Antarctic islands of Antipodes Island (99% of the population) and Campbell Island (a few pairs). Endemic to New Zealand.

A full census of nesting pairs was made on Antipodes Island in 1994, 1995 and 1996, with 4 635, 5 757 and 5 148 pairs counted respectively (Clark *et al.* 1995; J. Amey, in litt.; K. Walker and G. Elliot, unpubl. data) indicating a total mature population of about 17 000 birds for this biennial breeding species. No earlier comparable data exist (Warham and Bell 1979), so recent trends in the population size of this species are not known.

Occasional checks on the population on Campbell Island since the 1960s recorded between three and six pairs nesting there each year.

No measures of recruitment and adult survival of the Antipodean Albatross are yet available. This species was a relatively frequent bycatch in the New Zealand longline fishery between 1987 and 1992 (Murray *et al.* 1993). Both breeding (K. Walker and G. Elliot, unpubl. data) and non-breeding (D. G. Nicholls, in litt.) birds forage in waters where longline fisheries operate.

Prepared by: K. Walker and G. Elliot, Department of Conservation, Private Bag 5, Nelson, New Zealand.

Gibson's Albatross *Diomedea gibsoni*

Status: Vulnerable (D2).

Rationale: The restricted range (fewer than five breeding locations) qualifies this species as Vulnerable.

Text: Breeds only on the sub-Antarctic islands of Adams, Disappointment and Auckland Islands (Auckland Islands group). Endemic to New Zealand.

Feral cats and pigs apparently restrict breeding success on the main Auckland Island, and 94% of the population now nest on Adams Island and 5% on tiny Disappointment Island (both predator-free).

A series of four counts between 1991 and 1995 found a total annual breeding population of, at most, 6 200 pairs (K. Walker and G. Elliot, unpubl. data). This suggests a total mature population of about 22 000 birds for this biennial breeding species. There are no comparable earlier data (Robertson 1975), so population trends are unknown.

A detailed study of part of the Adams Island population between 1991 and 1996 found adult female survival to be significantly lower than adult males (93.8% cf. 98.5%), and an average breeding success of 69% (K. Walker and G. Elliot, unpublished data). No estimates of recruitment are available.

Gibson's Albatrosses were a significant bycatch in the longline tuna fishery in New Zealand waters between 1987 and 1992 (Murray *et al.* 1993). During breeding, Gibson's Albatrosses forage in waters also used by longline fisheries (Elliot *et al.* 1995; Walker *et al.* 1995).

Prepared by: K. Walker and G. Elliot, Department of Conservation, Private Bag 5, Nelson, New Zealand.

Southern Royal Albatross *Diomedea epomophora*

Status: Vulnerable (D2).

Rationale: The restricted breeding range (fewer than five breeding locations) qualifies this species as Vulnerable.

Text: Breeds at Campbell Island, Adams, Enderby and Auckland Island (Auckland Island group) and Taiaroa Head (South Island, New Zealand). Endemic to New Zealand.

The main population (99%) is at Campbell Island where the counts between 1978 and 1983 indicated an annual breeding population of 4 200–4 600 pairs, with published annual breeding success rates varying from 58–75% (though estimated by different methods) (note that breeding is biennial if a chick is successfully reared in the previous year, so the annual breeding population will vary according to the breeding success of the previous season). Annual counts of 6 300 and 7 800 pairs were obtained in 1995 and 1996 (but search effort in these years may have been more intense than previously). There has been an increase in the numbers of birds in study plots during the eight years up to 1995/96. The numbers at other breeding locations are small — Adams Island and Auckland Island have a combined total being no more than 20 breeding pairs annually, and Enderby Island has a mean annual population of 40 pairs, including two mixed pairs with *D. sanfordi*. No pure-bred *D. epomophora* are currently at Taiaroa Head on the New Zealand mainland, but five hybrids with *D. sanfordi* are successfully breeding with other hybrids or pure *D. sanfordi*. The total world breeding population is therefore about 7 870 pairs, equivalent to approximately 26 000 mature individuals.

There are no available data for juvenile recruitment or mean annual adult survival. The population at Enderby Island is currently increasing rapidly after extirpation in the 1800s. There is evidence of bycatch of this species in tuna fisheries in the Australasian region (Gales *et al.*, unpubl. data; C. J. R. Robertson, unpubl. data). Southern Royal Albatrosses may circumnavigate the southern hemisphere and are therefore potentially exposed to threats in high seas far from their breeding sites.

Prepared by: C. J. R. Robertson, Department of Conservation, PO Box 10420, Wellington, New Zealand.

Northern Royal Albatross *Diomedea sanfordi*

Status: Endangered (B1 + B2c,e).

Rationale: On the basis of restricted range (fewer than five breeding locations) and the observed and projected population declines, this species qualifies as Endangered.

Text: Breeds at The Forty-Fours, and the Big and Little Sister Islands in the Chatham Islands, Taiaroa Head (New Zealand mainland) and Enderby Island (Auckland Islands). Endemic to New Zealand.

Breeding is confined to a total land area of less than 100 ha. The main population (99%) is in the Chatham Islands with an estimated total breeding population of 6 500–7 000 pairs (based on air photographic counts in 1970s and 1990s). In 1995 a count of 5 200 pairs was recorded, equivalent to a total mature population of about 17 000 birds. At Taiaroa Head (a heavily protected and managed mainland population) there is a total breeding population of 27 breeding pairs, including five individuals as hybrids of *D. epomophora* × *D. sanfordi*. At Enderby Island from 1993–95, two individuals of *D. sanfordi* were recorded breeding with *D. epomophora*.

Breeding is biennial (every two years) if a chick is successfully reared, so the size of the annual breeding population may vary extensively according to the breeding success of the previous season. At Taiaroa Head the age at first breeding is 8–10 years with a mean 69% of fledglings surviving to five years old (Robertson 1993a). At Taiaroa Head mean annual adult survival (1936–93) is 0.9456 for birds of unknown age and 0.9461 for birds of known age (Robertson 1993a). At the Little Sister (1973–1995) the current mean annual survival rates are 0.951 for birds banded as adults and 0.953 for known age birds from time of fledging. Mean annual productivity (chicks fledged from eggs laid) from 1936–1996 at Taiaroa Head was 46% but this was partly assisted by management, especially since 1970. In the Chatham Islands the estimated mean annual productivity was 48% from 1973–1976, but only an estimated 18% p.a. from 1989– 1996.

Historically, Northern Royal Albatrosses have been harvested by Chatham Islanders, but harvesting is now illegal and occurs only sporadically (Robertson 1991). Human depredation's, disturbance and introduced predators (cats, ferrets, stoats) have affected the Taiaroa Head population and are currently controlled by intensive management (Robertson 1993b, 1997). Regulated tourism there since 1972 has also had some effect on nesting distribution (Robertson 1993c), but remedial measures have been tested since 1994. Since the mid 1970s the habitat at both the Chatham Island colonies and Taiaroa Head has dried out and the ambient temperature has increased. At Taiaroa Head this has led to an increasing incidence of strike by introduced Blowflies *Lucilia sericata* which, since 1986, has become the major cause of mortality. At the Chatham Island colonies in 1985 a "hurricane" and severe storm events have modified the habitat and reduced the soil cover where albatrosses breed. Worst affected is the Forty-Fours where estimated mean annual albatross productivity has dropped from 54% (1973–1976) to 8% (1989–1996). The total Chatham Island breeding population is now predicted to drop by some 30% without any compensating changes in habitat, breeding or productivity during the first generation (20 years) from 1985, with a similar or greater reduction for the second generation.

Though there has been some incidence of fisheries bycatch outside New Zealand waters (this taxon is one which has a circumpolar dispersal), this seems not to have been large, as evidenced by the current mean annual survival rate. Of concern is a reduction of about 20% in mean egg shell thickness in the Chatham Island populations since the 1970s and an increased occurrence of chicks dead in the shell or immediately after hatching at the Chatham Islands and Taiaroa Head populations, which may indicate a pesticide or contaminant problem.

Prepared by: C. J. R. Robertson, Department of Conservation, PO Box 10420, Wellington, New Zealand.

Amsterdam Albatross
Diomedea amsterdamensis

Status: Critically Endangered (D).

Rationale: With a population of fewer than 50 mature individuals this species qualifies as Critically Endangered.

Text: Amsterdam Albatrosses breed only at Amsterdam Island (Indian Ocean) and have a total population of about 90 birds including 40 mature individuals, about 13 pairs breeding annually. Breeding, which is biennial when successful, is restricted to the central plateau of the island. All the birds are banded, and the population is censused and monitored every year. The population has increased since 1984 when it was first censused (Weimerskirch *et al.* 1997).

The population was probably larger when its range was more extensive on the slopes of the island (Jouventin *et al.* 1990). The small size of the current population most likely results from declines caused by degradation of the breeding sites by introduced cattle, which are now restricted by a fence (Micol and Jouventin 1995), and longline fishing activity around the island in the 1970s and early 1980s (Weimerskirch *et al.* 1997). Other potential threats include predation by feral cats and rats. During the breeding season adult birds forage around Amsterdam Island and up to 2 200 km from the island in subtropical waters (H. Weimerskirch, unpubl. data) so the potential exists for birds to interact with longliners targeting tropical tuna species. Extinction risks are high because of the small size of the population and the susceptibility of the population to a possible increase in adult mortality.

Prepared by: H. Weimerskirch, CEBC–CNRS, 79360, Beauvoir, France.

Short-tailed Albatross *Phoebastria albatrus*

Status: Vulnerable (D1 + D2).

Rationale: The small size of the breeding population (<1 000 mature individuals) and the restricted breeding range (<5 locations) qualifies this species as Vulnerable.

Text: Short-tailed Albatrosses once bred in huge numbers on at least 11 islands in the Izu-Bonin chains and in the Ryukyu Islands in Japan, in Taiwan and probably on several islands off China. Exploitation for their feathers from the late 19th century onwards almost wiped out most populations by 1930. The only places where breeding has recently been confirmed are the volcanic ash slopes of Torishima in the Izu Islands, and on a cliff on Minami-kojima in the Senkaku Islands (Hasegawa 1984, 1991). One unsuccessful breeding attempt has also been observed on Midway Atoll, Northwestern Hawaiian Islands (Richardson 1994).

The population on Torishima increased from perhaps less than 50 birds in 1951 to approximately 500 birds in 1990/91 when 108 eggs were laid, further increasing to 158 eggs being laid in 1995/96 (H. Hasegawa, pers. comm.). Breeding success has improved with revegetation of the nesting area to stabilize the soil. Although increasing at about 7% per year, this population remains vulnerable to volcanic eruptions and land slides, and efforts have been made to attract adults to create a new colony at a stable site on the other side of the island (Hasegawa 1984, 1991). The first chick hatched from an egg laid in this new colony in January 1996 (H. Hasegawa, pers. comm.).

On Minami-kojima, 12 adults were observed in 1971, but breeding was not confirmed until 1988 when chicks were seen: a population of 75 was estimated in 1991, with 15 breeding pairs (Hasegawa 1991). On Midway Atoll, one or two birds have been present intermittently over the last two decades. A single incubating bird was found in November 1993, but the egg was subsequently abandoned (Richardson 1994).

Since 1987 there have been three confirmed mortalities of Short-tailed Albatrosses in North Pacific longline fisheries (P. Gould, in litt.) and this could be a serious problem given that these birds are known to be attracted by fishing vessels. It is possible that the abundant rats on Torishima predate eggs and newly hatched young; plastic debris inadvertently fed to chicks by their parents may also be a problem (see Gales 1993).

Prepared by: Updated from existing assessment (authored by M. Crosby in Collar *et al.* 1994) by R. Gales and J. P. Croxall using supplementary information provided by: H. Hasegawa, Toho University, 2–1, Miyama 2-chome, Funbashi, Chiba, Japan and P. Gould, U.S. National Biological Service, 1011 East Tudor Road, Anchorage, Alaska 9950.

Waved Albatross *Phoebastria irrorata*

Status: Vulnerable (D2).

Rationale: The restriction of the breeding population to only two locations qualifies the species as Vulnerable.

Text: Almost all of the breeding by this species takes place on Isla Espanola (area 58 km^2), Galapagos Islands. Censuses by Harris (1973) in 1970 and 1971 put the breeding population at "at least 12 000 pairs". The only other census occurred in 1994, and different interpretations of the resulting data put the breeding population at 15 581 pairs (Anderson 1995) and approximately 17 000 pairs (Douglas 1995). Thus, no decline in population size is indicated on Isla Espanola. The only other breeding site is Isla de la Plata (area 14 km^2), off the Ecuador coast. The maximum estimate of the La Plata breeding population is 10 pairs, and up to six chicks have been recorded at one time (Ortiz-Crespo and Agnew 1992). The long-term data from this site are too sketchy to allow any realistic assessment of trends in population size and breeding rate.

Prepared by: D. J. Anderson, Department of Biology, Wake Forest University, Box 7325, Winston-Salem, North Carolina 2710 USA.

Laysan Albatross *Phoebastria immutabilis*

Status: Lower Risk — Least Concern.

Rationale: Based on the breeding population size and range, and as most of the world population nests in protected sites and shows signs of range expansion and re-occupation of former nesting sites, this species qualifies as Lower Risk at this time.

Text: Laysan Albatrosses are the most numerous of the North Pacific albatross species and nest on oceanic islands ranging from Japan to Hawaii to Mexico. Most birds breed in the northern portion of the Northwestern Hawaiian Islands on U.S. National Wildlife Refuge Islands. There are also recently established colonies on islands near Japan, the main Hawaiian Islands and Mexico which have shown increases in the last decade. There is also visitation by birds to former colony sites at Wake and Johnston Atolls in the Pacific Ocean (whose populations were extirpated in the last century).

The primary threat to this species is mortality from fishery interactions (Johnson *et al.* 1993; Pacific Seabird Group, unpubl. data). Levels of mortality may be have been reduced by the closure of the North Pacific high seas driftnet fisheries in 1992 but it is unknown how the shift from drift netting to longline fishing will affect this species (Gould and Hobbs 1993). Other threats to Laysan Albatross populations include contaminants (Ludwig *et al.* 1997), human disturbance, mammalian predation, oiling, and plastic ingestion (Harrison 1990; Auman 1994). Establishment of new colonies in the main Hawaiian Islands has necessitated management activities such as hazing and egg removing around airports on Oahu and Kauai and fencing and predator control at Kilauea Point National Wildlife Refuge. Managers are currently using albatross decoys and sound recordings to lure prospecting birds to a safe nesting site on an offshore islet near Oahu.

Laysan Albatrosses currently breed at 16 locations. The U.S. Fish and Wildlife Service conducts population estimates at Midway every five years and annually at French Frigate Shoals and Laysan Island. These monitoring sites account for 93% of the world population of *c.* 607 000 breeding pairs, 70% (427 500 pairs) of which is at Midway Atoll. Population estimates at Midway in 1992 revealed a 15.4% reduction in the number of breeding pairs. At French Frigate Shoals the population has increased by 15% over the last 10 years. Populations at Laysan Island have shown no statistically significant change in the last five years (USF&WS data). Despite deaths of more than 17 500 birds to the high seas driftnet fishery in 1990 alone (Johnson *et al.* 1993) and deaths of an estimated 1 500 to 5 000 birds per year in the U.S. based pelagic longline fisheries for tuna, billfish, and groundfish (Pacific Seabird Group, unpubl. data) major decreases in the size of breeding populations at the main colonies have not yet been detected.

Prepared by: E. Flint, U.S. Fish and Wildlife Service, Pacific Remote Islands National Wildlife Refuge Complex, PO Box 50167, Honolulu, HI 96850, USA.

Black-footed Albatross *Phoebastria nigripes*

Status: Vulnerable (A2b).

Rationale: A reduction of the population of ≥20% projected to be met within the next three generations (generation time: ≥15 years) based on observed rates of decline qualifies the species as Vulnerable.

Text: Most Black-footed Albatrosses breed at nine locations in the Northwestern Hawaiian Islands, all of which are part of the U.S. National Wildlife Refuge system or are State of Hawaii Seabird Sanctuaries. Three small colonies on Japanese islands appear to be increasing in size. Primary threats include mortality as a result of fishery interactions and loss of nests from wave action at colonies near sea-level. Other threats include contaminants (Ludwig *et al.* 1996), mammalian predators (Hasegawa 1984), oiling and plastic ingestion (Harrison 1990). In the past, high levels of mortality occurred from interaction with fishing gear associated with the North Pacific high seas squid and large mesh driftnet fisheries that lasted through 1992 (Johnson *et al.* 1993). Although these fisheries are now banned, it is unclear how long these losses of Black-footed Albatrosses will affect population processes. Black-footed Albatrosses are susceptible to interaction with longline fisheries for tuna, billfish and groundfish. The level of impact of these fisheries in the North Pacific is unknown but observed mortality of at least 3 000 Black-footed Albatrosses per year in U.S. based fisheries alone prompts concern (Gould and Hobbs 1993; Pacific Seabird Group, unpubl. data).

Black-footed Albatrosses breed at 12 locations and the world population is *c.* 58 500 breeding pairs. At least 80% of the world breeding population is censused annually by the U.S. Fish and Wildlife Service in the Northwestern Hawaiian Islands. The remaining 20% of the population is censused less frequently but all sites except one have been surveyed since 1991. At Midway Atoll and French Frigate Shoals, where the estimates are most reliable, Black footed Albatrosses have decreased in numbers by 2.5% over four years (Williamson and Nishimoto 1994; Seto 1995, 1996) and 46.8% over 10 years, respectively (USF&WS data). At the world's largest colony (Laysan Island *c.* 25 000 pairs) the population has remained apparently stable for five years, but the level of sampling does not allow detection of the levels of change being seen at the other two colonies.

Prepared by: E. Flint, U.S. Fish and Wildlife Service, Pacific Remote Islands National Wildlife Refuge Complex, PO Box 50167, Honolulu, HI 96850, USA.

Black-browed Albatross *Thalassarche melanophrys*

Status: Lower Risk — Near Threatened (A1a + A2d).

Rationale: Despite the large size and wide distribution of populations, decreases at the best studied sites suggest that this species is a serious candidate for threatened status and therefore merits Near Threatened classification at present.

Text: This is the most widespread of all albatross species, breeding at most sub-Antarctic island groups (South Georgia, Crozet, Kerguelen, Heard, Macquarie, Campbell, Antipodes, Snares), the Falkland Islands and southern South America. Its world breeding population is about 680 000 pairs (i.e., about 1.3 million mature individuals) with 86% of this total at the Falkland Islands, 10% at South Georgia and 3% in Chile.

The population at Bird Island, South Georgia (15% of the South Georgia total) has decreased at 6.95% p.a. since 1989/90, an overall reduction of 35% (Croxall *et al.* 1997). Populations at Chile and Kerguelen have also declined in the 1970s and 1980s (Schlatter 1984; Weimerskirch *et al.* 1989), the latter by 17% between 1978/79 and 1994/95 (Weimerskirch and Jouventin 1997). Populations at the Falkland Islands probably increased substantially during the 1980s and several colonies, especially at West Point, North, New and Steeple Jason Islands (50% of total) are still doing so (M. Bingham, I. J. Strange, pers. comm.). However, the population at Beauchene Island (24% of Falkland Islands total) has remained stable between 1981 and 1996 (Prince 1982; Thompson 1993; M. Riddy, pers. comm). Increases are probably attributable to abundant offal and discards from trawl fisheries (Thompson and Riddy 1995). Recent decreases may well be attributable to the development of new longline fisheries over much of the Patagonian Shelf and around South Georgia and off South Africa (Prince *et al.* 1997; Stagi *et al.* 1997; Schiavini *et al.* 1997); given current mortality rates there must be a strong likelihood of forthcoming declines in Falkland Islands Black-browed Albatross. The main contribution to the population decline at South Georgia is the low survival rate of juveniles (Prince *et al.* 1994); if this is, or becomes, the case elsewhere breeding population numbers are very unlikely to recover for a further 10–20 years.

For the best studied population (at Bird Island, South Georgia) and for important colonies in Chile and at Kerguelen there are clear indications of substantial declines in recent years. However, the bulk of the world population, at the Falkland Islands, is probably stable at present.

Prepared by: J. P. Croxall and P. A. Prince, British Antarctic Survey, Madingley Road, High Cross, Cambridge CB3 OET, United Kingdom.

Campbell Albatross *Thalassarche impavida*

Status: Vulnerable (A1a,d + D2).

Rationale: Based either on restricted breeding range (fewer than five locations), or on a past decline exceeding 20% over the last three generations, this species qualifies as Vulnerable. If the decline is shown, or inferred, to be continuing then it would qualify as Endangered (criteria B1 + B2).

Text: Breeds only at Campbell Island in a small number of colonies on the northern part of the island. Endemic to New Zealand.

Usually in mixed colonies with Grey-headed Albatrosses. Some 26 000 breeding pairs were estimated from a mix of counts and photographs from 1987–94. There is evidence of a 33% decrease in at least part of the population since the 1960s (Moore and Moffat 1990a,b; Moore 1995). Mean adult survival 1989–92 was 0.954 (S. Waugh, pers. comm., New Zealand Department of Conservation, unpubl. data). Mean age of first breeding was estimated at 10.4 years. Mean annual productivity was 65% with a range of 51–84% (Waugh *et al.* 1995).

At-sea distribution may be confined to southern Australian waters, Tasman Sea and western Pacific Ocean. There is evidence of bycatch in tuna fisheries, predominantly of juveniles in New Zealand seas, but also of breeding adults in Australian seas (Murray *et al.* 1993; R. Gales *et al.*, unpubl. data).

Prepared by: C. J. R. Robertson, Department of Conservation, PO Box 10420, Wellington, New Zealand.

Buller's Albatross *Thalassarche bulleri*

Status: Vulnerable (D2).

Rationale: Based on restricted range (two breeding sites) this species qualifies as Vulnerable.

Text: Breeds only at Snares and Solander Islands. Endemic to New Zealand.

The total annual breeding population was estimated at 13 760 pairs (Sagar *et al.* 1994), with an estimated 8 460 breeding pairs on the Snares (Sagar *et al.* 1994) and 4 300 to 5 300 breeding pairs on the Solander Islands (Cooper *et al.* 1986). On The Snares, where two counts were made 23 years apart, the number of breeding pairs increased by 64% between 1969 and 1992 (Sagar *et al.* 1994). On Solander Island 2 500 pairs were recorded in February 1996 (J. C. Stahl and P. M. Sagar, unpubl. data). Thus, these most recent counts at the Snares and Solander Islands give a revised estimated breeding population of 10 960 pairs. Substantial mortality of Buller's Albatross has been reported in trawl and longline fisheries (Bartle 1991; Murray *et al.* 1993).

Prepared by: P. Sagar, NIWA, PO Box 8602, Christchurch, New Zealand.

Pacific Albatross *Thalassarche* nov. sp.

Status: Vulnerable (D2).

Rationale: The restricted breeding range (fewer than five breeding locations) automatically qualifies this species as Vulnerable.

Text: Breeds at Forty-Fours, Big and Little Sister Islands (Chatham Islands) and Three Kings Islands (Dawson 1973). Endemic to New Zealand. Breeding confined to a total land area of less than 100 ha.

The main population is in the Chatham Islands, especially the Forty-Fours. No count has been made (except at Little Sister), but on the basis of area a possible 16 000 breeding pairs is estimated on the Forty-Fours and 1 500 on the Big Sister (Robertson 1974, 1985, 1991). Counts at Little Sister from 1994–1996 indicate 630–670 breeding pairs. Some 20 pairs are estimated at the Three Kings Islands.

Breeding is annual. Mean annual adult survival between 1974–95 is 0.935 for birds banded as breeding adults of unknown age. Annual productivity (chicks fledged from eggs laid) between 1994–96 is estimated as 57–60% on Little Sister (62% at end of guard stage). There have been no reported incidences of bycatch in tuna fisheries. The effects of drying of habitat and storms (see Northern Royal Albatross; Robertson 1997) in the Chatham Islands since the 1970s may now be starting to affect nesting habitat, but no great changes are yet evident other than a tendency for increased collapsing of nest structures in dry conditions. At-sea range is uncertain but probably confined to waters east of New Zealand to the South American coast across the Pacific Ocean (Spear *et al.* 1995). May be difficult to distinguish at sea from Buller's Albatross *T. bulleri* .

Prepared by: C. J. R. Robertson, PO Box 10420, Department of Conservation, Wellington, New Zealand.

Shy Albatross *Thalassarche cauta*

Status: Vulnerable (D2).

Rationale: The restricted range (three breeding locations) automatically qualifies this species as Vulnerable.

Text: Breeds on three small islands off the coast of Tasmania. These populations contain *c.* 7 000 pairs (Mewstone), *c.* 5000 pairs (Albatross Island), and 200 pairs (Pedra Branca), giving an overall total of *c.* 12 200 pairs.

The status of the Mewstone and Pedra Branca populations is unknown. For the Mewstone population fisheries-related mortality is known to compound the naturally low reproductive success. At Pedra Branca interspecific competition for nest sites with Australian gannets is a limiting factor and the effect of this is confounded by at-sea mortality associated with fishing activities. The Albatross Island population is recovering slowly to its late 19th century population estimate of 20 000 pairs (currently at one-quarter this level). However, deaths associated with longline fishing are retarding the recovery process.

Adults of all three populations are relatively sedentary and their foraging ranges overlap throughout the year with longline fisheries in which this species is well represented as bycatch (Gales and Brothers 1995; Brothers and Foster, in press).

Efforts to reduce interactions with longline fisheries are occurring (1988 to present) and will have to be maintained for the foreseeable future. Longline fishing activities within Shy Albatross foraging areas are anticipated to continue expanding.

Prepared by: N. Brothers, Tasmanian Parks and Wildlife Service, GPO Box 44A, Hobart, Tasmania 7001, Australia.

White-capped Albatross *Thalassarche steadi*

Status: Vulnerable (D2).

Rationale: The restricted range (fewer than five confirmed breeding locations) automatically qualifies this species as Vulnerable.

Text: Breeds at Disappointment, Auckland and Adams Islands (Auckland Islands group), Bollon's Island (Antipodes Islands group), Forty-Fours Island (Chatham Islands group; unconfirmed). Endemic to New Zealand.

The main population is on Disappointment Island where estimates based on partial or complete counts from air and ground photographs between 1972 and 1994 indicate 70 000–80 000 breeding pairs. Comparable areas have shown a moderate increase during this period. Estimates for Auckland and Adams Islands are 3 000 and 100 pairs respectively. Less than 100 pairs breed at Bollon's Island. One pair occupied an empty nest mound at Forty-Fours Island in 1993 (Robertson 1975; Robertson 1985; Clark and Robertson 1996). No data on survival or breeding parameters are available.

On Auckland Island, only the cliff areas that are inaccessible to feral pigs have remained intact. Reduction of nesting was noted in an area accessible to pigs between 1972 and 1982. In spite of its occurrence as bycatch in a squid trawl fishery near the Auckland Islands in the 1980s (until fishing methods were modified), and continued bycatch in tuna fisheries (Bartle 1991), there have been no observed changes to the number of birds in the main breeding colonies. This species appears to be confined to New Zealand seas.

Prepared by: C. J. R. Robertson, Department of Conservation, PO Box 10420, Wellington, New Zealand.

Salvin's Albatross *Thalassarche salvini*

Status: Vulnerable (D2).

Rationale: The restricted range (fewer than five breeding locations) qualifies this species as Vulnerable.

Text: Breeds at Bounty Islands, Western Chain of the Snares Islands and possibly The Pyramid (Chatham Islands), Penguin Island (Crozet Islands). Virtually endemic to New Zealand.

Breeding is almost certainly annual. Breeding is primarily on small bare rocks with total land area less than 100 ha. The main population is at the Bounty Islands where in November 1978 an estimated 76 000 pairs bred (Robertson and van

Tets 1982). At the Western Chain an estimated 600 pairs bred in 1984 (Miskelly 1996). At The Pyramid two nests (without eggs) were occupied in October 1995. Four breeding pairs were recorded at Penguin Island in the 1980s. In 1985 1 000 fledglings were banded at the Bounty Islands; there have been no recoveries to date. No survival or breeding parameters are available. At sea, there is a possible dispersal westward as far as Africa and South America. There are also suggestions of up to 30% of population in Eastern Pacific along the Peru Current on the continental slope; associated with *D. eremita* in more pelagic conditions (Spear *et al.* 1995). Currently no reported interaction with fishing practices.

Prepared by: C. J. R. Robertson, Department of Conservation, PO Box 10420, Wellington, New Zealand.

Chatham Albatross *Thalassarche eremita*

Status: Critically Endangered (B1 + B2c).

Rationale: With breeding confined to a single location and an inferred/observed decrease in extent and/or quality of breeding habitat, this species qualifies as Critically Endangered.

Text: Breeds solely at The Pyramid (Chatham Islands). Endemic to New Zealand.

Breeding is confined to an area of less than 10 ha. Numbers based on air photographs estimated between 3 200–4 200 breeding pairs. Breeding site difficult to access and study. No breeding parameters currently available. An estimated 1 200–1 500 chicks fledged per annum 1992–1995. Permanent study plot installed 1995 for annual monitoring and further studies to be undertaken.

There is a history of significant harvesting of fledglings by Moriori and other early human settlers on Chatham Islands (Robertson 1974, 1991). Still rumoured to be small illegal annual "take" of up to 20 birds. There has been a significant recent (since 1980s) reduction in vegetation on the island (probably linked to hurricane and changed climatic conditions — see *D. sanfordi* and Robertson 1997) which may with lowered moisture retention increase the probability of nest collapse in dry periods. Parts of the island exposed to storms have in some years had very low productivity (documented between 1992–95). There is currently no documented evidence of fisheries related bycatch in New Zealand waters, but one banded bird, within a month of fledging in 1995, has been reported as caught on a swordfish longline south of Juan Fernandez Island in Chile. At-sea distribution is confined mainly to Pacific Ocean with an estimated 60% of population reported from Peru Current area in winter (Spear *et al.* 1995).

Prepared by: C. J. R. Robertson, Department of Conservation, PO Box 10420, Wellington, New Zealand.

Atlantic Yellow-nosed Albatross *Thalassarche chlororhynchos*

Status: Data Deficient.

Rationale: Breeds on too many locations to qualify as Vulnerable on basis of restricted range. If more data confirm that the population on Gough Island is still declining and/or declines are detected at any other sites, then Vulnerable (B1 + B2) would be appropriate. At present, lack of appropriate data is probably best reflected by a Data Deficient status rather than Near Threatened.

Text: Breeds on five islands of the Tristan da Cunha group and Gough Island. No recent reliable data on overall population size (an estimated 27 000 to 46 000 breeding pairs in the 1970s; Richardson 1984; Fraser *et al.* 1988). No information on population trends for most populations. The recent downward trend in breeding numbers and seemingly low recruitment in a 14-year study colony on Gough Island is reason for serious concern (J. Cooper, pers. comm). There are some reports of longline mortalities (Neves and Olmos 1997; J. Cooper, pers. comm.) and an assessment of longline mortality is urgently required. This species is now protected throughout the Tristan group and no longer legally taken for food (including its eggs).

Prepared by: J. Cooper, Percy FitzPatrick Institute, University of Cape Town, Rondebosch 7700, RSA.

Indian Yellow-nosed Albatross *Thalassarche carteri*

Status: Vulnerable (A1a + D2).

Rationale: Based either on the restricted breeding range (fewer than five locations), or on the past population decrease exceeding 20% over the last three generations, this species qualifies as Vulnerable.

Text: Breeds on four island groups in the Indian Ocean (Crozet, Kerguelen, Prince Edward, Amsterdam/St. Paul), with Amsterdam Island being the major site (68% of total population). Present population size is 36 500 breeding pairs. A decline of at least 36% over the past 14 years has been observed on Amsterdam Island, as a result of increased adult mortality and poor recruitment (Weimerskirch and Jouventin 1997). Bycatch of these birds in the Southern Bluefin Tuna fishery is significant (Klaer and Polacheck 1995) and could account for the decrease (Weimerskirch and Jouventin 1997). Other threats are a disease, under survey, that affects chicks and in some years can kill up to 95% of the chicks in some colonies. Indian Yellow-nosed Albatrosses forage during breeding in subtropical waters and overlap with the long-line fisheries for tropical tuna species (H. Weimerskirch, unpubl. data). Further studies are needed to determine the cause of the decline of the Amsterdam Island population and to monitor the trends at other locations.

Prepared by: H. Weimerskirch, CNRS–CEBC, 79360, Beauvoir, France.

Grey-headed Albatross
Thalassarche chrysostoma

Status: Vulnerable (A1a).

Rationale: The overall population decline at Bird Island (South Georgia) since 1975/76 (less than two generations) has been between 19–29% (depending on dates and colonies used), indicating a status of Vulnerable. The level of recruitment (Prince *et al.* 1994) is such that the population is likely to continue decreasing for at least the next decade.

Text: Breeds in Chile (Diego Ramirez) and at six subantarctic (South Georgia, Crozet, Kerguelen, Prince Edward, Macquarie, Campbell) island groups in the Southern Ocean with a total annual breeding population of at most 92 300 pairs (equivalent to a total population of 250 000 mature individuals in this biennially breeding species). At South Georgia, which represents at least 56% of the world population, the Bird Island sub-population (15% of the South Georgia total) has declined at 1.4% p.a. over the last 20 years (Prince *et al.* 1994; Croxall *et al.* 1997). The Marion Island population (7% of total) declined at 0.7% p.a. until 1992, since when it has increased slightly (J. Cooper, pers. comm.; Gales 1993). The Campbell Island population (7.3% of total) has declined (P. Moore, pers. comm.). No data are available on trends in the other populations (comprising 29% of the world population) but it is likely that they are also decreasing, particularly at Diego Ramirez (at 11% of total the second most important site for the species) because of the proximity to longline fisheries in the region. Mortality in longline fisheries is the main suspected cause of population decrease (Prince *et al.* 1994 1997; Croxall *et al.* 1997).

Prepared by: J. P. Croxall and P. A. Prince, British Antarctic Survey, High Cross, Madingley Road, Cambridge CB3 0ET, United Kingdom.

Sooty Albatross *Phoebetria fusca*

Status: Vulnerable (A1a).

Rationale: Although a population decline of >50% within three generations would suggest a classification of Endangered (A1a), because these data are based on only one site, a Vulnerable status is probably most appropriate.

Text: Breeds at five island groups in the Indian and Atlantic Oceans. This biennially breeding species is estimated to have an annual breeding population of 12 500–19 000 pairs per year, equivalent to a total breeding population of about 42 000 individuals. Only one population, on Possession Island (Crozet Islands) is monitored. At this locality the population has declined by 58% between 1980 and 1995 (Weimerskirch and Jouventin 1997). The remainder of the population is not monitored, and data from these sites are urgently required.

The observed population decline is caused by increased adult and immature mortality (Weimerskirch and Jouventin 1997) but the causes of this are not clear. Sooty albatrosses constitute a minor bycatch in the Southern Bluefin Tuna longline fishery in the Australian sector (Gales and Brothers 1995), but the only available bycatch information is from areas where Sooty Albatrosses are rare. From Crozet they forage in subtropical waters during the breeding season (Weimerskirch 1997), and are thus in contact with several major longline fisheries.

Prepared by: H. Weimerskirch, CNRS–CEBC, 79360 — Beauvoir, France.

Light-mantled Albatross *Phoebetria palpebrata*

Status: Data Deficient.

Rationale: Given that the population decline at the only site with data is insufficient to justify Vulnerable status, and no other data are available, classification of Data Deficient is probably most appropriate.

Text: This species breeds at nine subantarctic islands in the Southern Ocean (South Georgia, Prince Edward Islands, Crozet, Kerguelen, Amsterdam/St. Paul, Macquarie, Auckland, Campbell, Antipodes). It is a biennial breeder and is estimated to have a total annual breeding population of 19 000–24 000 pairs, equivalent to 58 000 mature individuals. Only one population has been monitored more than once, so that the status of the species is very difficult to assess. In the Indian Ocean, the population on Possession Island has decreased by 13% in 15 years.

This species is one of the longest-lived of all birds, with very low fecundity (Weimerskirch *et al.* 1987), and so its population is very susceptible to an increase in adult mortality. Light-mantled Albatrosses are killed in the tuna longline fishery in the Australian reigon (Gales and Brothers 1995). However, during the breeding season they forage mainly in pelagic Antarctic and sub-Antarctic waters (Weimerskirch and

Robertson 1994; Weimerskirch 1997) and are so not in contact with longliners operating in subtropical waters (tuna fisheries) or in neritic Antarctic waters (Patagonian toothfish fisheries).

Prepared by: H. Weimerskirch, CNRS–CEBC, 79360, Beauvoir, France.

DISCUSSION

Problems with assessing the conservation status of albatrosses

Given the relative complexity of the new IUCN criteria, the unfamiliarity of most evaluators with the process and the degree of inference often needed in applying the criteria (see Stattersfield 1996), there were inevitably some difficulties in the correct application of criteria. The independent evaluation by BirdLife International was an invaluable and essential part of the process. It suggested seven changes (out of 24 assessments) to the status proposed in the initial evaluation; five were a direct result of incorrect application of criteria, the other two involving re-assessments following consideration of additional data.

Particular difficulties that were experienced by evaluators included:

(a) Estimating the time-span equivalent to a generation. Generation is defined as the average age of parents in the population but this is probably unknown for almost all albatross populations. It is probably best estimated by calculating lifespan from mean annual adult survival rate, but this is only reliably known for a few species;

(b) Defining a location. Evaluator concepts of this ranged from a single colony or site clearly distinct from others in the general vicinity, to the whole of a sub-Antarctic island the size of Campbell Island. A realistic and comparable application of this criterion is of particular importance for albatrosses, given the number of species that are restricted to single islands or island-groups. In borderline cases it was considered essential that a specific known potential threat would be able to eliminate all individuals at a single location. On this basis, islands with several distinct and spatially very well separated colonies or groups of colonies would be regarded as having more than one location;

(c) The application of the Conservation Dependent category. The view was taken that Conservation Dependent should only be used for focused species-specific or habitat specific conservation programmes where, if they stopped, the species would rapidly (within about five years) move into threatened status. On this basis, programmes to promote the use of mitigating measures for reducing incidental mortality of albatrosses in longline fisheries were not thought to meet this criterion.

Despite these difficulties there was considerable unanimity over the assessments. The most difficult decisions, involving changes during the process of the present assessment, were for:

(a) Southern Royal Albatross (variously assessed as Vulnerable, or Near Threatened, depending on the interpretation of location criteria);

(b) Tristan Albatross (originally Vulnerable, reclassified as Endangered on the basis of very restricted range and lack of distinct viable subpopulations/locations);

(c) White-capped Albatross (assessed as Vulnerable or Near Threatened depending on location criteria);

(d) Buller's Albatross (originally Near Threatened, reclassified as Vulnerable on the basis of restricted range and locations);

(e) Shy Albatross (originally Conservation Dependent (see above), classified as Vulnerable on the basis of very few locations);

(f) Chatham Albatross (originally Vulnerable but if the category of inferred or projected population decline (B2c) is accepted then this single-island endemic automatically qualifies as Critically Endangered).

(g) Short-tailed Albatross (originally retained as Endangered but classification as C1 in Collar *et al.* (1994) was an error (M. Crosby, pers. comm.) and the continuing increase in population has now taken it above 250 mature individuals, which therefore only qualifies it for Vulnerable status).

Summary of assessments

The conservation status of albatrosses resulting from this assessment is summarized in Table 1, together with those criteria which triggered the category. This shows that 20 (83%) of 24 albatross species have threatened status, compared with 11% of bird species overall (Collar *et al.* 1994). This is the highest proportion of threatened species in any bird family that has more than a single species. Of these threatened albatrosses, two (10%) are Critically Endangered, two (10%) Endangered and the rest (80%) Vulnerable; these proportions are similar to those for all threatened species, being 15%, 21% and 63%, respectively (Collar *et al.* 1994). The criteria by which albatrosses qualify for threatened status are fairly evenly divided between known and projected population declines and small range with few breeding locations. Thus of the four most threatened

Table 1. Comparison between assessments of conservation status of albatrosses by Collar *et al.* (1994)[1] and in 1997[2].

	Collar *et al.* (1994)		Criteria[3]	1997		Criteria[3]
CRITICALLY ENDANGERED	Amsterdam Albatross	*Diomedea amsterdamensis*	D	Amsterdam Albatross	*Diomedea amsterdamensis*	D
				Chatham Albatross	*Thalassarche eremita*	B1 + B2c
ENDANGERED	Short-tailed Albatross	*Diomedea albatrus*	C1, D1, D2	Northern Royal Albatross	*Diomedea sanfordi*	B1 + B2c,e
				Tristan Albatross	*Diomedea dabbenena*	B1 + B2e
VULNERABLE	Wandering Albatross	*Diomedea exulans*	A1a,c, A2c	Southern Royal Albatross	*Diomedea epomophora*	D2
				Antipodean Albatross	*Diomedea antipodensis*	D2
				Gibson's Albatross	*Diomedea gibsoni*	D2
				Wandering Albatross	*Diomedea exulans*	A1a,d
				Waved Albatross	*Phoebastria irrorata*	D2
				Black-footed Albatross	*Phoebastria nigripes*	A2b
				Short-tailed Albatross	*Phoebastria albatrus*	D1, D2
				Campbell Albatross	*Thalassarche impavida*	A1a,d + D2
				Pacific Albatross	*Thalassarche* sp. nov.	D2
				Salvin's Albatross	*Thalassarche salvini*	D2
				White-capped Albatross	*Thalassarche steadi*	D2
				Indian Yellow-nosed Albatross	*Thalassarche carteri*	A1a, D2
				Buller's Albatross	*Thalassarche bulleri*	D2
				Grey-headed Albatross	*Thalassarche chrysostoma*	A1a
				Shy Albatross	*Thalassarche cauta*	D2
				Sooty Albatross	*Phoebetria fusca*	A1a
LOWER RISK						
— Near Threatened	Royal Albatross	*Diomedea epomophora*		Black-browed Albatross	*Thalassarche melanophrys*	A1a, A2d
	Buller's Albatross	*Diomedea bulleri*				
	Waved Albatross	*Diomedea irrorata*				
	Grey-headed Albatross	*Diomedea chrysostoma*				
	Sooty Albatross	*Phoebetria fusca*				
— Conservation Dependent						
— Least Concern	Black-footed Albatross	*Diomedea nigripes*		Laysan Albatross	*Phoebastria immutabilis*	
	Laysan Albatross	*Diomedea immutabilis*				
	Yellow-nosed Albatross	*Diomedea chlororhynchos*				
	Shy Albatross	*Diomedea cauta*				
	Black-browed Albatross	*Diomedea melanophrys*				
	Light-mantled Sooty Albatross	*Phoebetria palpebrata*				
DATA DEFICIENT				Atlantic Yellow-nosed Albatross	*Thalassarche chlororhynchos*	
				Light-mantled Albatross	*Phoebetria palpebrata*	

Note: 1. Based on taxonomy of Sibley and Monroe (1990, 1993)
2. Based on taxonomy of Robertson and Nunn (1997)
3. See Appendix 2c

species, population decline and restricted range together qualify Chatham, Tristan and Northern Royal Albatrosses, with Amsterdam Albatrosses qualifying on very small population size alone. For the Vulnerable species, half qualify through population declines and half by virtue of their small breeding range and few breeding sites.

The Lower Risk species comprise one that is Near Threatened (Black-browed Albatross; by virtue of declines in part of its range), one of Least Concern (Laysan Albatross) and two that are Data Deficient (Atlantic Yellow-nosed and Light-mantled Albatrosses), the latter for which population declines have recently been inferred from at-sea abundance data (Woehler 1996).

At first glance there would seem to be many changes between the evaluation by BirdLife International (Collar *et al.* 1994) — which recognized only three threatened species of albatross (out of 14; 21%) and the present exercise. However, most changes between the two assessments have resulted from improved knowledge of relationships within and between species, as reflected in the proposed new taxonomy. Otherwise, comparing the two evaluations:

(a) Three species (Amsterdam, Wandering and Laysan Albatrosses) have not changed status;

(b) More comprehensive and up-to-date data have warranted the reclassification of Short-tailed, Black-footed Albatrosses, Grey-headed and Sooty Albatrosses (two of which had already been recognized as Near Threatened by Collar *et al.* (1994)) and the treating of Light-mantled Albatross as Data Deficient;

(c) Waved Albatross was overlooked by Collar *et al.* (1994) (N. J. Collar, A. J. Stattersfield, pers. comm).

There can be little doubt, therefore, that albatrosses are the most threatened and vulnerable of all groups of marine birds. Only the Laysan Albatross, the second most abundant species, is not currently regarded as a subject of conservation concern.

BirdLife International will be incorporating these new assessments of the conservation status of albatrosses into their next edition of Birds to Watch, which will be the basis of the next IUCN Red Data List. It is very important, therefore, that BirdLife International are notified of any new data which have a bearing on population trend/threat assessment of albatrosses.

In the meantime, the very high conservation concern that albatrosses merit needs to be reflected in their treatment as a flagship group for the conservation of marine systems, as candidates for special protection in national and international conservation legislation and as subjects of priority action to substantially reduce, if not eliminate, the threats they face.

ACKNOWLEDGEMENTS

We thank everyone who contributed to the assessment process, whether as evaluators, reviewers or participants in the original Workshop sessions. We are very grateful to Alison Stattersfield and Nigel Collar for their critical reviews of all texts. Alison Stattersfield also gave much useful advice and comment on the introductory and summary sections.

APPENDIX 1

Structure of the IUCN Red List Categories

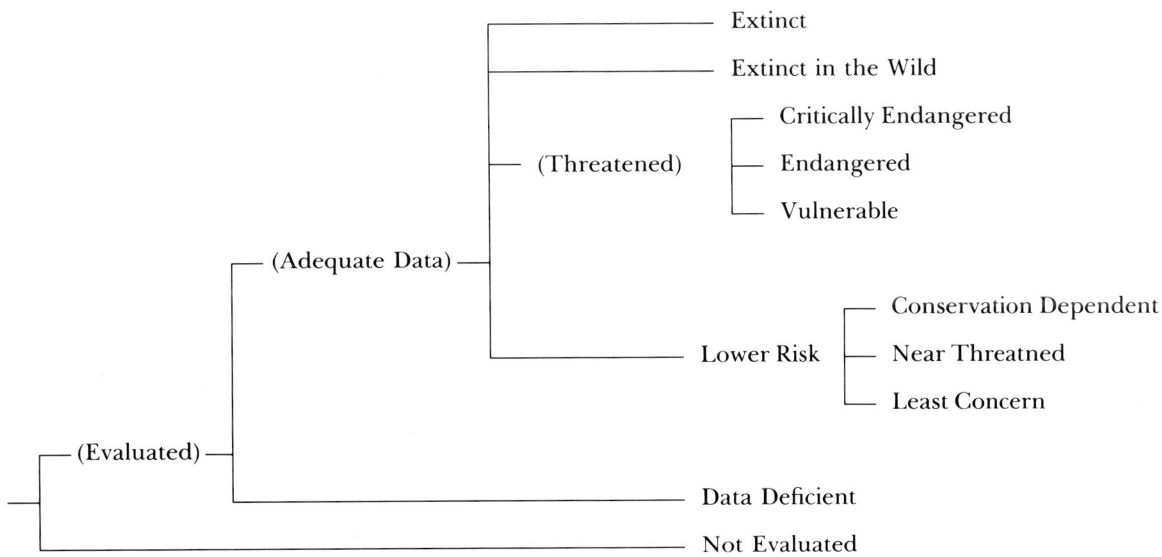

APPENDIX 2a

The new IUCN threatened category thresholds (in brief)

Criteria		Main numerical thresholds: Critical	Endangered	Vulnerable
A	RAPID DECLINE	>80% over 10 years or 3 generations	>50% over 10 years or 3 generations	>20% over 10 years or 3 generations
B	SMALL RANGE fragmented, declining or fluctuating	Extent of occurence <100 km^2 or area of occupancy <10 km^2	Extent of occurrence <5 000 km^2 or area of occupancy <500 km^2	Extent of occurrence <20 000 km^2 or area of occupancy <2 000 km^2
C	SMALL POPULATION declining	<250 mature individuals	<2 500 mature individuals	<10 000 mature individuals
D1	VERY SMALL POPULATION	<50 mature individuals	250 mature individuals	1 000 mature individuals
D2	VERY SMALL RANGE	—	—	<100 km^2 or <5 locations
E	UNFAVOURABLE PVA[1]	Probability of extinction >50% within 10 years	Probability of extinction >20% within 20 years	Probability of extinction >10% within 100 years

Note 1: Population Viability Assessment.

APPENDIX 2b

The IUCN categories

Extinct (EX)

A taxon is Extinct when there is no reasonable doubt that the last individual has died.

Extinct in the wild (EW)

A taxon is Extinct in the wild when it is known only to survive in cultivation, in captivity or as a naturalized population (or populations) well outside the past range. A taxon is presumed extinct in the wild when exhaustive surveys in known and/or expected habitat, at appropriate times (diurnal, seasonal, annual), throughout its historic range have failed to record an individual. Surveys should be over a time frame appropriate to the taxon's life cycle and life form.

Critically Endangered (CR)

A taxon is Critically Endangered when it is facing an extremely high risk of extinction in the wild in the immediate future, as defined by any of the criteria (A to E) on page 61.

Endangered (EN)

A taxon is Endangered when it is not Critically Endangered but is facing a very high risk of extinction in the wild in the near future, as defined by any of the criteria (A to E) on page 61.

Vulnerable (VU)

A taxon is Vulnerable when it is not Critically Endangered or Endangered but is facing a high risk of extinction in the wild in the medium-term future, as defined by any of the criteria (A to D) on page 61.

Lower Risk (LR)

A taxon is Lower Risk when it has been evaluated, does not satisfy the criteria for any of the categories Critically Endangered, Endangered or Vulnerable. Taxa included in the Lower Risk category can be separated into three subcategories:

1. *Conservation Dependent (cd).* Taxa which are the focus of a continuing taxon-specific or habitat-specific conservation programme targeted towards the taxon in question, the cessation of which would result in the taxon qualifying for one of the threatened categories above within a period of five years.
2. *Near Threatened (nt).* Taxa which do not qualify for Conservation Dependent, but which are close to qualifying for Vulnerable.
3. *Least Concern (lc).* Taxa which do not qualify for Conservation Dependent or Near Threatened.

Data Deficient (DD)

A taxon is Data Deficient when there is inadequate information to make a direct, or indirect, assessment of its risk of extinction based on its distribution and/or population status. A taxon in this category may be well studied, and its biology well known, but appropriate data on abundance and/or distribution is lacking. Data Deficient is therefore not a category of threat or Lower Risk. Listing of taxa in this category indicates that more information is required and acknowledges the possibility that future research will show that threatened classification is appropriate. It is important to make positive use of whatever data are available. In many cases great care should be exercised in choosing between DD and threatened status. If the range of a taxon is suspected to be relatively circumscribed, if a considerable period of time has elapsed since the last record of the taxon, threatened status may well be justified.

Not Evaluated (NE)

A taxon is Not Evaluated when it has not yet been assessed against the criteria.

APPENDIX 2c

The criteria for Critically Endangered, Endangered and Vulnerable

Critically Endangered (CR)

A taxon is Critically Endangered when it is facing an extremely high risk of extinction in the wild in the immediate future, as defined by any of the following criteria (A to E):

A. Population reduction in the form of either of the following:

1. An observed, estimated, inferred or suspected reduction of at least 80% over the last ten years or three generations, whichever is the longer, based on (and specifying) any of the following:
   (a) direct observation
   (b) an index of abundance appropriate for the taxon
   (c) a decline in area of occupancy, extent of occurrence and/or quality of habitat
   (d) actual or potential levels of exploitation
   (e) the effects of introduced taxa, hybridization, pathogens, pollutants, competitors or parasites.
2. A reduction of at least 80%, projected or suspected to be met within the next ten years or three generations, whichever is the longer, based on (and specifying) any of (b), (c), (d) or (e) above.

B. Extent of occurrence estimated to be less than 100 km^2 or area of occupancy estimated to be less than 10 km^2, and estimates indicating any two of the following:

1. Severely fragmented or known to exist at only a single location.
2. Continuing decline, observed, inferred or projected, in any of the following:
   (a) extent of occurrence
   (b) area of occupancy
   (c) area, extent and/or quality of habitat
   (d) number of locations or subpopulations
   (e) number of mature individuals.
3. Extreme fluctuations in any of the following:
   (a) extent of occurrence
   (b) area of occupancy
   (c) number of locations or subpopulations
   (d) number of mature individuals.

C. Population estimated to number less than 250 mature individuals and either:

1. An estimated continuing decline of at least 25% within 3 years or one generation, whichever is longer or
2. A continuing decline, observed, projected, or inferred, in numbers of mature individuals and population structure in the form of either:
   (a) severely fragmented (i.e., no subpopulation estimated to contain more than 50 mature individuals)
   (b) all individuals are in a single subpopulation.

D. Population estimated to number less than 50 mature individuals.

E. Quantitative analysis showing the probability of extinction in the wild is at least 50% within ten years or 3 generations, whichever is the longer.

Endangered (EN)

A taxon is Endangered when it is not Critically Endangered but is facing a very high risk of extinction in the wild in the near future, as defined by any of the following criteria (A to E):

A. Population reduction in the form of either of the following:

1. An observed, estimated, inferred or suspected reduction of at least 50% over the last ten years or three generations, whichever is the longer, based on (and specifying) any of the following:
   (a) direct observation
   (b) an index of abundance appropriate for the taxon
   (c) a decline in area of occupancy, extent of occurrence and/or quality of habitat
   (d) actual or potential levels of exploitation
   (e) the effects of introduced taxa, hybridization, pathogens, pollutants, competitors or parasites.
2. A reduction of at least 50%, projected or suspected to be met within the next ten years or three generations, whichever is the longer, based on (and specifying) any of (b), (c), (d) or (e) above.

B. Extent of occurrence estimated to be less than 5 000 km^2 or area of occupancy estimated to be less than 500 km^2, and estimates indicating any two of the following:

1. Severely fragmented or known to exist at no more than five locations.
2. Continuing decline, inferred, observed or projected, in any of the following:
   (a) extent of occurrence
   (b) area of occupancy
   (c) area, extent and/or quality of habitat
   (d) number of locations or subpopulations
   (e) number of mature individuals.
3. Extreme fluctuations in any of the following:
   (a) extent of occurrence
   (b) area of occupancy
   (c) number of locations or subpopulations
   (d) number of mature individuals.

C. Population estimated to number less than 2 500 mature individuals and either:

1. An estimated continuing decline of at least 20% within 5 years or 2 generations, whichever is longer, or
2. A continuing decline, observed, projected, or inferred, in numbers of mature individuals and population structure in the form of either:
   (a) severely fragmented (i.e., no subpopulation estimated to contain more than 250 mature individuals)
   (b) all individuals are in a single subpopulation.

D. Population estimated to number less than 250 mature individuals.

E. Quantitative analysis showing the probability of extinction in the wild is at least 20% within 20 years or 5 generations, whichever is the longer.

APPENDIX 2c — *continued*

Vulnerable (VU)

A taxon is Vulnerable when it is not Critically Endangered or Endangered but is facing a high risk of extinction in the wild in the medium-term future, as defined by any of the following criteria (A to E):

A. Population reduction in the form of either of the following:

1. An observed, estimated, inferred or suspected reduction of at least 20% over the last ten years or three generations, whichever is the longer, based on (and specifying) any of the following:
   (a) direct observation
   (b) an index of abundance appropriate for the taxon
   (c) a decline in area of occupancy, extent of occurrence and/or quality of habitat
   (d) actual or potential levels of exploitation
   (e) the effects of introduced taxa, hybridization, pathogens, pollutants, competitors or parasites.
2. A reduction of at least 20%, projected or suspected to be met within the next ten years or three generations, whichever is the longer, based on (and specifying) any of (b), (c), (d) or (e) above.

B. Extent of occurrence estimated to be less than 20 000 km^2 or area of occupancy estimated to be less than 2 000 km^2, and estimates indicating any two of the following:

1. Severely fragmented or known to exist at no more than ten locations.
2. Continuing decline, inferred, observed or projected, in any of the following:
   (a) extent of occurrence
   (b) area of occupancy
   (c) area, extent and/or quality of habitat
   (d) number of locations or subpopulations
   (e) number of mature individuals.
3. Extreme fluctuations in any of the following:
   (a) extent of occurrence
   (b) area of occupancy
   (c) number of locations or subpopulations
   (d) number of mature individuals.

C. Population estimated to number less than 10 000 mature individuals and either:

1. An estimated continuing decline of at least 10% within ten years or 3 generations, whichever is longer, or
2. A continuing decline, observed, projected, or inferred, in numbers of mature individuals and population structure in the form of either
   (a) severely fragmented (i.e., no subpopulation estimated to contain more than 1 000 mature individuals)
   (b) all individuals are in a single subpopulation.

D. Population very small or restricted in the form of either of the following:

1. Population estimated to number less than 1 000 mature individuals.
2. Population is characterized by an acute restriction in its area of occupancy (typically less than 100 km^2) or in the number of locations (typically less than 5). Such a taxon would thus be prone to the effects of human activities (or stochastic events whose impact is increased by human activities) within a very short period of time in an unforeseeable future, and is thus capable of becoming Critically Endangered or even Extinct in a very short period.

E. Quantitative analysis showing the probability of extinction in the wild is at least 10% within 100 years.

APPENDIX 3

Changes between the IUCN Criteria as used by BirdLife for Collar *et al.* (1994) and as published by IUCN (1994)

1. The extinction probability for the category Critical (CR) has been changed from 50% in five years to 50% in 10 years.
2. The relationship between the categories has changed such that Lower Risk (LR) category is sub-divided into Conservation Dependent (cd), Near Threatened (nt) and Least Concern (lc).
3. The main numerical threshold for the rapid decline criterion (A) for the Vulnerable category (VU) has changed from 50% over 20 years to 20% over the last 10 years. Declines in the future (A2) are specified as within the next 10 years or three generations whichever is the longer. An additional qualifier has been added (an index of abundance appropriate for the taxon) such that the qualifiers A1b, A2b used by BirdLife become A1c, A2c; A1c, A2c become A1d, A2d; and A1d, A2d become A1e, A2e.
4. The sub-criteria threshold for the small population and declining criterion (C) for the Vulnerable category (VU) has changed from 20% over 10 years to at least 10% within 10 years.

A. J. Stattersfield
November 1995

APPENDIX 4

Applying the IUCN Criteria to Seabirds

A total of 60 species of seabirds was classified as threatened by BirdLife in Birds to Watch 2 using the following criteria (species may qualify under more than one criterion):

A. 26 species (most commonly used code = A1a)
B. 12 (B1+2e)
C. 30 (C2b)
D. 54 (D2)

Seabirds have therefore been identified as threatened largely as a result of their tiny (breeding) ranges, small populations which are declining with all breeding individuals in a single subpopulation, and rapid declines which have been directly observed. NB. BirdLife used criteria A, B, C and D only in its analysis (E was excluded because formally published records of the results of Population Viability Analysis are virtually non-existent).

Some key points to bear in mind when assessing the status of seabirds include:

- use generation lengths (defined as the average age of parents in the population) rather than specified time periods for long-lived species (e.g., albatrosses).
- population thresholds refer to the numbers of mature individuals.
- severe fragmentation refers to small subpopulations which may go extinct with a reduced probability of recolonization.
- for the B or D2 criteria use area of occupancy (which refers to breeding range).
- for the B1 criterion use number of locations.
- a location is defined as a distinct area in which a single event (e.g., introduction of alien species, pollution) will soon affect all individuals. Therefore island groups may sometimes be treated as a single location.
- use Conservation Dependent with care. An active conservation programme should be targeted at the species in question (protection of breeding sites alone may not be enough) and conservation issues in the wider marine environment may need to be addressed.

A. J. Stattersfield
November 1995

REFERENCES

Alexander, K., Robertson, G. and Gales, R., 1997. The Incidental Mortality of Albatrosses in Longline Fisheries. Australian Antarctic Division, Tasmania. 44 pp.

Anderson, D. J., 1995. Census of the Waved Albatross, 1994. Unpublished Report to the Charles Darwin Research Station.

Auman, H. J., 1994. Plastic ingestion, biomarkers of health, PCBs and DDE in two species of albatrosses on Sand Island, Midway Atoll. Master's Thesis. Michigan State University, East Lansing, Michigan.

Bartle, J. A., 1991. Incidental capture of seabirds in the New Zealand and subantarctic squid trawl fishery, 1990. *Bird Conserv. Internat.* **1:** 351–59.

Brothers, N. and Foster, A., in press. Seabird catch rates: an assessment of causes and solutions in Australia's domestic tuna longline fishery. *Mar. Ornithol.*

Clark, G. and Robertson, C. J. R., 1996. New Zealand White-capped Mollymawks breeding with Black-browed Mollymawks at Antipodes Islands, New Zealand. *Notornis* **43.**

Clark, G., Amey, J. and McAlister, G., 1995. Unexpectedly large number of Wandering Albatrosses *(Diomedea exulans)* breeding on Antipodes Island, New Zealand. *Notornis* **42:** 42–46.

Collar, N. J., Crosby, M. J. and Stattersfield, A. J., 1994. Birds to Watch 2. The world list of threatened birds. BirdLife International: Cambridge.

Cooper, J., 1994. Seabird mortality from longline fisheries: evidence from Marion and Gough Islands. CCAMLR–WG–IMALF–94/18. 8 pp.

Cooper, W. J., Miskelly, C. M., Morrison, K. and Peacock, R. J., 1986. Birds of the Solander Islands. *Notornis* **33:** 77–89.

Croxall, J. P. and Prince, P. A., 1990. Recoveries of Wandering Albatrosses *Diomedea exulans* ringed at South Georgia, 1956–86. *Ringing and Migration* **11:** 43–51.

Croxall, J. P., Prince, P. A., Rothery, P. and Wood, A. G., 1997. Population changes in albatrosses at South Georgia. Pp. 69–83 *in* Albatross Biology and Conservation ed by G. Robertson and R. Gales. Surrey Beatty & Sons: Chipping Norton.

Croxall, J. P., Rothery, P., Pickering, S. P. C. and Prince, P. A., 1990. Reproductive performance, recruitment and survival of Wandering Albatrosses *Diomedea exulans* at Bird Island, South Georgia. *J. Anim. Ecol.* **59:** 775–95.

Dawson, E. W., 1973. Albatross populations in the Chatham Islands. *Notornis* **20:** 210–30.

de la Mare, W. K. and Kerry, K. R., 1994. Population dynamics of the Wandering Albatross *(Diomedea exulans)* on Macquarie Island and the effects of mortality from longline fishing. *Polar Biol.* **14:** 231–41.

Douglas, H. D., 1995. Population dynamics and ecology of the Waved Albatross *(Diomedea irrorata).* P. 44 *in* Abstracts of the First International Conference on the Biology and Conservation of Albatrosses, 28 August–1 September 1995, Hobart, Australia.

Elliot, G. P., Walker, K. J., Nicholls, D. G. and Murray, M. D., 1995. Foraging patterns of Wandering Albatross from the Auckland Islands. P. 19 *in* Abstracts of the First International Conference on the Biology and Conservation of Albatrosses, 28 August–1 September 1995, Hobart, Australia.

Fraser, M. W., Ryan, P. G. and Watkins, B. P., 1988. The seabirds of Inaccessible Island, South Atlantic Ocean. *Cormorant* **16:** 7–33.

Gales, R., 1993. Co-operative Mechanisms for the Conservation of Albatross. Australian Nature Conservation Agency: Hobart. 132 pp.

Gales, R. and Brothers, N., 1995. Characteristics of seabirds caught in the Japanese tuna longline fishery in the Australian region. P. 29 *in* Abstracts of the First International Conference on the Biology and Conservation of Albatrosses, 28 August–1 September 1995, Hobart, Australia.

Gould, P. J. and Hobbs, R., 1993. Population dynamics of the Laysan and other Albatrosses in the North Pacific. *Internat. Nth Pac. Fish. Bull.* **53:** 473–83.

Groombridge, B., (ed). 1993. IUCN Red List of Threatened Animals. IUCN: Gland, Switzerland and Cambridge, UK. lvi + 286 pp.

Harris, M. P., 1973. The biology of the Waved Albatross *Diomedea irrorata* of Hood Islands, Galápagos. *Ibis* **115:** 483–510.

Harrison, C., 1990. Seabirds of Hawaii: natural history and conservation. Cornell University Press: New York.

Hasegawa, H., 1984. Status and conservation of seabirds in Japan, with special attention to the Short-tailed Albatross. Pp. 487–500 *in* Status and Conservation of the World's Seabirds ed by J. P. Croxall, P. G. H. Evans and R. W. Schlatter. ICBP Technical Publication No. 2: Cambridge.

Hasegawa, H., 1991. Red data bird: Short-tailed Albatross. *World Birdwatch* **13:** 10.

IUCN. 1994. IUCN Red List Categories. IUCN: Gland, Switzerland.

Johnson, D. H., Shaffer, T. L. and Gould, P., 1993. Incidental catch of marine birds in the North Pacific high seas driftnet fisheries in 1990. *Internat. Nth Pac. Fish. Comm. Bull.* **53:** 473–83.

Jouventin, P., Martinez, J. and Roux, J. P., 1990. Breeding biology and current status of the Amsterdam Island Albatross. *Ibis* **131:** 171–89.

Klaer, N. and Polacheck, T., 1995. Japanese longline seabird bycatch in the Australian fishing zone April 1991–March 1994. CSIRO Report, Hobart.

Ludwig, J., Summer, C. L., Auman, H., Gauger, V., Bromley, D., Giesy, J. P., Rolland, R. and Colborn, T. 1997. The roles of organochlorine contaminants and fisheries bycatch in recent population changes of Black-footed and Laysan Albatrosses in the North Pacific Ocean. Pp. 225–38 *in* Albatross Biology and Conservation ed by G. Robertson and R. Gales. Surrey Beatty & Sons: Chipping Norton.

Mace, G. M. and Lande, R., 1991. Assessing extinction threats: towards a re-evaluation of IUCN threatened species categories. *Conserv. Biol.* **5:** 148–57.

Mace, G. M. and Stuart, S. N., 1994. Draft IUCN Red List Categories, Version 2.2. *Species* **21–22:** 13–24.

Mace, G. M., Collar, N. J., Cooke, J., Gaston, K., Ginsberg, J., Leader-Williams, N., Maunder, M. and Milner-Gulland, E. J., 1992. The development of new criteria for listing species on the IUCN Red List. *Species* **19:** 16–22.

Medway, D. G., 1993. The identity of the chocolate albatross *Diomedea spadicea* of Gmelin, 1789 and the wandering albatross of Linnaeus, 1758. *Notornis* **40:** 145–62.

Micol and Jouventin, P., 1995. Restoration of Amsterdam Island following control of feral cattle. *Biol. Conserv.* **73:** 199–206.

Miskelly, C., 1996. Biological notes on the Western Chain, Snares Islands 1984–85 and 1985–86. Conservation Advisory Science Notes No. 144. Department of Conservation, Wellington.

Moore, P. J. and Moffat, R. D., 1990a. Mollymawks on Campbell Island. Science and Research Internal Report No. 59. Department of Conservation: Wellington, New Zealand.

Moore, P. J. and Moffat, R. D., 1990b. Research and Management projects on Campbell Island 1987–1988. Science and Research Internal Report No. 57. Department of Conservation: Wellington, New Zealand.

Moore, P. J., 1995. Status and decline of mollymawks on Campbell Island, New Zealand. P. 7 *in* Abstracts of the First International Conference on the Biology and Conservation of Albatrosses, 28 August–1 September 1995, Hobart, Australia.

Murray, T. E., Bartle, J. A., Kalish, S. R. and Taylor, P. R., 1993. Incidental capture of seabirds by Japanese southern bluefin tuna long-line vessels in New Zealand waters, 1988–1992. *Bird Conserv. Internat.* **3:** 181–210.

Neves, T. and Olmos, F., 1997. Albatross mortality in fisheries off the coast of Brazil. Pp. 214–19 *in* Albatross Biology and Conservation ed by G. Robertson and R. Gales. Surrey Beatty & Sons: Chipping Norton.

Nunn, G. B., Cooper, J., Jouventin, P., Robertson, C. J. R. and Robertson, G. G., 1996. Evolutionary relationships among extant albatrosses (Procellariiformes: Diomedeidae) established from complete cytochrome-b gene sequences. *Auk* **113:** 784–801.

Ortiz-Crespo, F. and Agnew, P., 1992. The birds of La Plata Island, Ecuador. *Bull. Brit. Ornithol. Club* **112:** 66–73.

Prince P. A., Rothery, P., Croxall, J. P. and Wood, A. G., 1994. Population dynamics of Black-browed and Grey-headed Albatrosses *Diomedea melanophris* and *D. chrysostoma* at Bird Island, South Georgia. *Ibis* **136:** 50–71.

Prince, P. A., 1982. The Black-browed Albatross *Diomedea melanophris* population at Beauchêne Island, Falkland Islands. *Comité National Français des Recherches Antarctiques* **51:** 111–17.

Prince, P. A., Croxall, J. P., Trathan, P. N. and Wood, A. G., 1997. The pelagic distribution of South Georgia albatrosses and their relationships with fisheries. Pp. 137–67 *in* Albatross Biology and Conservation ed by G. Robertson and R. Gales. Surrey Beatty & Sons: Chipping Norton.

Prince, P. A., Rothery, P., Croxall, J. P. and Wood, A. G., 1994. Population dynamics of Black-browed and Grey-headed Albatrosses *Diomedea melanophris* and *D. chrysostoma* at Bird Island, South Georgia. *Ibis* **136:** 50–71.

Richardson, J., 1984. Aspects of the ornithology of the Tristan da Cunha group and Gough Island. *Cormorant* **12:** 122–201.

Richardson, S. A., 1994. Status of the Short-tailed Albatross on Midway Atoll. *'Elepaio* **54:** 35–37.

Richdale, L. E., 1950. The Pre-egg stage in the Albatross Family. Private Monograph No 3., Dunedin.

Richdale, L. E., 1952. Post-egg period in Albatrosses. Private Monograph No. 4., Dunedin.

Robertson, C. J. R., (ed.). 1985. The Complete Book of New Zealand Birds. Reader's Digest Services: Sydney. 319 pp.

Robertson, C. J. R., 1974. Albatrosses of the Chatham Islands. *Wildlife — a review* **5:** 20–22.

Robertson, C. J. R., 1975. Report on the distribution, status and breeding biology of the Royal Albatross, Wandering Albatross and White-capped Mollymawk on the Auckland Islands. Pp. 143–50 *in* Preliminary Results of the Auckland Island Expedition 1972–73 ed by J. C. Yaldwyn. New Zealand Department of Lands and Survey.

Robertson, C. J. R., 1991. Questions on the harvesting of Toroa in the Chatham Islands. Science and Research Series No. 35, Department of Conservation, Wellington. 105 pp.

Robertson, C. J. R., 1993a. Survival and longevity of the Northern Royal Albatross *Diomedea epomophora sanfordi* at Taiaroa Head 1937–1993. *Emu* **93:** 269–76.

Robertson, C. J. R., 1993b. Assessment of tinted glass for public observatory, Taiaroa Head Nature Reserve. Conservation Advisory Science Notes No. 42. Department of Conservation, Wellington. 9 pp.

Robertson, C. J. R., 1993c. Effects of nature tourism on marine wildlife. Pp. 53–60 *in* Proceedings of Marine Conservation and Wildlife Protection Conference. New Zealand Conservation Authority: Wellington.

Robertson, C. J. R. and van Tets, G. F., 1982. The status of birds at the Bounty Islands. *Notornis* **29:** 311–36.

Robertson, C. J. R., 1997. Factors influencing the breeding performance of the Northern Royal Albatross. Pp. 99–104 *in* Albatross Biology and Conservation ed by G. Robertson and R. Gales. Surrey Beatty & Sons: Chipping Norton.

Robertson, C. J. R. and Nunn, G. B., 1997. Towards a new taxonomy for albatrosses. Pp. 13–19 *in* Albatross Biology and Conservation ed by G. Robertson and R. Gales. Surrey Beatty & Sons: Chipping Norton.

Ryan, P. G., Dean, W. R. J., Moloney, C. L., Watkins, B. P. and Milton, S. J., 1990. New information on seabirds at Inaccessible Island and other islands in the Tristan da Cunha group. *Mar. Ornithol.* **18:** 43–54.

Sagar, P. M., Molly, J., Tennyson, A. J. D. and Butler, D., 1994. Numbers of Buller's Albatross breeding at the Snares Islands. *Notornis* **41:** 85–92.

Schiavini, A., Frere, E., Gandini, P., García, N. and Crespo, E., 1997. Albatross–fisheries interactions in Patagonian shelf waters. Pp. 208–13 *in* Albatross Biology and Conservation ed by G. Robertson and R. Gales. Surrey Beatty & Sons: Chipping Norton.

Schlatter, R. P., 1984. The status and conservation of seabirds in Chile. Pp. 261–69 *in* Status and Conservation of the World's Seabirds ed by J. P. Croxall, P. G. H. Evans and R. W. Schreiber. ICBP Technical Publication No. 2, Cambridge.

Seto, N., 1995. Black-footed Albatross population census on Midway Atoll, 1994–1995 breeding season. Administrative Report, US Fish and Wildlife Service, Honolulu: Hawaii.

Seto, N., 1996. Black-footed Albatross population census on Midway Atoll, 1995–1996 breeding season. Administrative Report, U.S. Fish and Wildlife Service, Honolulu: Hawaii.

Sibley, C. G. and Monroe, B. L., 1990. Distribution and Taxonomy of Birds of the World. Yale University Press: New Haven.

Sibley, C. G. and Monroe, B. L., 1993. A Supplement to Distribution and Taxonomy of Birds of the World. Yale University Press: New Haven.

Spear, L. B., Ainley, D. G. and Webb, S. W., 1995. Distribution, abundance and behaviour of Buller's, Chatham Island, and Salvin's albatrosses off Chile and Peru: Potential interaction with longliners. P. 25 *in* Abstracts of the First International Conference on the Biology and Conservation of Albatrosses, 28 August–1 September 1995, Hobart, Australia.

Stagi, A., Vaz-Ferreira, R., Marin, Y. and Joseph, L., 1997. The conservation of albatrosses in Uruguayan waters. Pp. 220–24 *in* Albatross Biology and Conservation ed by G. Robertson and R. Gales. Surrey Beatty & Sons: Chipping Norton.

Stattersfield, A. J., 1996. Applying the new IUCN threatened species categories. Pp. 67–70 *in* 1996 IUCN Red List of Threatened Animals ed. by B. Groombridge and J. Bailey. IUCN: Gland, Switzerland and Cambridge, United Kingdom.

Thompson, K. R., 1993. Falkland Islands seabird monitoring programme: summary of results 1989/90 to 1992/93. Falklands Conservation Report SMP/3, Stanley, Falkland Islands.

Thompson, K. R. and Riddy, M. D., 1995. Utilization of offal discards from "finfish" trawlers around the Falkland Islands by the Black-browed Albatross *Diomedea melanophris*. *Ibis* **137:** 198–206.

Walker, K. J. and Elliott, G. P., 1995. Albatross research on a shoestring: population biology of Auckland Island Wandering Albatross. P. 8 *in* Abstracts of the First International Conference on the Biology and Conservation of Albatrosses, 28 August–1 September 1995, Hobart, Australia.

Walker, K. J., Elliot, G. P., Nicholls, D. G., Murray, M. D. and Dilks, P., 1995. Satellite tracking of Wandering Albatross *(Diomedea exulans)* from the Auckland Islands: preliminary results. *Notornis* **42:** 127–37.

Warham, J. and Bell, B. D., 1979. The Birds of Antipodes Island, New Zealand. *Notornis* **26:** 121–69.

Watkins, B. P., 1987. Population sizes of King, Rockhopper and Macaroni penguins and Wandering albatrosses at the Prince Edward Islands and Gough Island, 1951–1986. *Sth Afr. J. Antarctic Res.* **17:** 155–62.

Waugh, S. M., Moore, P. J. and Weimerskirch, H., 1995. Population dynamics of Grey-head Mollymawks *Diomedea chrysostoma* and New Zealand Black-browed Mollymawks *D. melanophrys impavida* breeding at Campbell Island — a preliminary analysis. P. 60 *in* Abstracts of the First International Conference on the Biology and Conservation of Albatrosses, 28 August–1 September 1995, Hobart, Australia.

Weimerskirch, H. and Jouventin, P., 1987. Population dynamics of the Wandering Albatross *(Diomedea exulans)* of the Crozet Islands; causes and consequences of the population decline. *Oikos* **49:** 315–22.

Weimerskirch, H. and Jouventin, P., 1997. Changes in population sizes and demographic parameters of six albatross species breeding on the French sub-Antarctic islands. Pp. 84–91 *in* Albatross Biology and Conservation ed by G. Robertson and R. Gales. Surrey Beatty & Sons: Chipping Norton.

Weimerskirch, H. and Robertson, G., 1994. Satellite tracking of Light-mantled Sooty Albatrosses. *Polar Biol.* **14:** 123–26.

Weimerskirch, H., Clobert, J. and Jouventin, P., 1987. Survival in five southern albatross species and its relationship with their life history. *J. Anim. Ecol.* **56:** 1043–56.

Weimerskirch, H., Brothers, N. and Jouventin, P., 1997. Population dynamics of Wandering Albatross *Diomedea exulans* and Amsterdam Albatross *D. amsterdamensis* in the Indian Ocean and their relationships with longline fisheries: conservation implications. *Biol. Conserv.* **79:** 257–70.

Weimerskirch, H., 1997. Foraging strategies of Indian Ocean albatrosses and their relationships with fisheries. Pp. 168–79 *in* Albatross Biology and Conservation ed by G. Robertson and R. Gales. Surrey Beatty & Sons: Chipping Norton.

Weimerskirch, H., Zotier, R. and Jouventin, P., 1989. The avifauna of the Kerguelen Islands. *Emu* **89:** 15–29.

Williamson, D. and Nishimoto, M., 1994. Black-footed and Laysan Albatross population at Midway during 1991–1992 and 1992–1993 breeding seasons. Administrative Report. U.S. Fish and Wildlife Service: Honolulu, Hawaii.

Woehler, E. J., 1996. Concurrent decreases in five species of Southern Ocean seabirds in Prydz Bay. *Polar Biol.* **16:** 379–82.

The sun's last rays of the day highlight a lone incubating Wandering Albatross on wind swept Bird Island, South Georgia.
SIMON PICKERING

Population Studies

Contemporary studies of albatross populations are principally concerned with the numerical attributes of population — numbers, rates of change, sex ratio, survivorship and so on — together with the components of their environments that influence these attributes. Population studies lie at the heart of attempts to understand how albatrosses live and the biological and ecological processes that affect their lives. Given their life history strategies — the extended periods of immaturity, longevity and time between reproductive events long-term studies of albatrosses are essential. Very few, however, exist. Studies conducted at South Georgia (Chapter 5) and the French sub Antarctic islands (Chapter 6) over 20 years contribute greatly to our understanding of population processes in albatrosses. Studies of breeding schedules (Chapter 7), factors influencing reproductive performances (Chapter 8) and of habitat management regimes (Chapter 9) provide valuable insights into the life histories, constraints on breeding and habitat requirements of albatrosses. Knowledge on these issues is vital to our understanding of the status of albatross populations and the detection of influences which affect these populations.

Part of one of the world's largest albatross colonies, the 200 000 breeding pair-strong Black-browed Albatross colony on Steeple Jason, Falkland Islands.
GRAHAM ROBERTSON

CHAPTER FIVE

Population changes in albatrosses at South Georgia

J. P. CROXALL,[1] P. A. PRINCE,[1] P. ROTHERY[1,2] and A. G. WOOD[1]

At Bird Island, South Georgia, populations of Wandering, Grey-headed and Black-browed Albatrosses have been studied continuously since 1975 and are all decreasing. The magnitude and nature of these population changes are reviewed, together with an assessment of their demographic causes. For Wandering Albatrosses, the decrease was first recognized in 1979 and is still continuing. It reflects decreases of about 2–3% per annum in adult survival rate and 10% per annum in post-fledging juvenile survival rate, these contributing approximately equally to the 1% per annum reduction in population size. Breeding success is consistently high, and increased by 15% between 1975 and 1990, being stable currently. Average age of first breeding has decreased by about two years over the last 20 years, for both sexes. The decrease in Grey-headed Albatrosses is widespread amongst Bird Island colonies. However, unequivocal decreases were not evident until the late 1980s because of the magnitude of interannual fluctuations in population size in this biennially-breeding species. Adult survival may be lower since the 1970s than previously but the main cause of decrease is that recruitment to the breeding population nowadays is only 10% of that 15 years ago. Breeding success is variable and without obvious trends. In Black-browed Albatrosses, significant decreases were only evident for a proportion of the colonies at Bird Island by 1990; nowadays all colonies are decreasing. At the main study colonies recruitment is only 20% of that 15 years ago and adult survival has decreased by about 4% per annum since 1987. Breeding success is highly variable and a recent succession of poor years (bad weather, low krill availability) has caused lower breeding (especially fledging) success this decade compared to last. Interactions with fisheries are known, or believed, to be important in almost all these population changes and are briefly reviewed. Comparisons with conspecific populations at other islands indicate broad similarities in timing of population decreases but differences in the magnitude of these and in recovery rates. South Georgia albatross populations are unlikely to recover within the next 10–20 years and are potentially more seriously at risk than those of most other Southern Ocean islands.

Key words: South Georgia, Population Dynamics, Survival, Breeding Frequency, Recruitment, Wandering Albatross, Grey-headed Albatross, Black-browed Albatross.

INTRODUCTION

THERE is widespread evidence of recent decreases in populations of many albatross species (Gales 1993). The concern this generated, particularly in relation to mortality associated with longline fisheries, has focused considerable attention on the topic of population change in albatrosses. Many of these population changes were recognized initially only as differences between counts of breeding birds widely separated in time. Because albatross populations, especially those of biennially breeding species, show marked interannual fluctuations, continuous long-term studies are needed to establish population trends unequivocally. Such fundamental data are available for very few species and are urgently required for many species whose population status is uncertain but a cause for concern.

Detailed population monitoring studies should not be confined to decreasing species; they are also needed for albatross species that are, or appear to be, increasing (e.g., White-capped Albatross *Diomedea (Thalassarche) steadi* (C. J. R. Robertson, in litt.), Shy Albatross *D. (Thalassarche) cauta* (N. Brothers, in litt.), Buller's Albatross *D. (Thalassarche) bulleri* (Sagar *et al.* 1994)), in order to detect future decreases, or reductions in the rate of increase, of these populations in time to take appropriate action. In many circumstances, however, appropriate action will require knowledge (deriving from studies of survival rates) of which elements of the population are being affected, as well as when and where this might be happening. Comprehensive population studies are, therefore, fundamental to most aspects of albatross ecology and conservation — and vital for monitoring the effectiveness of conservation actions.

[1]British Antarctic Survey, Natural Environment Research Council, High Cross, Madingley Road, Cambridge CB3 0ET, United Kingdom.
[2]Present address: Institute of Terrestrial Ecology, Natural Environment Research Council, Monks Wood, Abbots Ripton, Huntingdon, Cambridgeshire PE17 2LS, United Kingdom.
Pages 69–83 *in* ALBATROSS BIOLOGY AND CONSERVATION, ed by G. Robertson and R. Gales. Surrey Beatty & Sons, Chipping Norton. 1997.

In this chapter we illustrate patterns of population and demographic change by comparing three species of albatross breeding at Bird Island, South Georgia — Wandering Albatross *D. exulans exulans*, Grey-headed Albatross *D. (Thalassarche) chrysostoma* and Black-browed Albatross *D. (Thalassarche) melanophris*. These species have been studied for more than 20 consecutive years at this site. Some of the population data have been analysed and published previously (Croxall *et al.* 1990; Prince *et al.* 1994), so we concentrate here on overall patterns and more recent, unpublished data. These are revealing some important changes in South Georgia populations, particularly of Black-browed Albatrosses, a species whose population was generally regarded as stable to increasing in the region (Gales 1993).

METHODS

The study site of Bird Island, South Georgia (54°S, 38°W) is described in detail in Hunter *et al.* (1982), which also shows the precise location of Black-browed and Grey-headed Albatross colonies and the principal breeding areas of the more dispersed Wandering Albatrosses. Most data for Grey-headed and Black-browed Albatrosses come from the long-term study colonies (E and H respectively), where every bird is checked and identified each year; supplementary data come mainly from colonies B and J respectively (see e.g., Table 1). Most data for Wandering Albatrosses come from the known-aged birds in the whole Bird Island population (such birds currently forming about 60% of the population); some supplementary data come from the Wanderer Ridge study population (150 pairs) where every bird is checked and identified each year. Details of the methods of the capture-mark-recapture studies are given in Croxall *et al.* (1990) and Prince *et al.* (1994); the annual monitored breeding populations of Wandering, Grey-headed and Black-browed Albatrosses are some 1 300, 350 and 120 pairs respectively, plus a new group of 200 pairs of Black-browed Albatrosses (colony J) added since 1988. Methods of organising and handling the data are described in Wood and Rothery (submitted); methods of analysing data on changes in population size, breeding success, breeding frequency, adult and juvenile survival are given in detail in Croxall *et al.* (1990), Prince *et al.* (1994) and Rothery and Prince (1990). There are two important differences in the data used here compared to our earlier published studies. For Wandering Albatrosses we previously estimated survival using only birds ringed in 1958, 1962 and 1963 (Croxall *et al.* 1990). We are now able to incorporate substantial numbers of birds ringed as chicks in each year from 1976 onwards. Comparing the same years, the overall effect is to produce a reduction in mean adult annual survival of 1.0% for males and 0.2% for females. For Grey-headed and Black-browed Albatrosses, mean annual adult survival rate was previously estimated using all adult birds, irrespective of status (Prince *et al.* 1994). For comparability with Wandering Albatrosses (and with other studies) here we only use data for known breeding birds. Comparing the same years, the difference in survival rate is about 1% (0.7–1.2%) for both sexes of both species. Methods are similar to those used in albatross studies at Possession Island, Iles Crozet (Weimerskirch and Jouventin 1997), so results should be comparable between the two studies and sites.

RESULTS

The main biological and ecological differences between the three study species of albatross are as follows. Wandering Albatrosses breed biennially (when successful), their breeding season lasting *ca.* 330 days (laying in December, chicks fledging in October–November); Grey-headed Albatrosses breed biennially (when successful), their breeding season lasting 238 days (laying in October, chicks fledging in late May); Black-browed Albatrosses breed annually, their season lasting 205 days (laying in October, chicks fledging in late April). Wandering Albatrosses take approximately equal proportions of fish, squid and carrion in their diet; Black-browed Albatrosses take similar proportions of krill, squid and fish, whereas Grey-headed Albatrosses mainly take squid and fish, with smaller amounts of krill (Prince 1980; Croxall *et al.* 1988a,b; Reid *et al.* 1996; Croxall *et al.* 1997). The foraging ranges and migrations of all three species are described in Prince *et al.* (1997). For each species we summarise data below on each of the main aspects that make up their demography, i.e., population size, breeding success, breeding frequency (equivalent to pregnancy rate in mammals) and adult and juvenile survival.

Wandering Albatross

Population size

Between 1961–63 and 1996 the Bird Island breeding population (61% of the South Georgia total) decreased by about 28% at an average annual rate of 0.8% (0.1% S.E.) ($t = 6.87$, $r^2 = 0.65$, $P = <0.001$; Fig. 1). The average annual rate of decrease since 1972 is also about 0.8%. However these rates are slightly lower than the estimate of 1.0% for 1961–1989 (Croxall *et al.* 1990), possibly reflecting some tendency towards stabilization.

Age of first breeding

As indicated in Croxall *et al.* (1990), data for cohorts from 1962 and 1963 compared with 1972 and 1973 suggested breeding at younger ages in

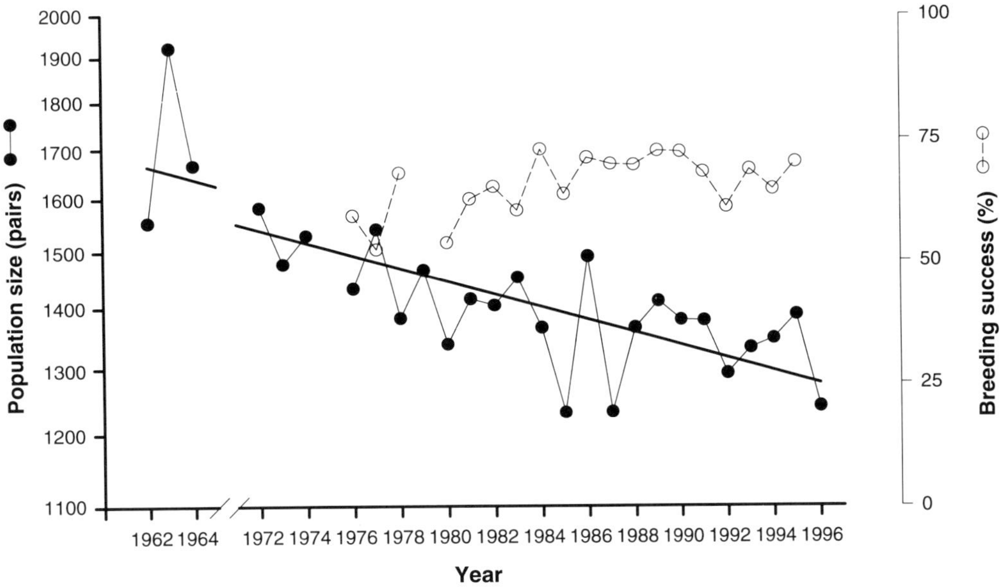

Fig. 1. Breeding population size (pairs; note log scale) and breeding success (chicks fledged per egg laid) for the Wandering Albatross population at Bird Island, South Georgia, 1962–1996. The solid line shows the fitted exponential decrease in population size with a rate of 0.8% per annum.

the more recent years but insufficient cohorts had been completely recruited to test this. Using data from the 1972 to 1980 cohorts (inclusive), the average age of first breeding is 10.4 years for females (n = 768) and 10.7 years for males (n = 734), a significant difference ($F_{1\ 500} = 10.0$, $P = 0.002$). In both sexes there has been a significant trend to reduce the mean age at first breeding, in females from 10.9 years for the 1972 cohort to 9.9 years for the 1980 cohort ($F_{7\ 767} = 5.2$, $P < 0.001$), and in males from 11.6 to 10.3 years ($F_{7\ 733} = 6.5$, $P < 0.001$). The overall change has been a decrease at a rate of 1.3–1.6 years per decade.

Breeding success

Between 1975 and 1990 breeding success increased by 15% (from 56% to 71%) at a rate of 1.2% per annum (0.4% S.E., $r^2 = 0.56$, $P < 0.05$). Between 1991 and 1995 breeding success has decreased slightly, median values being 68% compared with 71% over the previous five years ($U = 38$; $P = 0.04$, Mann-Whitney U-test); the overall rate of increase between 1975 and 1995 averages 0.6% annually (0.3% S.E., $r^2 = 0.31$, $P = 0.014$). Although hatching and fledging success contribute approximately equally to the overall breeding success and have both changed similarly in time (see also Croxall *et al.* 1990), correlations between the two are not statistically significant. Variation in breeding success can be partitioned into the individual components due to hatching and fledging success, together with a term due to covariation between these (Brown 1988). For Wandering Albatrosses hatching success contributes 40%, fledging success 45% and covariation 15% to year-to-year variation in breeding success.

Breeding frequency

No Wandering Albatross successful in fledging a chick has bred in the following (n + 1) year (the interval between mean fledging and laying dates averaging only about five weeks); theoretically all successful birds should therefore breed next in the n + 2 year. Unsuccessful birds should breed next in the n + 1 year, except those that fail relatively late in chick-rearing which are unable to do so. The proportion of birds deviating from this expectation is shown in Figure 2. On average 30% (2% S.E.) of successful and 35% (2% S.E.) of unsuccessful breeders defer breeding beyond the next year expected. Apart from 1977 for unsuccessful birds (25% more than usual postponing breeding to n+2 year) and 1979 for successful ones (25% more than usual breeding in n+2 year), there is remarkably little interannual variation, nor any indication of a trend with time.

Survival

Between 1976 and 1984 mean survival rates of adult males (94.8% (0.6% S.E.)) and females (93.1% (0.6% S.E.)) was significantly different, for the 1958, 1962 and 1963 cohorts combined (Croxall *et al.* 1990), and 93.9% (0.4% S.E.) and 92.7 (0.4% S.E.) for all cohorts. Including a further nine years in the time series (Figs 3a, b) gives overall values of 93.2% (0.4% S.E.) and 92.4% (0.4% S.E.) for males and females respectively (and 92.8% for both sexes combined). This difference between sexes is not statistically significant. There is no significant overall trend in survival rates but since 1987 they have averaged about 1% lower than previously estimated (92.0% versus 93.2%), though the

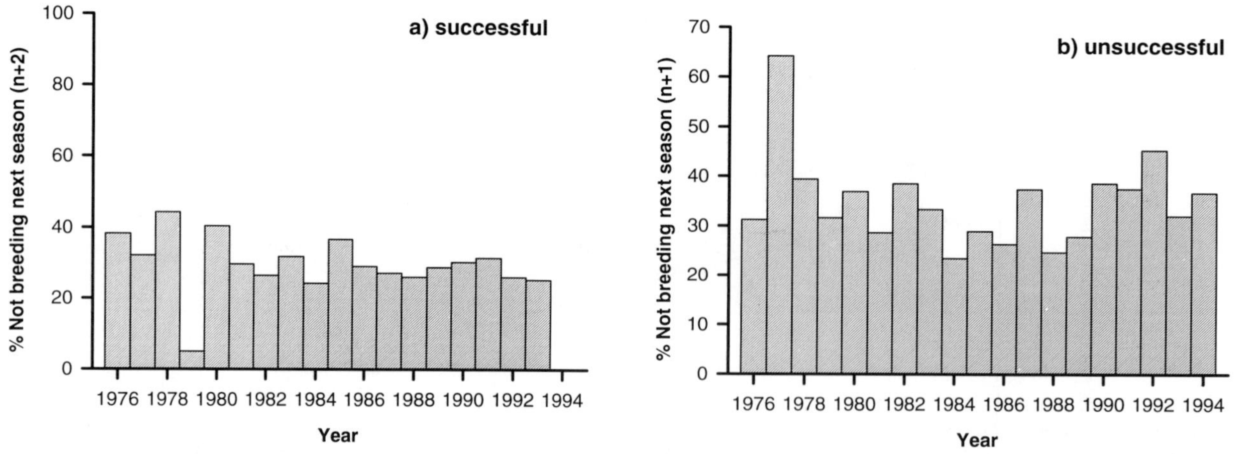

Fig. 2. Breeding frequency in Wandering Albatrosses at Bird Island, South Georgia, 1976–1993. (a) proportion of birds successful in season n deferring their next breeding attempt until after season n + 2; (b) proportion of birds unsuccessful in season n deferring their next breeding attempt until after season n + 1.

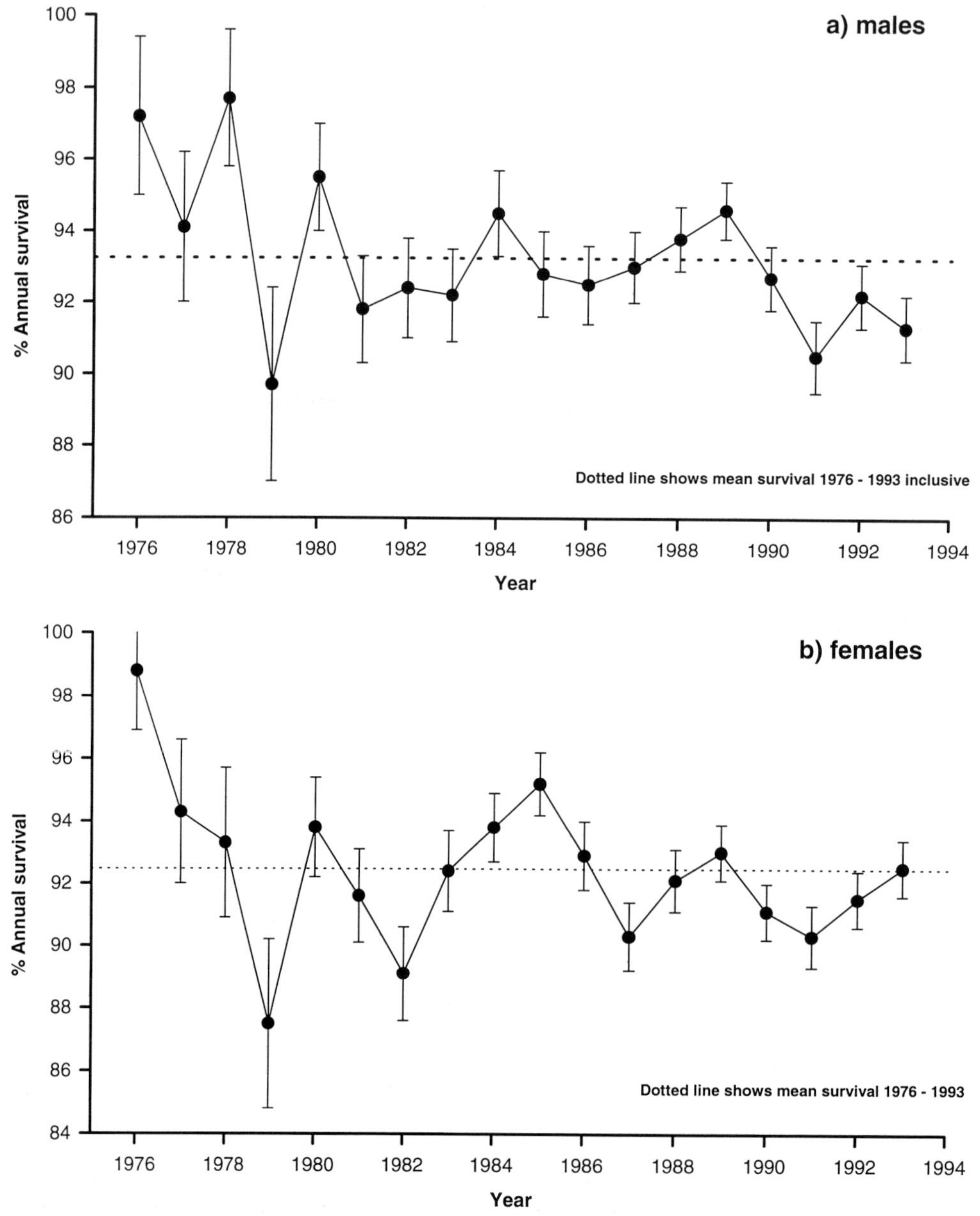

Fig. 3. Mean (± SE) estimated adult survival of male and female Wandering Albatrosses at Bird Island, South Georgia, 1976–1992.

Table 1. Average rates of recruitment to the breeding population in cohorts of albatrosses fledged at Bird Island, South Georgia. Values for the average proportion surviving to age five years are in parentheses.

	Proportion (%) of cohort surviving to breed				
	Black-browed Albatross		Grey-headed Albatross		Wandering Albatross
Cohorts	H	J	E	B	All
1959–64	21.8	26.6	35.5	35.6	35.8
1972–75	—	—	—	—	25.5 (45.1)
1976–81	11.0 (23.4)	14.4	8.3 (18.6)	4.4	32.9 (55.7)
1982–86	3.8 (15.0)	5.5	3.3 (19.1)	1.0	27.5 (44.4)

difference is not statistically significant ($t = 1.15$, $P = 0.1$). Overall there has been a decrease of nearly 4% in mean adult survival since Tickell's (1968) estimate for 1961–63 of 95.7% (compared with 93.2% and 91.9% for 1976–84 and 1985–92 respectively, using the population of exactly the same area as that studied by Tickell).

The six cohorts of chicks fledging in 1972–78 averaged 48.9% survival to five years of age (i.e., to 1977–83), with no significant difference between sexes (Croxall *et al.* 1990). For the 10 cohorts fledged between 1979–89, survival (to 1984–93) averaged 52.4%, again without significant difference between sexes and when compared to the earlier period. Comparison of the proportions surviving to breed (Table 1) shows a decrease from 36% (1.1% S.E.) for the 1959–64 cohorts to 32% (1.3% S.E.) for 1972–81 to 27% (2.5% S.E.) for 1982–86. This last value may increase by 1–2%, representing the relatively small number of birds still to recruit from these cohorts. Nevertheless it appears that about 25% fewer birds (a 9% reduction in annual recruitment rate) are currently being recruited to the breeding population than in the 1970s. Most of these losses are taking place between the time of first return to the breeding colony (at a modal age of five years) and recruitment (at a modal age of 10 years).

Grey-headed Albatross

Population size

Of the three species investigated here, this species shows the greatest inter-annual variation in breeding population size (Fig. 4). For the only study colony (E) where all breeding individuals are checked each year, between 1976/77 and 1995/96 the population decreased by 30% at an average annual rate of 1.9% (0.1% S.E.) ($t = 1.91$, $r^2 = 0.12$, $P = 0.07$; note, however, that biennial breeding produces negative serial autocorrelation which overestimates the S.E. of the estimated slope. Adjustment for this, using autocorrelations at lags 1, 2 and 3, approximately doubles the t-ratio and gives $P = 0.01$). For all colonies (totalling 65% of the Bird Island population) where regular counts have been made at the end of the laying period (and corrected using the known failure rate from colony E) (Table 2), the overall decrease is either 13.2% at 0.76% per annum (1976/77 to 1995/96), 18.1% at 1.42% per annum (comparing the demi-populations of 1976/77 plus 1977/78 with 1990/91 plus 1991/92), or 20.5% at 1.43% per annum (comparing the 1976/77 plus 1977/78 and 1992/93 plus 1993/94 demi-populations). At individual colonies based on similar estimates (Table 2), changes range from decreases of up to 31% to an increase of 5% at one colony (C). Colony C, however, has suffered the greatest reduction in the last two years so now all monitored colonies show clear evidence of population decrease.

Breeding success

Breeding success for the 21 years from 1976 to 1996 (Fig. 4) averages 39%. There is significant variation among years (see Prince *et al.* 1994) but no evidence of any overall trend. Mean hatching success is 60% (5% S.E.) and mean fledging success 65% (4% S.E.), both varying significantly among years (Prince *et al.* 1994). The hatching and fledging components of overall breeding success account for 65% and 48% of the total variation, with –13% due to covariation.

Breeding frequency

Departures from the breeding frequency patterns expected of a biennial species (Fig. 5), which average 36% (4% S.E.; CV 50%) and 47% (5% S.E.; CV 42%) in successful and unsuccessful birds, respectively, reveal variation much greater than in the Wandering Albatross (CV 26%) but much less than the Black-browed Albatross (CV 95%). There is, however, no evidence of any trend across years. Of the three most extreme years (1979, 1986, 1988, when 68–75% of birds deferred breeding beyond the $n + 2$ year) for successful birds, in two cases (1981, 1988) the $n + 2$ breeding season was one of very low breeding success, suggesting that poor conditions in these years may have induced more adults to defer breeding by another year. Of unsuccessful birds, those breeding in 1980 and 1987 were two of the three cohorts with greatest deferral of breeding beyond the $n + 1$ (1981, 1988) season. However, there was no general relationship

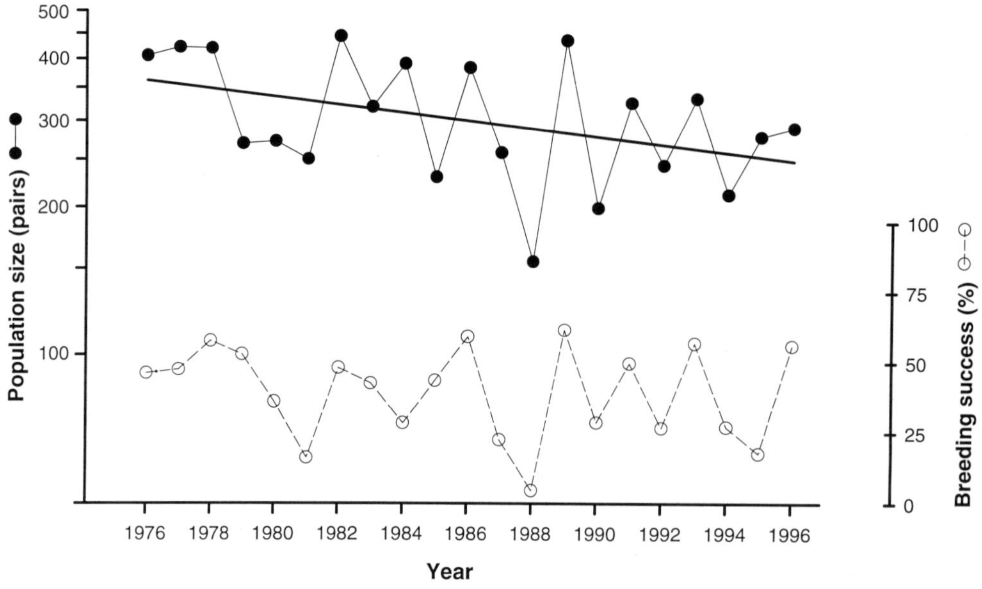

Fig. 4. Breeding population size (pairs; note log scale) and breeding success (chicks fledged per egg laid) for the Grey-headed Albatross population at colony E, Bird Island, South Georgia, 1976–1996. The solid line shows the fitted exponential decrease in population size, with a mean rate of 1.9% per annum.

Table 2. Breeding population (pairs) of Grey-headed Albatrosses at selected colonies at Bird Island, South Georgia. Values in parentheses for whole island assume equivalent change to that in combined total of colonies.

Year	Whole Island	Colonies								
		A/A1	B	C	D	E	K	L	M	Total
1976/77	9 053	332	752	369	373	408	2 854	80	625	5 793
1977/78	9 164	283	872	274	303	420	2 878	65	710	5 805
1989/90	5 632	251	595	222	164	198	2 067	50	291	3 838
1990/91	7 472	234	553	280	221	326	2 558	80	535	4 787
1991/92	(7 254)	272	689	374	280	243	2 404	73	472	4 707
1992/93	(8 954)	281	751	333	278	332	3 232	78	525	5 810
1993/94	(5 250)	232	451	241	199	211	1 666	57	350	3 407
1995/96	(7 752)	277	732	272	246	290	2 714	48	451	5 030

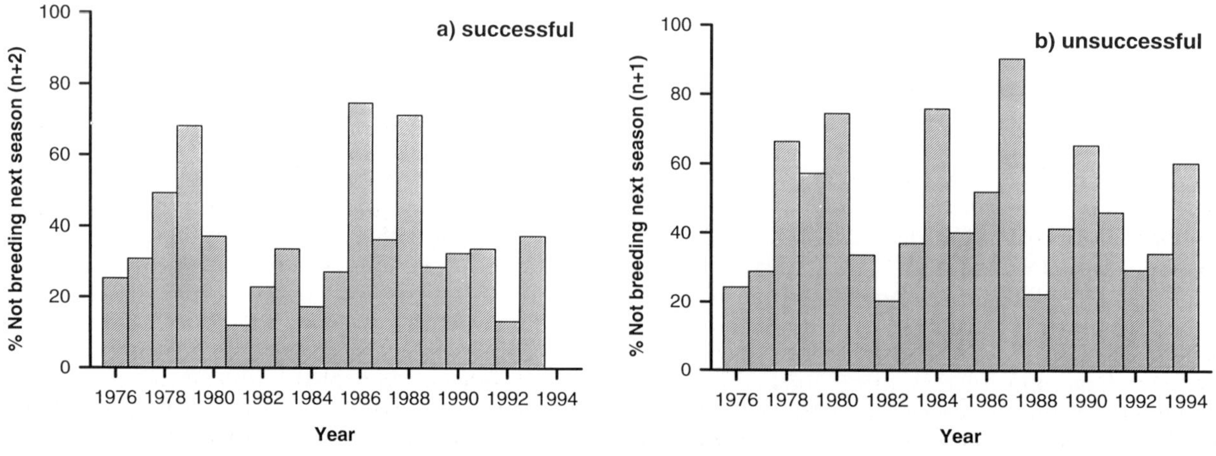

Fig. 5. Breeding frequency in Grey-headed Albatrosses at colony E, Bird Island, South Georgia, 1976–1994. (a) proportion of birds successful in each season n, deferring their next breeding attempt until after season n + 2; (b) proportion of birds unsuccessful in season n, deferring their next breeding attempt until after season n + 1.

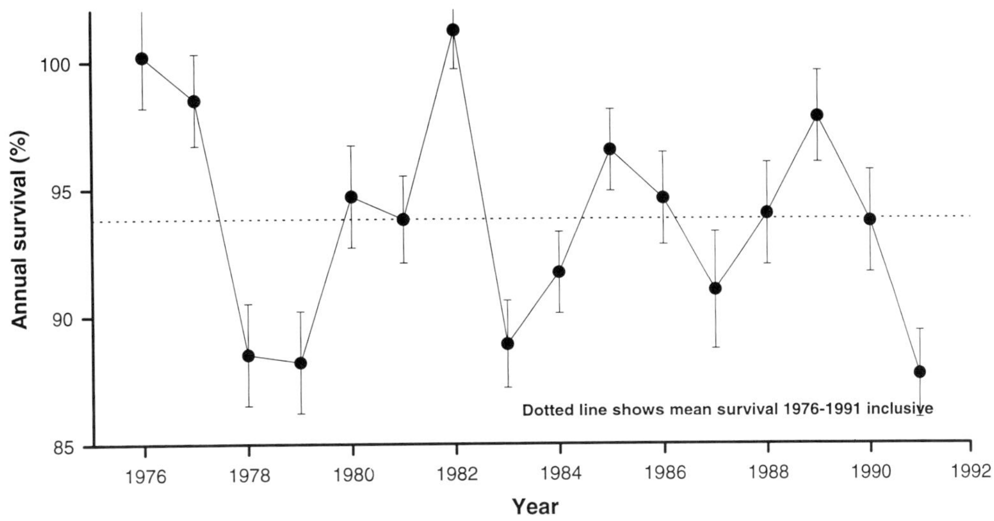

Fig. 6. Mean (± SE) estimated adult survival of Grey-headed Albatrosses (sexes combined) at colony E, Bird Island, South Georgia, 1976–1991.

between the extent of deferral by unsuccessful birds and breeding success in the n + 1 season. As Prince *et al.* (1994) showed, much of the variation in return rate to the next breeding attempt for unsuccessful birds relates to the timing of failure, with birds failing before and after March showing very different patterns. However, the relationship between failure date and likelihood of breeding clearly needs further study in conjunction with indices of potential food availability during future breeding seasons.

Survival

Re-analysis (see methods) of the data presented by Prince *et al.* (1994) shows that mean estimated annual survival of adults between 1977 and 1988 was 93.5% (0.8% S.E.) with no significant difference between the sexes. Adding three more years (1989–91) to the time series (Fig. 6) re-affirms the very strong and fairly regular pattern of fluctuation in survival rate but only changes the overall average to 93.4% (0.2% S.E.).

Until fairly recently we have assessed juvenile survival only over the whole period between fledging and recruitment to the breeding population. Data for 1959–83 were summarized in Prince *et al.* (1994); these are extended and summarized in Table 1. This indicates that juvenile survival decreased by 75% between values of 35–36% for the 1959–64 cohorts (recruiting in 1969–74) and of 7–8% for the 1976–81 cohorts (recruiting 1986–91) and shows signs of decreasing yet further (to <5%) for the 1982–86 cohorts (recruiting 1992–96), for which no more than an additional 2–3% of birds can be expected to return. Limited data for survival of birds to age five years (Table 1) give average survival rates of 18.6% and 19.1% for the 1976–81 and 1982–86 cohorts, respectively. These are double the values for overall juvenile survival of these same cohorts but still only half that of the earliest cohorts. This suggests that the overall decrease in juvenile survival results from approximately equal reductions in survival in the early years after fledging and in the years immediately prior to breeding (unlike the situation in Wandering Albatrosses).

Black-browed Albatross

Population size

At the only colony (H) where every bird is counted at laying, between 1976 and 1996 the population decreased by 64% at an average annual rate of 4.2% (0.1% S.E.) ($t = -5.55$, $r^2 = 0.40$, $P = 0.002$; note that for annual breeding species serial positive autocorrelation leads to underestimation of the slope S.E. and overestimation of t values. However, correction for this gives a t value of -3.33, $P < 0.01$, Fig. 7). This decrease may be atypically high because the colony is the only one (of those monitored) without direct shelter from prevailing winds and has suffered disproportionate erosion, somewhat reducing the amount of habitat suitable for breeding. Nevertheless the colony with the next most precise details (J) and opposite aspect decreased by 45% at a rate of 2.9% per annum over the same period (Table 3). The decreases in the other colonies ranged from 15% at 0.8% per annum to 47% at 3.3% per annum.

The overall population at all monitored colonies (10% of the total Bird Island population) has decreased by 30.8% at 1.8% per annum. It is possible, however, (see Table 3) that the overall Bird Island population may have been stable (or even increasing slightly) until about 1989/90 and then decreased sharply by about 35% at an average rate of 6.9% per annum.

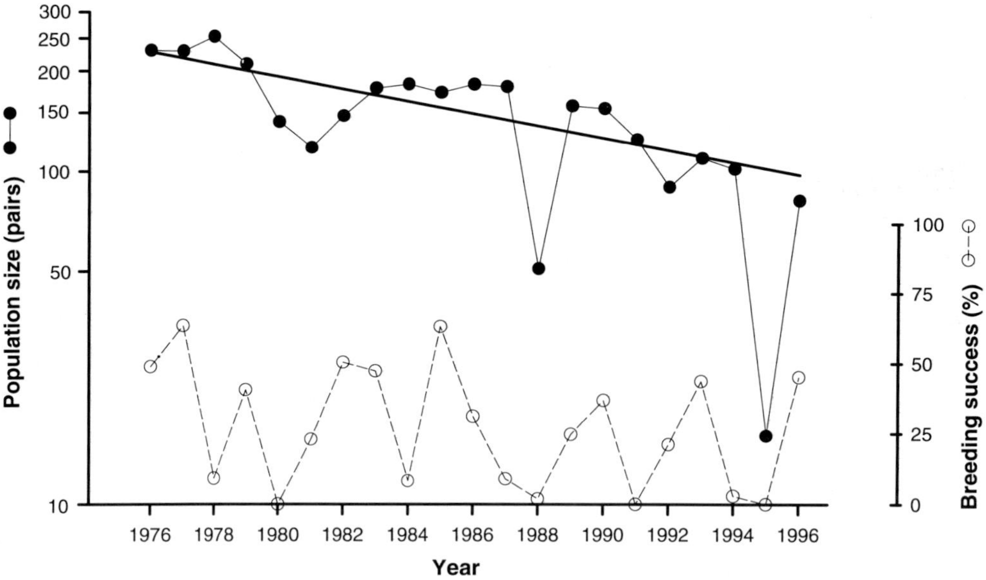

Fig. 7. Breeding population size (pairs; note log scale) and breeding success (chicks fledged per egg laid) for the Black-browed Albatross population at colony H, Bird Island, South Georgia, 1976–1996. The solid line shows the fitted exponential decrease in population size (excluding 1988 and 1995), with a rate of 4.1% per annum.

Table 3. Breeding population (pairs) of Black-browed Albatrosses at selected colonies on Bird Island, South Georgia. Values in parentheses for whole island assume equivalent change to that in combined total of colonies.

Year	Whole Island	Colonies								Mean % annual change
		H	I	J	K	K2	N	O	Total	
1976/77	13 328	227	1 252	533	731	150	1 283	1 080	5 256	
1989/90	14 695	157	1 345	502	640	104	1 441	1 417	5 606	+0.5
1992/93	(12 200)	110	1 135	375	538	164	1 196	1 136	4 654	−6.0
1993/94	(10 988)	102	1 005	378	422	159	1 095	1 031	4 192	−9.9
1995/96	(9 539)	82	881	294	390	82	992	918	3 639	−6.8

At colony H the rate of decrease for 1989–1996 averaged 9.4% (2.2% S.E.) ($r^2 = 0.75$, $P = 0.08$).

Breeding success

Breeding success is particularly variable in Black-browed Albatrosses (Prince *et al.* 1994), with occasional years of very poor reproductive performance typically associated with years of low krill availability (Prince 1985; Croxall *et al.* 1988a; Croxall *et al.* 1997). Such years were 1978, 1980, 1984, 1991 and 1994. Low breeding success in 1988 and 1995 was due to the effects of late cold winters with snow and ice still lying in the colonies at the time of breeding. Relatively few birds laid and most failed early in the incubation period. Overall breeding success averaged 27.2% (5% S.E.) between 1976 and 1996, and although there is no statistically significant trend across years, it averaged 34.9% (7% S.E.) between 1976 and 1986 compared with 18.5% (6% S.E.) between 1987 and 1996 ($t_{11,10} = 1.85$, $P = 0.08$). Even if years of poor performance due to physical conditions are excluded, the respective values are 38.4% and 23.1%, still markedly different.

Mean hatching success averaged 62% (5% S.E.) and fledging succcss 43% (8% S.E.), both varying significantly among years (Prince *et al.* 1994). Hatching success is very similar in both Grey-headed and Black-browed Albatrosses and neither shows significant decrease with time (Prince *et al.* 1994). Whereas hatching and fledging success account for similar proportions of variation in overall breeding success in Grey-headed Albatrosses, in Black-browed Albatrosses hatching success accounts for only 19%, fledging success for 85% and covariation –4%. Thus variation in breeding success in Black-browed Albatrosses is mainly due to variation in fledging success. Whereas hatching success has remained fairly constant across years, fledging success in the decade since 1986 was only one-half that between 1976 and 1986 and has decreased significantly between 1976 and 1996 (Pearson's $r = -0.52$, $P = 0.02$).

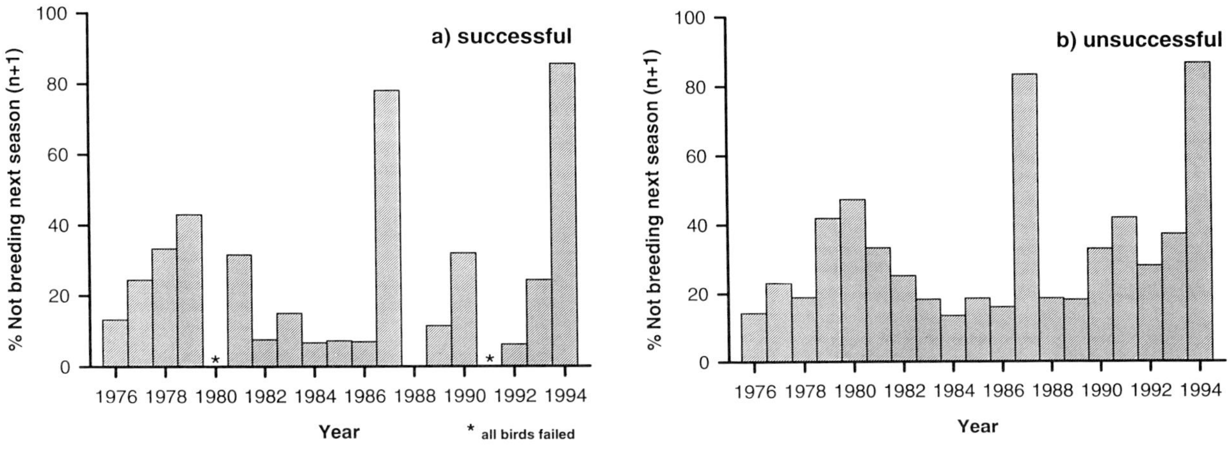

Fig. 8. Breeding frequency in Black-browed Albatrosses at colonies H and J, Bird Island, South Georgia, 1976–1994. (a) proportion of birds successful in season n, deferring their next breeding attempt until after season n + 1; (b) proportion of birds unsuccessful in season n, deferring their next breeding attempt until after season n + 1.

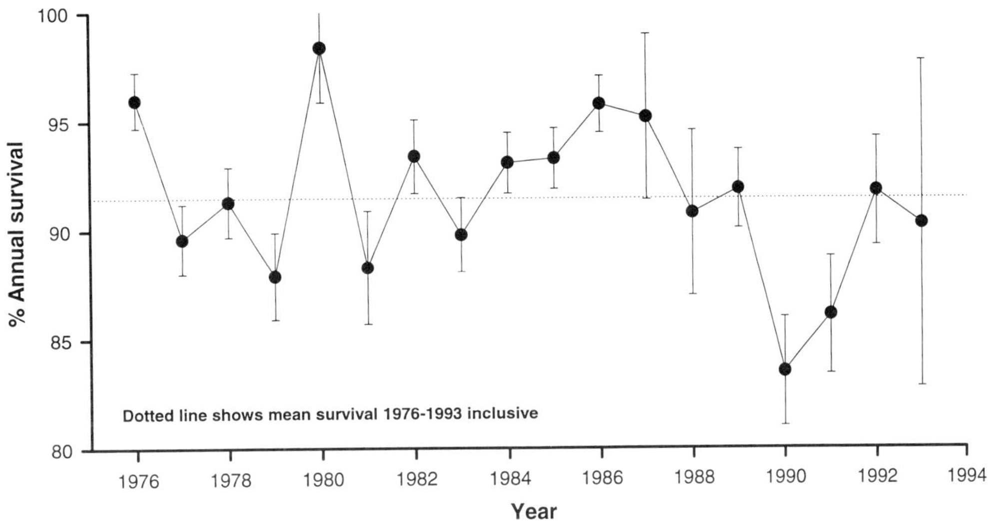

Fig. 9. Mean (± SE) estimated adult survival of Black-browed Albatrosses (sexes combined) at colony H, Bird Island, South Georgia, 1976–1991.

Breeding frequency

Black-browed Albatrosses are essentially annual breeders. Nevertheless 25% (5% S.E.; CV 95%) of birds successful in rearing a chick defer their next breeding attempt beyond the n+1 season (Fig. 8). This average excludes the years 1980 and 1991 when no birds were successful in rearing chicks. Exceptionally high levels of deferred breeding are evident for 1987 (78%) and 1994 (86%), both coinciding with years of low local krill availability and very reduced breeding success of those birds which did breed. If these years are also excluded then the overall average deferral of breeding by one year or more is 18%. Unsuccessful birds show an average deferral of breeding by one year or more of 33% (5% S.E.; CV 63%) (or 26% if 1987 and 1994 are excluded), neither significantly different from values for successful breeders. Once again 1987 and 1994 were years when exceptionally few birds (16% and 13%, respectively) bred in the following year.

Survival

Re-analysis of the data presented by Prince *et al.* (1994) gives an overall estimated mean adult annual survival rate of 92.5% (0.6% S.E.) for 1976–1988 with significant variation among years and significantly higher survival rates in females (95.1%) than in males (93.0%). Adding five more years data (Fig. 9) shows that survival rates in 1990 and 1991 were significantly lower than overall and than in any year since 1982, although some improvement occurred in 1992 and 1993. Nevertheless, the overall annual adult survival rate is now 91.5% (0.4% S.E.) and remains higher in females (92.9%) than in males (91.3%). Since 1987 adult survival rates have only averaged 89.1%, compared to 92.7% previously, but the difference is not significant ($t = 1.3$, $P = 0.10$).

Survival of fledglings to breeding age (Table 1) averaged 22–27% for the 1959–64 cohorts (recruited 1969–74) but had halved to 11–14%

Table 4. Differences in survival rates and annual rate of change of breeding populations of South Georgia albatrosses. Values are percentages.

Species	Adult survival		Breeding success		Recruitment rate		Population change	
	Then	Now	Then	Now	Then	Now	Then	Now
Wandering Albatross[1]	96	92	60	70	36	27	0	−1.0
Grey-headed Albatross[2]	(95)[3]	93	40	39	36	*c.*5	0	−1.4
Black-browed Albatross[2]	93	89	36	18	27	*c.*7	+0.5	−6.9

[1]Then: pre-1970; Now: since 1987. [2]Then: 1970s; Now: since 1987. [3]Estimated value.

for the 1976–81 cohorts (recruited 1986–91), with a clear indication of a further substantial drop to 4–5% for the 1982–86 cohorts (recruited 1992–96). Limited data for survival of birds to age five years give an average survival for the 1977–81 and 1982–86 cohorts of 23.4% and 15.0% respectively. This suggests that the survival of younger (<5 year old) juveniles dropped in the early 1970s but has since stabilized. Survival of older juveniles, however, has decreased progressively.

DISCUSSION

Demographic changes in South Georgia albatross populations

The main changes in breeding populations and the associated changes in survival rates since the 1970s of the three species of albatross studied at South Georgia are summarized in Table 4. Two parameters are omitted here. For breeding frequency we have no evidence that this has changed significantly for any species. For age of first breeding we have insufficient data to detect any change for Black-browed and Grey-headed Albatrosses. For Wandering Albatrosses there has been a significant reduction in the age of first breeding from a modal age of about 12 years for the 1962 and 1963 cohorts (Croxall *et al.* 1990) to a mean of 11 years for the 1972 cohort, and to 10 years for the 1980 cohort. However, reductions in age of first breeding have only a small effect on the overall population demography, a reduction of 4.2 years being needed to compensate for a 1% increase in adult mortality (Weimerskirch *et al.* 1997).

For Wandering Albatrosses prior to 1970, with adult survival of 95–96%, juvenile recruitment to breeding of 36% and breeding success of 60%, a stable or slightly increasing (by 0.5–1.1% per annum) population would be predicted (Croxall *et al.* 1990). By about 1985 adult survival was 94%, juvenile recruitment had decreased to 30% and breeding success increased to 64%, combining to produce an overall population decrease of 0.8% per annum, very similar to the actual rate of about 1.0% per annum (Croxall *et al.* 1990). Since 1987, however, annual adult survival has decreased by about another 1%, juvenile recruitment has decreased to 27% and breeding success has reached 70%, but may have decreased slightly recently. This would be predicted to result in a decrease in the breeding population of about 2% per annum, a rate somewhat higher than is evident in the overall population at present. It appears therefore, that there is little evidence of the population beginning to stabilize and there are even suggestions that the rate of population decrease may be about to increase. The overall population decrease has been caused mainly (about two-thirds) by the reduction in adult survival and by reduced recruitment (about one-third), notably of juveniles between the time of their first return to Bird Island and the commencement of breeding.

For Grey-headed Albatrosses we have a shorter time series of data. Nevertheless in the early 1970s annual rates of adult survival of 95% (assumed), recruitment of 36% and breeding success of 40% (both as estimated) would combine to produce a stable population and, despite large fluctuations, were consistent with the data on breeding population size from the only colony monitored in detail. Since 1977, however, breeding success has remained constant, annual adult survival has averaged only 93% and juvenile recruitment has decreased to 5% per annum (14% of levels in the 1970s), entirely consistent with a population decrease of 1.4–1.8% per annum. This population decrease since 1976 can be attributed mainly to the massive reduction in recruitment. This in turn is probably about equally divided between young birds prior to returning to Bird Island and mortality between then and the start of breeding.

For the annually breeding Black-browed Albatross in the 1970s, annual rates of adult survival of 92.5%, juvenile recruitment of 27% and breeding success of 36% were consistent with a population either stable or increasing at no more than 0.5% per annum. There is some indication that Black-browed Albatross populations had increased during (and prior to) this period and, in many colonies on Bird Island, continued to do so until about 1989. Other colonies, including the main study colony, were already decreasing by then and all colonies have decreased subsequently. Given that, at the study colony, annual adult survival rate between 1976 and 1992 averaged 91.5% (and only 89% since 1987), breeding success halved (to 18%) and juvenile recruitment decreased to 7% (26% of levels in the 1970s), an overall population decrease of about 7% per annum is not surprising. Prior to 1987 the decrease was caused partly (about one-third) by the reduction in adult survival and mainly (about two-thirds) by reduced recruitment, with lower breeding success also playing a small part; currently the reductions in adult and juvenile survival contribute about equally. Recently mortality of juveniles has mainly involved birds between their first return to Bird Island and the onset of breeding.

Causes of population change in South Georgia albatrosses

The topic of incidental mortality associated with commercial fisheries, in particular those using longline techniques, and its role in population decreases in albatrosses, has been extensively discussed, including in numerous papers in this book. In relation to the three species considered here, Prince *et al.* (1997) have discussed interactions with fisheries in some detail. Therefore this topic will be dealt with briefly here and principally in relation to the main demographic changes observed in the three species, and to interspecific differences in these.

Wandering Albatross

For Wandering Albatrosses at Bird Island in the 1960s there was no evidence of other than a stable population; indeed values for population variables suggest that the population may even have increased in the 1960s and early 1970s (Croxall *et al.* 1990). This could reflect benefit from the large-scale trawl fisheries which developed around South Georgia during the late 1960s (Kock 1992), because the Wandering Albatross is an accomplished opportunistic and scavenging forager (Croxall and Prince 1994), highly attracted to vessels and known to take many fish species characteristic of the by-catch of commercial fisheries around South Georgia (Croxall *et al.* 1988b).

For the subsequent population decrease, however, there has been no evidence, since the first assessment (Croxall 1979), of any causes other than incidental mortality associated with longline fishing operations (Croxall *et al.* 1990). The dependence of breeding (and immediately post-breeding) adults on foraging areas along the shelf slope of the South American continent and particularly in areas off southern Brazil and northern Argentina was entirely consistent with the greater impact on adults than juveniles and on females than males, because females spend more time foraging in more northerly areas (Weimerskirch and Jouventin 1987; Croxall and Prince 1990; Prince *et al.* 1992).

During the late 1980s and early 1990s longline fisheries, especially for Patagonian Toothfish *Dissostichus eleginoides*, have developed throughout the Patagonian Shelf, from Tierra del Fuego north to 30°S off Chile and 45°S off Argentina, around South Georgia (from 1988) and very recently (1994) around the Falkland Islands. These fisheries have much greater overlap with the distribution of male Wandering Albatrosses (see Prince *et al.* 1997, Fig. 2a) which may relate to the recent increase in mortality of adult males. At South Georgia adults of both sexes are particularly vulnerable during the brood-guard period when their foraging trips are mainly confined to the general area of the local longline fishery for *Dissostichus* (Croxall and Prince 1996). Because of their circum-Antarctic migrations (Prince *et al.* 1997), both sexes of adults are also substantially at risk from longline fishing for tuna in the Indian Ocean, especially around South Africa and Australia, although these fisheries are currently much smaller than in the 1970s and 1980s and may now use some measures to try to reduce mortality of albatrosses.

The pattern of juvenile mortality, with the progressive reduction in survival of about 10% since the early 1970s and the clear indication that birds older than age five years are primarily affected, is intriguing. It suggests that such birds are not greatly at risk (either through not frequenting the main areas of longline fishing or through being less able to compete with adults for access to baits) until the age at which they first return to Bird Island. They return to the breeding colonies in late summer, but progressively earlier and for longer as they get older (Pickering 1989) and presumably spend an increasing proportion of their year in the South Atlantic. This doubtless accounts for their being at greater risk of incidental mortality through interactions with the numerous longline fisheries in this region.

Grey-headed Albatross

We have fewer historical data on Grey-headed Albatrosses from South Georgia. However, all demographic data indicate that populations were stable in the 1960s and early 1970s. Juvenile survival has decreased markedly since the mid-1980s and adult survival may have decreased since 1977.

Grey-headed Albatrosses principally forage in association with deep water frontal systems (Prince *et al.* 1994, 1997; Rodhouse *et al.* 1996), which are prime areas for longline fishing for Southern Bluefin Tuna *Thunnus maccoyi* (Brothers 1991). During the breeding season adults from South Georgia do not visit the waters of the South American shelf; our very limited data on post-breeding movements indicate little overlap with longline fishery areas in the South Atlantic. However, the foraging range of adult birds from Bird Island encompasses that of the whole longline fishery for *Dissostichus* in the South Georgia area.

The more oceanic distribution of Grey-headed Albatrosses provides fewer data on their movements and interactions with fisheries. Nevertheless most recoveries are from longline fishing operations in the Australasian region (Prince *et al.* 1997) and, as reviewed in detail in Prince *et al.* (1994), their vulnerability to such mortality presumably accounts for much of the decrease in juvenile survival that has been particularly evident in the last decade. Why juveniles should be much more affected than adults is unclear, unless simply through their inexperience in avoiding capture on longlines. The apparent increased mortality of adult birds in some recent years may reflect the tendency of longline fishing to concentrate on higher latitude areas in the Australasian region, where Grey-headed Albatrosses are common in their non-breeding season, rather than on lower latitude areas in the Indian Ocean where Grey-headed Albatrosses are uncommon (Klaer and Polacheck, in press). The commencement of longline fishing for *Dissostichus* around South Georgia in 1988 may also have played a part. Although Grey-headed Albatrosses forage over deeper water than the areas favoured by longline fishing they have to cross the longline fishing zone on every foraging trip and may well interact with fishing vessels on these trips.

Black-browed Albatross

Demographic data for Black-browed Albatrosses for the 1970s from the only continuously studied colony suggest a stable or only slightly increasing population at this time. Nevertheless the Bird Island population as a whole may have continued to increase slowly up to about 1989. This presumably reflects the benefit from offal and discards associated with the trawl fisheries (*cf.* Thompson and Riddy 1995) which flourished around South Georgia in the late-1960s and 1970s and from similar operations in their wintering grounds in the Benguela Current area off South Africa, where South Georgia birds are abundant in association with trawlers (Prince *et al.* 1997).

Population data for the 1980s suggest that the survival of juveniles, but not adults, decreased substantially early in the decade. This difference is difficult to explain, but could reflect different movements and/or distributions of juveniles and adults in relation to the pelagic tuna fisheries off south-west and south-east South Africa. Different vulnerabilities of adults and juveniles to longline fishing may also be involved. The increased mortality of adults recently coincides with the start of longline fishing around South Georgia (in 1988), which overlaps very extensively with the foraging areas of adult Black-browed Albatrosses (Prince *et al.* 1997). This species is the main victim of longline vessels in the South Georgia area (Ashford *et al.* 1995). Birds of all ages are also substantially at risk from the very recently developed extensive on-shelf longline fisheries in South African coastal waters.

Black-browed Albatross populations at South Georgia also have suffered from substantially reduced breeding success (especially fledging success) this decade compared to last. The dependence of this species on krill during chick-rearing suggests that there have been more years of low krill availability and fewer years of high krill availability since 1987, compared with the 10 years previously. Whether this reflects any systematic change in the South Georgia marine environment and/or increased competition for food with the ever-expanding population of Antarctic Fur Seals *Arctocephalus gazella* (Boyd 1993), which also eat krill and have similar foraging areas, is unknown at present.

Comparisons with conspecific populations elsewhere

The causes of population change in albatrosses reflect the location of their interactions with their natural and human-influenced environment. Given that the three species studied at South Georgia have different patterns of distribution between breeding and non-breeding seasons and between adults and juveniles (and probably between different age classes of juveniles), it is not surprising that they should reveal distinct patterns of population change and different balances of the demographic factors that cause this. Equally, we should not necessarily expect populations of a species from different island groups to show similar patterns of change. This section briefly reviews data on trends in other populations of Wandering, Grey-headed and Black-browed Albatrosses.

Wandering Albatross

This discussion is confined to the taxon *Diomedea exulans exulans*, the large southern form of the Wandering Albatross, which breeds on six island groups in the Southern Ocean. Of major populations, that at the Marion and Prince Edward Islands (33% of the total) is least well documented but decreased substantially in the 1980s (Watkins 1987) and at Marion Island is currently decreasing by 0.7% per annum (J. Cooper in Gales 1993). At Macquarie Island the annual breeding population increased from about 10 pairs in 1956 to about 30 pairs in 1966 before decreasing at 8% per annum to about 10 pairs currently, primarily as a consequence of a reduction of adult survival rate from about 0.97

to 0.87, believed to be due to longline fishing for tuna (de la Mare and Kerry 1994). Macquarie Island is close to some important areas for longline fishing and, being a small population, may also be particularly vulnerable to stochastic environmental influences.

At the Indian Ocean sites, major decreases were recorded for Iles Crozet (population at Possession Island reduced by 54% to 1986) and Iles Kerguelen (population at Courbet Peninsula reduced by 60% to 1987: Weimerskirch and Jouventin 1987, 1997). At Possession Island the population was stable or slightly increasing between 1960 and 1969, decreased at 7% per annum from 1970 to 1976 and at 1.4% per annum from 1977 to 1985. Since 1986 it has increased at an average of 4% per annum. The main decrease reflected a reduction in adult survival of about 6.5% (together with reduced recruitment); between 1977 and 1985 adult survival rose by 4% and by a further 1.9% thereafter. During its main decline the foraging area of the population was coextensive with the main longline fishing area in the south Indian Ocean; subsequently the fishery diminished in intensity and was mainly located outside the albatrosses' core foraging area (Weimerskirch *et al.* 1997).

In broad terms, therefore, the period of main decrease of all populations of *D. e. exulans* coincided with the period of most intensive longline fishing for tuna within the foraging (and wintering) ranges of adult albatrosses; reduced adult survival was the main proximate cause of the population decrease. The magnitude of the decrease and the nature of any subsequent recovery mainly reflects the substantially reduced tuna fishing effort and changes in the location of this fishery, both combining to benefit populations of Wandering Albatrosses in the Indian Ocean. For South Georgia birds, the increase in other longline fisheries within their foraging range appears to have overshadowed any benefit from reduced fishing effort for tuna.

Grey-headed Albatross

There are few data from breeding locations other than South Georgia, which accounts for at least 56% of the world population. At Marion Island the population (7% of world total) decreased between the 1980s and 1992 at 0.7% per annum but is now increasing (J. Cooper in Gales 1997). At Campbell Island the population (7% of world total) has also decreased (P. Moore in Gales 1997). No data are available for the second largest population (11% of world total) at Diego Ramirez, but its proximity to the main longline fishing areas for Patagonian Toothfish off southern South America makes it likely that decreases have occurred. The drastic reduction in recruitment at South Georgia suggests that any population recovery will not occur for at least 10–20 years, unless there is an immediate increase in adult survival.

Black-browed Albatross

We deal here only with the taxon *D. m. melanophris*, which is the second most abundant of the world's albatrosses (after the Laysan Albatross *Diomedea (Phoebastria) immutabilis*), with a world population of around one million breeding individuals, 86% of which are at the Falkland Islands (Gales 1997). Black-browed Albatross populations around Cape Horn and at Iles Kerguelen decreased during the 1970s and 1980s (Schlatter 1984; Weimerskirch *et al.* 1989), the latter by 17% between 1978/79 and 1994/95 (Weimerskirch and Jouventin 1997). At Iles Kerguelen the main probable causes of mortality were collision with net-sonde cables during the breeding season (Bartle 1991; Weimerskirch and Jouventin 1997) and capture on longlines both locally and in Australasian waters (Weimerskirch and Jouventin 1997).

The situation at the Falkland Islands is complicated and few quantitative data exist. There are general indications that populations started to increase sometime between the 1960s and 1980s and those at several islands (totalling 50% of the Falklands population) are still increasing (Gales 1997). Population increases are readily explicable by the benefits to Black-browed Albatrosses, with their varied diet and strong association with vessels, of offal and discards from the major squid and fin-fishing operations around southern South America generally and the Patagonian Shelf in particular (Thompson 1992; Thompson and Riddy 1995). Decreases, however, may relate to the recent extensive development of longline fisheries in the region, given that Black-browed Albatrosses are known to be their main victim (Schiavini *et al.* 1997; Stagi *et al.* 1997) and that Black-browed Albatrosses from the Falkland Islands are probably mainly confined to the Patagonian Shelf year-round (Tickell 1967).

CONCLUSIONS

The plight of albatross populations at South Georgia is probably as serious as anywhere in the world. In terms of world populations of albatrosses, South Georgia is a key location, with 27% of *D. exulans exulans* (second most important island group, after the Prince Edward Islands), 56% of Grey-headed Albatrosses (most important site) and 12% of Black-browed Albatrosses *D. m. melanophris* (second most important island group, after the Falkland Islands). All three of these species are decreasing at South Georgia, Wandering Albatrosses at low rates over 20 years, Grey-headed Albatrosses at low (but possibly accelerating) rates fairly recently and Black-browed Albatrosses at higher rates, principally very recently. Decreases in both adult and juvenile survival are involved to different extents in different species. Arresting the decreases and restoring albatross populations at South Georgia will require elimination (or very substantial reduction) of incidental mortality of albatrosses in longline fisheries around South America, the Falkland Islands, South Georgia, South Africa and Australasia. This is a demanding task but ways of achieving it are already available (Alexander *et al.* 1997). The priority is to educate and persuade managers and practitioners of the fisheries concerned to employ effective mitigating measures, both for the sake of the birds themselves and also for the commercial benefits the fisheries will derive from minimising bait loss to birds. If this is not achieved then the future for albatrosses at South Georgia is not encouraging. They may only be saved by the collapse, through stock over-exploitation, of Southern Ocean longline fisheries!

ACKNOWLEDGEMENTS

We thank all those British Antarctic Survey staff who have assisted with the albatross population research at Bird Island and Ron Pinder and especially Lance Tickell for laying some of the foundations of this work. We are grateful to Dick Laws, Nigel Bonner and Andy Clarke for supporting the long-term programme, to Dirk Briggs for help with preparing the figures, to Julie Leland for typing the manuscript and to Knowles Kerry, Graham Robertson and Henri Weimerskirch for critically reviewing it.

REFERENCES

Alexander, K., Robertson, G. and Gales, R., 1997. The Incidental Mortality of Albatrosses in Longline Fisheries. Australian Antarctic Division, Tasmania. Pp. 44.

Ashford, J. R., Croxall, J. P., Rubilar, P. S. and Moreno, C. S., 1995. Seabird interactions with long-lining operations for *Dissostichus eleginoides* around South Georgia, April to May 1994. *CCAMLR Science* **2:** 111–22.

Bartle, J. A., 1991. Incidental capture of seabirds in the New Zealand subantarctic squid trawl fishery, 1990. *Bird Conserv. Internat.* **1:** 351–59.

Boyd, I. L., 1993. Pup production and distribution of breeding Antarctic Fur Seals *Arctocephalus gazella* at South Georgia. *Antarctic Sci.* **5:** 17–24.

Brothers, N. P., 1991. Albatross mortality and associated bait loss in the Japanese longline fishery in the Southern Ocean. *Biol. Conserv.* **55:** 255–68.

Brown, D., 1988. Components of lifetime reproductive success. Pp. 439–53 *in* Reproductive Success ed by T. H. Clutton-Brock. University of Chicago Press: Chicago and London.

Croxall, J. P., 1979. Distribution and population changes in the Wandering Albatross *Diomedea exulans* at South Georgia. *Ardea* **67:** 15–21.

Croxall, J. P. and Prince, P. A., 1990. Recoveries of Wandering Albatrosses *Diomedea exulans* ringed at South Georgia, 1956–86. *Ringing and Migration* **11:** 43–51.

Croxall, J. P. and Prince, P. A., 1994. Dead or alive, night or day; how do albatrosses catch squid? *Antarctic Sci.* **6:** 155–62.

Croxall, J. P., Prince, P. A., Trathan, P. N. and Wood, A. G., 1996b. Distribution and movements of South Georgia Black-browed Albatrosses: implications for Falkland Islands populations. *Warrah* **9:** 7–9.

Croxall, J. P., McCann, T. S., Prince, P. A. and Rothery, P., 1988a. Reproductive performance of seabirds and seals at South Georgia and Signey Island, South Orkney Islands 1976–1986: Implications for Southern Ocean monitoring studies. Pp. 261–85 *in* Antarctic Ocean and resources Variability ed by D. Sahrhage. Springer-Verlag: Berlin

Croxall, J. P., North, A. W. and Prince, P. A., 1988b. Fish prey of the Wandering Albatross *Diomedea exulans* at South Georgia. *Polar Biol.* **9:** 9–16.

Croxall, J. P., Prince, P. A. and Reid, K., 1997. Dietary segregation in South Georgia seabirds. *J. Zool., Lond.* **242:** 531–56.

Croxall, J. P., Rothery, P., Pickering, S. P. C. and Prince, P. A., 1990. Reproductive performance, recruitment and survival of Wandering Albatrosses *Diomedea exulans* at Bird Island, South Georgia. *J. Anim. Ecol.* **59:** 773–94.

de la Mare, W. K. and Kerry, K. R., 1994. Population dynamics of the Wandering Albatross *(Diomedea exulans)* on Macquarie island and the effects of mortality from longline fishing. *Polar Biol.* **14:** 231–41.

Gales, R., 1993. Co-operative Mechanisms for the Conservation of Albatross. Australian Nature Conservation Agency/Australian Antarctic Foundation: Hobart.

Gales, R., 1997. Albatross populations: status and threats. Pp. 20–45 *in*Albatross Biology and Conservation ed by G. Robertson and R. Gales. Surrey Beatty & Sons: Chipping Norton.

Hunter, I. Croxall, J. P. and Prince, P. A., 1982. The distribution and abundance of burrowing seabirds (Procellariformes) at Bird Island, South Georgia. I. Introduction and methods. *Brit. Antarctic Surv. Bull.* **56:** 49–67.

Klaer, N. and Polacheck, T., in press. Bycatch of albatrosses and other seabirds by Japanese longline fishing vessels in the Australian Fishing Zone from April 1992 to March 1995. *Emu.*

Kock, K.-H., 1992. Antarctic Fish and Fisheries. Cambridge University Press: Cambridge, England.

Pickering, S. P. C., 1989. Attendance patterns and behaviour in relation to experience and pair-bond formation in Wandering Albatrosses *Diomedea exulans* at South Georgia. *Ibis* **131:** 183–95.

Prince, P. A., 1980. The food and feeding ecology of Grey-headed Albatross *Diomedea chrysostoma* and Black-browed Albatross *D. melanophris. Ibis* **122:** 476–88.

Prince, P. A., 1985. Population and energetic aspects of the relationship between Black-browed and Grey-headed Albatrosses and the Southern Ocean marine environment. Pp. 473–77 *in* Antarctic Nutrient Cycles and Food Webs ed by W. R. Siegfried, P. R. Condy and R. M. Laws. Springer-Verlag: Berlin.

Prince, P. A., Croxall, J. P., Trathan, P. N. and Wood, A. G., 1997. The pelagic distribution of South Georgia albatrosses and their relationships with fisheries. Pp. 137–67 *in* Albatross Biology and Conservation ed by G. Robertson and R. Gales. Surrey Beatty & Sons: Chipping Norton.

Prince, P. A., Rothery, P., Croxall, J. P. and Wood, A. G., 1994. Population dynamics of Black-browed and Grey-headed Albatrosses *Diomedea melanophris* and *D. chrysostoma* at Bird Island, South Georgia. *Ibis* **136:** 50–71.

Prince, P. A., Wood, A. G., Barton, T. and Croxall, J. P., 1992. Satellite tracking of Wandering Albatrosses *(Diomedea exulans)* in the South Atlantic. *Antarctic Sci.* **4:** 31–36.

Reid, K., Croxall, J. P. and Prince, P. A., 1996. The fish diet of Black-browed and Grey-headed Albatrosses *Diomedea melanophris* and D. chrysostoma *at South Georgia. Polar Biol.* **16:** 469–77.

Rodhouse, P. G., Prince, P. A., Trathan, P. N., Hatfield, E. M. C., Watkins, J. L., Bone, D. G., Murphy, E. J. and White, M. G., 1996. Cephalopod community associated with mesoscale oceanographic features at the Antarctic Polar Front exploited by satellite tagged predators. *Mar. Ecol. Prog. Ser.* **136:** 37–50.

Rothery, P. and Prince, P. A., 1990. Survival and breeding frequency in albatrosses. *The Ring* **13:** 61–74.

Sagar, P. M., Molloy, J., Tennyson, A. J. D. and Butler, D., 1994. Numbers of Buller's Mollymawks breeding at the Snares Islands. *Notornis* **41:** 85–92.

Schiavini, A., Frere, E., Gandini, P., Garcia, N. and Crespo, E., 1997. Albatross–fisheries interactions in Patagonian shelf waters. Pp. 208–13 *in* Albatross Biology and Conservation ed by G. Robertson and R. Gales. Surrey Beatty & Sons: Chipping Norton.

Schlatter, R. P., 1984. The status and conservation of seabirds in Chile. Pp. 261–69 *in* Status and Conservation of the World's Seabirds ed by J. P. Croxall, P. G. H. Evans and R. W. Schreiber. ICBP Technical Publication No. 2. ICBP: Cambridge.

Stagi, A., Vaz-Ferreira, R., Marín, Y. and Joseph, L., 1997. The conservation of albatrosses in Uruguayan waters. Pp. 220–24 *in* Albatross Biology and Conservation ed by G. Robertson and R. Gales. Surrey Beatty & Sons: Chipping Norton.

Thompson, K. R., 1992. Quantitative analysis of the use of discards from squid trawlers by Black-browed Albatrosses *Diomedea melanophris* in the vicinity of the Falkland Islands. *Ibis* **134:** 11–12.

Thompson, K. R. and Riddy, M. D., 1995. Utilization of offal and discards from "finfish" trawlers around the Falkland Islands by the Black-browed Albatross *Diomedea melanophris. Ibis* **137:** 198–206.

Tickell, W. L. N., 1967. Movements of Black-browed and Grey-headed Albatrosses in the South Atlantic. *Emu* **66:** 357–67.

Tickell, W. L. N., 1968. The biology of the great albatrosses *Diomedea exulans* and *D. epomophora. Antarctic Res. Ser.* **12:** 1–55.

Watkins, B. P., 1987. Population sizes of King, Rockhopper and Macaroni Penguins and Wandering Albatrosses at the Prince Edward Islands and Gough Island, 1951–1986. *Sth Afr. J. Antarctic Res.* **17:** 166–62.

Weimerskirch, H. and Jouventin, P., 1987. Population dynamics of the Wandering Albatross *(Diomedea exulans)* of the Crozet Islands: causes and consequences of the population decline. *Oikos* **49:** 315–22.

Weimerskirch, H. and Jouventin, P., 1997. Changes in population sizes and demographic parameters of six albatross species breeding on the French sub-Antarctic islands. Pp. 84–91 *in* Albatross Biology and Conservation ed by G. Robertson and R. Gales. Surrey Beatty & Sons: Chipping Norton.

Weimerskirch, H., Brothers, N. and Jouventin, P., 1997. Population dynamics of Wandering Albatross *Diomedea exulans* and Amsterdam Albatross *D. amsterdamensis* in the Indian Ocean and their relationships with longline fisheries: conservation implications. *Biol. Conserv.* **79:** 257–70.

Weimerskirch, H., Zotier, R. and Jouventin, P., 1989. The avifauna of the Kerguelen Islands. *Emu* **89:** 15–29.

Wood, A. G. and Rothery, P., submitted. Database applications in long term studies of bird populations. *J. Avian Biol.*

CHAPTER SIX

Changes in population sizes and demographic parameters of six albatross species breeding on the French sub-Antarctic islands

HENRI WEIMERSKIRCH[1] and PIERRE JOUVENTIN[1]

Studies carried out during the past three decades on six albatross species on the French sub-Antarctic Islands indicate that at least five species have markedly decreased in number. Decreases in Wandering Albatrosses from Iles Crozet during the 1970s reflect high adult mortality and low juvenile recruitment rates; since the mid-1980s increases in the population reflect improvement in these parameters. Amsterdam Albatrosses at their sole breeding site on Ile Amsterdam are currently increasing in number. Sooty Albatrosses and Light-mantled Sooty Albatrosses have decreased since the early 1980s. Sooty Albatrosses are decreasing because of decreases in adult survival and juvenile recruitment rates. For Wandering Albatrosses and Sooty Albatrosses, breeding successes have increased throughout the study periods and are related statistically. The Black-browed Albatross population at Iles Kerguelen has been stable between 1979 and 1990, despite fluctuations in numbers, but since then the population has decreased due to low rates of juvenile recruitment. Yellow-nosed Albatrosses have decreased dramatically on Ile Amsterdam since 1981, because of a decrease in adult survival and low juvenile recruitment, low breeding success and low survival rates of immature birds. Although all decreases have been caused by low adult and immature survival, increased mortality occurred at different periods for each species. Subtropical waters appear more hazardous than sub-Antarctic and Antarctic waters for albatrosses when foraging due to the incidence of longline fishing vessels in sub-tropical waters. Although the Wandering Albatross populations of the Indian Ocean are presently recovering, a most worrying factor for the future of the small albatross species is the low rates of juvenile recruitment; these are likely to cause a dramatic decrease in the sizes of the respective populations in the future.

Key words: Albatrosses, French Sub-Antarctic Islands, Population Change, Demography, Adult Survival, Breeding Success, Recruitment.

INTRODUCTION

THE past decade has seen an increasing concern about the changes in the population size of large southern Procellariiformes, particularly albatrosses, and the possible effects of fisheries, especially longline fishing. In the early 1980s decreases in the size of Wandering Albatross *Diomedea exulans* populations were noted at South Georgia 55°S, 36°W (Croxall 1979) and at Iles Crozet 46°S, 52°E (Jouventin *et al.* 1984). The decrease of the Iles Crozet population was due to an increase in adult mortality, presumably as a result of fisheries, particularly longlining (Jouventin *et al.* 1984; Weimerskirch and Jouventin 1987). Data collected on Japanese tuna longliners in 1988 suggested that about 44 000 albatrosses were killed each year by this fishery alone, among them 9 600 Wandering Albatrosses (Brothers 1991). At South Georgia Wandering Albatross populations have also decreased due to high adult mortality, suspected to result, at least in part, from longline fishing off the coast of Brazil (Croxall *et al.* 1990; Croxall and Prince 1990; Croxall *et al.* 1997). Longline fishing may also explain the decrease since the 1960s in the Wandering Albatross population on Macquarie Island (54°S, 158°E) (de la Mare and Kerry 1994). Populations of Black-browed Albatrosses *Diomedea (Thalassarche) melanophrys* appear to have decreased at Iles Kerguelen (49°S, 69°E) (Weimerskirch *et al.* 1989) and possibly at Campbell Island (52°S, 169°E) (Moore and Moffat 1990). At South Georgia Black-browed Albatrosses and Grey-headed Albatrosses *D. (Thalassarche) chrysostoma* are decreasing, presumably as a result of mortality associated with longline fishing (Croxall *et al.* 1997).

With the recent expansion of longline fishing in Antarctic and sub-Antarctic waters (in particular for Patagonian Toothfish *Dissostichus eleginoides*) in addition to the existing Asian longline fisheries in subtropical waters, seabird populations are at risk more than ever in the

[1]CEBC–CNRS, 79360 Beauvoir sur Niort, France.

Pages 84–91 *in* ALBATROSS BIOLOGY AND CONSERVATION, ed by G. Robertson and R. Gales. Surrey Beatty & Sons, Chipping Norton. 1997.

Southern Ocean. Accurate information on albatross foraging zones, both during and outside the breeding seasons, are required to determine possible overlap between fisheries and seabirds, and to identify the possible threats to seabird populations. Information is also required on population sizes, population trends and demographics of albatrosses to determine which species are decreasing and to elucidate the proximate and ultimate causes of change. Information obtained on fishing vessels by specialist observers independent of the fishery is a vital complement to land-based studies, since it allows the development of measures to minimize mortality of albatrosses by fisheries (Alexander *et al.* 1997).

This chapter summarizes the information available on changes in numbers and demographic parameters of six species of albatrosses breeding on the French sub-Antarctic Islands to determine which demographic factors can be attributed to the observed changes in numbers, and to predict likely changes in future.

METHODS

For the six albatross species studied counts of eggs have been made since either the 1970s or 1980s. For Wandering Albatrosses the population at Possession Island, Iles Crozet, was counted, and at Iles Kerguelen the population at the Courbet Peninsula was counted; for Amsterdam Albatrosses *D. amsterdamensis* at Iles Amsterdam the entire population was counted; for Black-browed Albatrosses birds in the southern colony of Canon des Sourcils Noirs on the Jeanne d'Arc Peninsula, Iles Kerguelen were counted (see Weimerskirch *et al.* 1989); for Yellow-nosed Albatrosses *D. chlororhynchos (Thalassarche bassi)* birds in the southern colonies of Entrecasteaux, Iles Amsterdam 37°S, 77°E (see Roux *et al.* 1983) were counted; for Light-mantled Sooty Albatrosses *Phoebetria palpebrata* populations on the east coast of Possession Island were counted; and for Sooty Albatrosses *P. fusca* birds on the north coast of Possession Island were counted. Field and laboratory procedures to estimate breeding success, age at first breeding, adult and juvenile survival are described in Weimerskirch and Jouventin 1987; Weimerskirch *et al.* 1987; Weimerskirch *et al.* 1997).

RESULTS

Changes in the sizes of breeding populations

Wandering Albatrosses at Iles Crozet and Iles Kerguelen decreased steadily during the 1970s and early 1980s but have increased since 1986 (Fig. 1). Similarly, the tiny Amsterdam Albatross population has increased from five pairs breeding annually in 1984 to 17 pairs in 1996 (Fig. 1).

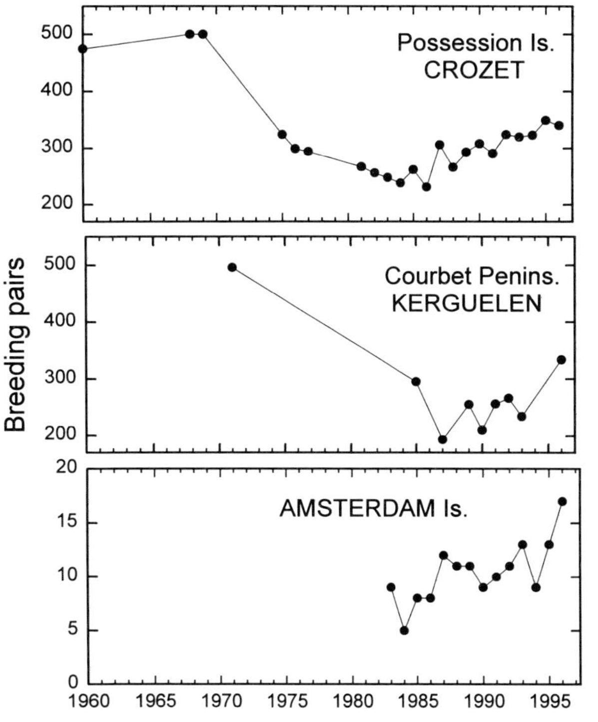

Fig. 1. Changes in the sizes of Wandering and Amsterdam Albatross populations at three breeding localities.

At Iles Kerguelen, Black-browed Albatrosses at the Jeanne d'Arc Peninsula have decreased slightly since the late 1970s (Fig. 2). The study population was stable during the period 1987–1994, varying only between 170 and 180 pairs. Yellow-nosed Albtrosses at Amsterdam Island have decreased substantially between 1981 and 1994 (Fig. 2). Sooty Albatrosses and Light-mantled Sooty Albatrosses at Possession Island decreased in number between 1980 and 1994, the former at a much higher rate than the latter (Fig. 2).

Demographic parameters

The decrease of the Iles Crozet Wandering Albatross population occurred in a period of low adult survival in the early 1970s. The stabilization and subsequent increase of the population was paralleled by an increase in adult survival rate and an increase in the survival rate of juvenile birds to the year of their first return to land. Breeding success was high over the entire period since the 1970s. For Sooty Albatrosses the survival of adults was high during the 1970s, dropped during the mid-1980s, and improved thereafter. The survival of juvenile birds to the year of their first return to land appears to have decreased, reaching its lowest level for the 1974/75 and 1975/76 cohort (i.e., those recruited in the mid-1980s). These changes have been paralleled by an increase in breeding success throughout the entire period. Adult survival of Yellow-nosed Albatrosses has decreased since the early 1980s, with low rates of juvenile

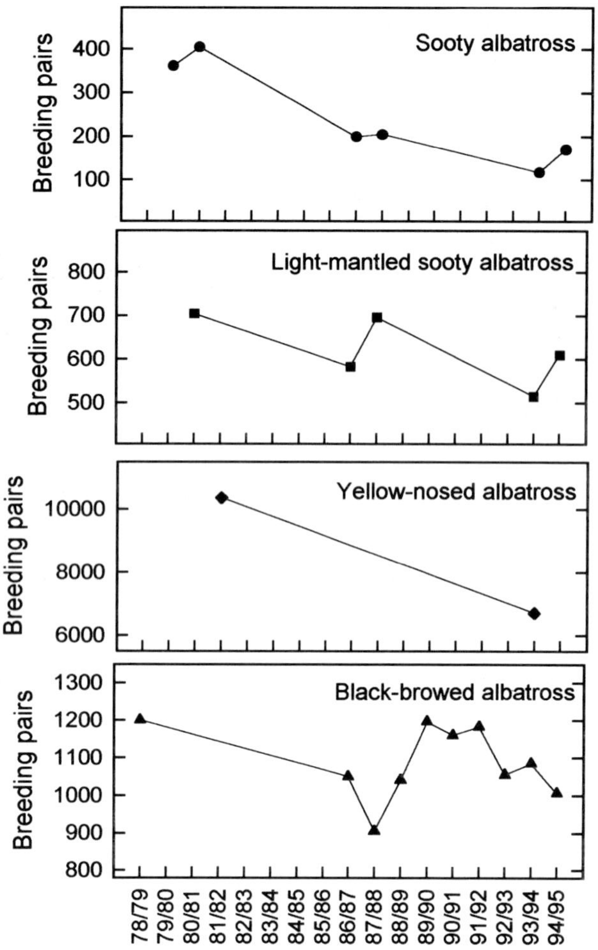

Fig. 2. Changes in the sizes of Sooty, Light-mantled Sooty, Yellow-nosed and Black-browed Albatross populations on French sub-Antarctic islands.

survival to their year of first return to land. Breeding success has been low, in general, and has varied greatly among years. Survival of adult Black-browed Albatrosses since the 1970s has been variable among years but appears stable overall. Breeding success has been high, on average, and has varied little among years. Survival of juvenile Black-browed Albatrosses has been low.

The breeding successes of Wandering and Sooty Albatrosses is correlated (r^2 = 0.303, P = 0.02, n = 17 breeding seasons), but the relationships between breeding successes of Sooty and Yellow-nosed Albatrosses (r^2 = 0.236, P = 0.09, n = 10), and Wandering and Yellow-nosed Albatrosses (r^2 = 0.282, P = 0.1, n = 9) is not statistically significant; the relationship for the other pairs of species is also not statistically significant (P > 0.2).

Average values for demographic parameters of the six albatross species are shown in Table 1, changes in adult survival and breeding success are shown in Figure 3 and juvenile survival is shown in Figure 4.

Table 1. Demographic parameters for six species of albatross breeding on the French sub-Antarctic Islands.

	Study period	Rate of change (%/year)	Adult survival (%/year) (±SD)	Juvenile survival (no. of cohorts) (±SD)	Minimum age at first breeding	Average (±SD) age at first breeding	Average breeding success % (no. of years)	Reference
Wandering Albatross	1966–1995	−1.4	93.1 ± 1.8	38.2 ± 13.9 (19)	7		68.5 ± 11.2 (26)	Weimerskirch *et al.* 1997
Amsterdam Albatross	1982–1995	+5	95.7 ± 1.8	70.4	7		71.6	Weimerskirch *et al.* 1997
Sooty Albatross	1966–1995	−6.9	89.8 ± 0.5	22.4 ± 11.7 (17)	9	12.7 ± 1.8 (55)	58.3 ± 18.2 (19)	this paper
Yellow-nosed Albatross	1978–1995	−3.6	85.7 ± 0.5	15.4 ± 7.3 (8)	5	8.7 ± 1.6 (67)	24.5 ± 18.3 (10)	this paper
Black-browed Albatross	1979–1995	−0.2	90.6 ± 0.5	13.7 ± 7.2 (11)	6	9.7 ± 1.8 (67)	62.7 ± 10.2 (16)	this paper
Light-mantled Sooty Albatross	1966–1995	−1.7	97.3		7	12	35.1 ± 18.6 (11)	H. W., unpubl.; Weimerskirch *et al.* 1987

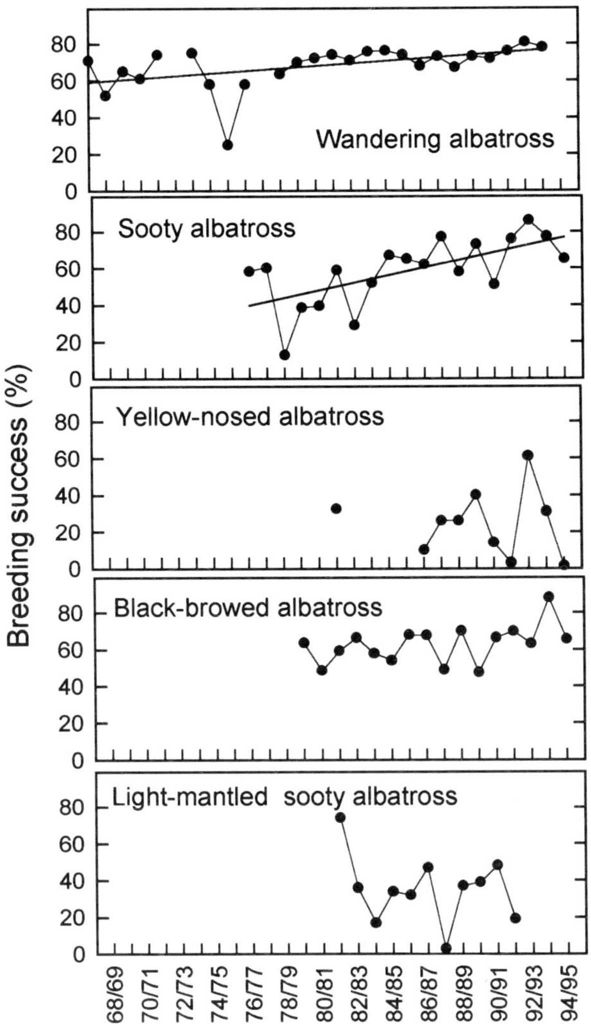

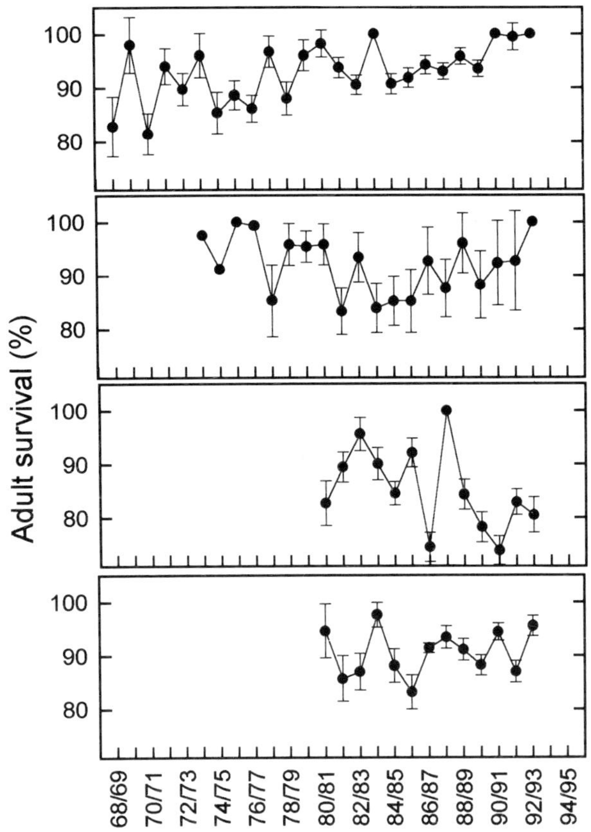

Fig. 3. Changes in adult survival (± 1 SD) and breeding success in five species of albatross. Wandering Albatrosses and Sooty Albatrosses show significant increases in breeding success over time (Wandering Albatross: $r^2 = 0.228$, $P = 0.02$, $y = 0.671x + 58.4$; Sooty Albatross: $r^2 = 0.411$, $P < 0.01$, $y = 2.07x + 18.9$).

Population trends

The changes in survival and breeding success that have occurred since monitoring programmes began are summarized in Table 2. Using demographic data for the most recent years and assuming adult survival is similar to that at present (Weimerskirch *et al.* 1997), a Leslie matrix model (Leslie 1964) predicts that in the next five years Black-browed, Yellow-nosed and Sooty Albatrosses will decrease in number, Sooty Albatrosses at much lower rates than observed since 1980. Yellow-nosed Albatrosses may decrease because of low adult and immature survival, and similarly for Black-browed Albatrosses, because of low immature survival rates.

DISCUSSION

Species analysis and evidence for albatross — fisheries interactions

Wandering Albatrosses

Previously, the decrease due to high mortality rate of adult birds, particularly females, was thought to result from mortality occurring in Asian longline fisheries (Weimerskirch and Jouventin 1987); this has recently been substantiated (Weimerskirch *et al.* 1997). Breeding Wandering Albatrosses (particularly females) from Iles Crozet forage in the sub-tropical and sub-Antarctic waters of the southwestern Indian Ocean (Weimerskirch *et al.* 1993) where longline fishing vessels have operated extensively (Ardill 1995). A significant proportion of breeding adults move from Iles Crozet to the western part of the Kerguelen shelf where longline vessels fishing for Patagonian toothfish operate (Weimerskirch 1997). In the non-breeding season immature and adult birds may forage in the same waters as those exploited by Asian longline vessels (Weimerskirch *et al.* 1997). The continuous increase in fishing effort by Japanese vessels in the Southern Ocean has paralleled the decrease of the Iles Crozet Wandering Albatross population (Weimerskirch *et al.* 1997). There is a poor relationship between fishing effort in the Iles Crozet sector and survival rates of adult birds, suggesting that mortality in the foraging area of breeding Iles Crozet birds is not the sole factor contributing to the low survival of adults. This is not surprising because adult birds only breed in alternate years and during the year off move in different waters (oceanic waters and waters of Australian continental shelf) to those exploited

Table 2. Present and projected changes in rates of change derived from adult and juvenile survival.

	Periods of comparison	Rate of change during last period (%/year)	Changes occurring between two periods			Projected rate of change from 1995–2000 (%/year)
			Adult survival	Immature survival	Breeding success	
Wandering Albatross	1966–1985/1986–1995	+4%	+6.4%	+40%	+8%	+4%
Amsterdam Albatross	/1986–1995	+4%	?	?	?	+4%
Black-browed Albatross	1979–1990/1991–1995	0	+ 3.0%	−7.4%	0	−2%
Yellow-nosed Albatross	1979–1985/1986–1995	−3.6%	−9.8%	−9.7%	0	−7%
Sooty Albatross	1979–1986/1987–1995	−6.9%	+6.4%	−1.1%	+23%	−3%
Light-mantled Sooty Albatross	1980–1995	−1.7%	?	?	?	?

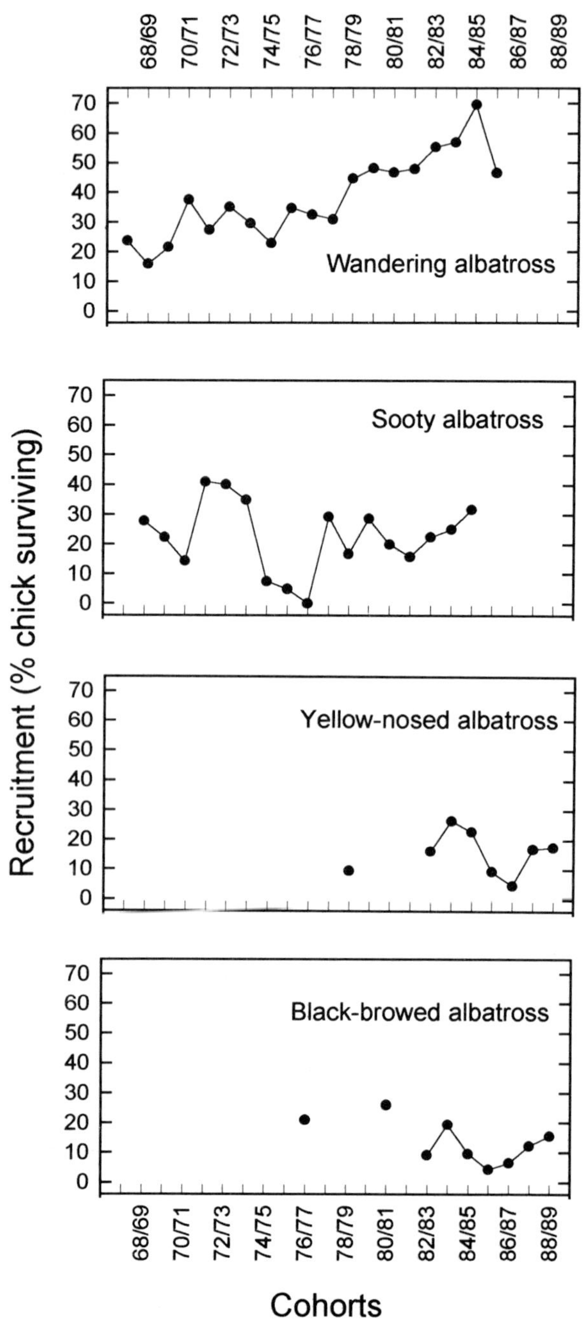

Fig. 4. Changes in the survival rates of cohorts of chicks from fledging to age of first return to land of four species of albatross.

by breeding birds (Weimerskirch *et al.* 1985; H. W., unpubl. data). Thus, both the breeding and the non-breeding adults may contact Asian longliners when foraging. Wandering Albatrosses form a major part of the seabird by-catch of the Japanese fishery (Brothers 1991), and several recoveries of banded Iles Crozet birds come from this source (Weimerskirch *et al.* 1985).

The increase of the population since 1986 has been paralleled by a decrease in fishing effort in the whole Southern Ocean, including near Iles Crozet (Weimerskirch *et al.* 1997). Since 1986 adult survival has improved and the increase in the population is the result of this improvement, although the improvement in juvenile survival and the decrease in the age at first breeding have also been a factor in the recovery. The recovery of the Iles Crozet Wandering Albatross population is very slow, as expected for a long lived seabird, and the population has not nearly recovered to its original numbers. The Iles Kerguelen population appears to have followed the same trends as the Iles Crozet population, but is likely to be more at risk from the Patagonian toothfish fishery than Iles Crozet birds, as this fishery operates closer to Iles Kerguelen, but less at risk from the Japanese tuna fishery, because Iles Kerguelen is located further from the main fishing grounds.

Amsterdam Albatrosses

This small population, discovered in 1981, has followed the same trend as that of the Wandering Albatross population at Iles Crozet. Its breeding ground is located in the fishing zone of Asian longliners, and it is potentially at high risk from this fishery. The population may have been depleted by this fishery in the 1970s when fishing effort was high close to Iles Amsterdam (Weimerskirch *et al.* 1997).

Sooty Albatrosses

The decrease of the population has mainly been caused by the low survival of adults and immature birds. The decrease is the greatest for the albatross species studied and is still occurring, but at a slower rate than during the early 1980s. Sooty Albatrosses forage in sub-tropical waters both during and outside the breeding season (Weimerskirch *et al.* 1986; Weimerskirch 1997;

Stahl *et al.*, in press) where they are likely to overlap with Asian longline vessels. However, no recaptures of banded birds have been obtained, and most recoveries from longliners where observers are present have been from Australasian waters where Sooty Albatrosses rarely occur. The lack of band recoveries in the southwestern Indian Ocean reflects the absence of observers on board fishing vessels rather than the lack of interaction between Sooty Albatrosses and longline vessels. The mortality rate of adults is related to the fishing effort in Sooty Albatross foraging waters (Weimerskirch *et al.*, unpubl. data.), suggesting longline fishing affects the survival prospects of this species of albatross.

Light-mantled Sooty Albatrosses

During the breeding season it is probably unlikely that Light-mantled Sooty Albatrosses contact oceanic longline fisheries as they forage in sub-Antarctic and Antarctic waters (Weimerskirch *et al.* 1986; Weimerskirch and Robertson 1994; Weimerskirch 1997). However, during the non-breeding season Light-mantled Sooty Albatrosses move northward from Iles Crozet (Stahl *et al.*, in press) where they constitute a significant bycatch in tuna longline fisheries (Brothers 1991). Assuming that the decreases of Sooty Albatrosses and Light-mantled Sooty Albatrosses is the result of fishing induced mortality, the difference in population change between the two species is probably best explained by their different foraging sectors, which result in Sooty Albatrosses being at risk year-round and Light-mantled Sooty Albatrosses being at risk only during the winter months.

Black-browed Albatrosses

Of the six albatross species studied Black-browed Albatrosses appear to be most at risk from fisheries. During the breeding season they forage over the Kerguelen shelf where fishing vessels trawl for Ice-fish *Champsocephalus gunnari* and longline for Patagonian toothfish. Birds from our study colony on southeastern Iles Kerguelen forage on the eastern part of the Kerguelen shelf (Weimerskirch 1997), which is not used by longline vessels. Birds from the northwestern colonies, however, forage over the western part of the shelf (Weimerskirch *et al.* 1988) and are likely to interact with longline vessels (see Cherel *et al.* 1995). In this fishery Black-browed Albatrosses and White-chinned Petrels *Procellaria aequinoctialis* form the major by-catch (Duhamel 1991; Cherel *et al.* 1995). Trawlers operate over the entire shelf, but during the 1980s they concentrated in the northeastern part of the shelf where Black-browed Albatrosses forage. In the past, mortality arising from net-sonde cables has probably been responsible for significant levels of mortality (Duhamel 1991; H. W., unpubl. data) and may have contributed to the relatively low survival of adults; this equipment is now forbidden in Iles Kerguelen waters. However, breeding success is strongly related to the take of fish by the trawl fishery (Weimerskirch and Jouventin 1988), suggesting a favourable affect on albatrosses. Thus, relationships between the fisheries and Black-browed Albatrosses during the breeding season are complex. Outside the breeding season Black-browed Albatrosses are caught in Australasian waters by the Asian longline tuna fishery. Survival of adult birds is lower than at South Georgia where the population is also decreasing (Prince *et al.* 1994; Croxall *et al.* 1997).

Yellow-nosed Albatrosses

The change in Yellow-nosed Albatross numbers between 1981 and 1994 may be explained by low adult survival in recent years. Surprisingly, only one recapture, in Australian fishing grounds, has been reported (H.W., unpubl. data). Yellow-nosed Albatrosses overwinter in Australasian waters (Weimerskirch *et al.* 1985) where large numbers of birds are caught in the Japanese longline fishery (Klaer and Polacheck 1995). Survival of immature birds appears to be very low, and breeding success approximates zero in some years; these factors result in low juvenile recruitment rates to the breeding population.

Predictions for future trends

Based on the above, in the absence of change to the demographic parameters recorded Wandering and Amsterdam Albtrosses will increase in number in the next several years and the Black-browed, Sooty and Light-mantled Sooty Albtrosses will decrease in number. The situation is worst for Yellow-nosed Albatrosses because of the effects on the population of low adult survival and low recruitment of juvenile birds; this could double the rate of decrease of the population if adult survival does not improve. Conversely, for Sooty Albatrosses the improvement in survival of adult and immature birds should reduce the rate of decrease in the future. For Black-browed Albatrosses the decrease which started three years ago should continue and may even accelerate in the future because of low survival of immature birds and an associated lower rate of recruitment of immature birds to the breeding population.

Comparison between species

In summary, all populations, with the possible exception of Amsterdam Albatrosses, have decreased over the past 30 years. Wandering Albatrosses decreased mainly between the early 1970s and the mid 1980s, and the other species since the early 1980s and possibly sooner. Increases in adult mortality rates have occurred at different periods for each. Survival of adult Black-browed Albatrosses have shown no clear trends in time. The increased mortality rate of immature birds is complex to explain as the effects of increase mortality during the first years of life is measured only at recruitment, i.e., 5–10 years later, according to the species (Table 1). Consequently it is impossible to determine precisely when each immature cohort was affected by increased mortality. These results suggest that the ultimate causes for population decreases could differ with species or, if similar, could have affected adults at different periods, or in different areas.

However, some similarities between species trends are evident. The long-term increase in breeding success and survival of adult Wandering and Sooty Albatrosses suggests improvement in foraging conditions for these two species and that factors affecting their breeding success may be similar. These two species forage in similar areas, mainly in sub-Antarctic and sub-tropical waters around Iles Crozet (Weimerskirch *et al.* 1986; Weimerskirch 1997). Yellow-nosed Albatrosses also forage in sub-tropical waters, north of Iles Amsterdam during breeding (H.W., unpubl. data). However, their survival rate has decreased over a different period (since the late 1980s) to the other two species. There could be two reasons for this. First, during the breeding season Yellow-nosed Albatrosses forage more in tropical waters where longline fishing for tuna is most common. Second, outside the breeding season Wandering and Sooty Albatrosses disperse in sub-tropical and sub-Antarctic waters of the Indian Ocean whereas Yellow-nosed Albatrosses move to waters of the continental shelf of Australia (Weimerskirch *et al.* 1985; Stahl *et al.*, in press) where they may interact with Australian and Asian longline vessels targeting Southern Bluefin Tuna.

There can be little doubt that fisheries, particularly longline fisheries targeting Southern Bluefin Tuna (Brothers 1991; Murray *et al.* 1993), are partly responsible for the observed decrease of albatross populations on the French sub-Antarctic islands. However, changes in food availability may have occurred concurrent with the development of fisheries, and this may have offset, to some degree, the effect of fisheries on albatross populations. For example, breeding success in Wandering and Sooty Albatrosses has improved concurrent with fisheries activity, and this suggests improvement in foraging success, and possibly in availability of squid, their principal food source (see Cherel and Klages 1997). However, low breeding success can be directly related to low adult survival (Weimerskirch and Jouventin 1987), because the disappearance of one member of a breeding pair results in breeding failure. This is possibly the reason for the parallel increase of breeding success and adult survival (of 6–8% each) in Wandering Albatrosses, but not in Sooty Albatrosses where breeding success has increased proportionally more than adult survival.

The results presented in this chapter indicate that although the demographic causes for the changes in numbers of albatross populations on the French sub-Antarctic islands have been identified, more data and analyses are needed to determine the exact causes for the changes in survival that have occurred, especially for the smaller albatross species. A major need is to obtain better information on the species, age, sex and provenance of albatrosses caught in the high sea longline fisheries to help integrate trends in data collection at colonies.

ACKNOWLEDGEMENTS

The long-term studies carried out on the French Antarctic Territories were funded by the Administration of Terres Australes et Antarctiques Françaises. We thank all the persons involved in the monitoring, banding and recoveries operations on Iles Amsterdam, Iles Kerguelen and Iles Crozet, and John Cooper and Graham Robertson for their helpful comments on the manuscript.

REFERENCES

Alexander, K., Robertson, G. and Gales, R., 1997. The incidental mortality of albatrosses in longline fisheries. Australian Antarctic Division, Tasmania. 44 pp.

Ardill, J. D., 1995. Atlas of industrial tuna fisheries in the Indian Ocean. Indo-Pacific tuna development and management programme/95/AT/3, 144pp.

Bartle, J. A., 1991. Incidental capture of seabirds in the New Zealand subantarctic squid trawl fishery, 1990. *Bird Conserv. Internat.* 351–59.

Brothers, N., 1991. Albatross mortality and associated bait loss in the Japanese longline fishery in the Southern Ocean. *Biol. Conserv.* **55:** 255–68.

Cherel, Y. and Klages, N., 1997. A review of the food of albatrosses. Pp. 113–36 *in* Albatross Biology and Conservation ed by G. Robertson and R. Gales. Surrey Beatty & Sons: Chipping Norton.

Cherel, Y., Weimerskirch, H. and Duhamel, G., 1995. Interactions between long-line vessels and seabirds in Kerguelen waters and a method to reduce seabird mortality. *Biol. Conserv.* **75:** 63–70.

Croxall, J. P., 1979. Distribution and population changes in the wandering albatross *Diomedea exulans* at South Georgia. *Ardea* **67:** 15–21.

Croxall, J. P., Rothery, P., Pickering, S. P. C. and Prince, P. A., 1990. Reproductive performance, recruitment and survival of wandering albatrosses *Diomedea exulans* at Bird Island, South Georgia. *J. Anim. Ecol.* **59:** 775–95.

Croxall J. P. and Prince, P. A., 1990. Recoveries of Wandering Albatrosses *Diomedea exulans* ringed at South Georgia 1958–1986. *Ringing and Migration* **11:** 43–51.

Croxall, J. P., Prince, P. A., Rothery, P. and Wood, A. G., 1997. Population changes in albatrosses at South Georgia. Pp. 69–83 *in* Albatross Biology and Conservation ed by G. Robertson and R. Gales. Surrey Beatty & Sons: Chipping Norton.

de la Mare, W. K. and Kerry, K. R., 1994. Population dynamics of the wandering albatross *Diomedea exulans* on Macquarie Island and the effects of mortality from longline fishing. *Polar Biol.* **14:** 231–41.

Duhamel, G., 1991.,Incidental mortality arising from fisheries activities around Kerguelen Islands (Division 58.5.1). Scientific Commission for the Conservation of Antarctic Marine Living Resources. SC-CCAMLR-X/ BG/14.

Jouventin, P., Stahl, J. C., Weimerskirch, H. and Mougin, J. L., 1984. The seabirds of the French subantarctic islands and Adélie Land, their status and conservation. Pp. 609–25 *in* Status and Conservation of the World's Seabirds ed by J. P. Croxall, P. G. H. Evans and R. W. Schreiber. I.C.B.P. Techn. Public. No. 2: Cambridge.

Klaer, N. and Polacheck, T., 1995. Japanese longline seabird bycatch in the Australian fishing zone April 1991–March 1994. CSIRO, Division of Fisheries Report, 95 pp.

Leslie, P. H., 1964. On the use of matrix in certain population mathematics. *Biometrika* **33:** 183–212.

Moore, P. J. and Moffat, R. D., 1990. Mollymawks on Campbell Island. Department of Conservation, Science and Research Division Internal Report No. 59. Wellington, New Zealand.

Murray, T. E., Bartle, J. A., Kalish, S. R. and Taylor, P. R., 1993. Incidental capture of seabirds by Japanese southern blue-fin longline vessels in New Zealand waters, 1988–1992. *Bird Conserv. Internat.* **3:** 181–210.

Prince, P. A., Rothery, P., Croxall, J. P. and Wood, A. G., 1994. Population dynamics of Black-browed and Grey-headed Albatrosses *Diomedea melanophris* and *D. chrysostoma* at Bird Island, South Georgia. *Ibis* **136:** 50–71.

Stahl, J. C., Bartle, J. A., Jouventin, P., Roux, J. P. and Weimerskirch H., in press. Atlas of seabird distribution in the south-west Indian ocean. *Mar. Ornithol.*

Weimerskirch, H., 1997. Foraging strategies of Indian Ocean albatrosses and their relationships with fisheries. Pp. 168–79 *in* Albatross Biology and Conservation ed by G. Robertson and R. Gales. Surrey Beatty & Sons: Chipping Norton.

Weimerskirch, H. and Jouventin, P., 1987. Population dynamics of the wandering albatross, *Diomedea exulans*, of the Crozet Islands: causes and consequences of the population decrease. *Oikos* **49:** 315–22.

Weimerskirch, H. and Jouventin, P., 1988. La démographie des oizeaux et mammiferes des Terres Australes et Antarctiques Françaises. *La Recherche dans les T.A.A.F., T.A.A.F. — M.R.* **Paris:** 95–108.

Weimerskirch, H. and Robertson, G., 1994. Satellite tracking of light-mantled sooty albatrosses. *Polar Biol.* **14:** 12–18.

Weimerskirch, H., Jouventin, P., Mougin, J. L., Stahl, J. C. and Van Beveren, M., 1985. Banding recoveries and the dispersion of seabirds breeding in the French austral and Antarctic territories. *Emu* **85:** 22–33.

Weimerskirch, H., Jouventin, P. and Stahl, J. C., 1986. Comparative ecology of the six albatross species breeding on the Crozet islands. *Ibis* **128:** 195–213.

Weimerskirch, H., Clobert, J. and Jouventin, P., 1987. Survival in five southern albatross species and its relationship with their life history. *J. Anim. Ecol.* **56:** 1043–56.

Weimerskirch, H., Bartle, J. A., Jouventin, P. and Stahl, J. C., 1988. Foraging ranges and partitioning of feeding zones in three species of southern albatrosses. *Condor* **90:** 214–19.

Weimerskirch, H., Zotier, R. and Jouventin, P., 1989. The avifauna of the Kerguelen Islands. *Emu* **89:** 15–29.

Weimerskirch, H., Salamolard, M., Sarrazin, F. and Jouventin, P., 1993. Foraging strategy of wandering albatrosses through the breeding season: a study using satellite telemetry. *Auk* **110:** 325–42.

Weimerskirch, H., Brothers, N. and Jouventin, P., 1997. Population dynamics of wandering albatrosses *Diomedea exulans* and Amsterdam albatross *Diomedea Amsterdamensis* in the Indian Ocean and their relationships with longline fisheries: conservation implications. *Biol. Conserv.* **79:** 257–70.

CHAPTER SEVEN

Breeding biology of Southern Buller's Albatrosses at The Snares, New Zealand

P. M. SAGAR[1,2] and JOHN WARHAM[1]

The breeding biology of Southern Buller's Albatrosses was studied at The Snares, New Zealand, from 1969 to 1977. Southern Buller's Albatrosses breed annually, with 90% of intact pairs returning to breed in consecutive years. Divorce was rare (3.8%) and about 60% of intact pairs used the same nest in consecutive years. Although there was considerable movement within the colony, no breeding birds moved between colonies. The first birds ashore were observed from mid-December, with males returning, on average, seven days earlier than females. The laying period lasted 43–62 days and the timing was consistent between years with peak laying in the third week of January. Incubation averaged 69 days, with both parents sharing incubation duties equally. Hatching occurred from mid-March to early May, and the chicks were brooded for 18–30 days, in shifts averaging 2.4 days. The average nestling period was 167 days, and chicks departed from late August to early October. In 1972, 57% of 89 eggs laid produced fledglings. Many aspects of the breeding biology of Southern Buller's Albatrosses are consistent with those of other small species of albatross. However, Southern Buller's Albatross is the only small albatross species to breed during winter, and the duration of the laying and nestling periods are longer than in other small albatrosses. Despite this extended commitment, Southern Buller's Albatrosses still manage to breed annually, which may reflect differences in the diet and at-sea distribution during the breeding and post-breeding periods.

Key words: Southern Buller's Albatross, The Snares, Breeding Biology.

INTRODUCTION

SOUTHERN Buller's Albatrosses *Diomedea bulleri bulleri (Thallassarche bulleri)* are endemic to New Zealand, breeding only at The Snares (48°02′S, 166°36′E) and Solander Islands (46°35′S, 166°36′E) (Turbott 1990). With a total population of less than 15 000 breeding pairs (Cooper *et al.* 1986; Sagar *et al.* 1994), Southern Buller's Albatrosses are rare compared with other Southern Hemisphere albatrosses. Despite the proximity of The Snares and Solander Islands to New Zealand, this species has been little studied, and many details of its biology are still not known. Most of what is known comes from studies of the pre-egg and incubation stages (Richdale 1949a, 1949b), survival, pair bond retention and nest-site tenacity (Richdale and Warham 1973), population size (Warham and Bennington 1983; Sagar *et al.* 1994), foods (West and Imber 1986) and vocalizations (Warham and Fitzsimons 1987; Warham 1997). However, detailed studies of such basic aspects of breeding as duration and timing of the incubation and nestling periods, length of incubation shifts, and breeding frequency have not been reported. Timing of the egg and nestling stages and duration of incubation shifts are linked to foraging ecology because maximum foraging range is limited by the time adults are able to spend away from the nest. Southern Buller's Albatrosses are one of the most frequently killed albatross species in fisheries operating in New Zealand waters (Bartle 1991; Murray *et al.* 1993). Consequently, through its links to foraging ecology and population dynamics, knowledge of the breeding biology has important implications for conservation of this albatross species.

In this chapter we provide detailed information on aspects of Southern Buller's Albatross breeding biology at The Snares, and compare the general characteristics of this population with those of other populations of *Diomedea*. If the general characteristics of the breeding biology of Southern Buller's Albatrosses are shown to be similar to those of other *Diomedea*, then the results of extensive studies of these other populations may be extrapolated to determine interim conservation measures for Southern Buller's Albatrosses.

[1]Department of Zoology, University of Canterbury, Private Bag, Christchurch, New Zealand.
[2]Present address: National Institute of Water and Atmospheric Research, PO Box 8602, Christchurch, New Zealand.

Pages 92–98 *in* ALBATROSS BIOLOGY AND CONSERVATION, ed by G. Robertson and R. Gales. Surrey Beatty & Sons, Chipping Norton. 1997.

STUDY AREA AND METHODS

Field work was conducted at The Snares during the 1968/69–1971/72, 1974–75 and 1976–77 summers (hereafter referred to as 1969, 1970, etc.). In all summers except 1972, observations were made from pre-laying to the late incubation stage; during 1972 observations extended throughout the entire season. During January 1969, a study colony was established at Mollymawk Bay on the north-east coast of The Snares. The site was uniform with respect to nest building material, ground slope, forest cover and aspect, being under the edge of the *Olearia lyalli* forest (6–8 m canopy) on terraces at the top of 50 m cliffs. Within this site all occupied nest mounds (n = 110) were marked with individually labelled wooden stakes, and their locations plotted on a map.

Study nests were visited daily from mid-December 1968 to late February 1969, from early December 1969 to mid-February 1970, and five visits were made to the study colony at six to 10 day intervals from mid-January to mid-February 1971. Observations were made throughout 1971–1973. During this period, daily visits were made during the laying (early January to mid-February 1972), hatching (mid-March to mid-April 1972) and fledging (late August to early October 1972) periods; inspections were made every three to five days between these periods. From early December 1974 to late February 1975, and early December 1976 to early March 1977, visits were made at three-day intervals.

In 1969, all unbanded breeding birds were fitted with numbered metal bands. Birds were sexed by observation of copulation, pre-laying attendance pattern, which bird undertook the first long incubation shift, or a combination of these. Band numbers of both parents attending a nest were recorded during incubation. Laying dates and attendance patterns were determined by recording the contents and identity of the bird(s) present at each nest respectively during daily visits in 1969, 1970 and 1972. Egg measurements were obtained using a spring balance and vernier callipers (measuring to 0.1 mm).

RESULTS

Return to the colony

The earliest a banded bird returned to the colony was 6 December (in 1971). Birds were observed consistently from mid-December, after which numbers increased rapidly. Males returned on average seven days before their partners. Of 51 pairs observed during 1968–69, the mean date of return of males was 24 December (range: 16 December–13 January) and that of females was 1 January (range: 16 December–19 January). During the 1970 season, males returned first at 28 of 31 nests; the mean difference between the arrival of males and females was seven days (range: 0–25 days, n = 31). The mean interval between the first return and laying date was 21 days (range: 0–35 days, n = 31) for males and 10 days (range: 0–22 days, n = 31) for females. In 1969, both parents were present together at the nest at three peak times before laying — 12–14 days, 7–9 days and 1–2 days (Fig. 1). Of 63 nests where banded birds were identified the day before laying, 34 (54.0%) had both parents present, the male alone was present at 16 (25.4%), and the female alone was present at a further 13 (20.6%).

Breeding frequency

Over all four years, 161 of 178 (90.4%) pairs which remained intact for at least two consecutive years bred annually (i.e., laid an egg in both years). Data collected over four consecutive summers (1969–1972) showed that the breeding frequency of intact pairs between consecutive years was similar (range: 89.6–90.8%, Table 1. There were insufficient data to compare the effect of mate change on breeding frequency between consecutive years. However, over the four years, five of nine (55.6%) females known to change mate between consecutive years laid in both years.

Mate retention

Between 1969 and 1972, there were 178 pair-years where the same partners were found breeding together. Seven instances of divorce, where both previous partners were known to be alive, were recorded in the same period — a rate of 3.8%. There were 16 instances when only the male was recorded again, 15 instances of only the female being recorded, and six instances of neither partner being seen again.

Nest site retention

Where the same partners were found breeding together in successive years, 99 of 148 pairs (66.9%, range: 62.2–71.2%, Table 1) also retained the same nest. A similar result was calculated for nine males that changed partners between consecutive years, when six (66.7%) also retained the same nest. However, of nine females that changed partner between two consecutive years, only two (22.2%) kept the same nest. Where the partnership remained the same between years, but the nest site changed (49/148) the mean distance moved was 3.3 ± 2.3 m. [SD] (range: 0.6–10.2 m, n = 49). Though the sample sizes are small, the distances moved by males and females with new partners were similar (new female: $\bar{x}$ = 5.5 ± 4.5 m, range: 1.0–10.0 m, n = 3; new male: $\bar{x}$ = 6.5 ± 6.2 m, range: 1.2–15.6 m, n = 7). The causes of some changes in nest sites included inaccessibility due to fallen

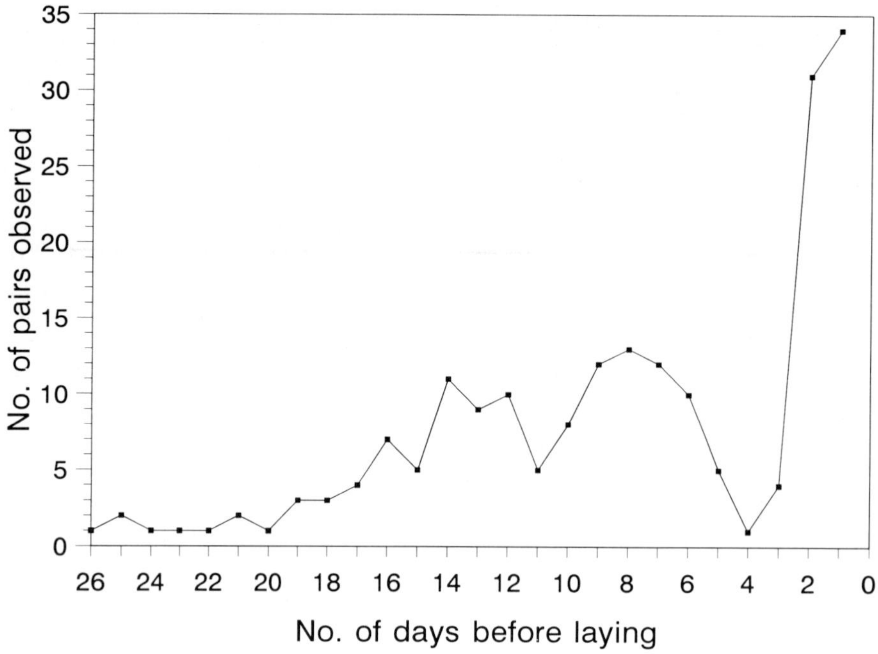

Fig. 1. Number of mated pairs observed together, before laying, at 34 nests of Southern Buller's Albatrosses, The Snares, January–February 1969.

Table 1. Breeding frequency of intact pairs and nest retention of Southern Buller's Albatrosses at The Snares, 1969–1972.

	Number (%) intact pairs laying		Intact pair			
			Same nest		Nest change	
Season	Both years	First year only	N	%	N	%
1969 and 1970	43 (89.6)	5 (10.4)	28	62.2	17	37.8
1970 and 1971	59 (90.8)	6 (9.2)	37	71.2	15	28.8
1971 and 1972	59 (90.8)	6 (9.2)	34	66.7	17	33.3
Overall			99	66.9	49	33.1

branches, and subsidence due to the burrowing activities of Sooty Shearwaters *Puffinus griseus.*

Egg laying, incubation and hatching

Mean laying date was during the third week of January and varied little between years. However, onset of laying varied by up to 12 days, and the appearance of the last egg by up to 19 days between years (Table 2). The laying period extended over 43–62 days (Table 2), and 80% of eggs were laid within the first 35 days (Fig. 2). Egg dimensions and masses at laying were similar in both 1969 and 1970 (see below). The incubation period averaged 69.1 days (± 1.3 days, range: 67–75 days, n = 61). Both adults shared incubation and after the relatively short first (laying) shift, the duration of shifts was similar for both parents, until at least 37 days of incubation (Table 3). Chicks took two to five days to emerge from the egg, but most hatched in three or four days. The mean hatching date in 1972 was 2 April (± 8.9 days; range: 14 March–30 April to 8 May; n = 73). The male was present in 31 of 57 (54%) cases where the duty parent was identified at hatching.

Chick rearing period

The mean duration of the fledging period was 167 ± 9.7 days (range: 147–189 days; n = 52). In 1972, the mean fledging date was 13 September (range: 22 August–2 October; n = 52). Males and females shared brooding in shifts of 1 to 12 days, the average length of 149 shifts being 2.4 days (± 2.1 days; n = 149). Shifts of males ($\bar{x}$ = 2.7 ± 2.3 days; range: 1–12 days; n = 75) were similar to those of females ($\bar{x}$ = 2.1 ± 1.9; range: 1–11 days; n = 74; t = 1.01, P > 0.2). The chicks were brooded for 15–30 days, usually 23–25 days. No chick was seen fledging in 1972. It is probable that they fledged at dawn or during the night, and immediately left the vicinity of the island. Some adults visited their chicks up to the day before the chick fledged. On four occasions a parent was found on the nest after the chick had fledged.

Breeding success and causes of losses

In 1972, 69 of 89 eggs (77.5%) laid in the study colony hatched and 51 (73.9%) of these chicks fledged, an overall breeding success of 57.3%. Of the 20 eggs that did not hatch, seven

Table 2. Egg laying periods for Southern Buller's Albatrosses at The Snares.

Year	No. eggs	Mean (s.d.)	Range	Approx. length of laying period (days)
1969	98	21 Jan. (9.0 d)	31 Dec. to 20 Feb.	51
1970	50	22 Jan. (6.6 d)	7 Jan. to 10 Feb.	43
1972	89	23 Jan. (9.2 d)	2 Jan. to 22–29 Feb.	51–58
1977	85	20–26 Jan.	26 Dec.–1 Jan. to 23–26 Feb.	51–62

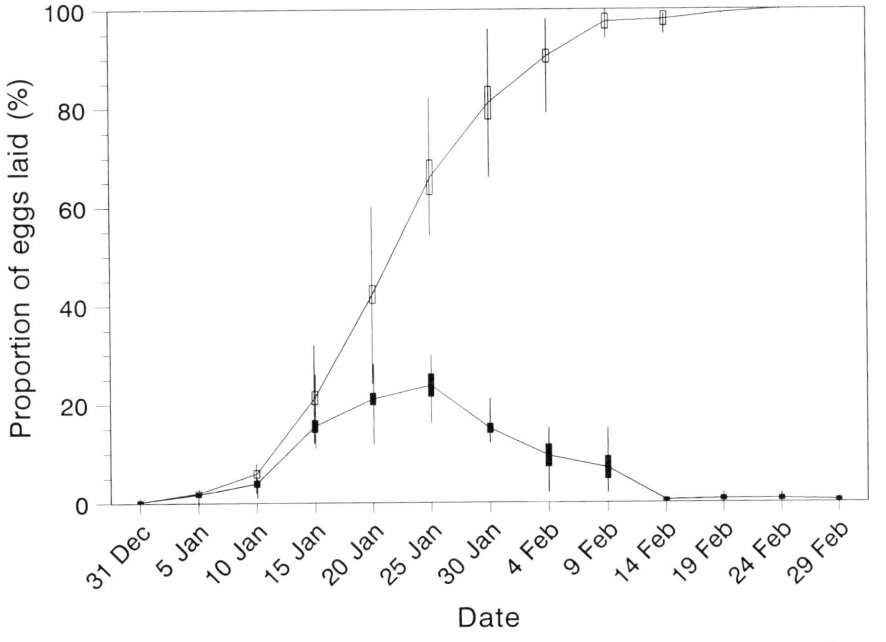

Fig. 2. Cumulative frequency (upper plot) and progression of laying (lower plot) of Southern Buller's Albatrosses at The Snares — 1969, 1970, 1972 and 1977. Mean, s.e. and range of values at five-day intervals are shown for the four years of observations.

Table 3. Duration (days) of completed incubation shifts in Southern Buller's Albatrosses at The Snares, 1948* and 1969–1972. *From Richdale (1994b).

		1948*				1969–1972			
Shift	Sex	Mean	S.D.	Range	n	Mean	S.D.	Range	n
1	F	6.0	4.7	0–24	60	4.5	4.1	1–19	86
2	M	11.0	4.1	6–21	23	10.7	4.0	3–22	72
3	F	10.8	5.0	5–20	13	11.3	4.4	2–21	49
4	M	9.3	0.5	9–10	3	10.9	2.8	5–18	22
5	F					7.1	3.8	1–12	7
6	M					10.0	1.0	9–11	2

were broken or rolled out of the nest within eight days of laying. Losses of the remaining eggs were spread evenly throughout the incubation period, although one infertile egg was incubated for 83 days before being broken. Northern Giant Petrels *Macronectes halli* were probably responsible for nine (50%) of the chick losses; unknown causes (five) and abandonment (four) accounted for the remainder. Although giant petrels were not seen killing chicks, their presence in the study colony in August and September coincided with the incidence of recently killed chicks.

DISCUSSION

Comparison with previous studies of Southern Buller's Albatrosses

Our information on pre-laying attendance at the nest, laying dates, duration of the laying period, and length of incubation shifts is consistent with Richdale (1949a, b), who found that laying extended over "fully 7 weeks" — from 16 January to early March, with peak laying between 22 January and 14 February. In our study, laying extended over 43–62 days, with peak laying also during the period late January to early February. Egg sizes in 1948, 1969 and 1970 were similar. In 1948, the mean size of the first 100 eggs laid was 102.2 ± 3.4 mm × 66.0 ± 1.8 mm (Richdale 1949b), compared to 103.1 ± 3.6 mm × 64.7 ± 1.8 mm in 1969 and 1970 (this study). The duration of the first four completed incubation shifts also varied little between years (Table 3).

Comparison with other albatross species

The timing of the breeding cycle of Southern Buller's Albatrosses, with chicks hatching during March and April and fledging during September, means that chick rearing is entirely within the austral winter. In this respect, Southern Buller's Albatrosses differ from all other small albatross species, where fledglings depart in the austral autumn (March–May). Breeding of the closely-related Northern Buller's Albatrosses *D. (Thallassarche) platei* is 2–3 months earlier than that of Southern Buller's Albatrosses, and fledglings of the northern taxon depart in June–July (Marchant and Higgins 1990).

Many aspects of the breeding biology of Southern Buller's Albatrosses are typical of the *Diomedea.* For example, they exhibit strong pair bond and nest site retention, and divorce is rare. The 3.8% divorce rate in Southern Buller's Albatrosses at The Snares is comparable to the 0% in Waved Albatrosses *D. irrorata* (Harris 1973) and the 2.1% for Laysan Albatrosses *D. (Phoebastria) immutabilis* (Rice and Kenyon 1962). A low divorce rate is also consistent with the long-term pair bonds (up to 23 years) reported for Southern Buller's Albatrosses at The Snares (Richdale and Warham 1973). Nest site tenacity is linked to mate retention (Warham 1990) and is well documented for small species of the albatrosses (e.g., Yellow-nosed Albatross *D. (Thalassarche) chlororhynchos* — Jouventin *et al.* 1983; Black-browed *D. (Thalassarche) melanophyrys* and Grey-headed *D. (Thalassarche) chrysostoma* Albatrosses — Prince *et al.* 1994) — no established breeding bird has been reported breeding at more than one colony.

Although none of 318 Southern Buller's Albatrosses banded as breeders at The Snares was found over a 23 year period at other than natal colonies (Richdale and Warham 1973), there may be considerable movement within colonies. Males usually return first and spend more time ashore than their partner (Warham 1990). Differences of 10 and 11 days are common in Southern Royal *D. epomophora* and Wandering *D. exulans* (Richdale 1950; Tickell 1968) Albatrosses, and eight and six days in Grey-headed and Black-browed Albatrosses respectively (Tickell and Pinder 1975), similar to the eight days difference in Southern Buller's Albatrosses. Similarly, the 20-day mean interval between male arrival and laying for Black-browed Albatrosses (Croxall 1984), 24 days for Southern Buller's Albatrosses (this study), and 25 days for Grey-headed Albatrosses (Croxall 1984) indicates similarity among the albatrosses. Despite size differences between species the mean incubation and brooding periods also vary little among the albatrosses. Mean incubation periods range from 68.1 to 72.4 days (Warham 1990) and brooding periods from 21.6 to 22.6 days (Tickell and Pinder 1975), so with a mean incubation period of 69.1 days and brooding period of 23–25 days, the values for Southern Buller's Albatrosses are of similar duration.

It is well established in albatrosses that the time either parent spends at the nest varies with the stage of the breeding cycle. For example, in six species of albatross breeding in the French sub-Antarctic islands brood shift lengths are shorter than those during incubation (Weimerskirch *et al.* 1986). We found that, with the mean duration of incubation shifts ranging from 7.1 to 11.3 days (after the first short shift) and brooding shifts averaging 2.4 days, Southern Buller's Albatrosses follow the typical pattern. However, the duration of the laying (43–62 days) and chick rearing periods of Southern Buller's Albatrosses is atypical, the laying period being 2–4 times longer than that for Sooty Albatrosses (12 days, Jouventin and Weimerskirch 1984), Grey-headed Albatrosses (17–19 days, Tickell and Pinder 1975; Jouventin *et al.* 1983), Black-browed Albatrosses (23 days, Tickell and Pinder 1975), and Light-mantled Sooty Albatrosses *P. palpebrata* (24 days, Mougin 1970). Even the laying periods of Northern Royal Albatrosses *D. epomophora sanfordi (D. sandfordi)* (36 days, Richdale 1950) and Wandering Albatrosses (38 days, Tickell 1968) are shorter: only that of the Waved Albatross *D. irrorata*, at 77 days (Harris 1973), is longer. In some petrel species that undertake trans-equatorial migrations a compressed laying period is necessary, but this is uncommon in migratory Procellariiformes (Warham 1990). In the albatrosses, Sooty Albatrosses, which have the shortest laying period, disperse widely after breeding (Marchant and Higgins 1990) and Waved Albatrosses, with the longest laying period, make only short post-nuptial flights (Warham 1990). The post-breeding movements of Southern Buller's Albatrosses are not well known, though they may move to waters off the western-coast of South America (Marchant and Higgins 1990). Clearly, without information on the body masses of breeding females when they return to the colonies before laying, the foods and foraging areas of females during the post-copulation exodus, and the timing and extent of post-breeding movements, it is difficult to assess the reasons for the very long period over which the eggs are laid.

The chick rearing period of Southern Buller's Albatrosses, at 167 days, is also longest among albatrosses of a comparable mass. The rearing periods of Sooty Albatrosses and Laysan Albatrosses at 164 days (Weimerskirch *et al.* 1986) and 160 days (Rice and Kenyon 1962), respectively, are similar to that of Southern Buller's Albatrosses. However, the rearing periods are only 115 days in Yellow-nosed Albatrosses (Weimerskirch *et al.* 1986) and 141 days in both Grey-headed Albatrosses (Tickell and Pinder 1975) and Light-mantled Sooty Albatrosses (Weimerskirch *et al.* 1986). A possible explanation for the long rearing period of this albatross species may be its dependence on squid as a source of food (West and Imber 1986). Squid were present in 78.4% of 14 food samples collected from adult and nestling Southern Buller's Albatrosses during three seasons at The Snares (West and Imber 1986). Squid are low energy foods and their dominance in the diet of Grey-headed Albatrosses compared with Black-browed Albatrosses (which feed on energy-rich Antarctic Krill *Euphausia superba*) at South Georgia may be implicated in the shorter nestling period of the latter species (Prince and Ricketts 1981; Ricketts and Prince 1981).

In Southern Buller's Albatrosses at The Snares, the commitment to the breeding colonies from the time of mean arrival of males (24 December) to mean departure of chicks (13 September) is 264 days. This compares with 205 days and 238 days in Black-browed and Grey-headed Albatrosses, respectively, breeding at South Georgia (Prince *et al.* 1994). Among other small species of albatross the commitment to the breeding colonies is about 230 days in Light-mantled Sooty Albatrosses (Thomas *et al.* 1983) and 250 days in Sooty Albatrosses (Weimerskirch 1982). Consequently, the annual breeding of Southern Buller's Albatrosses is surprising. Black-browed and Grey-headed Albatrosses are of similar mass and have similar timing and duration of breeding, yet breed annually and biennially, respectively (Croxall 1991). In Grey-headed Albatrosses an endocrine-mediated mechanism prevents females, but not males, from breeding in the season following successful chick rearing (Hector *et al.* 1986). Biennial breeding in males appears to be maintained by an inability to regain body mass and the advantages of breeding with the same partner (Croxall 1991). Our study indicates that in Southern Buller's Albatrosses no endocrine mechanism prevents annual breeding and that both males and females are able to gain sufficient condition between breeding seasons to breed annually.

As mentioned above a nutritionally poor diet probably accounts for the longer chick development period in Grey-headed Albatrosses (Ricketts and Prince 1981). Food may be one of the factors contributing to biennial breeding in Grey-headed Albatrosses, with adults not being able to attain breeding condition sufficiently fast enough during their post-nuptial time at sea in order to undertake annual breeding (Warham 1990). Apparently, with Southern Buller's Albatrosses, either loss of condition during the breeding season is proportionally less, or post-nuptial movements enable the birds to feed on higher energy foods or to obtain food more quickly than during the chick rearing period. In these cases they would attain condition in time to breed again the following year. There is still much to learn about the foods, feeding, and foraging areas of Southern Buller's Albatrosses.

ACKNOWLEDGEMENTS

The University of Canterbury Snares Islands Expeditions received the financial help of the Nuffield Foundation, the New Zealand University Grants Committee and the University of Canterbury. Transport was provided by the Royal New Zealand and United States' Navies, by the Department of Conservation and by a number of fishermen from Bluff and Stewart Island. For help in collecting field data we are grateful to C. H. Hay and S. L. Bennington and particularly to Carol Horning for her meticulous record keeping over a complete breeding season. Data analysis and writing of the manuscript by P. M. Sagar were funded by the Foundation for Research, Science and Technology (NZ) under contract CO1507.

REFERENCES

Bartle, J. A., 1991. Incidental capture of seabirds in the New Zealand subantarctic squid trawl fishery 1990. *Bird Conserv. Internat.* **1:** 351–59.

Cooper, W. J., Miskelly, C. M., Morrison, K. and Peacock, R. J., 1986. Birds of the Solander Islands. *Notornis* **33:** 77–89.

Croxall, J. P., 1984. Seabirds. Pp. 533–616 *in* Antarctic Ecology ed by R. M. Laws.

Croxall, J. P., 1991. Reproductive constraints on albatrosses. *Acta Congr. Int. Orn.* **XX:** 281–302.

Harris, M. P., 1973. The biology of the Waved Albatross *Diomedea irrorata* of Hood Island, Galapagos. *Ibis* **115:** 483–510.

Hector, J. A. L., Follett, B. K. and Prince, P. A., 1986. Reproductive endocrinology of the Black-browed Albatross *Diomedea melanophris* and the Grey-headed Albatross *D. chrysostoma*. *J. Zool. Lond.* **208:** 237–53.

Jouventin, P., Roux, J.-P., Stahl, J.-C. and Weimerskirch, H., 1983. Biologie et frequence de production chez lí Albatross a Bec Jaune *Diomedea chlororhynchos*. *Le Gerfaut* **73:** 161–71.

Jouventin, P. and Weimerskirch, H., 1984. Líalbatros fuligineux *Phoebetria fusca* exemple de strategie díadaptation extreme a la vie pelagique. *Rev. Ecol. Terre Vie* **39:** 401–29.

Marchant, S. and Higgins, P. J., 1990. Handbook of Australian, New Zealand and Antarctic Birds. Vol. 1. Oxford University Press: Melbourne.

Murray, T. E., Bartle, J. A., Kalish, S. R. and Taylor, P. R., 1993. Incidental capture of seabirds by Japanese Southern Bluefin tuna longline vessels in New Zealand waters, 1988–1992. *Bird Conserv. Internat.* **3:** 181–210.

Mougin, J.-L., 1970. Les albatros fuligineux *Phoebetria palpebrata* et *P. fusca* de líle de la Possession (Archipel Crozet). *Oizeau Rev. Fr. Orn. (No. Spec.)* **40:** 37–61.

Prince, P. A., 1985. Population and energetic aspects of the relationship between Black-browed and Grey-headed Albatrosses and the Southern Ocean marine environment. Pp. 473–77 *in* Antarctic Nutrient Cycles and Food Webs ed by W. R. Siegfried, P. R. Condy and R. M. Laws. Springer: Berlin.

Prince, P. A. and Ricketts, C., 1981. Relationships between food supply and growth in albatrosses: An interspecies chick fostering experiment. *Ornis Scand.* **12:** 207–10.

Prince, P. A., Rothery, P., Croxall, J. P. and Wood, A. G., 1994. Population dynamics of Black-browed and Grey-headed Albatrosses *Diomedea melanophris* and *D. chrysostoma* at Bird Island, South Georgia. *Ibis* **136:** 50–71.

Rice, D. W. and Kenyon, K. W., 1962. Breeding cycles and behavior of Laysan and Black-footed Albatrosses. *Auk* **79:** 517–67.

Richdale, L. E., 1949a. The pre-egg stage in Buller's Albatross. *Biol. Mongr.* 2 Author. Dunedin.

Richdale, L. E., 1949b. Buller's Albatross: incubation data. *Bird-banding* **20:** 127–41.

Richdale, L. E., 1950. The pre-egg stage in the albatross family. *Biol. Mongr.* 3 Author. Dunedin.

Richdale, L. E. and Warham, J., 1973. Survival, pair bond retention and nest-site tenacity in Buller's Albatross. *Ibis* **115:** 257–63.

Ricketts, C. and Prince, P. A., 1981. Comparison of growth of albatrosses. *Ornis Scand.* **12:** 120–24.

Sagar, P. M., Molloy, J., Tennyson, A. J. D. and Butler, D., 1994. Numbers of Buller's Albatrosses breeding at the Snares Islands. *Notornis* **41:** 85–92.

Thomas, G., Croxall, J. P. and Prince, P. A., 1983. Breeding biology of the Light-mantled Sooty Albatross *Phoebetria palpebrata* at South Georgia. *J. Zool. Lond.* **199:** 123–35.

Tickell, W. L. N., 1968. The biology of the great albatrosses, *Diomedea exulans* and *Diomedea epomophora. Antarct. Res. Ser.* **12:** 1–55.

Tickell, W. L. N. and Pinder, R., 1975. Breeding biology of the Black-browed Albatross *Diomedea melanophris* and Grey-headed Albatross *D. chrysostoma* at Bird Island, South Georgia. *Ibis.* **117:** 433–51.

Turbott, E. G., 1990. Checklist of the Birds of New Zealand. Ornithological Society of New Zealand: Wellington.

Warham, J., 1990. The Petrels: Their ecology and breeding systems. Academic Press: London.

Warham, J., 1996. The behaviour, population biology and physiology of the petrels. Academic Press: London.

Warham, J. and Bennington, S. L., 1983. A census of Buller's Albatross *Diomedea bulleri* at the Snares Islands, New Zealand. *Emu* **83:** 112–14.

Warham, J. and Fitzsimons, C. H., 1987. The vocalizations of Buller's Albatross, *Diomedea bulleri* (Aves: Diomedeidae), with comparative data on other albatrosses. *NZ J. Zool.* **14:** 65–79.

Weimerskirch, H., 1982. La strategie de reproduction de líAlbatros Fuligineux a dos sombre. *Comm. Natn. Fr. Rech. Antarct.* **51:** 437–47.

Weimerskirch, H., Jouventin, P. and Stahl, J.-C., 1986. Comparative ecology of six albatross species breeding on the Crozet Islands. *Ibis* **128:** 195–213.

West, J. A. and Imber, M. J., 1986. Some foods of Buller's Albatross *Diomedea bulleri. NZ. J. Zool.* **13:** 169–74.

CHAPTER EIGHT

Factors influencing the breeding performance of the Northern Royal Albatross

C. J. R. ROBERTSON[1]

The Northern Royal Albatross breeds only on three small islands in the Chatham Islands and at Taiaroa Head on the New Zealand mainland. At Taiaroa Head tourist activities over 25 years have changed the nesting distribution of the colony. Feral cats, ferrets, stoats and blowflies are the main predators and pests affecting the albatrosses. At the Chatham Islands modification of breeding habitat by severe weather events, and a trend towards higher temperatures and lower rainfall, together with egg shell thinning, has reduced annual productivity by 60% since 1985. If this trend continues at the current rate the breeding population will decrease by an estimated 40% in the next 20 years.

Key words: Northern Royal Albatross, Breeding Performance, Human Impacts, Mammalian Predators, Climate Change.

INTRODUCTION

THE Northern Royal Albatross *Diomedea epomophora sanfordi (D. sanfordi)* breeds on three small islands (the Big and Little Sister Islands and The Forty Fours) in the Chatham Islands (44°S, 176°30′W), 800 km east of New Zealand, and Taiaroa Head (45°46′S, 170°44′E), Otago Peninsula, on the New Zealand mainland. In 1995 the estimated world population of this albatross species was 6 500–7 000 breeding pairs; of these, only 27 pairs breed at Taiaroa Head. Although tiny, the Taiaroa Head colony is important not only because it allows the general public to develop an appreciation of albatrosses, but because of the continuous monitoring of the development of the colony over nearly 60 years. The results of this monitoring have allowed the identification of a range of factors, which would be difficult to determine at oceanic breeding sites, that affect albatross nesting behaviour and breeding performance. This chapter examines these factors and their effects on the survival prospects of the Northern Royal Albatross.

BREEDING BIOLOGY

The Northern Royal Albatross breeds biennially and takes about one year to raise a chick (Richdale 1950, 1952; Robertson 1991). Parents with eggs or chicks that fail before the end of the guard stage in March re-nest the following season. Pairing is usually for life and breeding usually takes place close to natal sites. Juvenile birds commence returning to the colony at three years of age and breeding commences from six years of age. About 70% of fledglings survive to five years of age. At Taiaroa Head the mean age of breeding birds is about 20 years, with one individual >61 years old. Mean annual adult survival is 94.5% at Taiaroa Head and 95.2% in the Chatham Islands (Robertson 1993a, C. J. R., unpubl. data).

HUMAN IMPACTS AT TAIAROA HEAD

General

Northern Royal Albatrosses first attempted nesting at Taiaroa Head between 1890 and 1920, when about 2 000 people lived there. Human interference with the birds' breeding attempts led to the Otago Heads area being reserved in 1938. In 1964 the tip of the Peninsula, where the birds currently breed, was made a Nature Reserve. This Reserve has contained most (but not all) albatrosses breeding on Taiaroa Head; some birds have bred up to 14 km from the Reserve. However, since 1960 all albatrosses nesting there have done so within the 9 ha Nature Reserve. Permanent staff have overseen the albatross breeding site since 1950 and have eliminated most direct visitor disturbance and associated egg and chick losses. In 1972, when managed viewing by the general public began, an observatory was built to allow viewing of the colony by the public, and although not directly affecting albatross

[1]Science and Research Division, Department of Conservation, PO Box 10420, Wellington, New Zealand.
Pages 99–104 *in* ALBATROSS BIOLOGY AND CONSERVATION, ed by G. Robertson and R. Gales. Surrey Beatty & Sons, Chipping Norton. 1997.

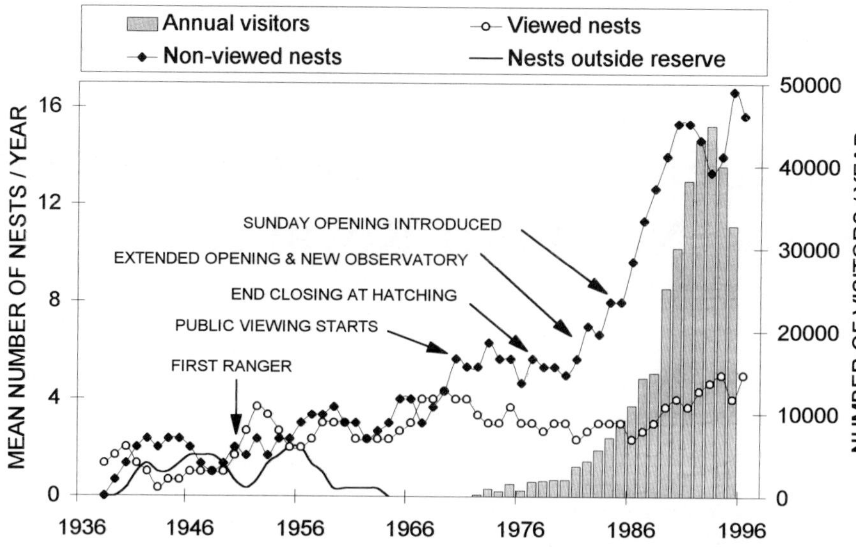

Fig. 1. Location of Royal Albatross nests (3-year running mean) in relation to the public observatory site and visitor numbers at Taiaroa Head Nature Reserve, 1936–1996. Visitor information courtesy of the Otago Peninsula Trust.

productivity, the visitors and viewing facilities have influenced the distribution of nesting sites on the headland (Robertson 1993b). With general sightseeing, coach tours, marine cruises, albatross and other nature tours, an estimated 350 000 people visit the Otago Heads annually (Fig. 1).

Interest by the public in the albatrosses has influenced the management of the Reserve (Robertson 1990). Initially, no viewing was allowed during the egg laying (Oct.–Nov.) or hatching (Jan.–Feb.) periods (Robertson 1993c), but as public demand increased, so did the number of allowable viewing days and hours per week, as no direct disturbance to nesting was observed. In the early 1980s visitors were allowed to view albatrosses from the observatory during the hatching period, and since 1987 viewing has been allowed daily until sundown, except during the pre-nuptial and egg laying periods (between mid-September and late-November).

Effects of tourist activities

There are at least four effects of tourist activity on the Northern Royal Albatrosses at Taiaroa Head. First, in the early 1990s it became clear that an increasing number of nests were being located out-of-view of the public observatory (Fig. 1). Birds abandoned a nesting area, within view of a walking track to the observatory having up to 100 000 visitor movements annually, in favour of sites furthest from visitor activity. Also, whereas the density of breeding birds away from the observatory has increased, the density of birds in nesting areas directly in front of the viewing windows of the observatory has remained stable. Since 1951, productivity (chicks fledged from eggs laid) of nests in areas out-of-view of the observatory has been lower (44%), than in areas in view of the observatory (61%).

Second, in the early years of monitoring, nesting attempts by maiden-breeding birds took place close to natal sites, and immigrant birds were distributed randomly throughout the Reserve. However, since 1975 fewer birds breeding for the first time have nested in view of the observatory compared to before 1975 ($\chi^2 = 16.78$, 2 d.f., $p \leq 0.0005$), and since 1975 fewer immigrant birds have bred in view of the observatory compared to before this date ($\chi^2 = 7.59$, 1 d.f., $p \leq 0.010$).

Third, in July, the first month when developing chicks are old enough to wander from the nest, chicks in view of the observatory moved on average 47 ± 20 m, compared to 12 ± 10 m by chicks from nests out-of-view of the observatory; these differences are statistically significant ($t_{33} = 8.784$; $p < 0.0001$).

Finally, the large un-tinted and un-screened windows in the observatory, while allowing clear viewing for visitors, also allows the albatrosses to see the visitors. Peak times of public visitation coincide with the main period of pair formation (December–March) which occurs for 2–4 years before nest site selection. Courtship behaviour and nest site selection is being increasingly conducted out-of-sight of the observatory windows, suggesting that young birds are disturbed by people inside the observatory. Recently, windows in the observatory were tinted in an attempt to conceal the public from the birds; this has resulted in an increase in the number of birds nesting in view of the observatory (Fig. 1).

EFFECTS OF MAMMALIAN PREDATORS AND HARVESTING BY PEOPLE

Unmanaged mainland albatross colonies are vulnerable to direct human disturbance (e.g., breakages and removal of eggs) and mammalian predators, which may take eggs and chicks. There will always be "natural" losses within a breeding colony, usually related to infertility, starvation or injury to the chick by the parent, especially when the chick is small. At Taiaroa Head the "natural" losses ranged from 5–60% per annum (Fig. 2).

In an effort to increase chick production at the colony, stoats *Mustela erminea*, ferrets *Putorius putorius* and cats *Felis catus*, all of which have been implicated in chick losses, have been trapped annually for over 45 years. As a result, the incidence of predation is low (Fig. 2). Predation is usually associated with periods of extreme heat and little wind; in these conditions incubating birds may become heat stressed, and may rise off the nest, allowing access to predators. Management of the colony has also included foster rearing of eggs and chicks, hand rearing of chicks (Robertson and Wright 1973), and minimizing the effects of dry, hot weather on the breeding birds. More recently, management practices include the control of blowflies *Lucilia sericata*, which strike hatchlings. In contrast to Taiaroa Head, the breeding sites at the Chatham Islands are remote and difficult to access, and this has led to an absence of mammalian predators. However, at the Chatham Islands pre-fledgling albatrosses have for 4–500 years been harvested, at varying levels of intensity, by people for cultural purposes (Robertson 1991; Bell and Robertson 1994). Though illegal since 1921, sporadic takes may still occur, and if continued this may ultimately affect the viability of the albatross population.

EFFECTS OF CLIMATE CHANGE AND UNUSUAL WEATHER EVENTS

High temperatures

Since 1970 the climate at both Taiaroa Head and at the Chatham Islands has become warmer and drier, consistent with the warming trend in New Zealand (Fig. 3). At Taiaroa Head, air temperatures as high as 55°C have been recorded between 1993 and 1996, with daily mean temperature ranges of up to 35°C during the incubation and hatching periods. This is especially stressful to adults restricted to the nest during incubation and chick guarding. In an extreme case, in January 1989 temperatures exceeded 50°C and four adult birds died from heat exhaustion when a lack of wind prevented them flying.

Since 1971, at both Taiaroa Head and the Chatham Islands, soil moisture available to maintain vegetation growth has diminished, especially in the summer, which is the nesting period for albatrosses. At Taiaroa Head soil dryness has caused problems in nest construction and attempts by chicks to leave the shell at hatching, since low relative humidity at hatching causes egg membranes to dry. In addition, warm conditions since 1986 have favoured the establishment of blowflies which strike hatchlings. Thus, the combination of fly-strike and heat stress, especially at hatching, has become the most

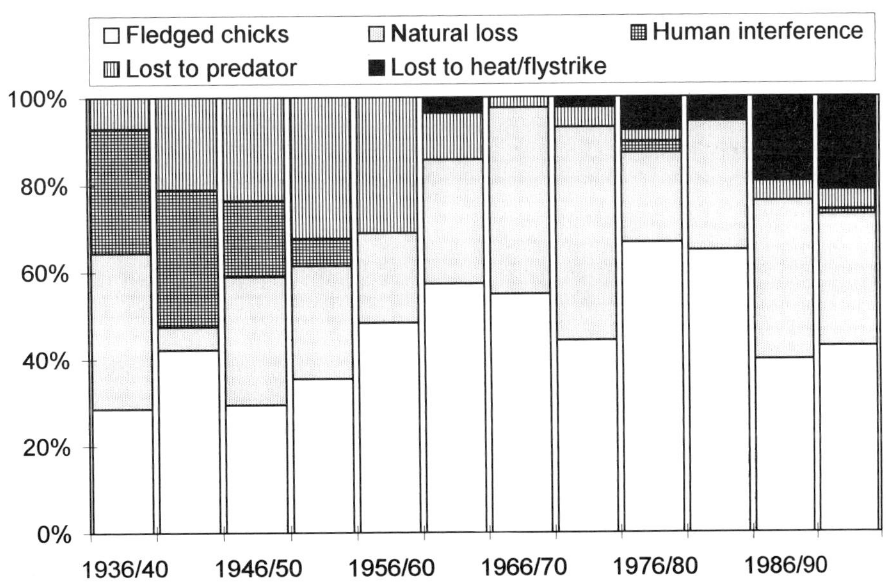

Fig. 2. Percentage productivity (fledged chicks from eggs laid) and cause of losses for Northern Royal Albatross eggs or chicks at Taiaroa Head from 1936 to 1995 (based on five-year totals), demonstrating the changing pattern of factors influencing productivity.

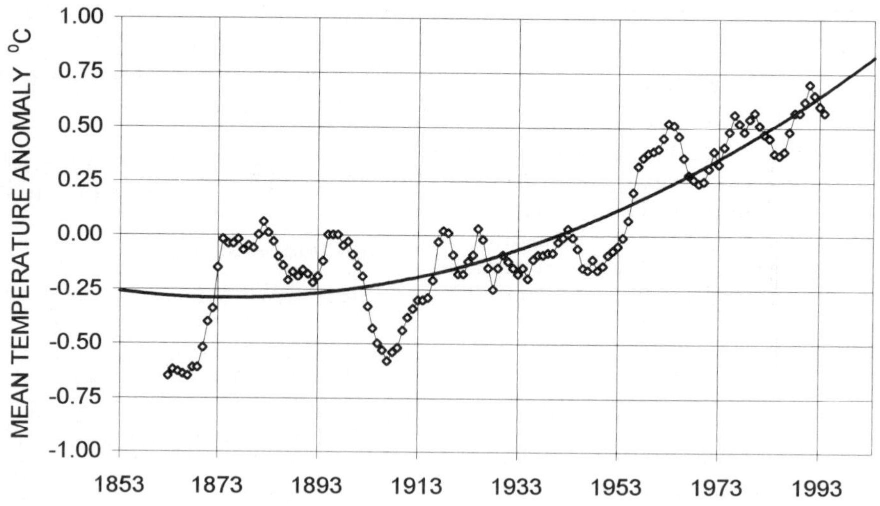

Fig. 3. Mean annual surface temperature anomaly (variation from the mean) and trend for New Zealand, 1853–1995 (10-year running mean). Information supplied by National Institute of Water and Atmospheric Research Ltd.

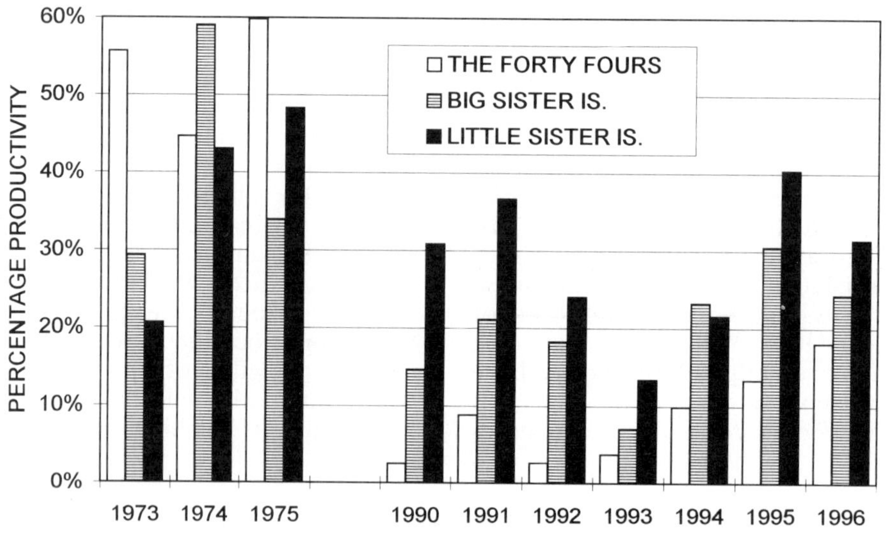

Fig. 4. Comparative rates of Northern Royal Albatross breeding success (fledglings produced from eggs laid) in 1970s and 1990s for the three breeding locations in the Chatham Islands. Total mean productivity in the 1970s was 48% compared with 18% in the 1990s. The 1996 data are provisional. Based on counts of air photographs by the Royal New Zealand Air Force and Department of Conservation.

important cause of egg and chick mortality at the Taiaroa Head colony (see Fig. 2). Fly strike has not been recorded at the Chatham Island colonies, although *L. sericata* occurs there.

Storms

The reduction in soil and vegetation cover (up to 80%) at the Chatham Island colonies (especially on The Forty Fours) has resulted in a rocky, desert-like landscape with poor water retention and little soil or vegetation for nest construction (Robertson and Sawyer 1994). Although the exact causes of the habitat changes are unknown there is evidence that weather events may have been primarily responsible.

In July 1985 a hurricane with wind speeds in excess of 170 kph and gusts of up to 280 km/h lasted for over 33 hours at the colonies (National Institute of Water and Atmospheric Research Ltd. [NIWA], S. Reid, pers. comm.). The calculated mean annual surface wind speed for the albatross islands is 53 km/h (NIWA), higher than the normal mean annual sea surface speed of 32 km/h (S. Reid, pers. comm.). The incidence of high wind speeds during and since 1985 is believed to be unusual for the Chatham Islands (S. Reid, pers. comm.). Six further storms with mean maximum wind speeds of over 140 km/h, and gusts to 280 km/h, occurred at the albatross islands 1988–1996.

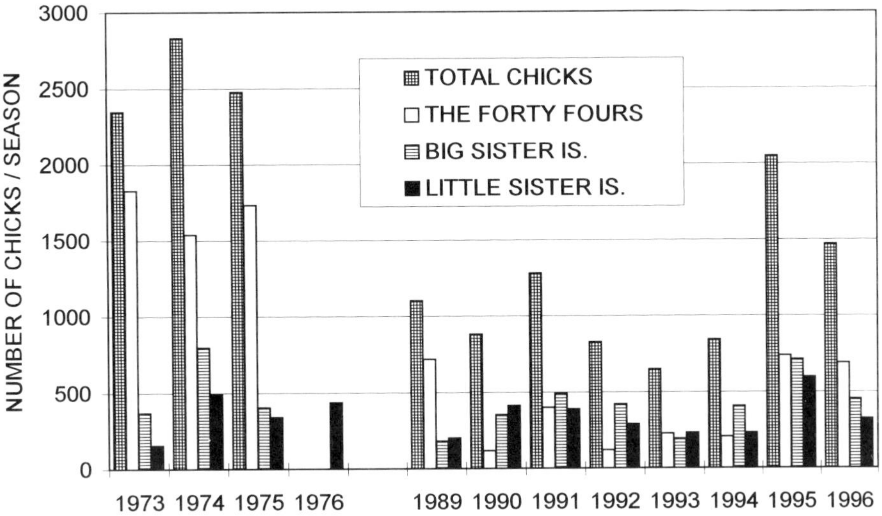

Fig. 5. Comparative numbers of Northern Royal Albatross chicks fledged at three breeding locations in the Chatham Islands during the 1970s and 1990s. Chick numbers in the 1990s are 50% of those recorded in the 1970s. The 1996 data are provisional. Based on counts of air photographs by the Royal New Zealand Air Force and Department of Conservation.

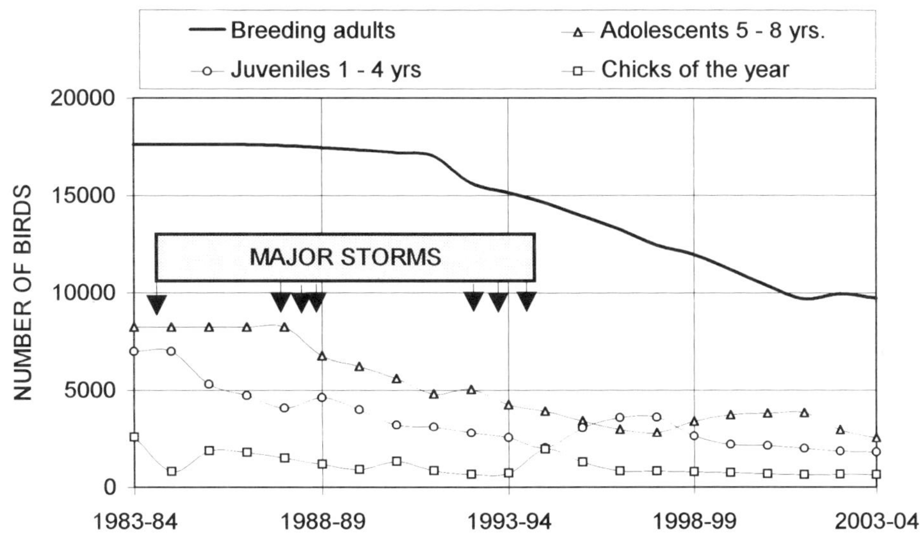

Fig. 6. Projected population of the Northern Royal Albatross breeding in the Chatham Islands, 1983–2004. Numbers of chicks of the year from aerial photograph counts 1989-96. Extrapolation from 1996 is based on an 18% mean annual productivity of fledglings (see text). Numbers of birds in other age classes are estimated using mean annual survival rates of 95.0% for juveniles, 97.5% for adolescents and 94.6% for breeding adults (modified from Robertson 1991).

In November 1994 the Chatham Islands experienced a storm (mean maximum speeds of 142 km/h and gust to 188 km/h) which caused the loss of eggs from 9% of albatross on the Little Sister Island. The incubating birds were blown off their nests. However, such short periodic events usually produce only a short-term fluctuation in nesting productivity and have probably been a regular occurrence in the past. Unlike the storm in 1994, the storm in 1985 destroyed the albatross nesting habitat, which will have a more long-lasting effect on the albatrosses than the event in 1994.

Following the 1985 and subsequent storm events, egg losses of over 50% have been recorded within the first month of incubation in the Chatham Islands. At The Forty Fours the percentage of nests producing fledglings has been as low as 3% in some years (Fig. 4). The Forty Fours, with some 60% of the total breeding population, has recently averaged only 8% mean productivity annually from 1990–1996 compared with 54% from 1973–76. The Little Sister Island, which has had the least habitat modification, shows the least long-term variation in numbers of chicks fledged (Fig. 5). Since

1990, mean annual productivity for the total population of the Northern Royal Albatross has dropped to only 18% compared with 48% from 1973–76 in the Chatham Islands and with 46% from 1938–95 at Taiaroa Head. At the Chatham Islands, if the low breeding success and production of fledglings continues into the future it is likely to cause the albatross population to decline (Fig. 6).

Storms may do more than destroy the albatrosses breeding habitat and cause egg losses; they also cause higher nesting densities in following breeding seasons. Higher nesting densities and associated stress are thought to have contributed to egg breakage by albatrosses breeding on the Little Sister (R. W. Risebrough, pers. comm.). Egg shell thickness measured at the Chatham Islands between 1993 and 1995 are at least 20% less than at the same location in the 1970s, when high density nesting also coincided with thinner shells. No similar change has occurred at Taiaroa Head where the density of nesting birds is much lower. Nesting failure from thin shells is compounded by a lack of vegetation, poor nest structures, low soil moisture and rocky substrates.

CONCLUSIONS

While much international concern has been directed towards the mortality of albatrosses in longline fisheries (e.g., Gales 1997), the factors affecting the Northern Royal Albatross reported here, highlight the importance of understanding both natural and human-induced effects on the albatrosses breeding performance and survival. Tourist activities may affect the breeding behaviour and nesting distribution of the Northern Royal Albatross, and severe weather events have the potential to cause long-term modification of nesting habitat, changes in population size, age structure and breeding success. Albatross taxa confined to single, or few, small nesting locations are especially vulnerable to direct human interference and sudden habitat modifications, and these factors have the capacity to exacerbate the effects of human-induced mortality at sea.

ACKNOWLEDGEMENTS

Long-lived and slow-breeding birds require long-term studies. I am indebted to Lancelot Richdale, Stan Sharpe and Alan Wright and their successors in the field at Taiaroa Head for an unsurpassed record of data collection and management expertise from 1937–96. I am further indebted, to those who have delivered or accompanied me (voluntarily or professionally) to the albatross islands and experienced the privations of tenting in wind speeds over 150 kph and the intricacies of landing or departure; to the owners of the islands or their agents who have graciously allowed access; to various staff at the National Institute of Water and Atmospheric Research Ltd. who analysed climatic conditions; to the Royal New Zealand Air Force who provided air photographs of the Chatham Island colonies in the 1970s; and to Arlbert Rebergen who has painstaking interpreted all air photographs used to provide the seasonal counts for the Chatham Island breeding colonies. The New Zealand Wildlife Service and its successor, the Department of Conservation, provided logistical and funding support for much of the field research. L. Clelland, R. D. Wooler, E. J. Woehler and G. G. Robertson critically reviewed and improved the manuscript with their suggestions.

REFERENCES

Bell, B. D. and Robertson, C. J. R., 1994. Seabirds of the Chatham Islands. Pp. 219–28 *in* Seabirds on Islands: Threats, Case Studies and Action Plans ed by D. N. Nettleship, J. Burger and M. Gochfeld. Bird Life International. Birdlife Conservation Series No. 1, Cambridge, United Kingdom.

Gales, R., 1997. Albatross populations: status and threats. Pp. 20–45 *in* Albatross Biology and Conservation ed by G. Robertson and R. Gales. Surrey Beatty & Sons: Chipping Norton.

Nicholls, D. G., Murray, M. D. and Robertson, C. J. R., 1994. Oceanic flights of the Northern Royal Albatross *Diomedea epomophora sanfordi* using satellite telemetry. *Corella* **18:** 50–52.

Richdale, L. E., 1950. The Pre-egg stage in albatrosses. Biological Monographs 3. Dunedin.

Richdale, L. E., 1952. The Post-egg stage in albatrosses. Biological Monographs 4. Dunedin.

Robertson, C. J. R., 1990. The Royal Albatrosses of Taiaroa Head. Pp. 150–51 *in* A Flying Start ed by B. J. Gill and B. D. Heather. Random Century, Auckland.

Robertson, C. J. R., 1991. Questions on the Harvesting of Toroa in the Chatham Islands. Science and Research Series No. 35, Department of Conservation, Wellington. 105 pp.

Robertson, C. J. R., 1993a. Survival and Longevity of the Northern Royal Albatross *Diomedea epomophora sanfordi* at Taiaroa Head 1937–1993. *Emu* **93:** 269–76.

Robertson, C. J. R., 1993b. Effects of Nature Tourism on Marine Wildlife. Pp. 53–60 *in* Proc. Marine Conservation and Wildlife Protection Conference. NZ Conservation Authority: Wellington.

Robertson, C. J. R., 1993c. Timing of egg laying in the Royal Albatross *Diomedea epomophora* at Taiaroa Head 1937–1992. Conservation Advisory Science Notes No. 50, Department of Conservation: Wellington. 8 p.

Robertson, C. J. R., 1993d. Assessment of tinted glass for public observatory, Taiaroa Head Nature Reserve. Conservation Advisory Science Notes No. 42. Department of Conservation: Wellington. 9 p.

Robertson, C. J. R. and Kinsky, F. C., 1972. Dispersal Movements of the Royal Albatross *Diomedea epomophora. Notornis* **19:** 289–301.

Robertson, C. J. R. and Richdale, L. E., 1993. The breeding phenology of the Royal Albatross *Diomedea epomophora sanfordi* 1937–1974. Conservation Advisory Science Notes No. 48, Department of Conservation: Wellington. 13 p.

Robertson, C. J. R., and Sawyer, S., 1994. Albatross research on (Motuhara) Forty-Fours Islands : 6–15 December 1993. Conservation Advisory Science Notes No. 70, Department of Conservation: Wellington. 10 p.

Robertson, C. J. R. and Wright, A., 1973. Successful Hand-rearing of an Abandoned Royal Albatross Chick. *Notornis* **20:** 49–58.

Thompson, C. S., 1983. The weather and climate of the Chatham Islands. NZ Meteorological Service: Wellington.

CHAPTER NINE

Biology and management of the Waved Albatross at the Galápagos Islands

DAVID J. ANDERSON[1] and FELIPE CRUZ[2]

The Waved Albatross has been the subject of few scientific studies. The breeding distribution is well characterized: the main breeding colony is on Isla Española, Galápagos Islands, Ecuador, and it is distributed at sea primarily near the Galápagos Islands and between the Islands and the Peruvian coast. The few population data that exist indicate that the Isla Española colony has not decreased in numbers over the past 25 years, and no serious anthropogenic threats have yet been recognized. Observations of behaviour at sea indicate that the birds avoid boats; however, more data are required in this area, particularly in relation to fishing vessels. The potential for interactions with longline fisheries, which are currently developing rapidly in the Galápagos region, deserves close attention. The species exhibits several traits that would repay field studies, including the tendency to transport their eggs from place to place and frequent attempts by males to force copulations on females that are not their mates.

Key words: Waved Albatross, Galápagos Islands, Peruvian Coast, Breeding Biology, Feeding Biology, Conservation.

INTRODUCTION

WE review knowledge on the biology of the Waved Albatross *Diomedea (Phoebastria) irrorata*, and briefly describe conservation and management issues affecting this species. Twenty-five years have passed since the only previous review, and only part of the scientific work that might have been expected after that review has eventuated. We hope to stimulate interest so that a review 25 years hence will provide a more complete understanding of the biology and management of this species than currently exists.

DISTRIBUTION AT SEA AND AT COLONIES

Since Salvin's (1883) collection off the Peruvian coast evidence from both ship- and land-based research has delimited the oceanic range of the Waved Albatross as a triangle with corners at the Galápagos Islands, the southern Peruvian coast, and the central Ecuadorian coast (Harris 1973; Anderson *et al.* 1997; Tickell, in press). Their distribution at sea is relatively restricted for an albatross: it includes the Peruvian upwelling region, southern Ecuadorian territorial waters, the waters near the Galápagos Islands, and the ocean between these areas, as judged by ship-based sightings (Tickell, in press). Satellite tracking has recently shown that breeding birds travelled from their colonies on the Galápagos Islands to forage in the Peruvian upwelling region during the incubation period of the breeding cycle (Anderson *et al.* 1997). These birds travelled rapidly between the Galápagos Islands and the upwelling region, and spent days to weeks in certain areas of the upwelling. All birds leave the Galápagos Islands after breeding in January and February, contracting their range to the waters of the Peruvian continental shelf.

Breeding occurs primarily at one site, Isla Española (1°S, 90°W), in the south-east corner of the Galápagos Islands. Adults have consistently been present on Isla Española during the breeding season (April–December), both historically (Salvin 1876; Rothschild and Hartert 1899; Snodgrass and Heller 1904; Loomis 1918; Lévêque 1963) and recently (Galápagos National Park Service, unpubl. data). Nests are located on the southern, windward half of the island, especially along the coast but also inland on the southern slopes of high hills. Owre (1976) recorded a second, much smaller breeding colony on Isla de La Plata (0°16′S, 81°06′W), off the western coast of Ecuador; this breeding site appears to be regularly occupied at present (Owre 1976; Nowak 1987; Ortiz-Crespo and Agnew 1992; Curry 1993). It is possible, but the evidence is weak, that this island has been recently colonized. Visitors in 1914 and 1925 failed to

[1]Department of Biology, Wake Forest University, Winston-Salem NC 27109, USA.
[2]Isla Santa Cruz, Isla Galápagos, Ecuador.
Pages 105–09 *in* ALBATROSS BIOLOGY AND CONSERVATION, ed by G. Robertson and R. Gales. Surrey Beatty & Sons, Chipping Norton. 1997.

report the species, but those visits may have occurred during the non-breeding season (Nowak 1987; Ortiz-Crespo and Agnew 1992). An island resident said that breeding began there around 1963 (Ortiz-Crespo and Agnew 1992), yet the lighthouse keeper collected a Waved Albatross skin in 1924 (Murphy 1936), so the history of this colony is unclear. No other breeding colonies are known, although non-breeding individuals and pairs are occasionally reported on Isla Genovesa in the Galápagos.

ABUNDANCE AND TRENDS

The total size of the population is unknown because we cannot say how many birds fail to attend breeding colonies where they can be counted easily. Complete censuses have been conducted on three occasions on Isla Española. Harris (1973) counted nesting pairs during the breeding seasons of 1970 and 1971. The estimate in 1970 was 10 600 breeding pairs. The 1971 census identified areas missed in the 1970 census and so represents the better estimate: Harris (1973) estimated "at least 12 000 pairs" from the 1971 census. The only other whole-island census occurred during the 1994 breeding season, using the same methods as Harris, with the number of breeding pairs estimated at 15 581 (Anderson 1995). Douglas (1995) used the same data to estimate the number of breeding pairs as "approximately 17 000", but the methods used to arrive at this number were not given, so the reasons for the discrepancy of around 1 500 pairs are unknown. Certainly the long-term trend in population size over the 24 years is not a decrease.

Published counts of pairs and/or nesting efforts of the Isla de la Plata population include five in May 1975 (Owre 1976); one, from an incomplete count, in May 1988 (Tickell, in Ortiz-Crespo and Agnew 1992); two in October 1985 (Nowak 1987); 10 in July–August 1990 (Ortiz-Crespo and Agnew 1992); and four in June 1991 (Curry 1993). Thirty individuals were reported on one occasion in 1981 (Hurtado 1981), and 22 adults on another in 1990 (Ortiz-Crespo and Agnew 1992). Clearly, the population at Isla de la Plata is small, and probably numbers less than 10 or 20 breeding pairs.

BREEDING BIOLOGY

Waved Albatross' breeding biology resembles that of other albatross species, including a well-defined season, single-egg clutch, high mate fidelity, and slow growth rate of nestling. Harris (1969, 1973, 1977) provided most of the available information on breeding biology, with additional material in Jungius and Hirsch (1979), Rechten (1985, 1986), Anderson and Fortner (1988), and Douglas (1995). Waved Albatrosses breed annually, with nest attendance at the colony beginning in March and egg laying lasting from April to June. The average nestling period is 167 days and the colony is vacated by parents and offspring by mid-January until the following March. Nesting success varies considerably with year and area, ranging from <9% to >80%. Complete breeding failure is not uncommon, occurring in four of the 11 years reported by Harris (1973). In that study most nestling mortality occurred by 30 days after hatching. Annual adult survival in the 1960s averaged 95%, and a minimum estimate of post-fledging, pre-reproductive annual survival was 93.4% (Harris 1973).

We draw attention to three areas of interest. The first is the phenomenon of forced copulation, in which males attack other birds (males and females) and attempt to copulate with them during the pre-egg stage of the breeding cycle (Harris 1973). In other bird species, extra-pair copulations yield increased reproductive success as an alternative reproductive tactic (Gibbs *et al.* 1990), which may explain the development of this behaviour by Waved Albatrosses. Harris (1973) interpreted the behaviour as territorial defence, yet only attack, and not copulation, is required for this purpose. On the other hand, Harris (1973) reported that males and females are both attacked and subjected to a copulation attempt, which is not considered to be an alternative mating tactic. Paternity studies using DNA fingerprinting would be a useful first step in evaluating this behaviour.

The second area of interest is the movement of eggs. In contrast to other albatross species that build substantial nests and maintain their egg within that structure, Waved Albatrosses normally do not build a nest, and regularly move their eggs around. Movement occurs when birds shuffle forward slowly with the egg held between the tarsi. Movement of up to 14 m/day (Rechten 1982) or 40 m over several days (Harris 1973) may be achieved. Egg movement frequently results in the egg's demise: it becomes lodged in a hole or under a bush. Hatching success is low (10–56% in 1970 and 1971), and over 80% of failures may be due to egg movement (Harris 1973).

With such high failures, one would expect some benefit to accrue to those parents that move their eggs. Vásquez (1981) tested hypotheses framing egg movement as an adaptation. Ticks may heavily infest chicks (Rechten 1982), so parents may move eggs away from tick infested areas. Vásquez sampled tick densities in the vicinity of (A) eggs that had not been moved in the preceding 10 d, (B) eggs in positions newly taken up, and (C) locations newly vacated. Reanalysis of her data reveals that counts from group A (80.4 ± 67.25, $n = 10$)

were significantly higher than those from group B (14.8 ± 19.57, n = 7; t_{11} = 2.91, $p < 0.02$) and also from group C (6.0 ± 11.60, n = 7; t_9 = 3.42, $p < 0.01$). Groups (B) and (C) did not differ (t_9 =1.02, $p > 0.20$). These results suggest that ticks congregate around parents that do not move their eggs and that ticks can be avoided by egg movement. However, experimental infestation of incubating adults did not increase egg movement compared to controls. Hence we are still uncertain about the role of egg movement in Waved Albatrosses.

Thermoregulation may also play a role in the movement of eggs. Waved Albatrosses nest within 150 km of the Equator, but vacate Isla Española during the hottest part of the year. Incubating birds frequently appear to be heat-stressed, as evidenced by laborious panting. Vásquez found that egg movement was no more likely during the day, when temperatures were higher, than during the night. During the day, frequency of movement was highest in the late afternoon, after temperatures had fallen from the midday high. Further, Vásquez could find no correlation between ground temperature and probability of movement. Females moved eggs more frequently than did males, and differences existed in both frequency and distance of movement between 1978 and 1981. Of greatest importance is the finding that hatching success was significantly higher in eggs that had been moved more frequently in one year, in apparent contrast to Harris' (1973) results and pointing to a dramatic positive effect of movement that is yet to be identified.

A third poorly understood aspect of the breeding biology of Waved Albatrosses is the occasional mass desertions of eggs on Isla Española. The desertions have been linked to the effects of rain: birds sometimes vacate the colony temporarily during heavy rains and flooding (Rechten 1985). Rain also may have an indirect effect: high mosquito densities and luxuriant vegetation following rain have both been implicated in mass desertions (Peterson 1967; Rechten 1985; Anderson and Fortner 1988). Waved Albatrosses nest when exposure to heavy rain is minimal (Rechten 1986), yet rainfall still appears to influence breeding success. Harris (1969) originally suggested that food shortage, occurring simultaneously with rain, provided an adequate explanation for mass desertions, but he later (Harris 1973) viewed food shortage as insufficient by itself to explain the phenomenon.

FEEDING BIOLOGY

Anderson *et al.*'s (1997) satellite-tracking study of Waved Albatross on foraging trips is the only published data on feeding biology since Harris (1973) summarized 259 regurgitation samples; these contained mostly squid, fish, and euphausiid crustaceans. Squid beaks occurred in 69% of the samples and the seven species of squid were from the Histioteuthidae and Octopodoteuthidae. Fish remains occurred in 40% of samples, with intact bodies ranging in size from 3–34 cm. The few fish identified were mainly flying fish (Exocetidae), the carangid *Decapterus scombrinus* and the clupeid *Etrumeus acumnatus*. A major gap in the information on feeding biology is direct observations of feeding behaviour at oceanic feeding sites in the Peruvian upwelling zone and waters around the Galápagos Islands. Merlen (1996) has made some progress in this respect with day-time observations of Waved Albatrosses scavenging regurgitated fish and other items from feeding aggregations of boobies and cetaceans. The relative importance of scavenging versus predation and day-time versus nocturnal foraging remains to be determined.

CONSERVATION CONCERNS

The Waved Albatross nests entirely within National Parks administered by the Galápagos National Park Service (Española) and the Machalilla National Park Service (la Plata), which are divisions of the Ecuadorian Institute of Forestry, Natural Areas, and Wildlife. Thus the species enjoys total protection at breeding colonies. The Galápagos Marine Resources Reserve was created by Presidential Decree in 1992, encompassing the waters within the Galápagos Archipelago and a 15 nautical mile (27 km) buffer zone around the islands (Merlen 1995). However, the marine reserve offers no protection for seabirds and there is no legal protection for Waved Albatrosses in continental Ecuadorian, Peruvian, or international waters in the eastern Pacific Ocean. On the basis of its restricted breeding range the Waved Albatross was considered "vulnerable" by Croxall and Gales (1997) according to the IUCN criteria for assigning threat status, and in 1997 it was added to Appendix 11 of the Convention on the Conservation of Migratory Species of Wild Animals. Appendix 11 includes migratory species of unfavourable conservation status that would require, or benefit from, international agreement on their conservation and management.

At least one Waved Albatross has been killed in the recent past by humans on Isla de la Plata (Nowak 1987), but the main population on Isla Española has not experienced known negative anthropogenic impacts. Poaching is unknown, and no introduced predators exist on the island. In contrast to the bleak situation on many other Galápagos islands, the only introduced species on Isla Española has been goats, and these were eradicated in 1978 (Hamann 1984). During their presence, which dated back at least to 1897 (Rothschild and Hartert 1899), the island's vegetation was substantially altered. Landscape photographs on Isla Española from 1905 show a widespread population of *Opuntia* cactus which is now absent. Since the eradication of the goats the vegetation has become more dense and closed, and this may ultimately cause a re-distribution of albatross nesting colonies as access to breeding sites inland from the coast decrease. We do not view a re-distribution to coastal areas (which have open vegetation due to salt spray) as a concern, because it probably represents a return to the distribution of albatrosses that existed before goats changed the vegetation. Large, open areas exist at the eastern, western and southern parts of the island, and these areas contain over half of the breeding sites (Harris 1973), yet appear to be well below their potential nest density (D. J. A., pers. obs.).

The only other current effect of humans on the Isla Española birds is the tourist visitation site at Punta Suarez, the western point of the island. A trail of several kilometres runs along the southern coast of the point, more than 30 m from most of the hundreds of albatross nests within view. The path was designed to minimize disturbance to the albatrosses, and the birds presently appear indifferent to presence of the well-regulated tourist groups, so no current negative influence has been detected. The primary anthropogenic risk at the Isla Española colony lies in the potential for further introductions of feral species, especially of rats *Rattus* spp. and arthropods such as the Fire Ant *Wasmannia auropuntata*. Such introductions have produced viable populations of these threats on other islands in the Galápagos Archipelago where boats can pull up to docks (Hoeck 1984). Such facilities do not presently exist on Isla Española, but tourism operators have in the past proposed construction of a landing dock for this island.

In the absence of concerns about this population at its Isla Española breeding site, the Galápagos National Park Service has not placed a high management priority on the Waved Albatross. Instead, it has viewed the albatrosses as an element of the entire Isla Española ecosystem. Until the eradication of the goat population, Isla Española was a top priority of the Park Service. After 1978, the Park's position with respect to Isla Española has been largely laissez-faire, allowing plant community succession and associated effects on albatrosses and other community members to proceed without intervention. The exception to this policy has been a re-introduction programme for the Giant Tortoise *Geochelone elephantopus hoodensis*, endemic to Isla Española and nearly extirpated earlier in this century by harvesting. The objective has been to return the island's biota to its original composition and then to allow that composition to determine, via its internal dynamics, the course of ecological succession. The albatross population holds no special position within that ecosystem context, and, given the healthy status of the population, we agree with this view.

Anderson *et al.* (1997) suggest that incidental bycatch in longline fishing operations (Brothers 1991; Alexander *et al.* 1997) has not affected Waved Albatrosses at sea, primarily because Waved Albatrosses tend not to associate with ships. This conclusion should be evaluated further by satellite tracking studies of immature birds and breeding birds in the non-breeding season, by observations of birds feeding at sea (especially in the vicinity of longline fishing operations) and by observations onboard longline fishing vessels; we view this gap in knowledge as the issue of highest priority for Waved Albatross conservation. Nonetheless, we are cautiously optimistic about the near-term viability of this species, given its protection at its main breeding site and its apparent low vulnerability to fishing activities at sea.

ACKNOWLEDGEMENTS

We thank Graham Robertson for inviting us to contribute this review; the National Geographic Society, Chicago Zoological Society, U.S. National Science Foundation, and Wake Forest University for funding our field work in the Galápagos Islands; the California Academy of Sciences for access to archival photographic material; and M. Harris and L. Tickell for comments on an earlier draft of the manuscript.

REFERENCES

Alexander. K., Robertson, G. and Gales, R., 1997. The incidental mortality of albatrosses in longline fisheries. Australian Antarctic Division, Tasmania. 44 pp.

Anderson, D. J., 1995. Census of Waved Albatross, 1994. Unpublished Report, Charles Darwin Research Station: Isla Galápagos, Ecuador.

Anderson, D. J. and Fortner, S., 1988. Waved Albatross egg neglect and associated mosquito ectoparasitism. *Condor* **90:** 727–29.

Anderson, D. J., Schwandt, A. J. and Douglas, H. D., 1997. Foraging ranges of Waved Albatrosses in the eastern Tropical Pacific Ocean. Pp. 180–85 *in* Albatross Biology and Conservation ed by G. Robertson and R. Gales. Surrey Beatty & Sons: Chipping Norton.

Brothers, N. P., 1991. Albatross mortality and associated bait loss in the Japanese longline fishery in the Southern Ocean. *Biol. Conserv.* **55:** 255–68.

Croxall, J. P. and Gales, R., 1997. An assessment of the conservation status of albatrosses. Pp. 46–65 *in* Albatross Biology and Conservation ed by G. Robertson and R. Gales. Surrey Beatty & Sons: Chipping Norton.

Curry, R. L., 1993. Update from Isla de La Plata. *Noticias de Galápagos* **52:** 24–25.

Douglas, H., 1995. Population dynamics and ecology of the Waved Albatross *(Diomedea irrorata)*. Abstracts of the First International Conference on the Biology and Conservation of Albatrosses. Hobart, Tasmania, Australia, August 1995.

Gibbs, H. L., Weatherhead, P. J., Boag, P. T., White, B. N., Tabak, L. M. and Hoysak, D. J., 1990. Realized reproductive success of polygynous red-winged blackbirds revealed by DNA markers. *Science* **250:** 1394–397.

Hamann, O., 1984. Changes and threats to the vegetation. Pp. 115–31 *in* Key Environments: Galápagos ed by R. Perry. Pergamon Press: New York.

Harris, M. P., 1969. Age at breeding and other observations on the Waved Albatross *Diomedea irrorata*. *Ibis* **111:** 97–98.

Harris, M. P., 1973. The biology of the Waved Albatross *Diomedea irrorata* of Hood Island, Galápagos. *Ibis* **115:** 483–510.

Harris, M. P., 1977. Comparative ecology of seabirds in the Galápagos Archipelago. Pp. 65–76 *in* Evolutionary Ecology ed by B. Stonehouse and C. Perrins. Macmillan Press: New York.

Hoeck, H. N., 1984. Introduced fauna. Pp. 233–45 *in* Key Environments: Galápagos ed by R. Perry. Pergamon Press: New York.

Hurtado, M., 1981. Informe preliminar sobre la Isla de La Plata. Instituto Nacional de Pesca, Guayaquil, Ecuador.

Jungius, H. and Hirsch, U., 1979. Herzfrequenzänderungen bei brutvölgeln in Galapagos als folge von störungen durch besucher. *J. für Ornith.* **120:** 299–310.

Lévêque, R., 1963. Le statut actuel des vertébrés rare et menacés de l'archipel de Galápagos. *Terre and Vie* **110:** 397–430.

Loomis, L. M., 1918. A review of the Albatrosses, petrels, and diving petrels. *Proc. Calif. Acad. Sci.* **2:** 1–187.

Merlen, G., 1995. Use and misuse of the seas around the Galápagos Archipelago. *Oryx* **29:** 99–106.

Merlen. G., 1996. Scavenging waved albatross around the Galapagos Archipelago: at risk from long-liners. *The Seabird Group Newsletter* **74:** 2–5.

Murphy, R. C., 1936. Oceanic Birds of South America. Am. Mus. Nat. Hist.: New York.

Nowak, J. B., 1987. Isla del La Plata and the Galápagos. *Noticias de Galápagos* **44:** 17.

Owre, O. T., 1976. A second breeding colony of Waved Albatross *Diomedea irrorata*. *Ibis* **118:** 419–20.

Ortiz-Crespo, F. and Agnew, P., 1992. The birds of La Plata Island, Ecuador. *Bull. B. O. C.* **112:** 66–73.

Peterson, R. T., 1967. The Galápagos. *Natl. Geog. Mag.* **131:** 541–85.

Rechten, C., 1982. The Waved Albatross of Española. Unpublished report, Charles Darwin Research Station: Galápagos, Ecuador.

Rechten, C., 1985. The Waved Albatross in 1983 — El Niño leads to complete breeding failure. Pp. 227–37 *in* El Niño in the Galápagos Islands: The 1982–1983 Event ed by G. Robinson and E. del Pino. Charles Darwin Foundation for the Galápagos Islands: Quito, Ecuador.

Rechten, C., 1986. Factors determing the laying date of the Waved Albatross *Diomedea irrorata*. *Ibis* **128:** 492–501.

Rothschild, W. and Hartert, E., 1899. A review of the ornithology of the Galápagos Islands. *Novitates Zoologicae* **6:** 85–205.

Salvin, O., 1876. On the avifauna of the Galápagos Archipelago. *Trans. Zool. Soc. London* **9:** 447–510.

Salvin, O., 1883. A list of birds collected by Cpt. A. H. Markham on the west coast of America. *Proc. Zool. Soc. Lond.* **1883:** 165–90.

Snodgrass, R. E. and Heller, E., 1904. Papers from the Hopkins-Stanford Galápagos Expedition 1898–99. 16, *Birds. Proc. Wash. Acad. Sci.* **5:** 231–372.

Tickell, W. L., in press. Galápagos Albatrosses at sea. *Sea Swallow*.

Vásquez, C., 1981. Why do Waved Albatrosses move their eggs? Unpublished report, Charles Darwin Research Station: Galápagos, Ecuador.

Watson, G. E. and Divoky, G. J., 1971. Identification of *Diomedea leptorhyncha* Coues 1866, an Albatross with remarkably small salt glands. *Condor* **73:** 487–89.

A pair of Waved Albatrosses bill clappering during pre-nuptial activities on Isla Española, Galápagos Islands.
TUI DE ROY/HEDGEHOG HOUSE.

An adult Sooty Albatross guards its two-month-old chick at its nest high on the cliffs of Iles Crozet, southern Indian Ocean.
HENRI WEIMERSKIRCH

Foraging and Fisheries Relationships

Albatrosses are an integral part of the marine environment, returning to land only to breed. The extensive oceanic migrations of albatrosses result in extended absence from land and until recently research on the lives of albatrosses was restricted to studies conducted at breeding sites. These studies have been important in shaping our understanding of the biology of albatrosses, including their dietary preferences and dietary overlap with prey species taken by commercial fisheries (Chapter 10). The recent miniaturization of satellite transmitters and data loggers has revolutionized our understanding of the at-sea ecology of albatrosses. Their use has enabled the measurement of foraging ranges, dive depths, foraging durations and meal sizes of albatrosses, as well as seasonal trends and inter-specific differences in these parameters (Chapters 11 and 12). The information gathered will not only help to explain the extent of spatial and temporal overlap with fisheries but the derivation of management options to reduce the impact of fisheries on albatrosses. Albatrosses favour ocean upwellings, frontal zones and continental shelves to forage (Chapter 13), areas that are also favoured by fisheries. The evidence to date suggests that all albatross species except the Waved Albatross are affected by longline fisheries.

Albatrosses and other petrel species are enthusiastic scavengers when fish and offal are made available to them, either as bait or when dumped by fishing vessels. This behaviour makes them prone to capture and death in longline fisheries.
TASMANIAN PARKS AND WILDLIFE SERVICE

A Chatham Albatross soars over the waters of the Pacific Ocean in preparation for landing at its breeding site at the Chatham Islands, New Zealand.
GARY NUNN.

CHAPTER TEN

A review of the food of albatrosses

YVES CHEREL[1] and NORBERT KLAGES[2]

The diets of nine species of albatrosses are dominated by cephalopods, fish and, to a lesser extent, crustaceans, while carrion and gelatinous zooplankton are minor components. Cephalopods (mainly the squid families Onychoteuthidae and Ommastrephidae) and fish rank first and second, respectively, by mass in the diets of Wandering, Royal, Laysan, Grey-headed, Sooty and Light-mantled Sooty Albatrosses. Fish dominates the diets of Black-footed, Black-browed and Yellow-nosed Albatrosses. Antarctic krill accounts for a high percentage by mass in the diets of Black-browed, Grey-headed and Light-mantled Sooty Albatrosses, and carrion in that of the Wandering, Sooty and Light-mantled Sooty Albatrosses. Albatrosses exploit the marine food chain at three levels: omnivorous zooplankton and small nekton, large carnivorous nektonic species, and top predators through scavenging. Association with human activities can be beneficial but is not without risk because of high mortality rates in fishing operations. Further studies on the foods and feeding ecologies of many albatross populations during and outside the breeding season are necessary to increase knowledge of these aspects of albatross ecology and to quantify the effects of human activities on albatrosses, especially the effects of commercial fishing activities.

Key words: Albatrosses, Diets, Cephalopods, Fish, Crustaceans, Fisheries.

INTRODUCTION

ALBATROSSES (Procellariiformes: Diomedeidae) are large seabirds with long wings adapted for gliding flight and powerful beaks adapted for seizing prey (Marchant and Higgins 1990). They are among the most oceanic of all seabirds, spending most of their lives foraging over vast tracts of ocean. Albatrosses have restricted abilities to obtain prey to any great depth and depend on finding food at or near the sea surface (Prince *et al.* 1994).

Increasing exploitation of marine resources by people throughout the world has focused attention on the management of marine ecosystems and the factors affecting them. Of concern are the effects of competition between marine predators such as seabirds, and fisheries, and the conservation of endangered species, including albatrosses (Alexander *et al.* 1997). In this context, accurate information on the diets and feeding ecologies of albatrosses is essential if the effects of resource depletion and incidental bycatch are to be minimized.

Until the end of the 1970s, knowledge of the foods of albatrosses was largely anecdotal, resulting from miscellaneous observations at sea, the collection of birds for museums and from prey remains at colonies. Following the pioneering work of Prince (1980), Thomas (1982) and Harrison *et al.* (1983), quantitative studies of albatross diets have been made of birds breeding in the Southern Ocean (e.g., at Marion Island, Iles Crozet and South Georgia). In total, data are available on the diets of nine of the 14 species of albatrosses in the current taxonomy (for revised taxonomy and nomenclature see Robertson and Nunn 1997), the nine species being the Wandering *Diomedea exulans*, Royal *D. epomophora*, Laysan *D. (Phoebastria) immutabilis*, Black-footed *D. (Phoebastria) nigripes*, Grey-headed *D. (Thalassarche) chrysostoma*, Black-browed *D. (Thalassarche) melanophris*, Yellow-nosed *D. (Thalassarche) chlororynchos (T. bassi)*, Sooty *Phoebetria fusca* and Light-mantled Sooty *Phoebetria palpebrata* Albatrosses. The aim of this chapter is to review all the available information on the diets of albatrosses. The review will examine both incomplete and comprehensive studies and, for a few species, temporal and spatial variations, and variations in the diets of sympatric albatross species. No such comprehensive work has been undertaken, but some information appeared in Prince and Morgan's (1987) review of the foods and feeding ecology of Procellariiformes, and in three recent reviews (Marchant and Higgins 1990; del Hoyo *et al.* 1992; Gales 1993).

[1]Centre d'Etudes Biologiques de Chizé, Centre National de la Recherche Scientifique, F79360 Villiers-en-Bois, France.
[2]Port Elizabeth Museum, P.O. Box 13147, Humewood 6013, South Africa.
Pages 113–36 *in* ALBATROSS BIOLOGY AND CONSERVATION, ed by G. Robertson and R. Gales. Surrey Beatty & Sons, Chipping Norton. 1997.

METHODS OF COLLECTING AND ANALYSING FOOD SAMPLES

There are three methods of obtaining diet samples from seabirds: by killing them, by sampling their stomach contents, and by collecting regurgitates at the nest. The difficulty of capturing albatrosses at sea usually necessitates the shooting of birds for studies of at-sea diet (e.g., Linkowski and Rembiszewski 1978; Ainley *et al.* 1984), if such studies are deemed to be scientifically necessary and morally defensible. In colony-based studies, however, almost all samples are obtained from regurgitates from chicks or adults, or from regurgitated casts (Berruti and Harcus 1978; Brooke and Klages 1986; Imber 1992). Regurgitations include the contents of the proventriculus only: flesh and hard parts of prey, such as squid beaks, are too large to pass into the ventriculus and may persist in a relatively uneroded condition in the stomachs of albatrosses for over one month (Furness *et al.* 1984). Special care needs to be taken in dietary analyses to separate fresh material from hard parts that may have accumulated over a long period.

Generally, the stomach contents of albatrosses are partially digested and their identification relies mainly on exoskeletons for crustaceans, beaks and gladii for cephalopods, and otoliths and bones (jaws, opercles and caudal skeletons) for fish. Such identifications are difficult and time-consuming, and require the use of specialized reference collections, atlases, and keys (e.g., Clarke 1986; Hecht 1987; Williams and McEldowney 1990; Smale *et al.* 1995). Although in the 1970s the paucity of descriptions of cephalopod beaks led to some misidentifications of prey (Imber 1992), the main problems today are the lack of up-to-date descriptions of many families of cephalopods, and the lack of otolith descriptions for many fish species.

As for other seabird species, the foods of albatrosses are poorly known outside the chick-rearing period; when foraging for self maintenance adult albatrosses may take different prey from that given to their chicks. The main difference between samples collected from chicks and those from adults feeding chicks is the number of cephalopod beaks with no flesh attached, because beaks accumulate in the chick's stomach during the nestling period. These beaks originate not only from fresh remains digested by chicks, but from prey previously digested by adults, which are passed to the chick during feeding. Further, because the mean number of items per sample is usually low (e.g., 1.7 items in the food of Wandering Albatross chicks at Iles Crozet [Ridoux 1994]) and prey items medium-to-large in size, biases may occur in estimates of the importance of various prey components in the diet. These points must be considered during dietary analyses and subsequent interpretations of the ecological significance of dietary data.

LIQUID AND SOLID CONSTITUENTS

The stomach contents of albatross chicks and adults usually comprise solid matter mixed with an oily liquid. The mean proportion of the liquid fraction constitutes 50–80% of sample mass (Table 1), but proportions vary greatly within species (Green 1974; Thomas 1982; Cooper *et al.* 1992); for example oil comprised 7–85% in the diet of Sooty Albatross chick samples from Marion Island (Cooper and Klages 1995). The solid fraction of albatross stomach contents includes both fresh items and those accumulated over many months. Fresh items typically consist of partially digested prey, while accumulated items are usually indigestible hard parts of prey and items of variable origin. In Wandering, Sooty and Light-mantled Sooty Albatrosses, cephalopod beaks free of body parts account for 0.6–1.3% and 1.8–2.8% of total prey mass and solid fraction, respectively (Cooper *et al.* 1992; Cooper and Klages 1995). Other accumulated remains of natural origin include cephalopod and fish lenses, cephalopod gladii and spermatophores, fish bones, feathers, crustacean exoskeletons, stones, pebbles and pumice, and vegetable and soil matter (Hagen 1952; Segonzac 1972; Green 1974; West and Imber 1986). Regurgitated pellets usually contain only indigestible material, mainly cephalopod beaks and feathers, but may include lenses, bones, spermatophores and stones (Berruti and Harcus 1978; Imber and Berruti 1981; Brooke and Klages 1986). Parasitic nematodes (Anisakidae in Buller's *D. (Thalassarche) bulleri* and Shy *D. (Thalassarche) cauta* Albatrosses) occur regularly in albatross food samples and in casts (Green 1974; Brooke and Klages 1986; West and Imber 1986).

STOMACH OIL AND DIGESTIVE PHYSIOLOGY

The Procellariiformes, with the exception of the diving petrels, are unique among birds in storing large volumes of oil in the proventriculus. Liquid-to-oil ratios vary with prey type eaten, fish and squid producing low lipid content (1% of the liquid fraction) and krill high lipid content (27% of the liquid fraction) (Clarke and Prince 1980; Prince 1980). Stomach oil is produced from the breakdown of ingested food and allows adult albatrosses to deliver to chicks meals of high energy density (40.2 kJ/g for oil; Warham *et al.* 1976) at low energy costs of transport (Prince and Morgan 1987). Stomach oil is the main source of energy and production of metabolic water (via lipid oxidation) for chicks, and complements nutrients and water provided

Table 1. Solid and liquid constituents of the diets of seven species of albatrosses.

Species	Location	*n*	Total mass (g)	Liquid mass (%)	Solid Mass (%)	Reference
Wandering	Prince Edward Islands	39–50 chicks	1 630	70.2	28.8	Cooper *et al.* 1992
	Iles Crozet	37 chicks	—	61.9	38.1	Ridoux 1994
Black-footed	Hawaii	172 (84% from chicks)	96[a]	*ca.* 10	*ca.* 90	Harrison *et al.* 1983
Laysan	Hawaii	183 (>95% from chicks)	64[a]	*ca.* 10	*ca.* 90	Harrison *et al.* 1983
Black-browed	South Georgia	138 adults	—	47.7	52.3	Prince 1980
	Falkland Islands	9 adults	—	80	20	Thompson 1992
Grey-headed	Prince Edward Islands	88 adults	343	55.4	44.6	Hunter and Klages 1989
	South Georgia	132 adults	—	51.7	48.3	Prince 1980
Sooty	Prince Edward Islands	37 chicks	549	52.5	47.5	Cooper and Klages 1995
Light-mantled Sooty	Prince Edward Islands	32 chicks	639	68.7	31.3	Cooper and Klages 1995
	South Georgia	9–10 adults	557	52	48	Thomas 1982

[a]in ml.

by the fresh and liquid fractions of the diet (Warham *et al.* 1976; Warham 1977; Prince 1980; Jacob 1982).

Little is known of the digestive physiology of albatrosses, as only the Sooty Albatross has been studied. Time to excretion of 50% of the total faecal mass and mean retention times of digesta in the birds' guts is lower for squid than for fish and crustaceans, despite squid being relatively indigestible *in vitro* (Jackson 1992). Sooty albatrosses are able to retain a substantial proportion (47%) of ingested chitin from crustacean exoskeletons, and to assimilate (91% efficiency) wax esters (Jackson and Place 1990; Jackson *et al.* 1992). Sooty Albatross chicks excrete lipids more slowly than adults, which enables them to eke out food between visits by parents (Jackson and Place 1990).

PREY CLASSES

General

Cephalopods, fish and, to a lesser extent, crustaceans are important food for albatrosses, while carrion and gelatinous plankton are of minor importance (Tables 2 and 3). Cephalopods rank first (>40% by mass), and fish second, in the diet of Wandering, Royal, Laysan, Grey-headed, Sooty and Light-mantled Sooty Albatrosses, while fishes predominate in the diets of Black-footed, Black-browed and Yellow-nosed Albatrosses. Crustaceans exceed 10% by mass in the diets of only the Black-browed, Grey-headed and Light-mantled Sooty Albatrosses, and carrion only in those of the Wandering, Sooty and Light-mantled Sooty Albatrosses (Table 3). Jellyfish comprise about 4% by volume and 5% by mass of the diets of Laysan and Black-browed Albatrosses, respectively (Harrison *et al.* 1983; Thompson 1992).

Cephalopods

Cephalopods play a major role in the diets of the nine albatross species investigated (Table 3). A total of 70 taxa of cephalopods in 27 families has been identified (Appendix 1). Squids (Teuthoidea, 21 families) predominate over cuttlefishes (Sepioidea, one family) and octopuses (Octopoda, four families) (Table 4). Only one species of non-squid cephalopod, the Octopus *Octopus maorum*, has been recorded as an important prey: this accounts for a significant proportion (52% by mass) of the cephalopods in the diet of Royal Albatrosses and presumably is obtained as discards by seals or from fishing boats (Imber 1991).

The squid families Ommastrephidae and Onychoteuthidae are the most important in the diets of albatrosses, other major families being Histioteuthidae, Cranchiidae, Loliginidae and Psychroteuthidae (Table 4). Albatrosses consume ommastrephids in the waters of the North Pacific, near the equator and in the Southern Hemisphere, the squid species eaten varying with location. The species of ommastrephids eaten by Laysan, Black-footed and Waved *D. (Phoebastria) irrorata* Albatrosses are poorly known (Harris 1973; Harrison *et al.* 1983). The neritic *Nototodarus* spp. and *Illex argentinus* are taken by albatrosses in Australasian waters and the Falkland Islands, respectively (West and Imber 1986; Thompson 1992), while *Martialia hyadesi* is taken by albatrosses from South Georgia, Kerguelen, Macquarie and Campbell Islands, and possibly also from Iles Crozet (Appendix 1). Among Southern Ocean albatrosses, the small species of albatrosses are the main ommastrephid consumers: at South Georgia Grey-headed and Black-browed Albatrosses take an estimated 20 000 tonnes of *M. hyadesi* annually (Rodhouse

Table 2. Frequency of occurrence of broad prey classes in the diets of albatrosses.

Species	Location	Frequency of occurrence (%)				Reference
		Fish	Cephalopod	Crustacean	Other	
Wandering	Prince Edward Islands	60.0	96.0	4.0	22.0	Cooper *et al.* 1992
	Iles Crozet	27.7–34.4	68.8[a]–91.7	0–2.8	3.1–25.0	Cooper *et al.* 1992; Ridoux 1994
Royal	Chatham Islands	19	?	19	6–68	Imber 1991
	New Zealand	16	?	32	22	Imber 1991
Waved	Galapagos Islands	41.3	52.9[a]	10.4–45.9	?	Harris 1973
Black-browed	Iles Crozet	58.3	41.7[a]	0	41.7	Weimerskirch *et al.* 1986
	South Georgia	47.8–72.2	30.0[a]–85.2[a]	2.9–91.7	0–18.5	Tickell 1964; Prince 1980
	Falkland Islands:					
	Beauchêne	60.0	93.3	20.0	20.0	Thompson 1992
	New Island	87.5	21.9	40.6	9.4	Thompson 1992
	Steeple Jason	60.0	65.0	5.0	10.0	Thompson 1992
Buller's	Snares/Chatham Islands	48.1	29.6 (85.1[a])	44.4	18.5–25.9	West and Imber 1986
Shy	Australia	58.7	53.3[a]	34.7	32.0	Green 1974
Yellow-nosed	Iles Crozet	100.0	22.0	55.0	0	Ridoux 1994; Weimerskirch *et al.* 1986
Grey-headed	Prince Edward Islands	59.1	63.6	14.8	3.4–5.7	Hunter and Klages 1989
	Iles Crozet	56.8	58.8[a]	25.5	13.7	Weimerskirch *et al.* 1986
	Campbell Island	29	100[a]	0	14	Marchant and Higgins 1990
	South Georgia	36.4–80.0	62.1[a]–94.1[a]	11.7–86.7	1.7–71.6	Tickell 1964; Prince 1980
Sooty	Prince Edward Islands	56.8	94.6	35.1	48.6	Cooper and Klages 1995
	Iles Crozet	0–35.0	50.0–95.1[a]	0–50.0	11–55.0	Mougin 1970; Weimerskirch *et al.* 1986; Ridoux 1994
	Tristan da Cunha	32	97[a]	2	74	Richardson 1984
Light-mantled Sooty	Prince Edward Islands	65.6	84.4	62.5	56.3	Cooper and Klages 1995
	Iles Crozet	10–31.1	75.9–95.6[a]	10–42.2	10–40.0	Mougin 1970; Weimerskirch *et al.* 1986; Ridoux 1994
	South Georgia	53.3	95.6[a]	87.0	8.7	Thomas 1982

[a]Including accumulated squid beaks, or not indicated.

Table 3. Proportion by fresh mass of broad prey classes in the diet of albatrosses.

Species	Location	Percentages by fresh mass				Reference
		Fish	Cephalopod	Crustacean	Other	
Wandering	Prince Edward Islands	36.5	58.6	0.1	4.8	Cooper *et al.* 1992
	Iles Crozet	14.9[a]	76.7[a]	0.1[a]	8.4[a]	Ridoux 1994
	South Georgia	41.5	39.5	0.2	18.8	Prince and Morgan 1987
Royal	Campbell Island	21[a]	75[a]	3[a]	1[a]	Marchant and Higgins 1990
	Chatham Islands	14[a]	85[a]	0[a]	1[a]	Marchant and Higgins 1990
	New Zealand	15[a]	80[a]	3[a]	2[a]	Marchant and Higgins 1990
Black-footed	Hawaii	50[b]	32[b]	5[b]	13[b]	Harrison *et al.* 1983
Laysan	Hawaii	9[b]	65[b]	9[b]	17[b]	Harrison *et al.* 1983
Black-browed	South Georgia	27.1–72.5	13.2–37.3	4.5–41.6	0–5.9	Prince 1980; Croxall *et al.* 1988b; Rodhouse and Prince 1993; Reid *et al.* 1996
	Falkland Islands:					
	Beauchêne	25.9	74.0	<0.1	0.1	Thompson 1992
	New Island	81.5	0.7	9.4	8.4	Thompson 1992
	Steeple Jason	41.0	56.7	0.7	1.6	Thompson 1992
Yellow-nosed	Iles Crozet	58.2	38.0	3.8	0	Ridoux 1994; Weimerskirch *et al.* 1986
Grey-headed	Prince Edward Islands	58.0	34.2	3.0	4.9	Hunter and Klages 1989
	Iles Crozet	traces	89.3	10.7	0	Ridoux 1994
	South Georgia	8.7–60.8	37.7–74.5	1.5–32.3	0–10.4	Prince 1980; Croxall *et al.* 1988b; Reid *et al.* 1996

Table 3 — *continued.*

Species	Location	Percentages by fresh mass				Reference
		Fish	Cephalopod	Crustacean	Other	
Sooty	Prince Edward Islands	33.1	41.8	0.5	24.7	Cooper and Klages 1995
	Iles Crozet	5.5[a]	40.5[a]	2.7[a]	51.2[a]	Ridoux 1994
Light-mantled Sooty	Prince Edward Islands	45.7	33.9	3.9	16.5	Cooper and Klages 1995
	Iles Crozet	10.9[a]	56.3[a]	16.0[a]	16.7[a]	Ridoux 1994
	South Georgia	10.9	45.5	39.5	4.1	Thomas 1982

[a]Percentages by reconstituted mass and not fresh mass. [b]Percentages by volume.

Table 4. Relative importance of cephalopod families by cephalopod reconstituted mass identified from lower beaks regurgitated by albatrosses. +, <1% by mass; ++, <10%; +++, <25%; and ++++, >25% by mass.

	Wandering	Royal	Waved	Black-footed	Laysan	Black-browed	Buller's	Yellow-nosed	Grey-headed	Sooty	Light-mantled Sooty
Sepiidae	–/+	–	–	–	–	–	–	–	–	–	–
Loliginidae	–/+	–	–	–	–	–/++++	–	–	–	–	–
Lycoteuthidae	–/+	–	–	–	–	–	–	–	–/+	–/+	–/+
Enoploteuthidae	+/++	–	++	–	+	–/++	+	–	–	–/+	–
Octopoteuthidae	+/+++	–	+++	++	–	–	+	–	–/++	+/++	–/+
Onychoteuthidae	++++	+++/++++	++	–	++	–/++	++	–/++++	++/++++	++++	+++/++++
Gonatidae	+/++	–	–	–	–	–/++	–	–	–/++	+/++	–/++
Histioteuthidae	++/++++	++	++++	–	++	–/++	++	–	+/+++	++/+++	–/++
Psychroteuthidae	–/++	–	–	–	–	–/+	–	–	–/++	–/++	++/++++
Architeuthidae	–/+++	–/++	–	–	–	–	–	–	–	+	–
Neoteuthidae	+/++	–	–	–	–	–	–	–	–/++	+/++	++/+++
Brachioteuthidae	–/+	–	–	–	–	–	–	–	–/+	–	–
Ommastrephidae	–/+++	+/++++	++++	++++	++++	+++/++++	++++	–/++++	–/++++	–/++	–/+++
Lepidoteuthidae	–/++	–	–	–	++	–	–	–	–	–/++	–/++
Pholidoteuthidae	–/+	–	+	–	–	–	–	–	–	–	–/+
Batoteuthidae	–/+	–	–	–	–	–	–	–	–/++	+/++	–/++
Cycloteuthidae	–/++	–	–	–	–	–	–	–	–/+	+/++	–/++
Chiroteuthidae	–/++	–	++	–	–	–/++	+	–	–/+	+/+++	–/++
Mastigoteuthidae	–/+	–	–	–	++	–	+	–	–/+	+/++	–/++
Promachoteuthidae	–/+	–	–	–	–	–	–	–	–	–	–
Grimalditeuthidae	–/+	–	–	–	–	–	–	–	–	–	–
Cranchiidae	+/+++	–	–	++	–	–/+++	+	–/++	+/++	+++	++/+++
Vampyroteuthidae	–/+	–	–	–	–	–	–	–	–	–	–
Octopodidae	–/+	+/++++	–	++	–	–/++	+	–	–/+	–	–/+
Alloposidae	–/++	–	–	–	–	–	–	–	–/+	–/++	–
Ocythoidae	–/+	–	–	–	–	–	+	–	–	–	–
Argonautidae	–/+	–	–	–	–	–	++	–	–	–	–

References: Berruti and Harcus 1978[a]; Clarke and Prince 1981; Cooper and Klages 1995; Cooper *et al.* 1992; Brooke and Klages 1986[b]; Harris 1973[b]; Harrison *et al.* 1983[c]; Hunter and Klages 1989[b]; Imber 1991; Imber 1992[d]; Prince 1980; Prince and Morgan 1987[e]; Ridoux 1994[b]; Rodhouse *et al.* 1990; Rodhouse and Prince 1993; Thompson 1992[f]; West and Imber 1986[f]. [a]Data modified according to Imber (1992). [b]Calculated from original data. [c]Percentages by volume. [d]This review paper includes data from Clarke *et al.* (1981); Imber and Russ (1975); Imber and Berruti (1981); and Rodhouse *et al.* (1987). [e]Including Thomas (1982). [f]Estimated from proportions by number.

et al. 1993). The majority of ommastrephids eaten are juveniles (West and Imber 1986; Rodhouse *et al.* 1993; Cherel and Weimerskirch 1995) taken from surface waters (Barton 1979; Imber and Berruti 1981; Croxall and Prince 1994).

Onychoteuthid squid form the bulk of the diets of Wandering and Royal Albatrosses, being less important for the other Southern Ocean species, except those at Prince Edward Islands, and are minor prey for Northern Hemisphere albatrosses (Table 4). The main onychoteuthids taken by albatrosses are *Kondakovia longimana*, *Moroteuthis knipovitchi* and *M. ingens* (Table 5); these are deep-water species associated with islands and continental masses (Nesis 1987), and are frequently found as fresh remains in albatross food samples (Cooper *et al.* 1992; Ridoux 1994). Most onychoteuthids eaten by albatrosses are large adults (Imber 1991, 1992).

Histioteuthid squids are major prey of Waved Albatrosses (Harris 1973), while two other species of this family are frequently caught by Southern Ocean albatrosses: *Histioteuthis eltaninae*, by birds from the Prince Edward Islands (Berruti and Harcus 1978; Hunter and Klages 1989; Cooper and Klages 1995), and *H. atlantica* by Wandering Albatrosses at various localities

Table 5. Percentages by number of main cephalopod species eaten by Southern Ocean Albatrosses. +, <1% by number; ++, <10%; +++, <25%; and ++++, >25% by number.

	Wandering	Royal	Black-browed	Buller's	Yellow-nosed	Grey-headed	Sooty	Light-mantled Sooty
Loligo gahi	–/+	–	–/++++	–	–	–	–	–
Kondakovia longimana	+/++++	–	–/++	–	–/++++	+/++++	++	++/+++
Moroteuthis knipovitchi	+/++++	–	–	–	–/++	–/+++	+/++	–/++
M. ingens	+/++	++/++++	–	++	–	–	–/+	–/+
Histioteuthis eltaninae	–/+++	–	–/++	–	–	–/++++	++/++++	–/++++
H. atlantica	+/++++	–	–/++	++	–	–/++	+	–/+
Psychroteuthis spp.	–/++	–	–/+++	–	–	–/++	–/+	++/+++
Alluroteuthis antarcticus	+/++	–	–	–	–	–/+	+/++	++/+++
Martialia hyadesi[a]	–/++	–	+/++++	–	–/++++	–/++++	–/+	–/+++
Nototodarus spp.	–	+/++++	–	++++	–	–	–/+?	–
Illex argentinus	–/+++	–	–/+++	–	–	–	–	–
Chiroteuthis spp.	–/++	–	–/+++	+	–	–/++	+/++++	–/+++
Taonius spp.	+/+++	–	–	+	–	–/+	+/++	–/++
Galiteuthis glacialis	++/+++	–	–/++++	–	–	+/++++	+++/++++	+++/++++
Octopus maorum	–	+/++++	–	++	–	–	–	–

References: Berruti and Harcus 1978[b]; Clarke and Prince 1981; Cooper and Klages 1995[c]; Cooper *et al.* 1992; Brooke and Klages 1986[c]; Hunter and Klages 1989; Imber 1991[d]; Imber 1992[c,e]; Prince 1980; Prince and Morgan 1987[f]; Ridoux 1994[c]; Rodhouse and Prince 1993; Rodhouse *et al.* 1990; Thompson 1992[c]; West and Imber 1986[c]. [a]Including *Todarodes filippovae* (Ridoux 1994). [b]Data modified according to Imber (1992). [c]Calculated from original data. [d]Estimated from proportions by mass. [e]This review paper includes data from Imber and Russ (1975); Clarke *et al.* (1981); Imber and Berruti (1981); and Rodhouse *et al.* (1987). [f]Including Thomas (1982).

(Imber 1992). Other major squid prey include the Cranchiid *Galiteuthis glacialis* taken by most of the Southern Ocean albatrosses (Table 5), *Psychroteuthis glacialis* by Light-mantled Sooty Albatrosses at Iles Crozet and South Georgia (Prince and Morgan 1987; Ridoux 1994), *Chiroteuthis* sp. by Sooty Albatrosses at Marion Island (Cooper and Klages 1995), and *Loligo gahi* by Black-browed Albatrosses at the Falkland Islands (Thompson 1992).

The size of squid taken by albatrosses varies greatly in mantle length (30–1 000 mm) and in mass (15–9 000 gm), over all species of albatrosses (Rodhouse *et al.* 1987; Hunter and Klages 1989; Ridoux 1994). Estimated mantle lengths range from 29–822 mm (15–5 354 g) and from 42–980 mm (10–8 253 g) for Sooty and Wandering Albatrosses, respectively (Cooper *et al.* 1992; Cooper and Klages 1995). Because the mean mass of the prey is positively related to the mass of the predator (Imber and Berruti 1981), Wandering Albatrosses eat on average larger prey (694 g) than the two *Phoebetria* spp. (137–340 g) (Cooper *et al.* 1992; Cooper and Klages 1995).

Despite the importance of cephalopods in the diets of albatrosses there are few direct observations at sea of albatrosses eating cephalopods (Croxall and Prince 1994). The best documented fact is the regular winter occurrence of large numbers of Wandering and Black-browed Albatrosses feeding on floating moribund cuttlefishes (*Sepia apama* = *Amplisepia verreauxi*) off eastern Australia (Gibson and Sefton 1959; Wood 1992). Mixed flocks of albatrosses, other seabirds and sea mammals have also been observed catching the loliginid *Sepioteuthis australis* in surface waters of the Tasman Sea (Barton 1979).

Given the paucity of direct observations, how albatrosses catch squid (by scavenging or predation) remains open to conjecture. Adult onychoteuthids taken by albatrosses are so large and powerful that it is doubtful whether albatrosses are able to subdue them when still alive. Hence albatrosses may scavenge dead or moribund squids near the surface (Weimerskirch *et al.* 1986; Cooper *et al.* 1992; Cooper and Klages 1995). Some squid taxa float after death making them accessible to surface-feeding seabirds, whereas others sink (Lipinski and Jackson 1989). "Floater" squid are common in the diet of Wandering Albatrosses, suggesting that scavenging plays an important role in the behaviour of this species (Cooper *et al.* 1992). Nevertheless, the relative importance of predation and scavenging in the diets of albatrosses is uncertain. In a recent study (Croxall and Prince 1994) the foraging ranges, methods and timing of feeding of South Georgia albatrosses was examined in relation to the size, distribution, buoyancy characteristics, bioluminescence and prey of the squid, and the location and potential contribution of fishery waste. Wandering and Light-mantled Sooty Albatrosses were found to depend on scavenged squid for a substantial part of their diets (see also Cooper *et al.* 1992; Ridoux 1994; Cooper and Klages 1995), but the evidence for Black-browed and Grey-headed Albatrosses was less clear. Hence the roles of scavenging and predation in the diet of albatrosses requires critical re-evaluation.

Fish

A total of 78 fish taxa including 54 identified species, have been found in the diets of 13 of the 14 albatross species; there is no information available for the Amsterdam Albatross

D. amsterdamensis (Appendix 2). Fish prey comprise epipelagic, mesopelagic and benthic species. The epipelagic fishes include fish in the Engraulidae, Clupeidae, Scomberesocidae, Exocoetidae and Carangidae families. The Exocoetidae are especially important in the diets of Laysan, Black-footed, Short-tailed *D. (Phoebastria) albatrus* and Waved Albatrosses (Harris 1973; Hasegawa and deGange 1982; Harrison *et al.* 1983). Black-footed Albatrosses are the only species to exploit flying fish eggs, which account for 44% by volume of their diet (Harrison *et al.* 1983). Shy and Yellow-nosed Albatrosses have been observed feeding on schools of *Trachurus declivis, Sardinops neopilchardus* and *Engraulis australis* in Tasmanian waters (Barton 1979). The paucity of epipelagic schooling fish in the Southern Ocean (Rodhouse and White 1995) explains why their importance to albatrosses is greater in the Northern Hemisphere.

Oceanic mesopelagic species form a minor part of albatross diets and include myctophid fish and the paralepidid *Magnisudis prionosa* (Cooper and Klages 1995; Reid *et al.* 1996). Benthic fish form the bulk of the fish component of the Southern Ocean albatross species. Benthic fish in the neritic zone belong to the Muraenolepididae, Nototheniidae and Channichthyidae, and form the main fish prey of Wandering, Black-browed and, to a lesser extent, of Grey-headed Albatrosses at South Georgia (Croxall *et al.* 1988a; Reid *et al.* 1996). Surprisingly, deep-water fish of the Gadiformes (Macrouridae, Moridae and Macruronidae) occur regularly in food samples of several species of albatrosses (Appendix 2). Presumably these fish species are taken by albatrosses scavenging from fishing operations (Reid *et al.* 1996) and/or float when dead (e.g., macrourid fish) and are eaten as carrion at the surface (Y. Cherel, unpubl. data).

At South Georgia, Grey-headed Albatrosses and, to a lesser extent, Black-browed Albatrosses also eat parasitic lampreys *Geotria australis* (Tickell 1964; Prince 1980). Lampreys are probably found in large groups at sea where they swim near the surface during their adult marine life cycle (Potter *et al.* 1979).

Crustaceans that parasitize fish ingested with their hosts are regularly found in the diet of albatrosses (e.g., Green 1974; Harris 1973; West and Imber 1986). Since the exoskeletons are more resistant to digestion than fish, they are useful indicators of fish in the diet. For example, the Copepod *Sphyrion lumpi*, which occurs in the food of the Wandering Albatross at Iles Crozet, parasitizes fish in the Scorpaeniformes and Gadiformes and was once found associated with *Halargyreus johnsonii* (Ridoux 1994).

The lengths of fish eaten by albatrosses range from 50–600 mm. Myctophids of <100 mm are the commonest small species of fish taken by Grey-headed and Black-browed Albatrosses (Reid *et al.* 1996). Wandering Albatrosses catch fish 225–634 mm, equivalent to fish weighing 74–3 027 g (Croxall *et al.* 1988a). Fish in excess of 400 mm may also be eaten (e.g., by Shy Albatrosses, Green 1974) but fish of this size are likely dismembered before being ingested (see Thompson and Riddy 1995).

Crustaceans

Crustaceans are a minor component by mass of the diets of albatrosses. Notable exceptions are the Antarctic Krill *Euphausia superba* and the Lobster-krill *Munida* sp. in the Southern Ocean. Lobster-krill accounts for about 5% by mass of the diet of Black-browed Albatross chicks at the Falkland Islands (Thompson 1992). Antarctic Krill is important in the diets of Black-browed (40% by mass), Grey-headed (17%) and Light-mantled Sooty Albatrosses (40%) at South Georgia (Prince 1980; Thomas 1982). Studies at sea confirm the importance of krill to Light-mantled Sooty Albatrosses (Linkowski and Rembiszewski 1978; Ainley *et al.* 1984; Harper 1987) and to the small albatross species of the Southern Hemisphere (Harrison *et al.* 1991). Albatrosses usually take adult krill 40–50 mm in length (Prince 1980; Thomas 1982; Ridoux 1994).

Giant mysids of the genus *Gnathophausia* (80–150 mm length) and amphipods of the genus *Eurythenes* (40–85 mm length) are common prey (but always few in number per sample) of several albatross species (Green 1974; Harrison *et al.* 1983; West and Imber 1986; Ridoux 1994). These crustaceans, together with *Notostomus, Anuropus* and *Chalaraspidum*, occur from 1 000–3 500 m depth (Holthuis and Sivertsen 1967; Ledoyer 1995) and their occurrence in the diet of albatrosses is surprising. Secondary consumption may explain the presence of copepods, amphipods and prawns in the diets of the small Southern Ocean Albatrosses (Appendix 3). Although Buller's Albatrosses have been observed diving into plankton swarms (Fenwick 1978), it is not known if albatrosses were preying directly on crustaceans or on planktivorous fish that were also present.

Other dietary material

Dead seal and gelatinous plankton are occasionally eaten by albatrosses, but in general they account for <10% by mass of the diet (Appendix 4–5). The proportion of carrion is medium to high in the diets of Wandering (5–19% by mass), Light-mantled Sooty (4–17%) and Sooty Albatrosses (25–51%) (Prince and Morgan 1987; Cooper *et al.* 1992; Ridoux 1994; Cooper and Klages 1995). Marine mammal is common in the diet of Wandering Albatrosses, and seabirds in the diet of the two *Phoebetria* species. Blubber with skin attached is the most frequent remains of cetaceans and seals (Cooper *et al.* 1992; Ridoux 1994), but seal hairs and skin, an elephant seal testicle, and large nematod parasites of fur seals have all been found in albatross stomachs (Paulian 1953; Tickell 1964; Mougin 1970; Segonzac 1972).

Seventeen bird taxa of two orders (Sphenisciformes and Procellariiformes) have been identified in albatross food samples (Appendix 4). For the Southern Ocean Albatrosses, penguins occur most frequently. Penguin feathers are typically found as accumulated items in stomach contents, and some casts are almost entirely made of them (Y.C. and N.K., unpubl. data). The penguin species eaten are generally unknown, but due to their huge populations, Macaroni Penguins *Eudyptes chrysolophus* probably form the bulk of the penguin prey in the sub-Antarctic region, while Rockhopper Penguins *E. chrysochome* are likely to be eaten in the northern localities (where Macaroni Penguins do not breed) of the Southern Hemisphere. Prions *Pachyptila* spp. and Diving Petrels *Pelecanoides* spp. also occur regularly in the diets of albatrosses (Appendix 4). Some petrels are ingested whole (Richardson 1984; Ridoux 1994; Cooper and Klages 1995) and are either captured alive or scavenged.

Jellyfish and tunicates are also taken by albatrosses but these form only a minor component of their diets overall (Appendix 5). Jellyfish and tunicates are probably quickly digested by albatrosses and are thus likely to be underestimated in diet analyses. Wandering and Black-browed Albatrosses are often seen feeding on jellyfish over the Iles Crozet shelf (Weimerskirch *et al.* 1986), but no remains have been found in chick stomach samples (Ridoux 1994). Salps have been recorded in a highly digested condition in food samples of Wandering Albatrosses (up to 90 individuals in one sample) and of the small Southern Ocean Albatrosses (Tickell 1964; West and Imber 1986; Imber 1992). At South Georgia Albatrosses tend to select salps that are asexual and solitary in preference to the sexual aggregate forms (Tickell 1964).

TEMPORAL AND SPATIAL VARIATIONS

Temporal variation

Information on seasonal variations in albatross diet is limited to the reproductive period; there are no available data on dietary changes between the breeding and non-breeding periods. At Hawaii, only minor differences were found in the major prey items consumed by albatross chicks between the hatching and fledging periods (Harrison *et al.* 1983). Studies over five years into the food of albatross chicks at South Georgia show annual fluctuations in broad prey classes (Fig. 1). Krill ranged from 5–42% by mass in the diet of Black-browed Albatrosses, and fish

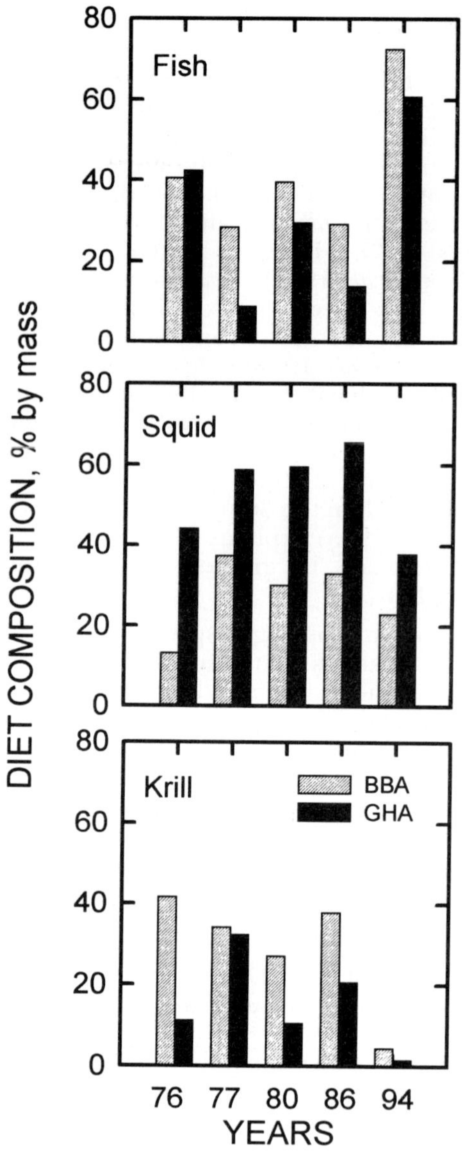

Fig. 1. Interannual variations in the proportion by mass of broad prey classes in the diet of Black-browed (BBA) and Grey-headed (GHA) Albatross chicks from South Georgia (data from Croxall *et al.* 1988b; Reid *et al.* 1996).

comprised 9–61% by mass in the diet of Grey-headed Albatrosses (Croxall *et al.* 1988b; Reid *et al.* 1996). A dietary shift from krill to fish is obvious in some years of suspected poor krill availability (Fig. 1), while no changes in the relative proportions of cephalopod were found (Rodhouse *et al.* 1987; Rodhouse and Prince 1993). Wandering Albatrosses may respond to "abnormal" years, marked by a decrease in Antarctic squid abundance in their diet, by extending their foraging range to exploit a cephalopod community in warmer waters farther north (Rodhouse *et al.* 1987; Rodhouse 1989). At such times, the overall proportion of squid in the diet decreases, and is compensated by increased contribution of benthic fish (Rodhouse *et al.* 1987; Croxall *et al.* 1988a).

Spatial variation

Only two studies have investigated dietary differences between sites at the same locality. At Hawaii, few differences in the major prey items consumed by Black-footed and Laysan Albatrosses were detected among three breeding island groups (Harrison *et al.* 1983). In contrast the relative proportions of squid prey of Black-browed Albatrosses from the Falkland Islands varied markedly between sites, from <1% to 74% by mass, suggesting that birds from different colonies may have different feeding grounds (Thompson 1992), as described by Weimerskirch *et al.* (1988) for albatrosses at Iles Kerguelen.

For five species of Southern Ocean albatrosses the food has been described for two to seven breeding localities, thus allowing comparisons among localities. Proportions by mass of prey classes varied little (Royal Albatross), moderately (Wandering and the two *Phoebetria* Albatrosses), or greatly (Black-browed and Grey-headed Albatrosses) (Table 3). Surprisingly, large differences occurred in the diet of Grey-headed Albatrosses breeding at Prince Edward and Iles Crozet, both being located in similar water masses of the Indian Ocean (Table 3) (Hunter and Klages 1989; Ridoux 1994); however, these differences may be an artefact of the small sample sizes collected from albatrosses at Iles Crozet and/or interannual variation in the diet (Weimerskirch *et al.* 1986). The proportion of squid in the cephalopod component of albatross diet may differ with locality Imber (1992): many squid species decrease in frequency southwards, others increase, and several are most frequent at sub-Antarctic localities. Onychoteuthids form the bulk of the diet by mass (46–96%) of Wandering Albatrosses south and north of the Antarctic Polar Front, except at the northern locality, Gough Island where they are partially replaced by histioteuthids (Cooper *et al.* 1992; Ridoux 1994) (Fig. 2). Since Amsterdam and Gough Islands are both located near the Subtropical Front, histioteuthids can be expected to be important in the food of the endangered Amsterdam Albatross.

Dietary differences among birds breeding in different localities probably reflect the prey availability in the foraging areas of the respective populations. Antarctic krill is abundant in the Scotia Sea and, consequently, it accounts for a higher percentage by mass in the diet of Black-browed, Grey-headed and Light-mantled Sooty Albatrosses at South Georgia than elsewhere in the Southern Ocean (Prince 1980; Thomas 1982). The virtual absence of ommastrephids in the diet of albatrosses, including ommastrephid-eating albatrosses, at the Prince Edward Islands (Hunter and Klages 1989; Cooper *et al.* 1992; Cooper and Klages 1995), suggests that these squids do not occur near the islands, while being abundant at other localities (Imber 1991; Rodhouse *et al.* 1993; Ridoux 1994; Cherel and Weimerskirch 1995).

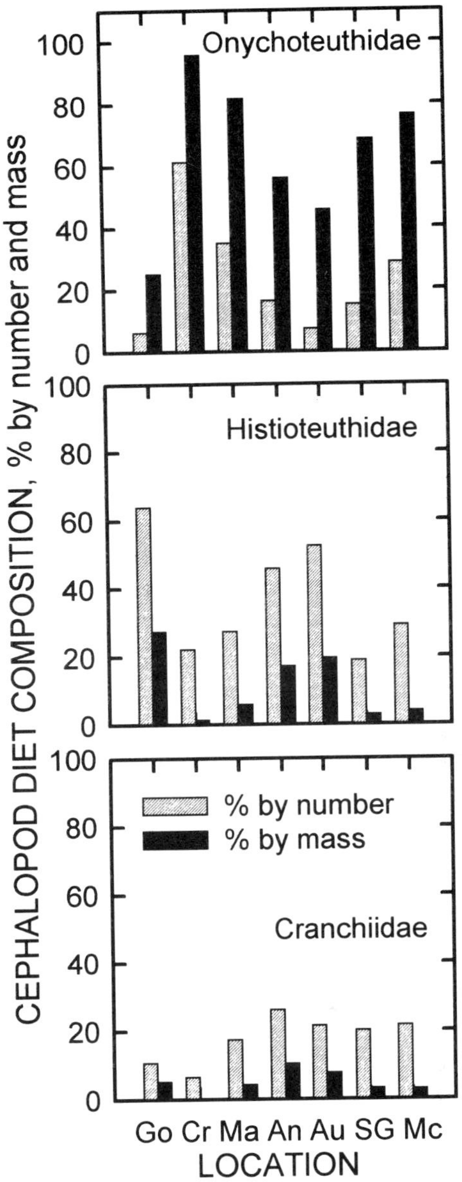

Fig. 2. Spatial variations in the proportions by number and mass of the three main families of squids eaten by Wandering Albatrosses (data from Cooper *et al.* 1992; Imber 1992; Ridoux 1994). Abbreviations: Go, Gough (40°S); Cr, Crozet (46°S); Ma, Marion (47°S); An, Antipodes (50°S); Au, Auckland (51°S); SG, South Georgia (54°S); Mc, Macquarie (55°S).

DIETARY SEGREGATION AMONG SYMPATRIC ALBATROSSES

Partitioning of food resources among sympatric albatross species has been studied at four sites: Hawaii (two species), Iles Crozet (six species), Prince Edward (five species) and South Georgia (four species). There is considerable overlap among sites in the broad prey classes, but the proportions by number and by mass of the prey

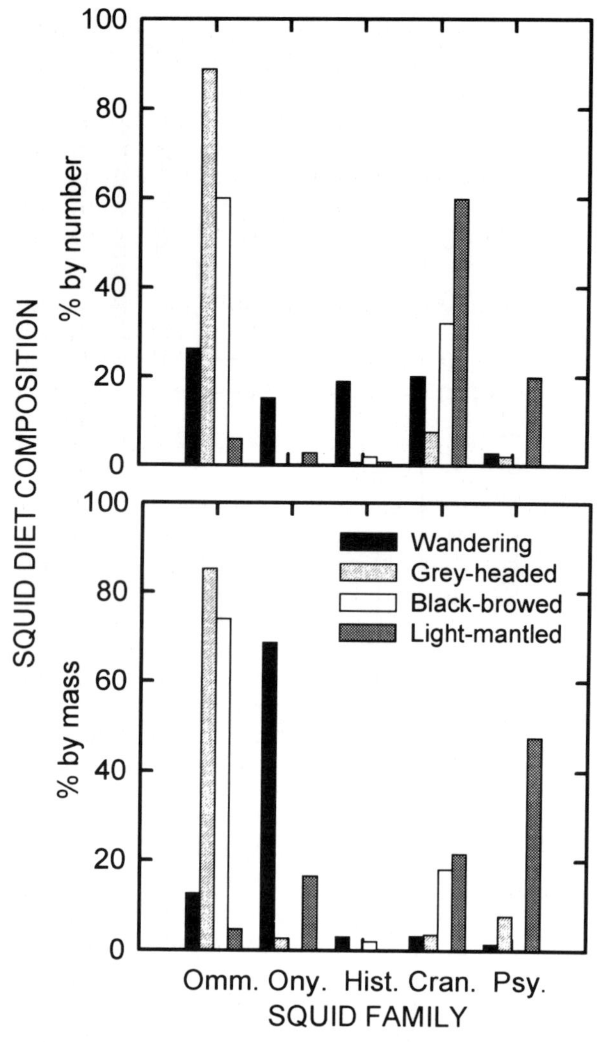

Fig. 3. Interspecific variations in the proportions by number and by mass of the main squid families in the diet of four sympatric albatross species from South Georgia (data from Prince and Morgan 1987; Croxall *et al.* 1988b; Imber 1992; Rodhouse *et al.* 1990; Rodhouse and Prince 1993). Abbreviations: Omm, Ommastrephidae; Ony, Onychoteuthidae; Cran, Cranchiidae; Psy, Psychroteuthidae.

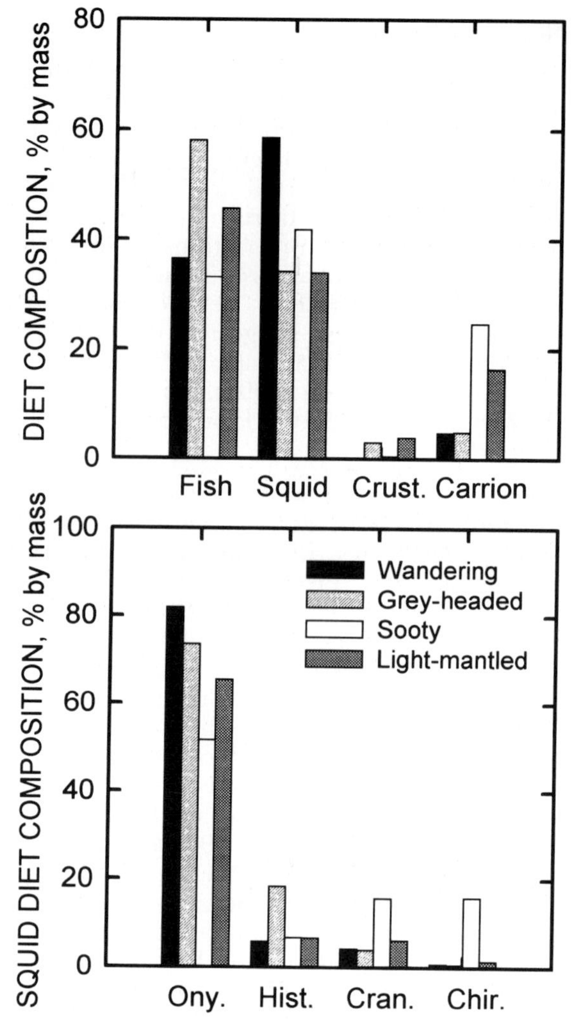

Fig. 4. Interspecific variations in the proportion by mass of broad prey classes and of the main squid families in the diet of four sympatric albatross species from Marion Island (data from Hunter and Klages 1989; Cooper *et al.* 1992; Cooper and Klages 1995). Abbreviations: Ony, Onychoteuthidae; Hist, Histioteuthidae; Cran, Cranchiidae; Chir, Chiroteuthidae.

differ among albatross species. At Hawaii, both the Black-footed and Laysan Albatrosses forage in cool oceanic waters; however, Laysan Albatrosses take more squids (65 versus 32% by volume) and Black-footed Albatrosses more flying fish eggs (44 versus 4%). Such differences may result from temporal segregation, Laysan Albatrosses generally feeding at night and Black-footed Albatrosses feeding during the daylight hours (Harrison *et al.* 1983).

Differences and similarities in diet composition are consistent among Southern Ocean Albatrosses (Figs 1, 3 and 4). At South Georgia Black-browed Albatrosses eat greater amounts of krill and krill-eating fish, while Grey-headed Albatrosses feed more on squid, lamprey, and myctophids and deep-water fish (Tickell 1964; Prince 1980; Reid *et al.* 1996). Juvenile ommastrephids are, however, the main cephalopod prey of all Black-browed, Grey-headed and Yellow-nosed Albatrosses at Iles Crozet and South Georgia, while adult onychoteuthids predominate either alone for the Wandering Albatross, or in association with cranchiids, histioteuthids or psychroteuthids for the other species (Fig. 3) (Croxall and Prince 1994; Ridoux 1994). Observations of albatrosses at sea and knowledge of the biogeography of prey suggest that dietary differences result more from spatial segregation of albatrosses at sea than from specialization on different prey within the same area. At the Iles Crozet, the separation of foraging zones occurs not only between the different albatross species, but between birds at different colonies, between immatures and adults, and between the two sexes (Weimerskirch *et al.* 1986, 1988). Each species of albatross is associated with a distinct marine habitat: Black-browed Albatrosses are restricted to insular shelves and slopes and Yellow-nosed and

Grey-headed Albatrosses to oceanic waters north and south of the subtropical convergence, respectively. Wandering Albatrosses, which rear their chicks during the austral winter, forage over extensive oceanic areas and in slope waters (Weimerskirch *et al.* 1986, 1993, 1994). Sooty and Light-mantled Sooty Albatrosses overlap greatly in their diets, but Sooty Albatrosses consume more squid north of the Antarctic Polar Front than do Light-mantled Sooty Albatrosses (Ridoux 1994; Cooper and Klages 1995). The prevalence of the squid *Psychroteuthis glacialis* in the diet of Light-mantled Sooty Albatrosses indicates that this albatross feeds closer to Antarctica than any other albatross species (Prince and Morgan 1987; Ridoux 1994).

When albatrosses congregate at sea differences in their flight manoeuvrability and dominance behaviour may explain food partitioning among species (Wood 1992). When competing for the same prey, larger species of albatrosses tend to exclude smaller species (Weimerskirch *et al.* 1986; Wood 1992). This foraging hierarchy is reversed when albatrosses scavenge discards from moving fishing vessels which favours smaller, more manoeuvrable albatross species (Wood 1992).

RELATIONSHIPS BETWEEN ALBATROSSES AND HUMAN ACTIVITIES

Fisheries

Important effects of fishing operations on albatrosses are incidental mortality through entanglement in nets, collision with net sonde cables (Bartle 1991), and drowning on longline hooks (Alexander *et al.* 1997), as well as competition for the common resources, and the provision of new ones through discards and bait. Fisheries may affect albatross diet by increasing the availability of food items and by providing prey species unavailable to the birds under natural conditions. At the Falkland Islands, Black-browed Albatross chicks are fed extensively on commercially exploited species of squid and fish, including *Loligo gahi*, nototheniids, *Micromesistius australis* and *Merluccius* spp., scavenged by adult birds (Thompson 1992; Thompson and Riddy 1995). Without a fishery some of these taxa (e.g., *M. australis*) would be unavailable to albatrosses, while for others (e.g., *L. gahi*) the availability has increased. The quantity of waste generated by the *Loligo* trawl fishery alone amounts to about 5% of the reported catch. Just over 50% of this waste is scavenged by adult Black-browed Albatrosses nesting on Beauchêne Island, Falkland Islands, and between February and May waste provides 10–15% of the total food requirement of the breeding population. In addition, about 5% of the annual energy requirements of the breeding population at the Falkland Islands are obtained from discards of finfish and squid fisheries operating in the vicinity of the island. Fisheries in Falkland Islands waters are greater predators of finfish and squid than are Black-browed Albatrosses, and the benefit gained by the birds from scavenging on fisheries waste may be more than offset by potential depletion of their natural prey stocks (Thompson 1992; Thompson and Riddy 1995), at least in the short term.

Considerably less detailed information is available for the other species of albatrosses regarding the contributions of fisheries' discards and scavenged bait to their diet. For example, it seems likely that the Falkland Islands fisheries provide squid prey for Wandering Albatrosses nesting at South Georgia (Imber 1992). Wandering Albatrosses, as long-range oceanic feeders, are potentially in contact year-round with longline fishing fleets targeting Tuna *Thunnus* spp. and Toothfish *Dissosticus eleginoides* (Prince *et al.* 1997). Wandering Albatrosses are successful in competition with other species of seabirds for baits on hooks set by these fisheries, and are likely to benefit from the supplementary food provided from dislodged baits and offal discards. They are also highly susceptible to being hooked and to drown in the process (Brothers 1991; Prince *et al.* 1997).

The limited diet data available for Buller's Albatrosses (Appendix 1–4) indicate that *Nototodarus* sp. is a key prey item for this species. We suspect that the trawl fishery targeting this squid off New Zealand may have led to an artificially elevated increase of the availability of *Nototodarus* to Buller's Albatrosses. The same may also apply to Shy Albatrosses in that region which are second only to Wandering Albatrosses in their aggression towards other seabirds in obtaining food from trawler discards (Brothers 1991). Short-tailed, Black-footed and Laysan Albatrosses were attracted to the North Pacific pelagic gill net fisheries before this fishery was banned at the end of 1992 (Gould and Hobbs 1992; Gould *et al.* 1997). Although the dietary data available for these species are limited, the squid *Ommastrephes bartrami*, scavenged from the drift net fishery, comprised, respectively, 41% and 39% of the diet of Laysan and Black-footed Albatrosses feeding from gill nets (P. Gould, unpubl. data).

Other human activities

Albatrosses are well-known ship followers and there are numerous observations of albatrosses eating ship's garbage (Harris 1973; Barton 1979; Thomas 1982; Harper 1987). However, some species, such as Waved, Grey-headed and both *Phoebetria* Albatrosses, associate with ships less often than do the other species (Harris 1973; Croxall and Prince 1994).

Plastic particles and other artefacts have been found in stomach contents and casts of all the species of albatrosses investigated, including the endangered Short-tailed Albatross (Gales 1993; Auman *et al.* 1997). Most foreign material floats, suggesting that the particles are picked up at sea (Ryan 1987). In Hawaiian waters, floating objects may provide a substrate for flying fish ova and plastic particles may be ingested incidentally with the fish ova (Harrison *et al.* 1983). The size of plastic debris ingested by albatrosses is related to their body size, and birds show a strong colour selectivity, favouring orange and red particles that probably resemble some of their natural prey, such as crustaceans (Ryan 1987).

CONCLUSIONS AND PERSPECTIVES

Among seabirds, albatrosses belong to the guild of surface feeders that prey on organisms from 0.1 cm to over 100 cm (Croxall and Prince 1980; Harrison *et al.* 1983; Ridoux 1994). Within this guild, albatrosses are the largest species and feed on larger prey than other seabirds. They are opportunistic and probably take any squid and fish and, to a lesser extent, crustaceans of appropriate size that occur in epipelagic waters. Some albatrosses feed on organisms in excess of their lifting abilities, indicating scavenging (Croxall and Prince 1994; Ridoux 1994). Albatrosses exploit the marine food chain at three levels: omnivorous zooplankton and small nekton, larger carnivorous nektonic species, and top predators through scavenging. Association with marine mammals and human activities can be important, but the latter is not without risk because of the potential for high mortality rates during fisheries activities.

There is still considerable basic information to be gathered on the feeding ecologies of albatrosses. The diets of some species or populations are unstudied, such as the endangered Short-tailed and Amsterdam Albatrosses, Waved, Shy and Yellow-nosed Albatrosses, and most of the populations from New Zealand, the world's richest area for albatrosses. Determinations of prey and foraging grounds during and outside the breeding season are of paramount importance for albatross conservation because they permit the clarification of the relationships between seabirds and fisheries. An integrative approach to feeding ecology is therefore necessary, including the simultaneous measurements of foraging ranges, feeding rates and behaviour at sea (Weimerskirch *et al.* 1993; Wilson *et al.* 1995a,b), supplemented by stomach contents analysis of instrumented birds.

ACKNOWLEDGEMENTS

The authors thank Vincent Ridoux for his help in improving the manuscript, and Mike Imber and Keith Reid for their comments as referees.

APPENDIX 1

Cephalopod taxa recorded in the diets of albatrosses.

Taxa	Wandering	Royal	Waved	Short-tailed	Black-footed	Laysan	Black-browed	Buller's	Shy	Yellow-nosed	Grey-headed	Sooty	Light-mantled Sooty
Sepiidae													
Sepia sp.	An	—	—	—	—	—	Aus, SA	—	—	—	—	—	—
S. apama	Aus	—	—	—	—	—	—	—	—	—	—	—	—
Loliginidae													
Sepioteuthis australis	Aus	—	—	—	—	—	Aus	—	Aus	Aus	—	—	—
Loligo sp./*gahi*	SG	—	—	—	—	—	F	—	—	—	—	—	—
L. vulgaris reynaudi	—	—	—	—	—	—	SA	—	—	—	—	—	—
Lycoteuthidae													
Lycoteuthis (Oregoniateuthis) sp.	Cr	—	—	—	—	—	—	—	—	—	Cr	Cr	Cr, SG
Enoploteuthidae sp.	—	—	Ga	—	—	—	SG	—	—	—	—	—	—
Enoploteuthis sp.	—	—	—	—	—	—	SG	—	—	—	—	—	—
Ancistrocheirus sp./*lesueuri*	An, Auc, Cr, Go, Mc, PE, SG	—	—	—	—	H	SG	Ch, Sn	—	—	—	Cr, PE	—
Octopoteuthidae sp.	—	—	Ga	—	H	H	—	—	—	—	—	—	—
Taningia danae	An, Auc, Cr, Go, Mc, PE, SG	—	—	—	—	—	—	—	—	—	PE	Cr, PE	PE
Octopoteuthis sp./*rugosa*	An, Auc, Cr, Go, Mc, PE, SG	—	—	—	—	—	—	Ch, Sn	—	—	—	Cr, PE	—
Onychoteuthidae sp.	—	—	Ga	—	—	H	—	—	—	—	PE		PE
Onychoteuthis sp.	An, Auc, Cr, Go, Mc, PE, SG	—	Ga	—	—	—	—	—	—	—	—	Cr	Cr
O. banksi	An, Go	—	—	—	—	—	—	—	—	—	—	—	—
Moroteuthis sp.	—	—	Ga	—	—	—	—	—	—	—	PE	PE	—
M. ingens	An, Auc, Cr, Go, Mc, PE, SG	Ca, Ch, NZ	—	—	—	—	—	Ch, Sn	—	—	—	PE	Cr
M. robsoni	An, Auc, Cr, Go, Mc, PE, SG	—	—	—	—	—	—	Ch, Sn	—	PE	PE	Cr, PE	PE
M. knipovitchi	An, Auc, Cr, Go, Mc, PE, SG	—	—	—	—	—	—	—	—	PE	Cr, PE, SG	Cr, PE	Cr, PE, SG
Kondakovia longimana	An, Auc, Cr, Go, Mc, PE, SG	Ca	—	—	—	—	SG	—	—	PE	Ca, Cr, PE, SG	Cr, PE	Ca, Cr, Mc, PE, SG
Gonatidae													
Gonatus sp./*antarcticus*	An, Auc, Cr, Go, Mc, PE, SG	Ch	—	—	—	—	F, SA, SG	—	—	—	Cr, PE, SG	Cr, PE	Cr, Mc, PE, SG
G. phoebetriae	An, SG	—	—	—	—	—	—	—	—	—	—	PE	—
Histioteuthidae													
Histioteuthis sp.	An, Cr, PE	Ch, NZ	Ga	—	—	H	SA, SG	Ch, Sn	—	—	Ca, Cr, PE, SG	Cr, PE	PE
H. bonnellii corpuscula	An, Auc, Go, PE, SG	—	—	—	—	—	—	—	—	—	—	PE	PE
H. macrohista	An, Auc, Go, Mc, PE, SG	—	—	—	—	—	—	Ch, Sn	—	—	—	PE	—
H. atlantica	An, Auc, Go, Mc, PE, SG	Ch, NZ	—	—	—	—	—	Ch, Sn	—	—	—	PE	PE

Appendix 1 — *continued.*

Taxa	Wandering	Royal	Waved	Short-tailed	Black-footed	Laysan	Black-browed	Buller's	Shy	Yellow-nosed	Grey-headed	Sooty	Light-mantled Sooty
Histioteuthidae — *continued*													
H. miranda	An, Auc, Go, PE, SG	—	—	—	—	—	—	Ch, Sn	—	—	—	PE	PE
H. meleagroteuthis	An, Auc, Go, PE	—	—	—	—	—	—	—	—	—	—	—	—
H. dofleini (= *hoylei*)	An, Auc, Go, Mc, PE, SG	—	—	—	—	—	—	Ch, Sn	—	—	—	PE	—
H. eltaninae	An, Auc, Cr, Mc, PE, SG	—	—	—	—	—	SG	—	—	—	Cr, PE, SG	Cr, PE	Cr, PE, SG
H. celetaria	An	—	—	—	—	—	—	—	—	—	—	—	—
Psychroteuthidae													
Psychroteuthis sp./*glacialis*	An, Cr, Mc, PE, SG	—	—	—	—	—	Cr	—	—	—	Cr, SG	PE	Ca, Cr, Mc, PE, SG
Architeuthidae													
Architeuthis sp.	An, Auc, Cr, Go, Mc, PE, SG	Ch	—	—	—	—	—	—	—	—	—	Cr, PE	—
Neoteuthidae													
Alluroteuthis antarcticus	An, Auc, Cr, Go, Mc, PE, SG	—	—	—	—	—	SG	—	—	—	Cr, SG	Cr, PE	Ca, Cr, Mc, PE, SG
Brachioteuthidae													
Brachioteuthis sp.	An, Cr	—	—	—	—	—	—	—	—	—	SG	—	—
Ommastrephidae sp.	Auc, Mc, PE	—	Ga	—	H	H	—	Ch, Sn	—	—	—	—	—
Illex argentinus	SG	—	—	—	—	—	F	—	—	—	—	—	—
Todarodes sp.	—	—	—	—	—	—	K	—	—	—	—	—	—
T. filippovae	—	—	—	—	—	—	Cr	—	—	Cr	Cr	—	—
T. pacificus	—	—	—	Ja	—	—	—	—	—	—	—	—	—
Nototodarus sp./*sloani*	—	Ch, NZ	—	—	—	—	—	Ch, Sn	Ca	—	—	PE?	—
N. gouldi	—	—	—	—	—	—	—	—	Aus	—	—	—	—
Martialia hyadesi	An, Mc, SG	—	—	—	—	—	F, K, SG	—	—	—	Ca, SG	PE	Ca, Mc, PE, SG
Ommastrephes bartrami	—	—	—	—	—	—	SA	—	SA	—	—	—	—
Symplectoteuthis sp./*oualaniensis*	—	—	Ga	—	H	H	—	—	—	—	—	—	—
Hyaloteuthis pelagica	—	—	—	—	H	—	—	—	—	—	—	—	—
Lepidoteuthidae sp.	—	—	—	—	—	H	—	—	—	—	—	—	—
Lepidoteuthis sp./*grimaldii*	An, Cr, Go, PE, SG	—	—	—	—	—	—	—	—	—	—	Cr	SG
Pholidoteuthidae													
Pholidoteuthis sp./*boschmai*	An, PE, SG	—	Ga	—	—	—	—	—	—	—	—	—	PE
Batoteuthidae sp.	Cr	—	—	—	—	—	—	—	—	—	Cr	Cr	Cr
Batoteuthis skolops	An, Auc, Mc, PE, SG	—	—	—	—	—	—	—	—	—	—	PE	—
Cycloteuthidae													
Cycloteuthis sp./*akimushkini*	An, Auc, Cr, Go, Mc, PE	—	—	—	—	—	—	—	—	—	—	Cr, PE	Cr
Discoteuthis sp./*laciniosa*	PE	—	—	—	—	—	—	—	—	—	PE	PE	PE

Appendix 1 — *continued.*

Taxa	Wandering	Royal	Waved	Short-tailed	Black-footed	Laysan	Black-browed	Buller's	Shy	Yellow-nosed	Grey-headed	Sooty	Light-mantled Sooty
Chiroteuthidae sp.	An, Go	—	Ga	—	—	—	—	—	—	—	—	—	—
Chiroteuthis sp.	An, Auc, Cr, Go, PE, SG	—	—	—	—	—	Cr, SA, SG	Ch, Sn	—	—	Cr, PE, SG	Cr, PE	Cr, PE, SG
C. capensis	An, Go	—	—	—	—	—	—	—	—	—	—	—	—
C. veranyi	—	—	—	—	—	—	SG	—	—	—	SG	PE	—
C. imperator (= *picteti*)	Cr, SG	—	—	—	—	—	—	—	—	—	Cr	Cr	—
Mastigoteuthidae													
Mastigoteuthis sp.	An, Cr, Go, PE	—	—	—	—	H	SA	Ch, Sn	—	—	PE, SG	Cr, PE	Cr, PE
M. psychrophila	An, Mc, SG	—	—	—	—	—	—	—	—	—	SG	—	—
Promachoteuthidae													
Promachoteuthis sp.	SG?	—	—	—	—	—	—	—	—	—	—	—	—
Grimalditeuthidae													
Grimalditeuthis bonplandi	An	—	—	—	—	—	—	—	—	—	—	—	—
Cranchiidae sp.	An, Go, PE, SG	—	—	—	H	—	—	—	—	PE	—	—	—
Cranchia scabra	An, Go, Mc, PE, SG	—	—	—	—	—	—	—	—	—	—	PE	—
Galiteuthis sp.	An, Auc, Go, Mc, PE, SG	Ca, Ch	—	—	—	—	—	—	—	—	PE	PE	—
G. glacialis	An, Auc, Cr, Go, Mc, PE, SG	Ch	—	—	—	—	SG	—	—	—	Ca, Cr, PE, SG	Cr, PE	Ca, Cr, K, Mc, PE, SG
Mesonychoteuthis hamiltoni	An, Cr, Mc, PE, SG	—	—	—	—	—	—	—	—	—	SG	Cr, PE	Cr, PE
Megalocranchia sp.	Cr	Ch	—	—	—	—	—	—	—	—	—	—	Cr
Taonius sp./*pavo*	An, Auc, Cr, Go, Mc, PE, SG	Ca	—	—	—	—	SA	Ch, Sn	—	—	Cr, SG	Cr, PE	PE
Teuthowenia sp./*pellucida*	An, Auc, Cr?, Go, Mc, PE, SG	Ch	—	—	—	—	—	Ch, Sn	—	—	Cr, PE	PE	PE
Bathothauma sp./*lyromma*	An, Auc, Cr	—	—	—	—	—	—	—	—	—	SG	Cr, PE	Cr, Mc
Vampyroteuthidae													
Vampyroteuthis infernalis	An, Go, SG	—	—	—	—	—	—	—	—	—	—	—	—
Octopodidae sp.	SG	—	—	—	H	—	SG	—	—	—	SG	—	PE
Octopus dofleini	SG	—	—	—	—	—	—	—	—	—	—	—	—
O. maorum	—	Ch, NZ	—	—	—	—	—	Ch, Sn	—	—	—	—	—
Pareledone polymorpha	—	—	—	—	—	—	SG	—	—	—	—	—	—
Alloposidae													
Alloposus mollis	An, Go, PE, SG	—	—	—	—	—	—	—	—	—	SG	PE	—
Ocythoidae													
Ocythoe tuberculata	Auc, Mc, SG	—	—	—	—	—	—	Ch. Sn	—	—	—	—	—
Argonautidae													
Argonauta nodosa	Auc	—	—	—	—	—	—	Ch, Sn	—	—	—	—	—

Abbreviations: An, Antipodes; Auc, Auckland; Aus, Australia; Ca, Campbell; Ch, Chatham; Cr, Crozet; F, Falklands; Ga, Galapagos; Go, Gough; H, Hawaii; Ja, Japan; K, Kerguelen; Mc, Macquarie; NZ, New Zealand; PE, Prince Edward and Marion; SA, South Africa; SG, South Georgia; Sn, Snares.

References: Barton 1979; Berruti and Harcus 1978; Brooke and Klages 1986; Cherel and Weimerskirch 1995; Clarke and Prince 1981; Clarke *et al.* 1981; Cooper and Klages 1995; Cooper *et al.* 1992; Croxall and Prince 1994; Gales 1993; Gibson and Sefton 1959; Green 1974; Harper 1987; Harris 1973; Harrison *et al.* 1983; Hunter and Klages 1989; Imber 1991; Imber 1978; Imber 1992; Imber and Berruti 1981; Imber and Russ 1975; Lipinski and Jackson 1989; Marchant and Higgins 1990; Prince 1980; Prince 1982; Prince and Morgan 1987; Ridoux 1994; Rodhouse and Prince 1993; Rodhouse *et al.* 1987; Rodhouse *et al.* 1990; Thomas 1982; Thompson 1992; West and Imber 1986; Wood 1992.

APPENDIX 2

Fish taxa recorded in the diets of albatrosses.

Taxa	Wandering	Royal	Waved	Short-tailed	Black-footed	Laysan	Black-browed	Buller's	Shy	Yellow-nosed	Grey-headed	Sooty	Light-mantled Sooty
Petromyzontidae													
Geotria australis	An	—	—	—	—	—	SG	—	—	—	SG	—	—
Engraulidae													
Engraulis australis	—	—	—	—	—	—	—	—	Aus	Aus	—	—	—
E. ingens	—	—	PC	—	—	—	—	—	—	—	—	—	—
Clupeidae													
Sardinops neopilchardus	—	—	—	—	—	—	—	—	Aus	Aus	—	—	—
Etrumeus acuminatus	—	—	Ga	—	—	—	—	—	—	—	—	—	—
Microstomatidae													
Nansenia antarcticus	—	—	—	—	—	—	—	—	—	—	SG	—	—
Bathylagidae													
Bathylagus sp.	—	—	—	—	—	—	—	—	—	—	—	PE	—
Gonostomatidae													
Vinciguerria sp.	—	—	—	—	—	H	—	—	—	—	—	—	—
Sternoptychidae sp.	—	—	—	—	H	—	—	—	—	—	—	—	—
Argyropelecus sp.	—	—	—	—	—	H	—	—	—	—	—	—	—
Paralepididae													
Magnisudis prionosa	—	—	—	—	—	—	SG	—	—	—	SG	PE	PE
Anotopteridae													
Anotopterus pharao	—	—	—	—	—	—	—	—	—	—	SG	—	—
Myctophidae sp.	—	—	—	—	—	H	SG	—	—	—	SG	—	SG
Electrona antarctica	—	—	—	—	—	—	—	—	—	—	SG	—	—
E. carlsbergi	—	—	—	—	—	—	—	—	—	—	—	PE	PE
Gymnoscopelus bolini	—	—	—	—	—	—	SG	—	—	—	—	—	—
G. nicholsi	—	—	—	—	—	—	—	—	—	—	SG	—	—
G. fraseri	—	—	—	—	—	—	—	—	—	—	SG	—	—
G. microlampas	—	—	—	—	—	—	—	—	—	—	SG	—	—
G. piabilis	—	—	—	—	—	—	—	—	—	—	—	PE	—
Gymnoscopelus sp.	—	—	—	—	—	—	—	—	—	—	—	PE	PE
Krefftichthys anderssoni	—	—	—	—	—	—	SG	—	—	—	SG	—	—
Protomyctophum bolini	—	—	—	—	—	—	—	—	—	—	SG	—	—
P. choriodon	—	—	—	—	—	—	—	—	—	—	SG	—	—
Ophidiidae													
Genypterus blacodes	—	—	—	—	—	—	F	—	—	—	—	—	—
Spectrunculus grandis	PE	—	—	—	—	—	—	—	—	—	—	—	—
?*Ophidiidae* sp.	PE	—	—	—	—	—	—	—	—	—	—	—	—

Appendix 2 — *continued.*

Taxa	Wandering	Royal	Waved	Short-tailed	Black-footed	Laysan	Black-browed	Buller's	Shy	Yellow-nosed	Grey-headed	Sooty	Light-mantled Sooty
?Macrouridae sp.	PE, SG	—	—	—	—	—	—	—	—	—	—	—	PE
Coelorhynchus sp.	—	—	—	—	—	—	F	—	—	—	—	—	—
?*Lionurus* sp.	—	—	—	—	—	—	—	—	—	—	—	—	PE
?*Nezumia* sp.	—	—	—	—	—	—	—	—	—	—	—	PE	—
Moridae													
?*Antimora rostrata*	SG	—	—	—	—	—	—	—	—	—	—	—	—
Halargyreus johnsonii	Auc	—	—	—	—	—	—	—	—	—	—	—	—
Salilota australis	—	—	—	—	—	—	F	—	—	—	—	—	—
Pseudophycis bacchus	—	—	—	—	—	—	—	Ch, Sn	—	—	—	—	—
Melanonidae													
Melanonus gracilis	—	—	—	—	—	—	—	—	—	—	—	PE	PE
Macruronidae													
Macruronus magellanicus	—	—	—	—	—	—	F	—	—	—	—	—	—
M. novaezelandiae	Auc	Ca	—	—	—	—	—	—	—	—	—	—	—
Muraenolepididae													
Muraenolepis microps	SG	—	—	—	—	—	—	—	—	—	SG	—	—
Merlucciidae													
Merluccius sp.	—	—	—	—	—	—	F	—	—	—	—	—	—
Gadidae													
Micromesistius australis	—	—	—	—	—	—	F	—	—	—	—	—	—
Scomberesocidae													
Scomberesox saurus	—	—	—	—	—	—	—	—	—	T	—	—	—
Cololabis saira	—	—	—	—	—	H	—	—	—	—	—	—	—
Exocoetidae sp.	—	—	Ga	Ja	H, sea	H	—	—	—	—	—	—	—
Cypselurus sp.	—	—	—	—	—	H	—	—	—	—	—	—	—
Cheilopogon furcatus	—	—	—	—	—	—	—	—	—	Am	—	—	—
Hemiramphidae sp.	—	—	—	—	—	H	—	—	—	—	—	—	—
Euleptorhamphus viridis	—	—	—	—	—	H	—	—	—	—	—	—	—
Melamphaidae													
Melamphaes sp.	—	—	—	—	—	—	—	—	—	—	—	—	PE
Sio nordenskjöldii	—	—	—	—	—	—	—	—	—	—	SG	—	—
Berycidae													
Trachichthodes gerrardi	—	—	—	—	—	—	—	—	Aus	—	—	—	—
?Oreosomatidae sp.	PE	—	—	—	—	—	—	—	—	—	—	—	—
Macroramphosidae													
Notopogon lilliei	—	—	—	—	—	—	T	—	—	—	—	—	—

Appendix 2 — *continued.*

Taxa	Wandering	Royal	Waved	Short-tailed	Black-footed	Laysan	Black-browed	Buller's	Shy	Yellow-nosed	Grey-headed	Sooty	Light-mantled Sooty
Scorpaenidae sp.	—	—	—	—	Al	—	—	—	—	—	—	—	—
Priacanthidae													
Priacanthus cruentatus	—	—	Ga	—	—	—	—	—	—	—	—	—	—
Carangidae													
Trachurus declivis	—	—	—	—	—	—	—	—	Aus	—	—	—	—
Decapterus sp.	—	—	—	—	H	—	—	—	—	—	—	—	—
D. scombrinus	—	—	Ga	—	—	—	—	—	—	—	—	—	—
Mullidae sp.	—	—	—	—	—	H	—	—	—	—	—	—	—
Pentacerotidae													
Pseudopentaceros richardsoni	—	—	—	—	—	—	—	Ch, Sn	—	—	—	—	—
Nototheniidae sp.	SG	Ca	—	—	—	—	F, SG	—	—	—	SG	—	SG
Dissostichus eleginoides	—	—	—	—	—	—	F	—	—	—	—	—	—
Gobionotothen gibberifrons	SG	—	—	—	—	—	—	—	—	—	—	—	—
Lepidonotothen squamifrons	—	—	—	—	—	—	—	—	—	—	—	Cr	—
Patagonotothen guntheri	—	—	—	—	—	—	SG	—	—	—	SG	—	—
Trematomus hansoni	SG	—	—	—	—	—	—	—	—	—	—	—	—
Channichthyidae													
Chaenocephalus aceratus	SG	—	—	—	—	—	—	—	—	—	—	—	—
Champsocephalus gunnari	SG	—	—	—	—	—	SG	—	—	—	SG	—	—
Channichthys rhinoceratus	—	—	—	—	—	—	—	—	—	—	PE	—	—
Pseudochaenichthys georgianus	SG	—	—	—	—	—	SG	—	—	—	SG	—	—
Chiasmodontidae													
Dysalotus alcocki	—	—	—	—	—	—	—	—	—	—	PE	—	—
Pinguipedidae													
Parapercis colias	—	—	—	—	—	—	—	—	Ca	—	—	—	—
Antennariidae													
Antennarius sp.	—	—	Ga	—	—	—	—	—	—	—	—	—	—
Gempylidae sp.	—	—	—	—	H	H	—	—	—	—	—	—	—
Paradiplospinus gracilis	—	—	—	—	—	—	SG	—	—	—	—	Cr, PE	—
Thyrsites atun	—	—	—	—	—	—	—	—	—	Am	—	Am	—
Trichiuridae													
Benthodesmus elongatus	PE	—	—	—	—	—	—	—	—	—	—	—	—
Centrolophidae													
Icichthys australis	—	—	—	—	—	—	SG	—	—	—	—	—	—
Molidae sp.	—	—	—	—	—	H	—	—	—	—	—	—	—
Ranzania laevis	—	—	—	—	—	H	—	—	—	—	—	—	—

Abbreviations: Al, Aleutian Islands; Am, Amsterdam; An, Antipodes; Auc, Auckland; Aus, Australia; Ca, Campbell; Ch, Chatham; Cr, Crozet; F, Falklands; Ga, Galapagos; H, Hawaii; Ja, Japan; PC, Peruvial Current; PE, Prince Edward and Marion; sea, observation at sea; SG, South Georgia; Sn, Snares; T, Tristan da Cunha.

References: Barton 1979; Cooper and Klages 1995; Cooper *et al.* 1992; Cottam and Knappen 1939; Croxall *et al.* 1988a; Duffy 1983; Fisher 1945; Green 1974; Hagen 1952; Harris 1973; Harrison *et al.* 1983; Hasegawa and deGange 1982; Hunter and Klages 1989; Imber 1992; Imber and Russ 1975; Marchant and Higgins 1990; Murphy 1936; Potter *et al.* 1979; Prince 1980; Reid *et al.* 1996; Ridoux 1994; Rodhouse and Prince 1993; Segonzac 1972; Thomas 1982; Thompson 1992; Thompson and Riddy 1995; Tickell 1964; West and Imber 1986.

APPENDIX 3

Crustacean taxa recorded in the diets of albatrosses.

Taxa	Wandering	Royal	Waved	Black-footed	Laysan	Black-browed	Buller's	Shy	Yellow-nosed	Grey-headed	Sooty	Light-mantled Sooty
Stomatopoda												
Pseudosquilla sp.	—	—	—	—	H	—	—	—	—	—	—	—
Mysidacea												
Gnathophausia sp.	—	—	—	H	H	—	Ch, Sn	Aus	T	—	—	—
G. ingens	—	—	—	H	H	—	—	—	T	—	—	—
G. gigas	—	—	—	H	H	—	—	—	Cr, T	Cr	Cr	Cr
Chalaraspidum sp.	—	—	—	—	—	—	—	—	T	—	—	—
Mysidacea sp.	—	—	—	H	H	—	—	—	Cr	—	—	—
Isopoda												
Anuropus sp.	—	—	—	H	H	SG?	—	—	—	SG?	—	—
A. branchiatus	—	—	—	H	H	—	—	—	—	—	—	—
A. pacifica	—	—	Ga	—	—	—	—	—	—	—	—	—
Meinertia gaudichaudii	—	—	Ga	—	—	—	—	—	—	—	—	—
Livoneca sp./*raynaudi*	—	—	—	—	—	—	Ch, Sn	Aus	—	—	—	—
Isopoda sp.	—	Ch, NZ	—	H	H	SG	—	—	—	—	—	PE, SG
Amphipoda sp.	—	—	—	H	H	SG	—	—	—	SG	—	K
Themisto gaudichaudii	—	—	—	—	—	SG	—	—	—	Cr, PE, SG	Cr, PE	Cr, PE, SG
Hyperia sp.	—	—	—	—	—	—	—	—	—	—	Cr	—
Cyllopus ?lucasii	—	—	—	—	—	—	—	—	—	—	—	SG
Alicella sp.	—	—	—	H	—	—	—	—	—	—	—	—
Eurythenes sp.	—	—	—	—	—	—	—	—	—	—	PE	PE
E. obesus	—	—	—	—	—	—	—	—	—	—	—	Cr
E. gryllus	—	—	—	—	H	SG?	—	—	T	Cr, SG?	Cr	—
Hyale grandicornis	—	—	—	—	—	—	—	—	T	—	—	—
Gammaridea sp.	—	—	—	Al	—	—	—	—	—	—	—	PE
Euphausiacea sp.	sea	—	—	H	H	sea	Ch, Sn	Aus?, Ca	—	PE	—	sea
Euphausia superba	SG	—	—	—	—	SG	—	—	—	SG	Cr, PE	Cr, PE, RS, SG, sea
E. frigida	—	—	—	—	—	SG	—	—	—	SG	—	—
E. vallentini	—	—	—	—	—	—	—	—	—	—	Cr, PE	Cr
E. spinifera	—	—	—	—	—	—	Ch, Sn	—	—	—	—	—
Thysanopoda monacantha	—	—	Ga	—	—	—	—	—	—	—	—	—
Nyctiphanes capensis	—	—	—	—	—	—	Sn	—	—	—	—	—
N. australis	—	—	—	—	—	—	Sn	—	—	—	—	—
Benthophausia sp. (?)	—	—	Ga	—	—	—	—	—	—	—	—	—
Natantia (Decapoda) sp.	—	—	—	H	H	—	—	—	—	—	PE	PE
Chorismus antarcticus	—	—	—	—	—	SG?	—	—	—	SG?	—	—
Nematocarcinus longirostris	PE	—	—	—	—	—	—	—	—	—	—	—
Notostomus sp.	—	—	—	H	H	—	—	—	T	—	—	—
N. japonicus	—	—	—	H	—	—	—	—	—	—	—	—
Acanthephyra sp.	—	—	—	—	H	SG	—	—	—	SG	—	SG
A. exima	—	—	—	—	H	—	—	—	—	—	—	—
Oplophoridae sp.	—	—	—	H	—	—	—	—	—	—	—	—

Appendix 3 — *continued.*

Taxa	Wandering	Royal	Waved	Black-footed	Laysan	Black-browed	Buller's	Shy	Yellow-nosed	Grey-headed	Sooty	Light-mantled Sooty
Natantia (Decapoda) sp. — *continued*												
Pasiphaea scotiae/longispina	—	—	—	—	—	—	—	—	—	Cr	Cr	Cr, PE
Pasiphaeidae sp.	—	—	—	H	H	—	—	—	—	—	—	—
Austropandalus grayi	—	—	—	—	—	—	—	—	—	—	—	PE
Pandalidae sp.	—	—	—	—	—	—	—	Aus	—	—	—	—
Reptantia (Decapoda)												
Munida sp./*gregaria*	F	Ch, NZ	—	—	—	F	Ch, Sn	—	—	—	—	—
Lithodes sp.	PE	—	—	—	—	—	—	—	—	—	—	—
Planes cyaneus	—	—	—	H	H	—	—	—	—	—	—	—
Nectocarcinus integrifrons	—	—	—	—	—	—	—	Aus	—	—	—	—
Brachyura sp.	—	—	—	H	H	—	—	—	—	—	—	—
Decapoda sp.	—	—	Ga	—	—	SG	Ch, Sn	—	—	PE, SG	—	SG
Leptostraca												
Nebaliopsis typica	—	—	—	H	H	—	—	—	—	—	—	—
Copepoda												
Sphyrion lumpi	Cr	—	—	—	—	—	—	—	—	—	—	—
Penella sp.	—	—	—	H	—	—	—	—	—	—	—	—
Trifur lotellae	—	—	—	—	—	—	Ch, Sn	—	—	—	—	—
Copepoda sp.	—	—	—	—	—	—	—	—	Cr	—	—	—
Cirripedia thoracica												
Lepas sp.	—	Ch, NZ	—	—	—	—	Ch, Sn	Aus	—	PE	—	—

Abbreviations: Al, Aleutian Islands; Aus, Australia; Ca, Campbell; Ch, Chatham; Cr, Crozet; F, Falklands; Ga, Galapagos; H, Hawaii; K, Kerguelen; NZ, New Zealand; PE, Prince Edward and Marion; RS, Ross Sea; sea, observation at sea; SG, South Georgia; Sn, Snares; T, Tristan da Cunha.

References: Ainley *et al.* 1984; Bourne 1975; Cooper and Klages 1995; Cooper *et al.* 1992; Cottam and Knappen 1939; Falla 1937; Fenwick 1978; Green 1974; Hagen 1952; Harper 1987; Harris 1973; Harrison *et al.* 1983; Harrison *et al.* 1991; Holthuis and Sivertsen 1967; Hunter and Klages 1989; Imber 1991; Imber 1992; Ledoyer 1995; Linkowski and Rembiszewski 1978; Marchant and Higgins 1990; Murphy 1936; Prince 1980; Prince 1982; Ridoux 1994; Thomas 1982; Thompson 1992; Tickell 1964; West and Imber 1986.

APPENDIX 4

Details of carrion eaten by albatrosses.

	Wandering	Royal	Black-footed	Black-browed	Buller's	Grey-headed	Sooty	Light-mantled Sooty
Marine mammals (whale, seal)	Cr, K, PE, SG	—	—	SG	—	PE, SG	?Am, Cr	Cr, sea
Birds								
Spheniscidae sp.	Cr, Mc, SG	Ch	—	Cr, K, SG	—	Ca, Cr, PE, SG	Am, Cr, Go, PE, T	Ca, Cr, PE, SG
Eudyptes chrysolophus	—	—	—	—	—	—	PE	—
E. chrysocome	—	—	—	—	—	—	PE, T	PE
Procellariidae								
Pterodroma mollis	—	—	—	—	—	—	PE	—
P. hypoleuca	—	—	H	—	—	—	—	—
Pachyptila sp.	—	—	—	—	—	—	Cr, PE	PE, SG
P. vittata	—	—	—	—	—	—	T	—
P. salvini	—	—	—	—	—	NZ	Cr, PE	—
P. desolata	—	—	—	—	—	—	—	K
P. turtur	—	—	—	—	—	—	Cr	—
Procellaria aequinoctialis	PE	—	—	—	—	—	—	—
Puffinus sp.	—	—	H	—	—	—	—	—
P. pacificus	—	—	H	—	—	—	—	—
P. gravis	—	—	—	—	—	—	T	—
Hydrobatidae								
Fregetta sp.	—	—	—	—	—	—	T	—
Pelecanoididae								
Pelecanoides sp.	—	—	—	Auc	—	—	—	—
P. urinatrix	—	—	—	—	—	NZ	T	—
Undetermined birds	PE, ?	—	—	SG	Ch, Sn	SG	PE	PE

Abbreviations: Am, Amsterdam; Auc, Auckland; Ca, Campbell; Ch, Chatham; Cr, Crozet; Go, Gough; H, Hawaii; K, Kerguelen; Mc, Macquarie; NZ, New Zealand; PE, Prince Edward and Marion; sea, observation at sea, SG, South Georgia; Sn, Snares; T, Tristan da Cunha.

References: Berruti and Harcus 1978; Cooper and Klages 1995; Cooper *et al.* 1992; Falla 1937; Hagen 1952; Harrison *et al.* 1983; Hunter and Klages 1989; Imber 1991; Imber 1992; Linkowski and Rembiszewski 1978; Marchant and Higgins 1990; Mougin 1970; Murphy 1936; Paulian 1953; Prince 1980; Prince and Morgan 1987; Richardson 1984; Ridoux 1994; Segonzac 1972; Thomas 1982; Tickell 1964; Weimerskirch *et al.* 1986; Weimerskirch *et al.* 1988; West and Imber 1986.

APPENDIX 5

Details of minor prey items eaten by albatrosses.

	Wandering	Royal	Waved	Black-footed	Laysan	Black-browed	Buller's	Shy	Grey-headed	Sooty	Light-mantled Sooty
Hydrozoa (Cnidaria)											
Physalia sp.	?	—	—	—	—	—	—	—	—	—	—
Vellela vellela	—	—	—	H	H	—	—	—	—	Am	—
Cnidaria sp. (jellyfish)	Cr, PE	—	—	—	—	Cr, F, sea	—	—	—	Am	—
Ctenophora											
Ctenophora sp.	—	—	—	—	—	—	—	—	PE	—	—
Mollusca											
Janthina sp. (ova)	—	—	—	—	H	—	—	—	—	—	—
Nudibranchia	—	—	Ga	—	—	—	—	—	—	—	—
Unidentified mollusca	—	—	—	—	H	—	—	—	PE		
Insecta											
Halobates sericeus (Gerridae)	—	—	—	H	H	—	—	—	—	—	—
Insecta sp. (Caterpillar)	—	—	—	H	—	—	—	—	—	—	—
Echinodermata											
Strongylocentrotus dröbachiensis	—	—	—	Al	—	—	—	—	—	—	—

Appendix 5 — continued

	Wandering	Royal	Waved	Black-footed	Laysan	Black-browed	Buller's	Shy	Grey-headed	Sooty	Light-mantled Sooty
Tunicata											
Thaliacea											
Pyrosoma sp.	An, SG	Ch, NZ	—	H	H	sea	Ch, Sn	—	—	—	—
?*Iasis zonaria*	—	—	—	—	—	—	Ch, Sn	—	—	—	—
?*Metcalfina hexagona*	—	—	—	—	—	—	Ch, Sn	—	—	—	—
?*Salpa thompsoni*	—	—	—	—	—	SG	—	—	SG	—	—
Salps	—	—	—	—	—	sea	—	Aus	—	—	—
Ascidiacea											
?*Cnemidocarpa stewartensis*	—	—	—	—	—	—	Ch, Sn	—	—	—	—
Tunicata sp.	—	—	—	—	—	—	Ch, Sn	—	—	—	PE

Abbreviations: Al, Aleutian Islands; Am, Amsterdam; An, Antipodes; Aus, Australia; Ch, Chatham; Cr, Crozet; F, Falklands; Ga, Galapagos; H, Hawaii; NZ, New Zealand; PE, Prince Edward and Marion; sea, observation at sea, SG, South Georgia; Sn, Snares.

References: Cooper and Klages 1995; Cooper *et al.* 1992; Cottam and Knappen 1939; Green 1974; Harper 1987; Harris 1973; Harrison *et al.* 1983; Hunter and Klages 1989; Imber 1991; Imber 1992; Marchant and Higgins 1990; Murphy 1936; Prince 1980; Segonzac 1972; Thompson 1992; Tickell 1964; Weimerskirch *et al.* 1986; West and Imber 1986.

REFERENCES

Ainley, D. G., O'Connor, E. F. and Boekelheide, R. J., 1984. The marine ecology of birds in the Ross Sea. *Orn. Monogr.* **32:** 1–97.

Alexander, K., Robertson, G. and Gales, R., 1997. The incidental mortality of albatrosses in longline fisheries. Australian Antarctic Division, Tasmania. 44 pp.

Auman, H., J., Ludwig, J. P., Giesy, J. P. and Colborn, T., 1997. Plastic ingestion by Laysan Albatross chicks on Sand Island, Midway Atoll, in 1994 and 1995. Pp. 239–44 *in* Albatross Biology and Conservation ed by G. Robertson and R. Gales. Surrey Beatty & Sons: Chipping Norton.

Bartle, J. A., 1991. Incidental capture of seabirds in the New Zealand sub-Antarctic squid trawl fishery 1990. *Bird Conserv. Int.* **1:** 351–59.

Barton, D., 1979. Albatrosses in the western Tasman Sea. *Emu* **79:** 31–35.

Berruti, A. and Harcus, T., 1978. Cephalopod prey of the sooty albatrosses *Phoebetria fusca* and *P. palpebrata* at Marion Island. *S. Afr. J. Antarct. Res.* **8:** 99–103.

Bourne, W. R. P., 1975. Birds feeding on lobster-krill off the Falkland Islands. *Sea Swallow* **24:** 22–23.

Brooke, M. De L. and Klages, N., 1986. Squid beaks regurgitated by greyheaded and yellownosed albatrosses, *Diomedea chrysostoma* and *D. chlororhynchos* at the Prince Edward Islands. *Ostrich* **57:** 203–06.

Brothers, N., 1991. Albatross mortality and associated bait loss in the Japanese longline fishery in the Southern Ocean. *Biol. Conserv.* **55:** 255–68.

Cherel, Y. and Weimerskirch, H., 1995. Seabirds as indicators of marine resources: black-browed albatrosses feeding on ommastrephid squids in Kerguelen waters. *Mar. Ecol. Prog. Ser.* **129:** 295–300.

Clarke, M. R., 1986. A handbook for the identification of cephalopod beaks. Oxford University Press: Oxford.

Clarke, A. and Prince, P. A., 1980. Chemical composition and calorific value of food fed to mollymauk chicks *Diomedea melanophris* and *D. chrysostoma* at Bird Island, South Georgia. *Ibis* **122:** 488–94.

Clarke, M. R. and Prince, P. A., 1981. Cephalopod remains in regurgitations of black-browed and grey-headed albatrosses at South Georgia. *Br. Antarct. Surv. Bull.* **54:** 1–7.

Clarke, M. R., Croxall, J. P. and Prince, P. A., 1981. Cephalopod remains in regurgitations of the wandering albatross *Diomedea exulans* L. at South Georgia. *Br. Antarct. Surv. Bull.* **54:** 9–21.

Cooper, J. and Klages, N. T. W., 1995. The diets and dietary segregation of sooty albatrosses (*Phoebetria* spp.) at sub-antarctic Marion Island. *Ant. Sci.* **7:** 15–23.

Cooper, J., Henley, S. R. and Klages, N. T. W., 1992. The diet of the wandering albatross *Diomedea exulans* at sub-antarctic Marion Island. *Polar Biol.* **12:** 477–84.

Cottam, C. and Knappen, P., 1939. Food of some uncommon North American birds. *Auk* **56:** 138–69.

Croxall, J. P. and Prince, P. A., 1980. Food, feeding ecology and ecological segregation of seabirds at South Georgia. *Biol. J. Linnean Soc.* **14:** 103–31.

Croxall, J. P. and Prince, P. A., 1994. Dead or alive, night and day: how do albatrosses catch squid? *Ant. Sci.* **6:** 155–62.

Croxall, J. P., North, A. W. and Prince, P. A., 1988a. Fish prey of the wandering albatross *Diomedea exulans* at South Georgia. *Polar Biol.* **9:** 9–16.

Croxall, J. P., McCann, T. S., Prince, P. A. and Rothery, P., 1988b. Reproductive performance of seabirds and seals at South Georgia and Signy Island, South Orkney Islands, 1976–1987: implications for Southern Ocean Monitoring studies. Pp. 261–85 *in* Antarctic Ocean and Resources Variability ed by D. Sahrhage. Springer-Verlag: Berlin.

del Hoyo, J., Elliott, A. and Sargatal, J., 1992. Handbook of the Birds of the World. Vol. 1. Lynx Ediciones: Barcelona.

Duffy, D. C., 1983. The foraging ecology of Peruvian seabirds. *Auk* **100:** 800–10.

Falla, R. A., 1937. Birds. *B.A.N.Z.A.R.E. 1929–31, Ser. B* **2:** 1–304.

Fenwick, G. D., 1978. Plankton swarms and their predators at the Snares Islands. *N.Z. J. Mar. Freshwater Res.* **12:** 223–24.

Fisher, H. I., 1945. Black-footed albatrosses eating flyingfish. *Condor* **47:** 128–29.

Furness, B. L., Laugksch, R. C. and Duffy, D. C., 1984. Cephalopod beaks and studies of seabird diets. *Auk* **101:** 619–20.

Gales, R., 1993. Co-operative mechanisms for the conservation of albatross. Australian Nature Conservation Agency. Tasmanian Government Printer: Hobart.

Gibson, J. D. and Sefton, A. R., 1959. First report of the New South Wales Albatross Study Group. *Emu* **59:** 73–82.

Gould, P. and Hobbs, R., 1992. Population dynamics of the Laysan and other albatrosses in the North Pacific. *In* Proceedings of a symposium on the biology, distribution and stock assessment of species caught in high seas driftnet fisheries in the North Pacific Ocean ed by W. Shaw, R. Burger and J. Itô. International North Pacific Fisheries Commission: Tokyo.

Gould, P., Ostrom, P., Walker, W. and Pilichowski, K., 1997. Laysan and Black-footed Albatrosses: trophic relationships and driftnet fisheries associations of non-breeding birds. Pp. 199–207 *in* Albatross Biology and Conservation ed by G. Robertson and R. Gales. Surrey Beatty & Sons: Chipping Norton.

Green, R. H., 1974. Albatross Island, 1973. *Rec. Queen Victoria Mus.* **51:** 1–17.

Hagen, Y., 1952. Birds of Tristan da Cunha. *Res. Norwegian Sci. Exped. Tristan da Cunha 1937–1938* **20:** 1–248.

Harper, P. C., 1987. Feeding behaviour and other notes on 20 species of Procellariiformes at sea. *Notornis* **34:** 169–92.

Harris, M. P., 1973. The biology of the waved albatross *Diomedea irrorata* of Hood Island, Galapagos. *Ibis* **115:** 483–510.

Harrison, C. S., Hida, T. S. and Seki, M. P., 1983. Hawaiian seabird feeding ecology. *Wildl. Monogr.* **85:** 1–71.

Harrison, N. M., Whitehouse, M. J., Heinemann, D., Prince, P. A., Hunt, G. A. and Veit, R. R., 1991. Observations of multispecies seabird flocks around South Georgia. *Auk* **108:** 801–10.

Hasegawa, H. and deGange, A. R., 1982. The short-tailed albatross, *Diomedea albatrus*, its status, distribution and natural history. *Amer. Birds* **36:** 806–14.

Hecht, T., 1987. A guide to the otoliths of Southern Ocean fishes. *S. Afr. J. Antarct. Res.* **17:** 1–87.

Holthuis, L. B. and Sivertsen, E., 1967. The Crustacea Decapoda, Mysidacea and Cirripedia of the Tristan da Cunha Archipelago, with a revision of the "frontalis" subgroup of the genus *Jasus*. *Res. Norwegian Sci. Exped. Tristan da Cunha 1937–1938* **52:** 1–55.

Hunter, S. and Klages, N. T. W., 1989. The diet of grey-headed albatrosses *Diomedea chrysostoma* at the Prince Edward Islands. *S. Afr. J. Antarct. Res.* **19:** 31–33.

Imber, M. J., 1978. The squid families Cranchiidae and Gonatidae (Cephalopoda: Teuthoidea) in the New Zealand region. *N.Z. J. Zool.* **5:** 445–84.

Imber, M. J., 1991. Feeding ecology of antarctic and subantarctic Procellariiformes. Pp. 1402–412 *in* Acta XX Congr. Int. Orn. New Zealand Ornithological Congress Trust Board: Wellington.

Imber, M. J., 1992. Cephalopods eaten by wandering albatrosses *Diomedea exulans* L. breeding at six circumpolar localities. *J. R. Soc. N. Z.* **22:** 243–63.

Imber, M. J. and Russ, R., 1975. Some foods of the wandering albatross *Diomedea exulans*. *Notornis* **22:** 27–36.

Imber, M. J. and Berruti, A., 1981. Procellariiform seabirds as squid predators. Pp. 43–61 *in* Proceedings of the Symposium on Birds of the Sea and Shore, 1979 ed by J. Cooper. African Seabird Group: Cape Town.

Jackson, S., 1992. Do seabird gut sizes and mean retention times reflect adaptation to diet and foraging method? *Physiol. Zool.* **65:** 674–97.

Jackson, S. and Place, A. R., 1990. Gastrointestinal transit and lipid assimilation efficiencies in three species of subantarctic seabird. *J. Exp. Zool.* **255:** 141–54.

Jackson, S., Place, A. R. and Seiderer, L. J., 1992. Chitin digestion and assimilation by seabirds. *Auk* **109:** 758–70.

Jacob, J., 1982. Stomach oils. Pp. 325–40 *in* Avian Biology, Vol. VI ed by D. S. Farner, J. R. King and K. C. Parkes. Academic Press: London.

Ledoyer, M., 1995. Mysidacés (Crustacea) de Kerguelen, Crozet et Bouvet (Océan Austral) récoltés par la Japonaise, le Marion-Dufresne (1972–82) et dans des contenus stomacaux d'oiseaux. *J. Nat. Hist.* **29:** 601–18.

Linkowski, T. B. and Rembiszewski, J. M., 1978. Distribution of sea birds of Argentina coast and the feeding habits of the birds fauna in the Drake Passage and Scotia Sea. *Pol. Arch. Hydrobiol.* **25:** 717–27.

Lipinski, M. R. and Jackson, S., 1989. Surface-feeding on cephalopods by procellariiform seabirds in the southern Benguela region, South Africa. *J. Zool. Lond.* **218:** 549–63.

Marchant, S. and Higgins, P. J., 1990. Handbook of Australian, New Zealand and Antarctic birds. Volume 1: Ratites to Ducks. Part A: Ratites to Petrels. Oxford University Press: Melbourne.

Mougin, J. L., 1970. Les albatros fuligineux *Phoebetria palpebrata* et *P. fusca* de l'Ile de la Possession (Archipel Crozet). *L'Oiseau et R.F.O.* **40**(no. spécial)**:** 37–61.

Murphy, R. C., 1936. Oceanic Birds of South America. The American Museum of Natural History: New York.

Nesis, K. N., 1987. Cephalopods of the world. Squids, cuttlefishes, octopuses and allies. T.F.H. Publications: Neptune City, New Jersey.

Paulian, P., 1953. Pinnipèdes, cétacés, oiseaux des Iles Kerguelen et Amsterdam. Mission Kerguelen 1951. *Mém. Inst. Sci. Madagascar, Sér. A* **8:** 111–234.

Potter, I. C., Prince, P. A. and Croxall, J. P., 1979. Data on the adult marine and migratory phases in the life cycle of the southern hemisphere lamprey, *Geotria australis* Gray. *Environ. Biol. Fish.* **4:** 65–69.

Prince, P. A., 1980. The food and feeding ecology of grey-headed albatross *Diomedea chrysostoma* and black-browed albatross *D. melanophris*. *Ibis* **122:** 476–88.

Prince, P. A., 1982. The Black-browed albatross *Diomedea melanophris* population at Beauchêne Island, Falkland Islands. *Com. Nat. Fr. Rech. Antarct.* **51:** 111–17.

Prince, P. A. and Morgan, R. A., 1987. Diet and feeding ecology of Procellariiformes. Pp. 135–71 *in* Seabirds: Feeding Ecology and Role in Marine Ecosystems ed by J. P. Croxall. Cambridge University Press: Cambridge.

Prince, P. A., Huin, N. and Weimerskirch, H., 1994. Diving depths of albatrosses. *Ant. Sci.* **6:** 353–54.

Prince, P. A., Croxall, J. P., Trathan, P. N. and Wood, A. G., 1997. The pelagic distribution of South Georgia albatrosses and their relationships with fisheries. Pp. 137–67 *in* Albatross Biology and Conservation ed by G. Robertson and R. Gales. Surrey Beatty & Sons: Chipping Norton.

Reid, K., Croxall, J. P. and Prince, P. A., 1996. The fish diet of black-browed albatross *Diomedea melanophis* and grey-headed albatross *D. chrysostoma* at South Georgia. *Polar Biol.* **16:** 469–77.

Richardson, M. E., 1984. Aspects of the ornithology of the Tristan da Cunha Group and Gough Island, 1972–1974. *Cormorant* **12:** 123–201.

Ridoux, V., 1994. The diets and dietary segregation of seabirds at the subantarctic Crozet Islands. *Mar. Orn.* **22:** 1–192.

Robertson, C. J. R. and Nunn, G. B., 1997. Towards a new taxonomy for albatrosses. Pp. 13–19 *in* Albatross Biology and Conservation ed by G. Robertson and R. Gales. Surrey Beatty & Sons: Chipping Norton.

Rodhouse, P. G., 1989. Cephalopods in the diet of wandering albatrosses and sea-surface temperatures at the Sub-antarctic Front. *Scient. Mar.* **53:** 277–81.

Rodhouse, P. G. and Prince, P. A., 1993. Cephalopod prey of the black-browed albatross *Diomedea melanophrys* at South Georgia. *Polar Biol.* **13:** 373–76.

Rodhouse, P. G., Clarke, M. R. and Murray, A. W. A., 1987. Cephalopod prey of the wandering albatross *Diomedea exulans*. *Mar. Biol.* **96:** 1–10.

Rodhouse, P. G. and White, M. G., 1995. Cephalopods occupy the ecological niche of epipelagic fish in the Antarctic Polar Frontal Zone. *Biol. Bull.* **189:** 77–80.

Rodhouse, P. G., Croxall, J. P. and Prince, P. A., 1993. Towards an assessment of the stock of the ommastrephid squid *Martialia hyadesi* in the Scotia Sea: data from predators. Pp. 433–40 *in* Recent Advances in Fisheries Biology ed by T. Okutani, R. K. O'Dor and T. Kubodera. Tokai University Press: Tokyo.

Rodhouse, P. G., Prince, P. A., Clarke, M. R. and Murray, A. W. A., 1990. Cephalopod prey of the grey-headed albatross *Diomedea chrysostoma*. *Mar. Biol.* **104:** 353–62.

Ryan, P. G., 1987. The incidence and characteristics of plastic particles ingested by seabirds. *Mar. Environ. Res.* **23:** 175–206.

Segonzac, M., 1972. Données récentes sur la faune des Iles Saint-Paul et Nouvelle Amsterdam. *L'Oiseau et R.F.O.* **42**(no spècial): 3–68.

Smale, M. J., Watson, G. and Hecht, T., 1995. Otolith atlas of Southern African marine fishes. *Ichthyol. Monogr. J.L.B. Smith Inst. Ichthyol.* **1:** 1–253.

Thomas, G., 1982. The food and feeding ecology of the light-mantled sooty albatross at South Georgia. *Emu* **82:** 92–100.

Thompson, K. R., 1992. Quantitative analysis of the use of discards from squid trawlers by black-browed albatrosses *Diomedea melanophris* in the vicinity of the Falkland Islands. *Ibis* **134:** 11–21.

Thompson, K. R. and Riddy, M. D., 1995. Utilization of offal and discards from "finfish" trawlers around the Falkland Islands by the black-browed albatross *Diomedea melanophris*. *Ibis* **137:** 198–206.

Tickell, W. L. N., 1964. Feeding preferences of the albatrosses *Diomedea melanophris* and *D. chrysostoma* at South Georgia. Pp. 383–87 *in* Biologie Antarctique ed by R. Carrick, M. W. Holdgate and J. Prévost. Hermann: Paris.

Warham, J., 1977. The incidence, functions and ecological significance of petrel stomach oils. *Proc. N.Z. Ecol. Soc.* **24:** 84–93.

Warham, J., Watts, R. and Dainty, R. J., 1976. The composition, energy content and function of the stomach oils of petrels (Order, Procellariiformes). *J. Exp. Mar. Biol. Ecol.* **23:** 1–13.

Weimerskirch, H., Doncaster, C. P. and Cuénot-Chaillet, F., 1994. Pelagic seabirds and the marine environment: foraging patterns of wandering albatrosses in relation to prey availability and distribution. *Proc. R. Soc. Lond. B* **255:** 91–97.

Weimerskirch, H., Jouventin, P. and Stahl, J. C., 1986. Comparative ecology of the six albatross species breeding on the Crozet Islands. *Ibis* **128:** 195–213.

Weimerskirch, H., Bartle, J. A., Jouventin, P. and Stahl, J. C., 1988. Foraging ranges and partitioning of feeding zones in three species of southern albatrosses. *Condor* **90:** 214–19.

Weimerskirch, H., Salamolard M., Sarrazin, F. and Jouventin, P., 1993. Foraging strategy of wandering albatrosses through the breeding season: a study using satellite telemetry. *Auk* **110:** 325–42.

West, J. A. and Imber, M. J., 1986. Some foods of Buller's mollymawk *Diomedea bulleri*. *N.Z. J. Zool.* **13:** 169–74.

Williams, R. and McEldowney, A., 1990. A guide to the fish otoliths from waters off the Australian Antarctic Territory, Heard and Macquarie Islands. *ANARE Res. Notes* **75:** 1–173.

Wilson, R. P., Pütz, K., Gremillet, D., Culik, B. M., Kierspel, M., Regel, J., Bost, C. A., Lage, J. and Cooper, J., 1995a. Reliability of stomach temperature changes in determining feeding characteristics of seabirds. *J. Exp. Biol.* **198:** 1115–135.

Wilson, R. P., Weimerskirch, H. and Lys, P., 1995b. A device for measuring seabird activity at sea. *J. Avian Biol.* **26:** 172–75.

Wood, K. A., 1992. Seasonal abundance and spatial distribution of albatrosses off central New South Wales. *Aust. Bird Watchers* **14:** 207–25.

CHAPTER ELEVEN

The pelagic distribution of South Georgia albatrosses and their relationships with fisheries

P. A. PRINCE[1], J. P. CROXALL[1], P. N. TRATHAN[1] and A. G. WOOD[1]

At South Georgia populations of Wandering, Black-browed and Grey-headed Albatrosses are all decreasing. Interactions with fisheries are known, or believed, to be important in all these population changes. Such interactions are reviewed, using recoveries of banded birds and recent data from satellite-tracking studies. Wandering Albatrosses range particularly widely. While breeding, birds from South Georgia range over vast areas of the South Atlantic, from 67°S off the Antarctic Peninsula to 26°S off southern Brazil, and from 17°W to 85°W off the Pacific coast of southern Chile. During the breeding season birds are frequently caught in subtropical areas over the Patagonian Shelf by vessels using longlines to fish for tuna; recoveries from other longline fisheries further south are presently increasing. In the non-breeding season birds from South Georgia disperse across the South Atlantic Ocean, spending time off South Africa before crossing the Indian Ocean to winter around Australia. They are vulnerable to longline fishing at all stages of this migration. In the breeding season, Grey-headed Albatrosses are particularly associated with the Antarctic Polar Frontal Zone (APFZ) over a wide arc to the west and north of South Georgia. Few interactions with fishing vessels are known and information on their distribution in the non-breeding season is scarce; mortality derives from interactions with longline fishing in the Indian and South Pacific Oceans, particularly in the vicinity of Australia. Black-browed Albatrosses breeding at South Georgia also frequent the APFZ, but are most commonly associated with the shelf-slope areas around South Georgia and the South Orkney Islands. After the breeding season, birds from South Georgia migrate across the South Atlantic Ocean to winter in South African waters, where they commonly associate with fishing vessels in the Benguela Current; some birds also reach Australian waters. The rapid development of longline fisheries in many regions of the Southern Ocean, particularly at the edge of the continental shelves of South America and South Africa and in deeper waters of the Indo-Pacific region, poses substantial actual and potential threats to South Georgia albatross species. Whereas Black-browed and Grey-headed Albatrosses are currently mainly at risk outside the breeding season, Wandering Albatrosses are vulnerable year-round throughout most of the shelf and oceanic waters of the South Atlantic and Indian Oceans east to New Zealand. They and Grey-headed Albatrosses may also be at risk in the South Pacific Ocean.

Key words: South Georgia, Fisheries, Banding Recoveries, Satellite-tracking, Wandering Albatross *Diomedea exulans*, Grey-headed Albatross *D. chrysostoma*, Black-browed Albatross *D. melanophrys*.

INTRODUCTION

THERE is world-wide concern nowadays over the conservation status of albatrosses. Populations of many species, including those of Wandering *Diomedea exulans*, Black-browed *D. melanophrys* and Grey-headed *D. chrysostoma* Albatrosses at South Georgia, are in decrease. Interactions with fisheries, particularly those using longline techniques, are believed to contribute significantly to these decreases (Gales 1993). Much work has been undertaken trying to monitor population trends and processes in albatrosses, to assess the nature of their interactions with fisheries and to devise solutions to some of the problems (Croxall 1979; Weimerskirch and Jouventin 1987, 1997; Weimerskirch *et al.* 1997; Croxall *et al.* 1990; Bartle 1991; Brothers 1991; Vaske 1991; Murray *et al.* 1993; Robertson 1993; de la Mare and Kerry 1994; Prince *et al.* 1994; Brothers and Foster, in press).

However, obtaining accurate data on the number of albatrosses killed by longline fishing is a complex and difficult process, and the results are not easily related to the populations being studied demographically. The rapid expansion (both in terms of fishing effort and area) of longline fishing and the often unreliable nature of data on incidental mortality of albatrosses (especially in fisheries lacking comprehensive observer programmes), compound the difficulty of determining the actual or potential impact of new and developing longline fisheries on albatross populations.

[1]British Antarctic Survey, Natural Environment Research Council, High Cross, Madingley Road, Cambridge CB3 0ET United Kingdom.
Pages 137–67 *in* ALBATROSS BIOLOGY AND CONSERVATION, ed by G. Robertson and R. Gales. Surrey Beatty & Sons, Chipping Norton. 1997.

Hitherto most attention has been focused on interactions between albatrosses and fishing for Tuna *Thunnus* spp. (e.g., Brothers 1991; Murray *et al.* 1993), but it is clear that the recent development of longline fisheries targeting a wider range of deep-water and/or pelagic fish species poses problems of considerable potential significance for albatrosses. It is important to use all available data to characterize the at-sea distributions and preferred foraging areas of albatrosses in order to understand the full range of threats they face and to influence policy and practice in relation to the development of new fisheries in such areas. A priority is to determine the movements and foraging areas of albatrosses outside the breeding season. This chapter uses data on recoveries of Wandering, Grey-headed and Black-browed Albatrosses banded at Bird Island, South Georgia over the last 40 years, together with data from satellite tracking studies over the last five years, to assess the threats to the South Georgia populations of these species from current and developing Southern Ocean fisheries. More detailed treatment of some aspects of this topic, especially relating to albatrosses during their breeding seasons, can be found in Croxall and Prince (1990), Croxall *et al.* (1990), Prince *et al.* (1994) and Croxall and Prince (1996); this chapter provides an overview of available data, but concentrates on events and interactions outside the breeding seasons of each species.

METHODS

Satellite telemetry

Deployment methods were the same as described in Prince *et al.* (1992), except that smaller (60–80 g) Microwave and Toyocom T2038 platform terminal transmitters (PTT) were used on Black-browed and Grey-headed Albatrosses. All PTT's were deployed on birds breeding at the time of instrumentation. For Wandering Albatrosses this included incubating, brooding and chick-rearing adults. PTT's on Black-browed and Grey-headed Albatrosses were deployed during the chick-rearing periods only. Analytical methods are as described in Prince *et al.* (1992); discrimination between uplinks relating to foraging and commuting in Grey-headed and Black-browed Albatrosses are derived from independent objective assessment of distance and time between successive uplinks (Prince *et al.*, unpubl. data).

Banding

Materials and methods concerning banded birds and recoveries are as described in Croxall and Prince (1990), Croxall *et al.* (1990) and Prince *et al.* (1994).

RESULTS

Wandering Albatross

Recoveries

Since the detailed analysis of data up to 1988 (Croxall and Prince 1990), a further 57 recoveries from 7 276 banded birds have been reported (Appendix 1). This proportion, 0.78%, is double that (0.39%) reported previously (Croxall and Prince 1990). While this may reflect improvements in reporting of recoveries, especially in recent years, it may also represent a genuine increase in the rate of birds dying. We can now partition recoveries since 1975 into two decadal blocks (Appendix 1), showing that recovery rates since 1985 have been substantially higher than in the previous decade.

The location of recoveries (excluding the many birds caught and released by the New South Wales Albatross Study Group, Australia (Gibson 1963, 1973), illustrated in Figure 1, is chiefly (60%) from the South Atlantic (90% of which were between 24–42°S off the coasts of southern Brazil, Uruguay and northern Argentina) with a few, mainly recent, recoveries from southern and western Chile and the Falkland Islands. The remaining recoveries are split equally (17% each) between the Indian Ocean and Tasman Sea, with the remainder (6%) from the southern Pacific Ocean. Overall, birds banded as chicks and recovered when younger than the modal age of first breeding (10 years old) are no more likely to be recovered than birds banded as adults (0.42% and 0.48% respectively), though in the last decade more juveniles have been recovered than adults (0.37% and 0.24% respectively).

Recoveries of birds in the South Atlantic Ocean (Table 1) indicate the high proportion (53%) in the period 1985–96 known to come from longline fishing, particularly off the coast of sub-tropical South America. For many birds reported as "at-sea" and "associated with fishing vessels" details are often inadequate to determine the precise circumstances; however, the involvement of longline fishing cannot be excluded. The total absence, until very recently, of scientific observers aboard fishing vessels in the South Atlantic Oceans means that the number, nature and location of recoveries of albatrosses is very open to bias, the most important of these being the wide-scale failure, of vessels without observers, to report the capture of banded birds. In contrast, interest in and reporting of birds on beaches has at least been maintained, if not increased substantially, over recent decades. Therefore it is probably safe to assume that the pattern and nature of recoveries indicate the increasing importance of interactions between fisheries, especially those using longlines, and albatrosses as a cause of mortality and source of recoveries.

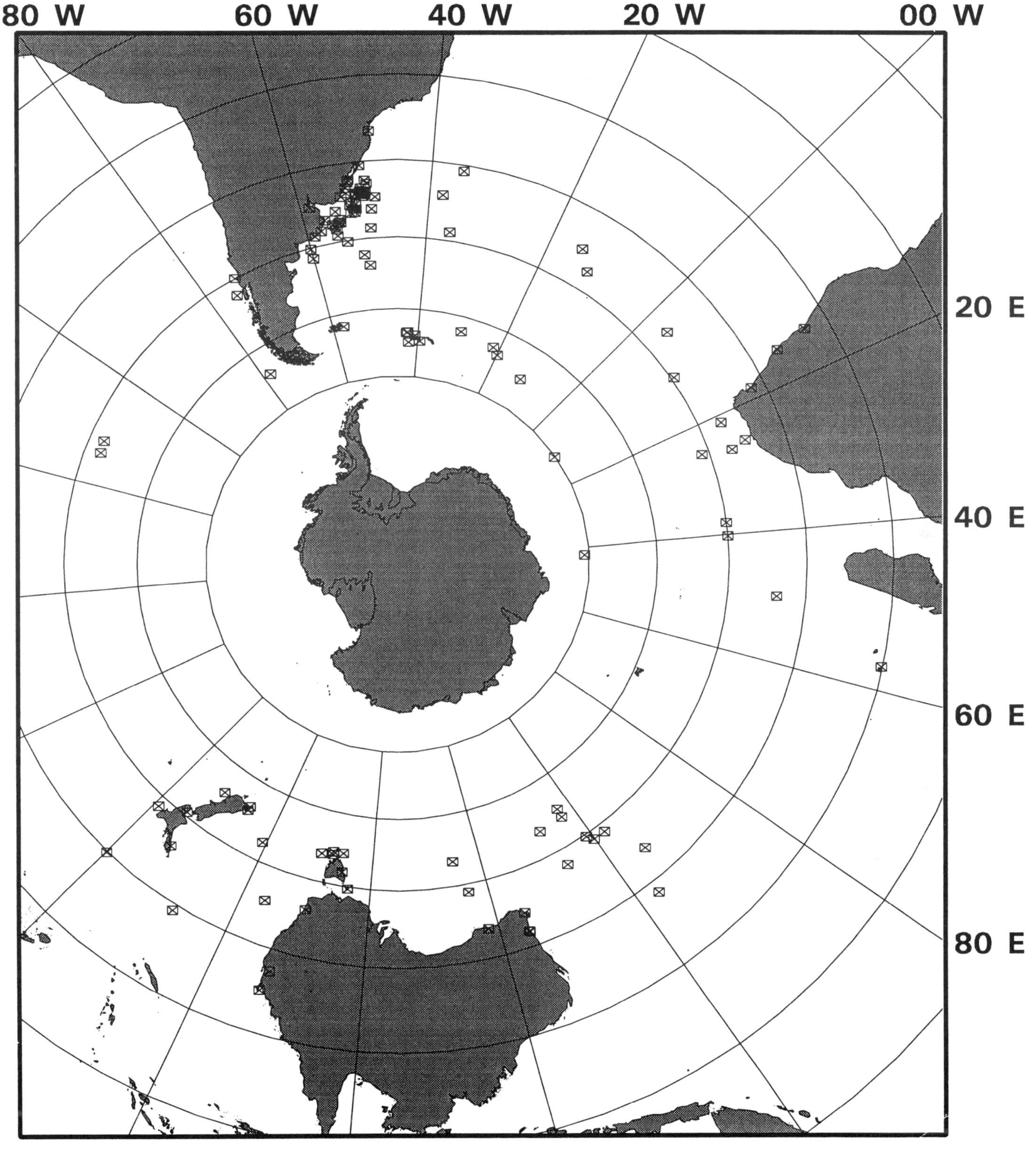

Fig. 1a. Distribution of recoveries of Wandering Albatrosses (n = 138) banded at Bird Island, South Georgia (54°S, 38°W). Additional details in Appendix 1.

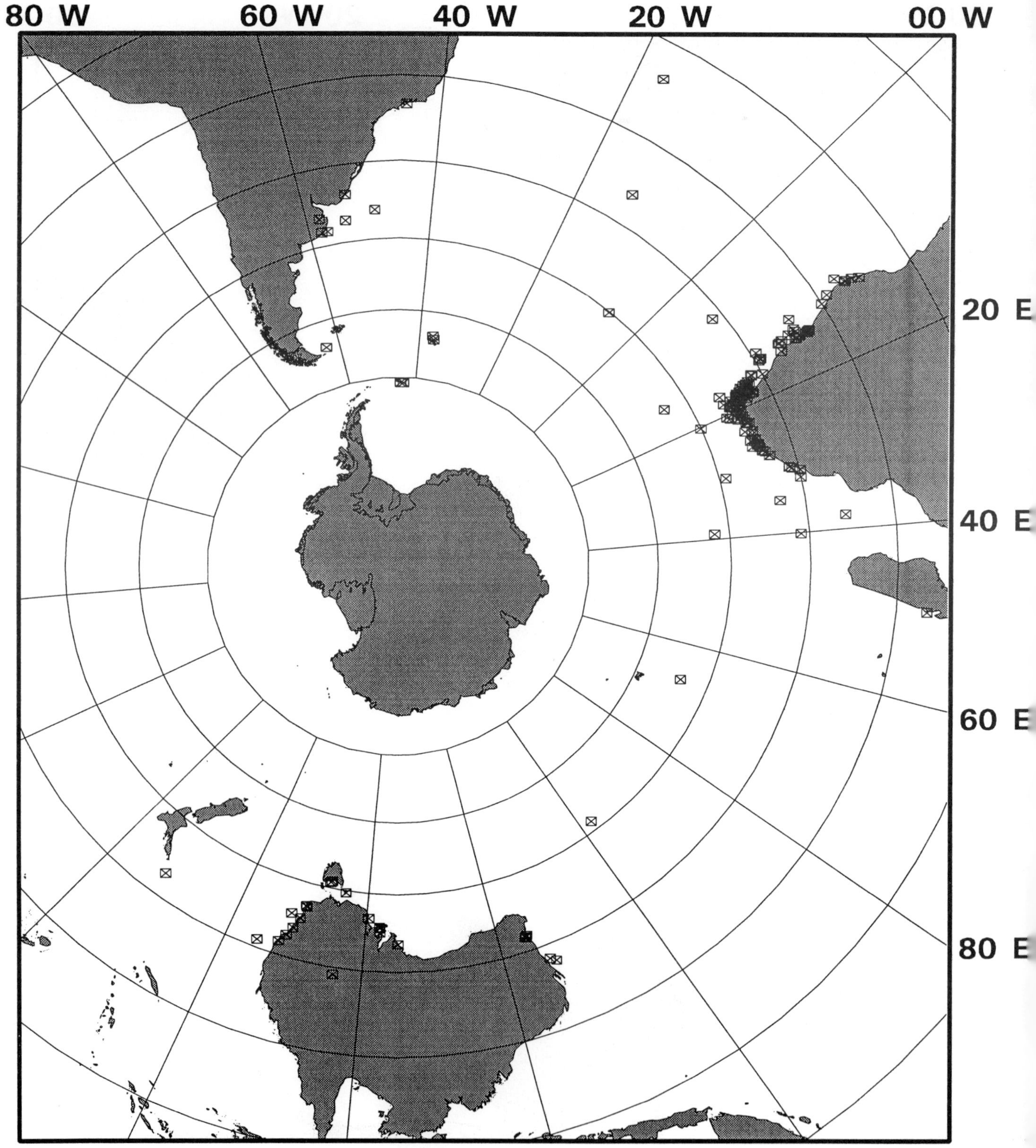

Fig. 1b. Distribution of recoveries of Black-browed Albatrosses (n = 281) banded at Bird Island, South Georgia (54°S, 38°W). Additional details in Appendix 1.

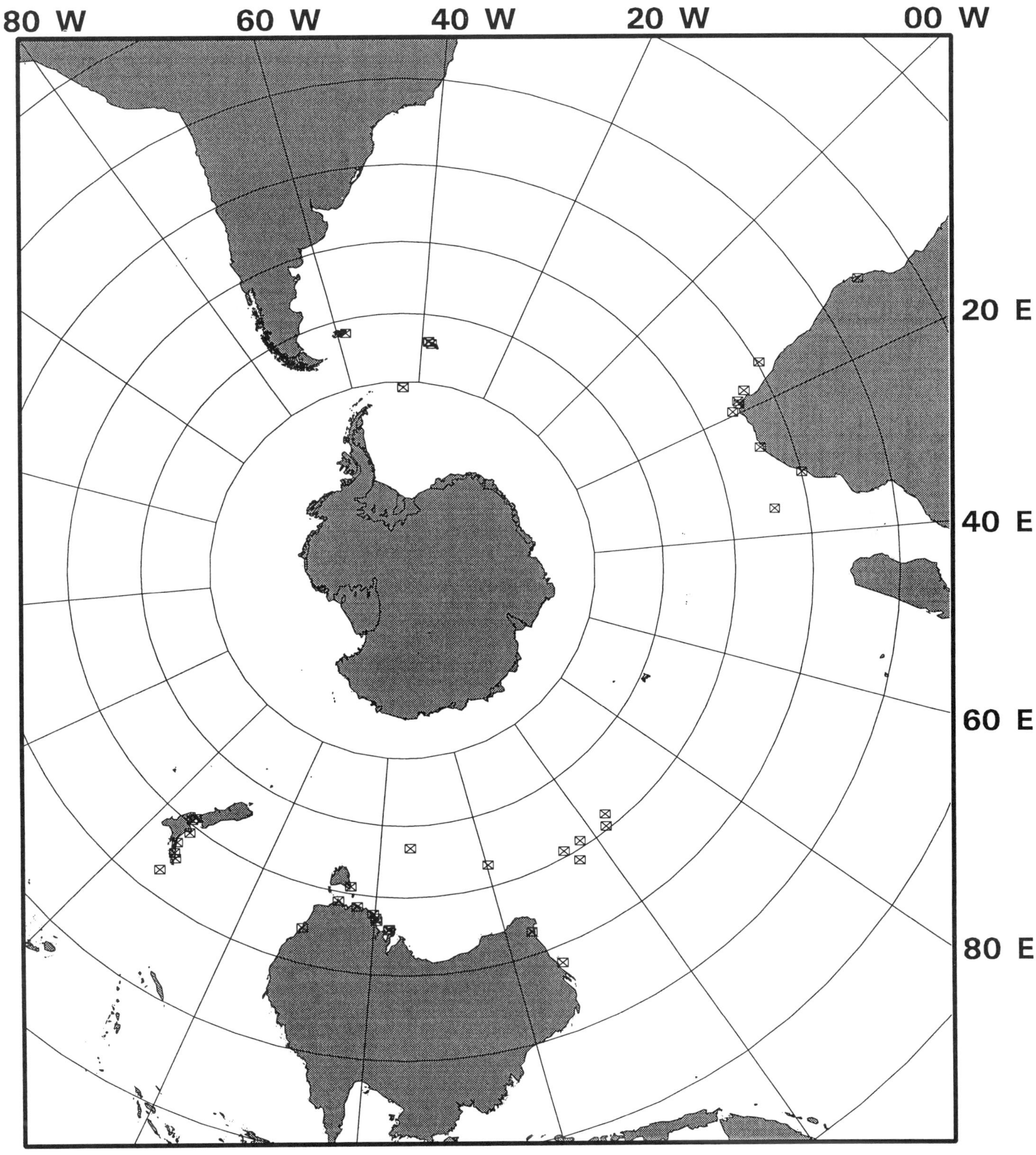

Fig. 1c. Distribution of recoveries of Grey-headed Albatrosses (n = 42) banded at Bird Island, South Georgia (54°S, 38°W). Additional details in Appendix 1.

Table 1. Location and circumstances of Wandering Albatrosses banded at Bird Island, South Georgia and recovered in the South Atlantic Ocean region.

	South America (n = 64)						South Georgia (n = 16)		
	26–43°S			51–55°S			53–55°S		
	1960–74	1975–84	1985–96	1960–74	1975–84	1985–96	1960–74	1975–84	1985–96
Longline	0	2	22	0	0	0	0	0	2
Fishing	5[1]	2	10	0	0	2[2]	3[3]	0	3
At sea	4	1	3	0	0	0	3	5	0
Ashore	1	2	5	0	2[1]	3[4]	0	0	0
Total	10	7	40	0	2	5	6	5	5

Notes: [1]Includes one bird from Chile. [2]Includes one bird from each of Chile and the Falkland Islands. [3]There are also three recoveries from the vicinity of Tristan de Cunha. [4]Includes two birds from Chile.

In the Australasian region the use of scientific observers in the Japanese longline fishery for Southern Bluefin Tuna *Thunnus maccoyi* has, since 1988, been responsible for many recoveries. Of banded birds, the Wandering Albatross has been the commonest species recorded and Bird Island, South Georgia, the primary source (Gales and Brothers 1995). Indeed, almost all South Georgia Wandering Albatrosses recovered in the Australasian region since 1988 have been caught by longline fishing vessels. Overall, therefore, recoveries of Wandering Albatrosses banded at South Georgia confirm the wide geographical range of this population. They indicate aggregation of recoveries, principally associated with longline fishing activities. However, banding records are an imperfect guide to the distribution of marine seabirds and their existing and potential vulnerability to interactions with fisheries.

Satellite tracking

The distribution at sea of South Georgia Wandering Albatrosses during their breeding season is shown in Figure 2. Overall, birds ranged from 28°S off southern Brazil to 67°S off the southwestern Antarctic Peninsula, east to 17°W (near Tristan da Cunha) in the central South Atlantic Ocean, and west to 85°W in the south-east Pacific Ocean and to 75°W off the coast of south-west Chile. There are many similarities between the sexes, particularly in the west and north-west axis of travel from South Georgia, with little activity to the east or south. For most birds, important destinations for their foraging trips were the coastal shelf-slope of South America, particularly between 33–38°S, and between 43–45°S, and the Patagonian Shelf around the Falkland Islands. A particularly interesting feature is how seldom Wandering Albatrosses foraged close inshore and the extent to which uplinks were clustered, close to the 1 000 m isobath, in certain localities. This is shown more clearly in Figure 3, which also includes the location of Wandering Albatrosses recovered in the area, mainly between 1985 and 1993. The similarity between the locations of live satellite-tracked albatrosses and of birds recovered dead from longline fishing vessels is remarkable. Furthermore, the locations at which Wandering Albatrosses were recovered dead ashore strongly suggest that these birds were not the victims of natural environmental events at sea (as assumed by Croxall and Prince (1990)), but more likely drifted ashore after being cut free or thrown overboard after capture by fishing vessels.

Only 13 (23%) birds, mostly females, visited the area shown in Figure 3. They occurred there during some of the longer trips recorded (15.3 ± 2.2 days, covering 9 052 ± 2 629 km on average) and all occurred between late July and late September (towards the end of chick-rearing). All 18 recoveries from this area (again mainly females) occurred between August and October (Croxall and Prince 1990; unpubl. data). Low numbers of both sexes visited the shelf-slope at 35°S, the Patagonian Shelf around the Falkland Islands and various sea-mounts to the south-east of the Falklands (Fig. 2). In this region the main difference between the sexes is that only one female ventured south of 55°S (into the eastern Drake Passage), whereas several males foraged in the Drake Passage, one travelling south to 67°S off the Antarctic Peninsula and then north to 53°S on the west (Pacific) coast of Chile, moving along the 1 000 m isobath once in shelf-slope waters. Only males frequented areas, again along the 1 000 m isobath, east of Staten Island and off the south coast of Tierra del Fuego.

These distribution patterns apply to birds throughout the incubation (Jan.–Feb.) and main chick-rearing (June–Sept.) periods. However, during the brooding period (March to mid-May) foraging trips were much shorter than at all other times (average: 2.6 days; Tickell 1968) and the area covered is greatly reduced (Fig. 4). This was especially so for males, with all trips confined to areas within a 350 km radius of Bird Island and 74% of uplinks within the South Georgia shelf and shelf-slope areas. Females tended to range somewhat further afield at this time with two individuals reaching the nearest point of the Patagonian Shelf, one spending appreciable time around the Falkland Islands.

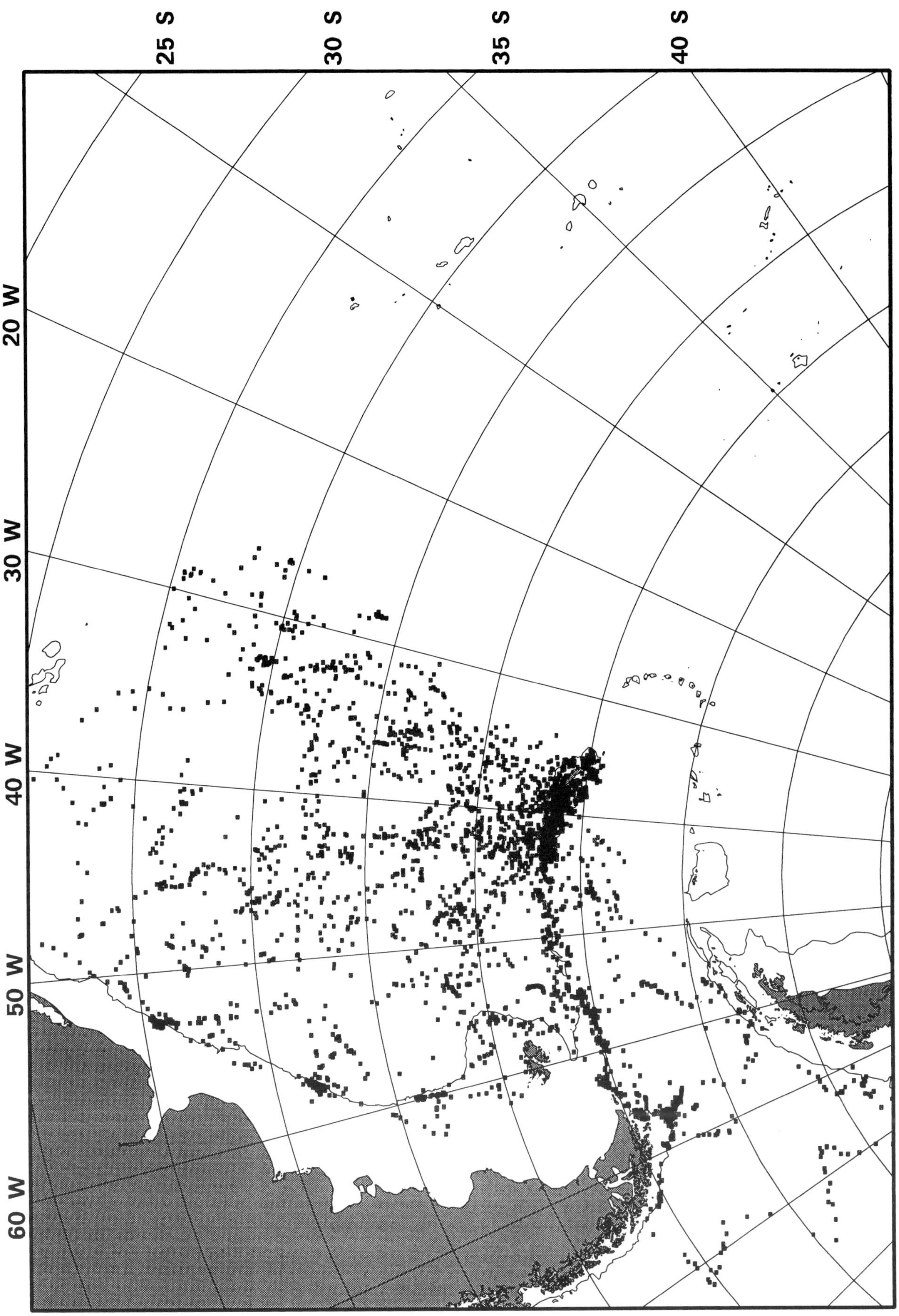

Fig. 2a. Distribution of uplinks from satellite-tracked male Wandering Albatrosses breeding at Bird Island, South Georgia. Based on 4 843 uplinks from 19 birds tracked for a total of 212 days between August 1990 and September 1994. The position of the 1 000 m isobath is shown.

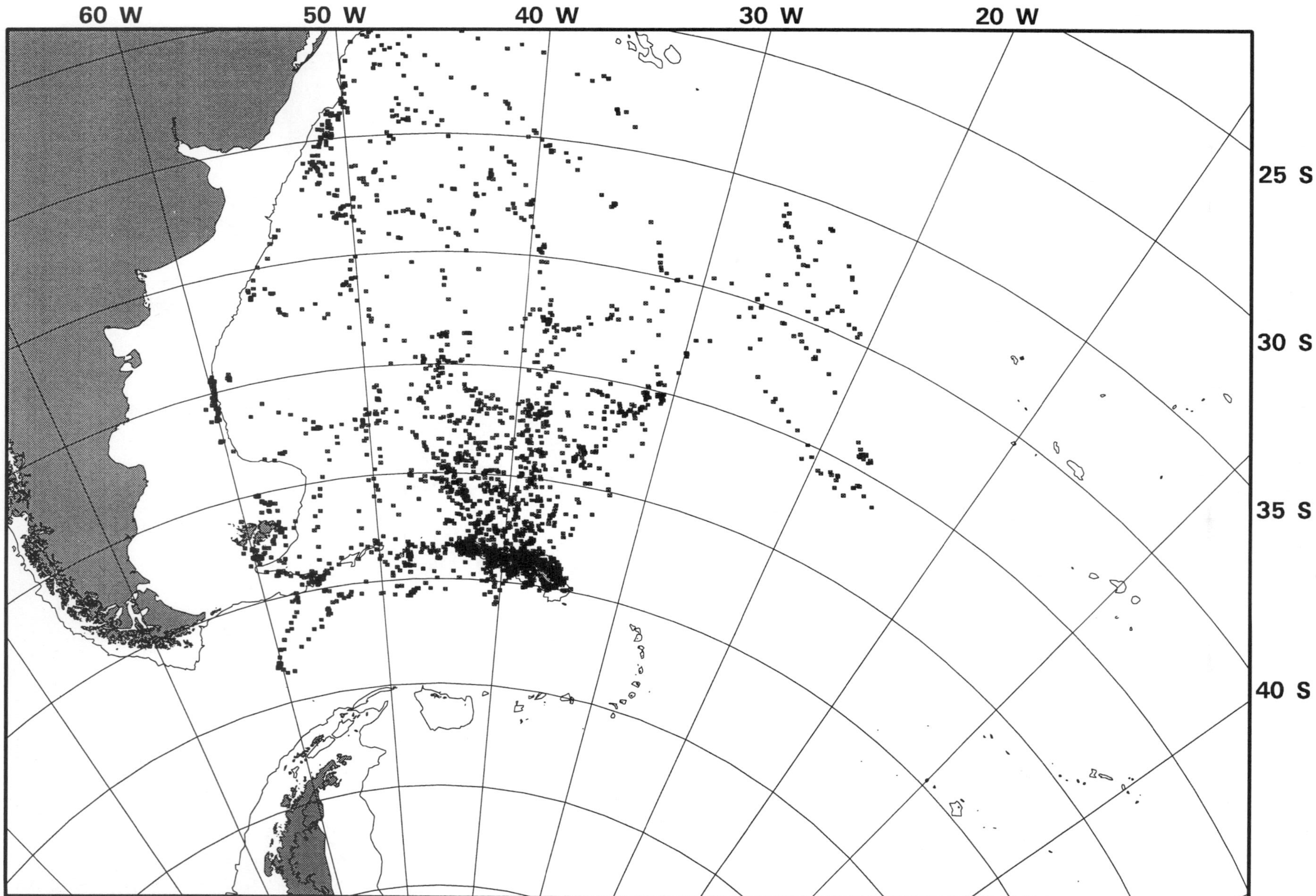

Fig. 2b. Distribution of uplinks from satellite-tracked female Wandering Albatrosses breeding at Bird Island, South Georgia. Based on 3 583 uplinks from 14 birds tracked for a total of 190 days between August 1990 and September 1994. The position of the 1 000 m isobath is shown.

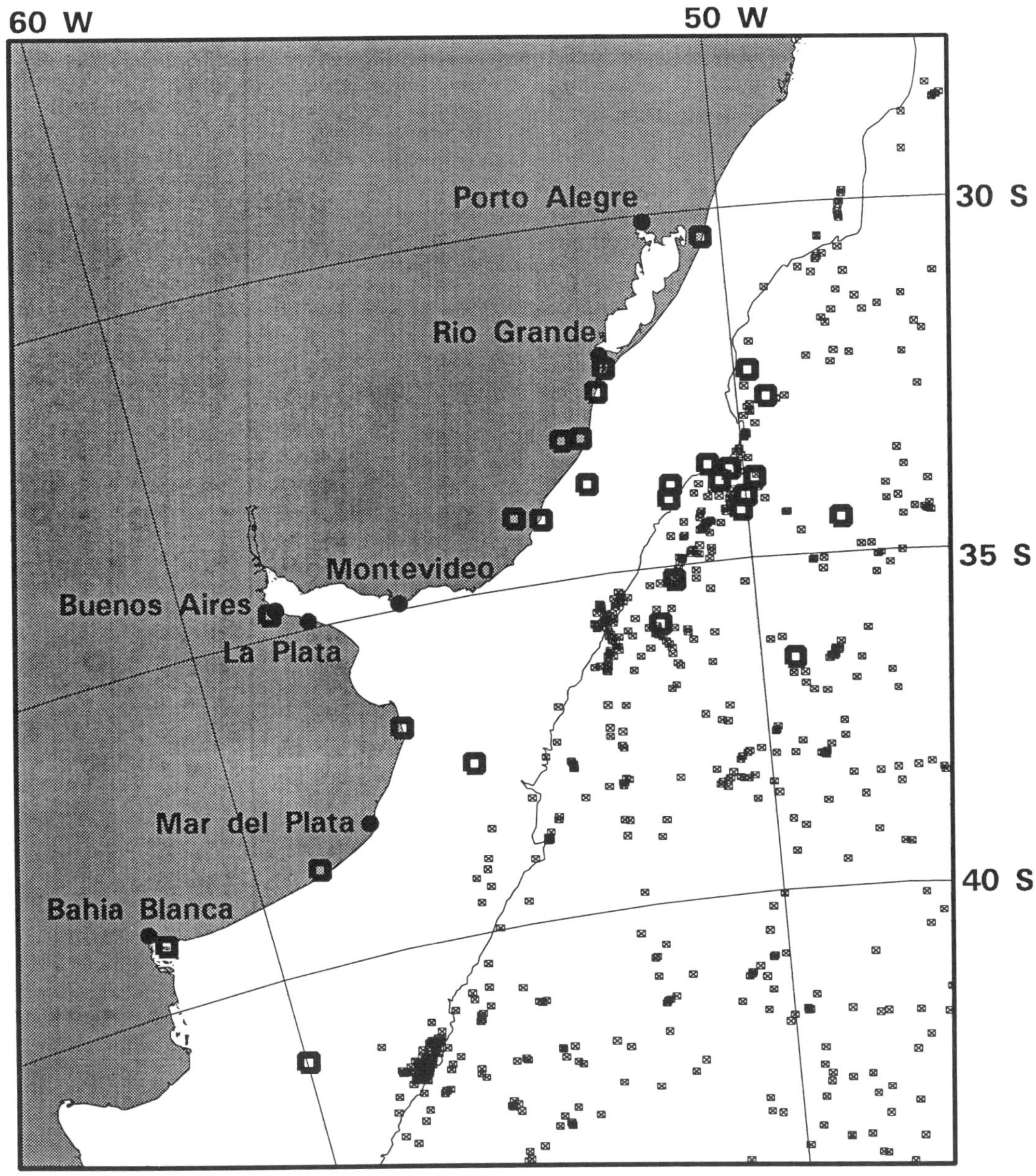

Fig. 3. Distribution of uplinks from satellite-tracked Wandering Albatrosses breeding at Bird Island, South Georgia (data sources as in Fig. 2) off the coast of South America between southern Brazil and northern Argentina, together with locations (large open squares) of Wandering Albatrosses from South Georgia recovered in this area. The position of the 1 000 m isobath is shown.

Nevertheless, 56% of uplinks from females were confined to the South Georgia shelf and shelf-slope (Croxall and Prince 1996). The restriction of Wandering Albatrosses to the South Georgia shelf is of particular concern because of the much enhanced potential for interaction with the local longline fishery for Patagonian Toothfish *Dissostichus eleginoides*. This operates at the same time over the South Georgia shelf-slope in water depths between 500–2 000 m (Fig. 4; Ashford *et al.* 1995; Croxall and Prince 1996).

To assess the movements of South Georgia birds after completing breeding we have data from only two birds (both females) in 1992 (Fig. 5). Both birds showed broadly similar patterns, at least initially, and therefore may be reasonably typical of the nature of the post-breeding migration away from South Georgia towards Australia, the east coast of which is an important wintering area for South Georgia birds (Gibson 1963, 1973; Tickell 1968). Female 9288 (Fig. 5a) left Bird Island on 2 July (day 183) and travelled north-west to latitude 37°S off the Brazilian coast (covering 1 155 km in 40 h at 28.9 km h^{-1}), spending three days in this area before moving west (days 187–197; covering 8 081 km at an average speed of 34.7 km h^{-1}) to reach the southern tip of the South African coastal shelf. She then spent nine days (travelling intermittently) in the region (and to the south-east) of the Agulhas Current before moving

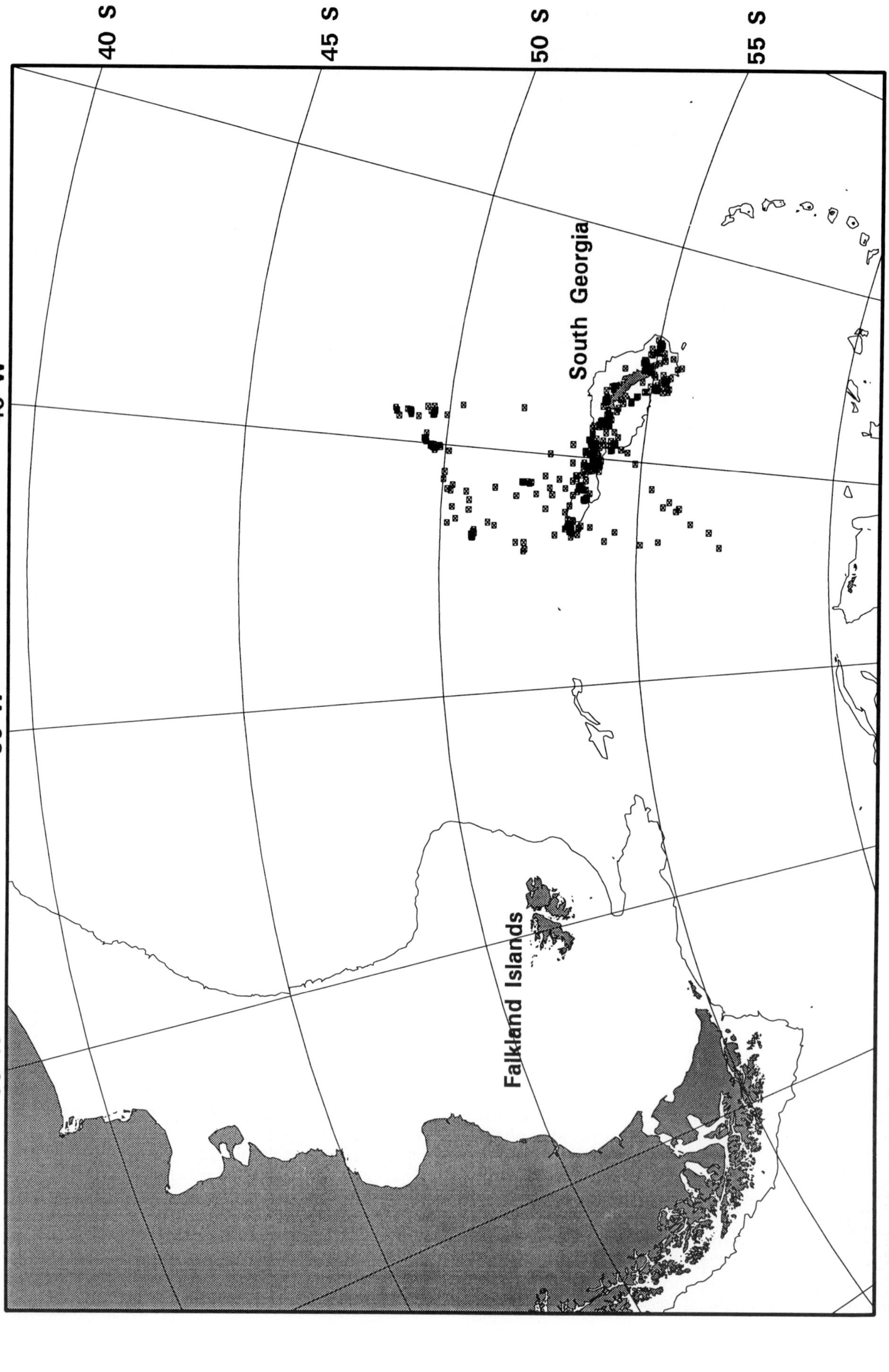

Fig. 4a. Distribution of uplinks from satellite-tracked male Wandering Albatrosses breeding at Bird Island, South Georgia during their brood-guard period. Based on 498 uplinks from seven birds tracked for a total of 31 days between March and mid-May of 1991 and 1992. The position of the 1 000 m isobath is shown.

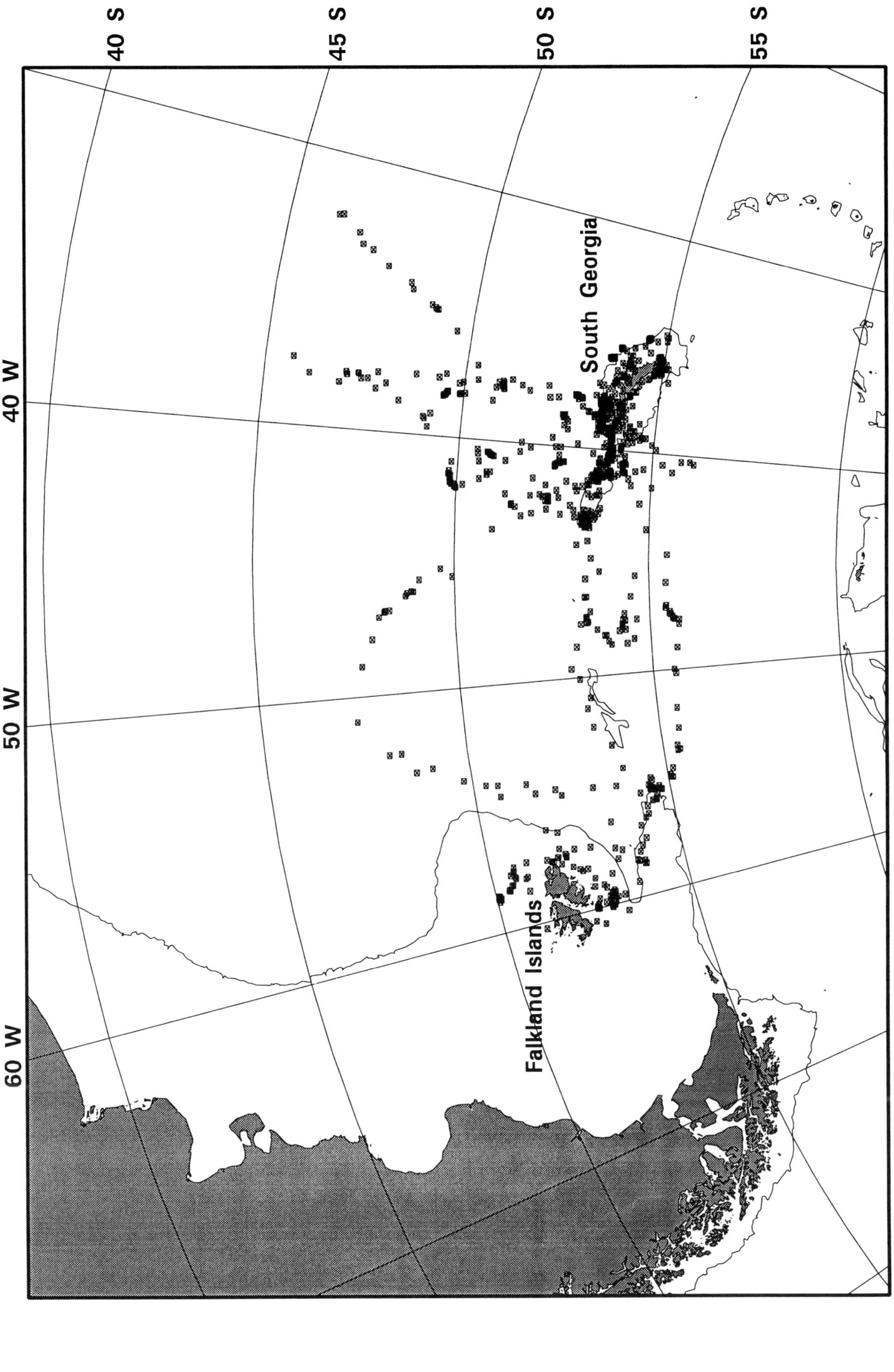

Fig. 4b. Distribution of uplinks from satellite-tracked female Wandering Albatrosses breeding at Bird Island, South Georgia during their brood-guard period. Based on 565 uplinks from four birds tracked for a total of 37 days between March and mid-May of 1991 and 1992. The position of the 1 000 m isobath is shown.

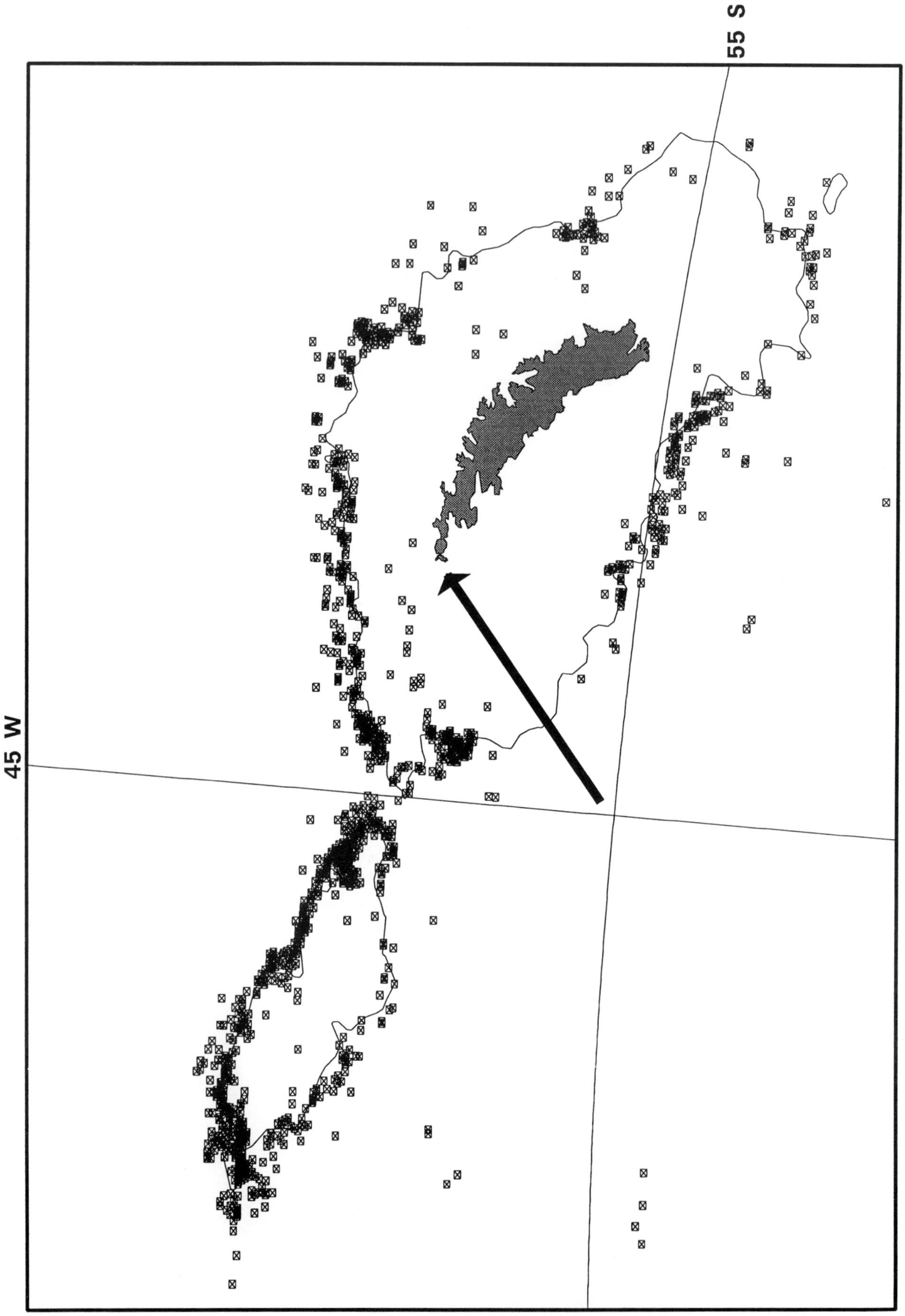

Fig. 4c. The location of hauls by vessels longlining for Patagonian Toothfish *Dissostichus eleginoides* between December and September of 1990/91 to 1993/94 (see Croxall and Prince 1996 for further details). The position of the 1 000 m isobath is shown. The location of Bird Island is indicated.

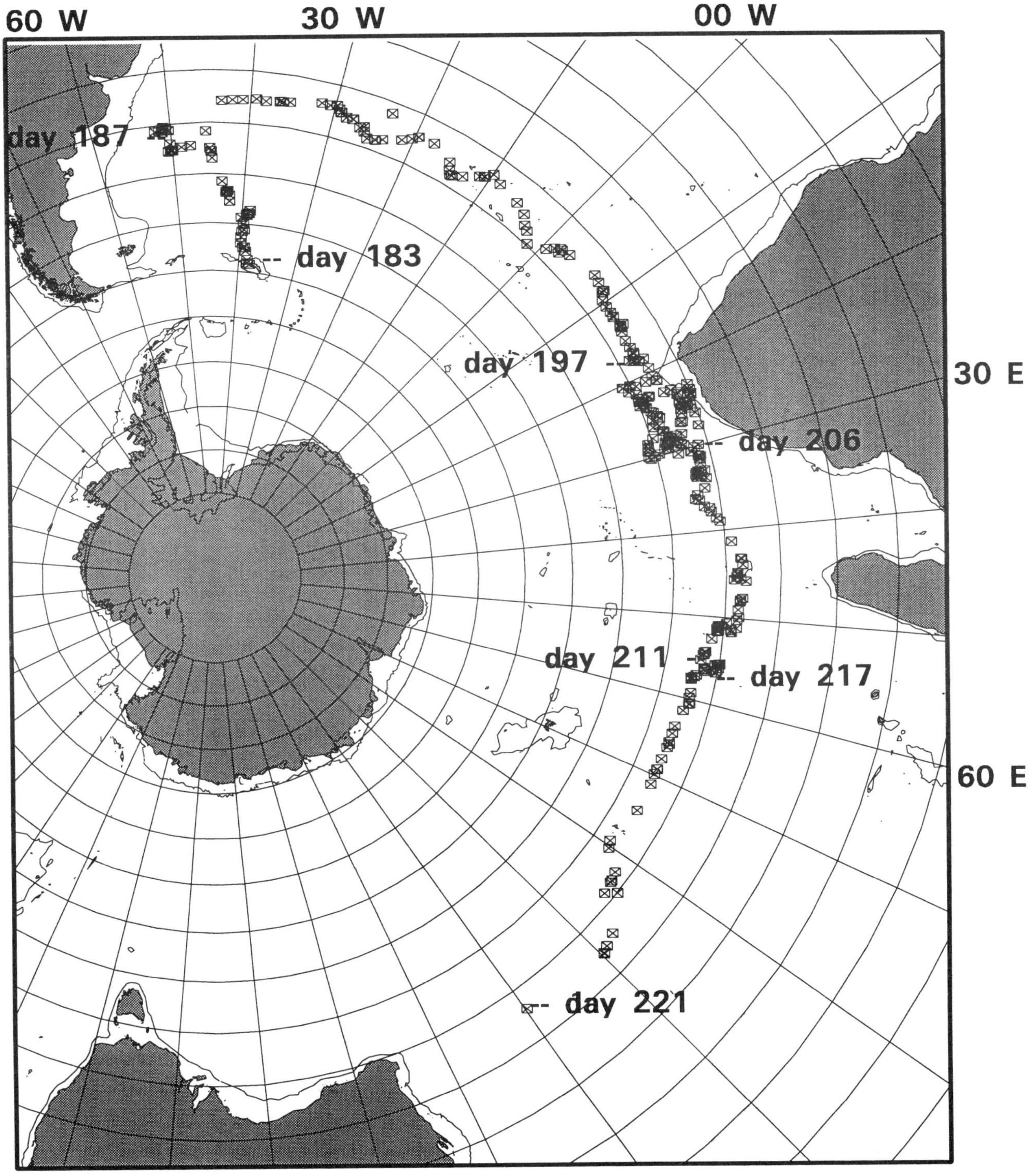

Fig. 5a. Track of a Wandering Albatross in July–August 1992 after concluding breeding activities at Bird Island, South Georgia. Shown is female 9288 (514–6112); distance travelled 25 075 km in 36.3 days. Numbers on map refer to Julian days (day 183 = 2 July, day 226 = 14 August). The position of the 1 000 m isobath is shown.

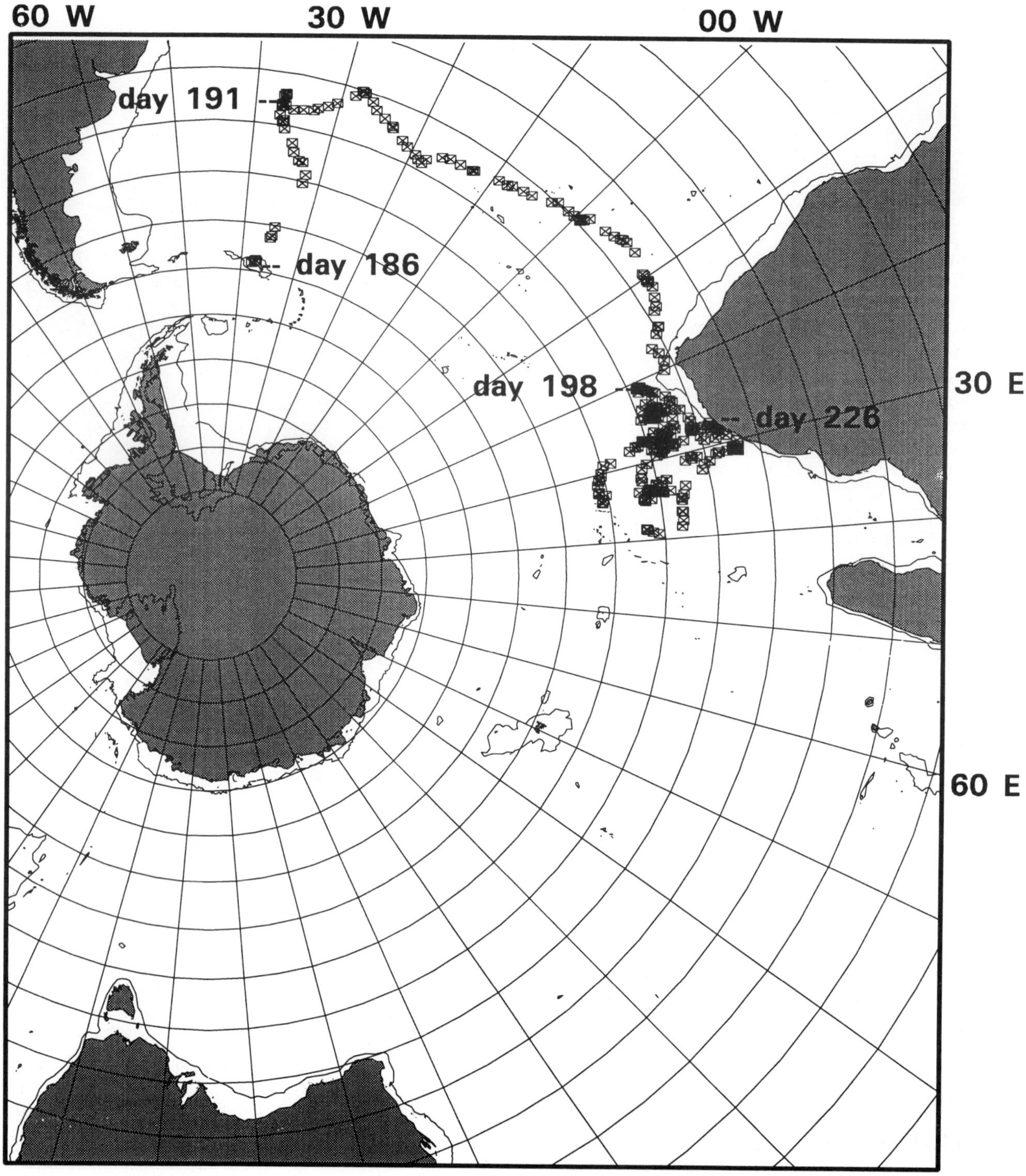

Fig. 5b. Track of a Wandering Albatross in July–August 1992 after concluding breeding activities at Bird Island, South Georgia. Shown is 11-year-old female 9280 (510–8646); distance travelled 20 619 km in 40.1 days. Numbers on map refer to Julian days (day 183 = 2 July, day 226 = 14 August). The position of the 1 000 m isobath is shown.

Table 2. Circumstances of recovery of Black-browed and Grey-headed Albatrosses banded at Bird Island, South Georgia.

	Black-browed Albatross			Grey-headed Albatross		
Source	1960–74	1975–84	1985–96	1960–74	1975–84	1985–96
Longline	0	4	1	0	0	2
Fishing	118	10	3	2	2	2
Ashore/at sea	62	10	3	10	6	4
Unknown	60	10	0	8	2	2
Total	240	34	7	20	10	10

(days 206–211) to an area 800 km NNE of the Iles Crozet, where she spent six days travelling short distances (1 455 km at an average speed of 9.5 km h^{-1}). On 5 September (day 217) she moved rapidly east covering 4 447 km at an average speed of 53.6 km h^{-1} in the next 3.5 days until her transmitter failed on 8 September, some 800 km west of Australia. At this stage she was in an area where several South Georgia birds have been recovered (see Fig. 1) and within one week for direct travel from the traditional wintering grounds off eastern Australia. Throughout her journey the latitudinal movement of this bird had been remarkably small, seldom straying outside latitudes 35–40°S.

Female 9280 (Fig. 5b) left Bird Island three days after bird 9288 (5 July), spent two days on the South Georgia shelf and then headed due north to latitude 36–37°S (covering 1 790 km at an average speed of 45.9 km h^{-1}). She crossed the South Atlantic Ocean between 33–40°S in 6.5 days (days 191–198, covering 6 005 km at 38 km h^{-1}), arriving off Cape Agulhas one day after bird 9288. Subsequently, however, she spent 28 days ranging widely between the coast of Natal at 33°S and areas to the west of the Prince Edward Islands at 46°S (covering 12 108 km at 18 km h^{-1}). She was still in South African waters when her transmitter failed on 14 July (day 226), by which time bird 9288 would have reached Australia. Although the chicks of both females fledged successfully at Bird Island in October–November 1992, only female 9288 returned to breed (in December 1993); we assume that female 9280 died at some stage in the intervening year.

Black-browed Albatross

Recoveries

The distribution of recoveries of Black-browed Albatrosses banded at Bird Island between 1959 and 1964 and recovered to the end of 1965 was analysed by Tickell (1967). These and additional recoveries from the large scale banding in these years and from subsequent banding at Bird Island are summarized in Appendix 1 and Figure 1. Overall, the proportion of recoveries of Black-browed Albatrosses banded as adults is about one-quarter that of birds banded as chicks (0.31% and 1.12% respectively). Of the nine recoveries of adults to date, six are from South Africa, two from near South Georgia and, one from Australia. Of the 20 464 chicks banded between 1959–64, there had been 186 (0.91%) recoveries by the end of 1965 (Tickell 1967) with 95% from South Africa, 4% from Australia and 2% from South America. Subsequently there have been 76 more recoveries from these cohorts (giving an overall recovery rate of 1.25%), and 16 recoveries from chicks banded since 1974 (0.42%). These latter recoveries are more widely dispersed than those before 1965, with 44% from South Africa, 31% from Australia, 6% from South America and the rest from the central Atlantic and Indian Oceans. The overall distribution of recoveries is 85% from South Africa, 9% from Australia and 4% from South America. Whether the lower proportion of recoveries from cohorts after 1975 (0.51% for 1975–84; 0.19% for 1985–94) reflects fewer birds in these areas, lower mortality rates or less reporting of recoveries, is unknown.

The distribution of recoveries of Black-browed Albatrosses from South Georgia and the Falkland Islands differs greatly, despite the proximity of the two populations. Thus, of 66 recoveries of 8 793 chicks (0.75%) banded in the Falkland Islands (Tickell 1967), 58 (88%) were from South America and 8 (12%) from South Africa, compared with 4% and 85% respectively for Black-browed Albatrosses from South Georgia.

The circumstances of Black-browed Albatross recoveries (Table 2) are difficult to interpret. Until 1974 almost all detailed recoveries from South Africa involved birds reported by fishing trawlers, mainly operating in the Benguela Current. Many other recoveries probably had similar provenance but birds were also found dead ashore along the coasts of Southern Africa from Namibia in the west to Natal in the east. Most recoveries from Australia (mainly on the east coast) were of birds found dead ashore. Since 1975 most reports have been from fishing vessels at sea, the majority in South African waters (Morant *et al.* 1983; Croxall *et al.* 1984), but include several birds caught on longlines by Japanese vessels fishing for tuna around Australia and a single such record from New Zealand.

Satellite tracking

Black-browed Albatrosses rearing chicks at Bird Island, South Georgia are distributed at sea over a relatively limited area, bounded by 30–40°W at 50°S and 48–50°W at 62°S (Fig. 6a). This area encompasses the continental shelf of South Georgia (including the Shag Rocks area to the west) and areas to the north and north-east mainly between 30–40°W, the shelf of the South Orkney Islands (at 61°S; 500 km to the south of South Georgia) and much of the area in between these two shelves. If uplinks deriving from travel between Bird Island and feeding areas (characterized by lower travel speeds and longer residence times) are excluded, then four main foraging areas can be recognized (Fig. 6b). These are an area to the north of South Georgia associated with the Antarctic Polar Frontal Zone (APFZ), areas of the South Georgia shelf-slope to the north and south of Bird Island and around Shag Rocks, the north-west edge of the continental shelf of the South Orkney Islands and a poorly defined area between 56–58°S and 44–47°W, which lacks any obvious bathymetric features.

To illustrate the post-breeding migration of Black-browed Albatrosses, we have two tracks from birds which lost their chicks late in the rearing period (about three weeks before the mean fledging date). Bird 1440 (Fig. 7a) left the breeding colony for the last time on 19 March 1992 (day 79) and travelled 300 km (at 28 km h^{-1}) to the edge of the continental shelf to the east of South Georgia. It spent eight days in this area before travelling 690 km (at 10 km h^{-1}) to the APFZ to the north-east of South Georgia where it stayed for a further 18 days. On 19 April (day 109) it started to cross the South Atlantic Ocean, passing between Tristan da Cunha and Gough Island and reached the coastal shelf off South Africa at the Cape of Good Hope eight days later (covering 4 213 km at an average speed of 22 km h^{-1}). It remained in a relatively restricted area near the edge of the continental shelf in the Benguela Current for a further 19 days until the transmitter failed on 16 May (day 136) after 57 days of continuous transmission.

The following year, bird 3247 left the breeding colony on 24 March 1993 (day 83) and spent 1.6 days in the Shag Rocks area (covering 464 km at 11.9 km h^{-1}) before travelling 508 km (at 42.3 km h^{-1}) north-west to the APFZ. It spent the next 10 days travelling intermittently in this area (covering 2 915 km at 12.9 km h^{-1}). Starting on 5 April (day 95), it crossed the South Atlantic Ocean in four days (covering 4 533 km at an average speed of 47.7 km h^{-1}) along a similar track (also passing between Tristan da Cunha and Gough Islands) to that of bird 1440 the previous year. Bird 3247 spent the next 13 days in the Benguela Current near the shelf edge off south-west Africa until its transmitter failed on 22 April (day 112). Both albatrosses foraged in an area of the Benguela Current where local fishing vessels were operating at the same time (D. G. Miller, pers. comm.).

Grey-headed Albatross

Recoveries

The distribution of recoveries of Grey-headed Albatrosses banded at Bird Island between 1959 and 1964 and recovered to the end of 1965 was analysed by Tickell (1967). All recoveries from all banding at Bird Island are summarized in Appendix 1 and Figure 1. Recoveries of birds banded as adults are few and are about one-half the proportion of those from birds banded as chicks (0.09% and 0.17% respectively), recoveries for both age groups being many fewer than for the other two species of albatross breeding at South Georgia. This suggests that Grey-headed Albatrosses spend much less time in inshore waters than the other two species reviewed here. Recoveries come principally from the Australasian region (65%) and South Africa (23%), with the remainder (12%) in the South Atlantic Ocean, divided between South Georgia, near the South Orkney Islands and the Falkland Islands. This distribution is very different from Black-browed Albatrosses and presumably reflects the much greater importance of the Australasian area to Grey-headed Albatrosses from South Georgia.

There are few details of recovery circumstances of Grey-headed Albatrosses (Table 2). Before 1985 most recoveries with adequate details were of birds found ashore, particularly in South Africa, south-east Australia and northern New Zealand. Since then, most recoveries have been from fishing (including longline) vessels, particularly in areas to the south and south-west of Australia.

Satellite tracking

Grey-headed Albatrosses rearing chicks at Bird Island, South Georgia, were tracked over a wide area, bounded by 30–50°W at 47–50°S and 35–55°W at 60–62°S (Fig. 6c). If uplinks deriving from commuting between the breeding colony and feeding areas are excluded (Fig. 6d), then the main foraging areas consist of the APFZ to the north and west of South Georgia, an area around 45°W, 56°S with no discernable bathymetric or hydrographic features, part of the northern and southern shelf edge of the South Orkney Islands and a variety of small scattered oceanic areas in the southern and western Scotia Sea. Although at first glance the foraging distribution of Black-browed and Grey-headed Albatrosses look similar, differences are evident in the lack of foraging by Grey-headed Albatrosses over the South Georgia shelf, their

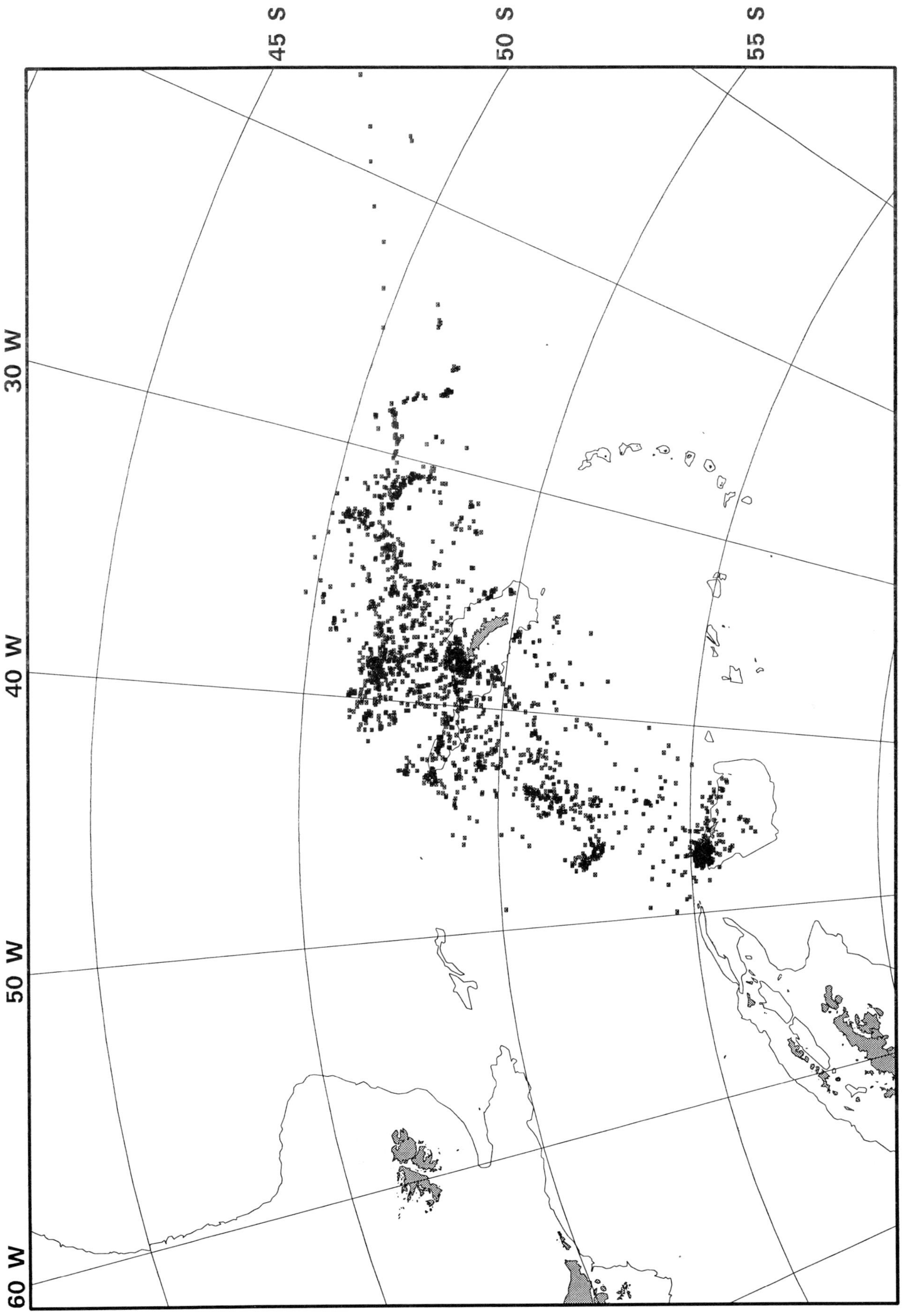

Fig. 6a. Distribution of uplinks from satellite-tracked Black-browed Albatrosses breeding at Bird Island, South Georgia in January–March 1993. Shown are all 1 819 uplinks for 19 birds tracked for 77 days. The position of the 1 000 m isobath is shown.

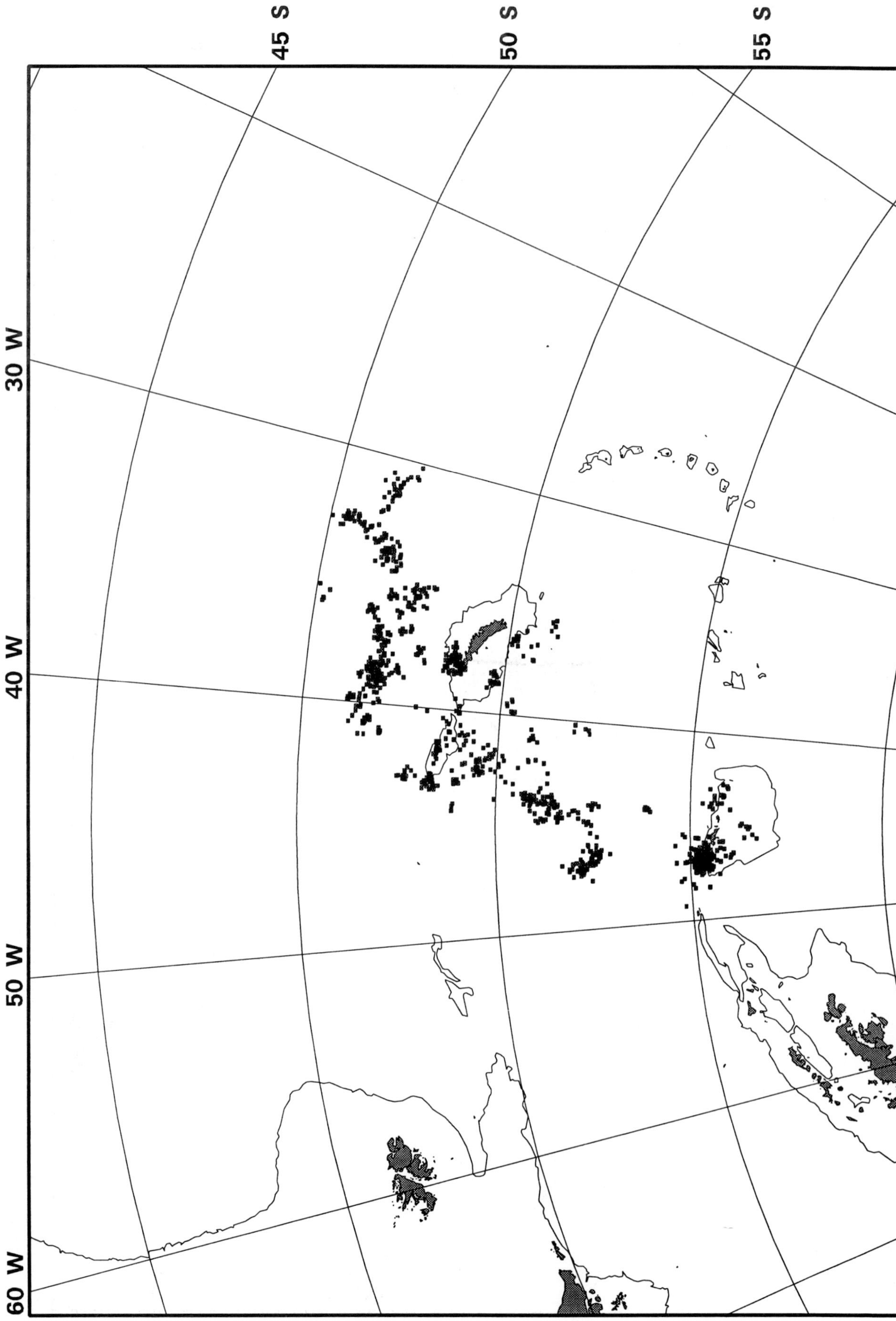

Fig. 6b. Distribution of uplinks from satellite-tracked Black-browed Albatrosses breeding at Bird Island, South Georgia in January–March 1993. Shown are uplinks as for Figure 6a but with transit movements excluded.

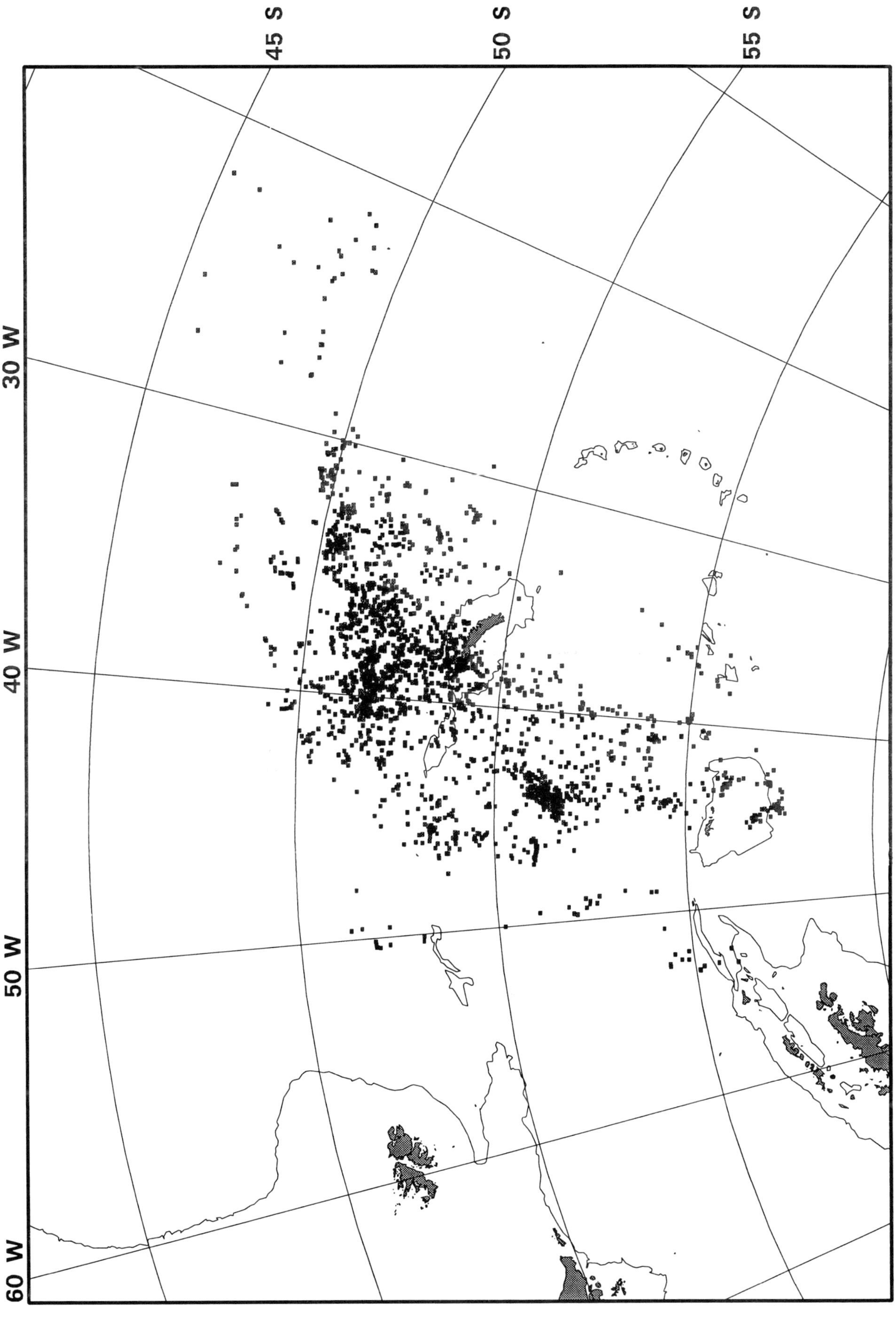

Fig. 6c. Distribution of uplinks from satellite-tracked Grey-headed Albatrosses breeding at Bird Island, South Georgia in January–March 1993. Shown are all 1 919 uplinks for 10 birds tracked for 90 days. The position of the 1 000 m isobath is shown.

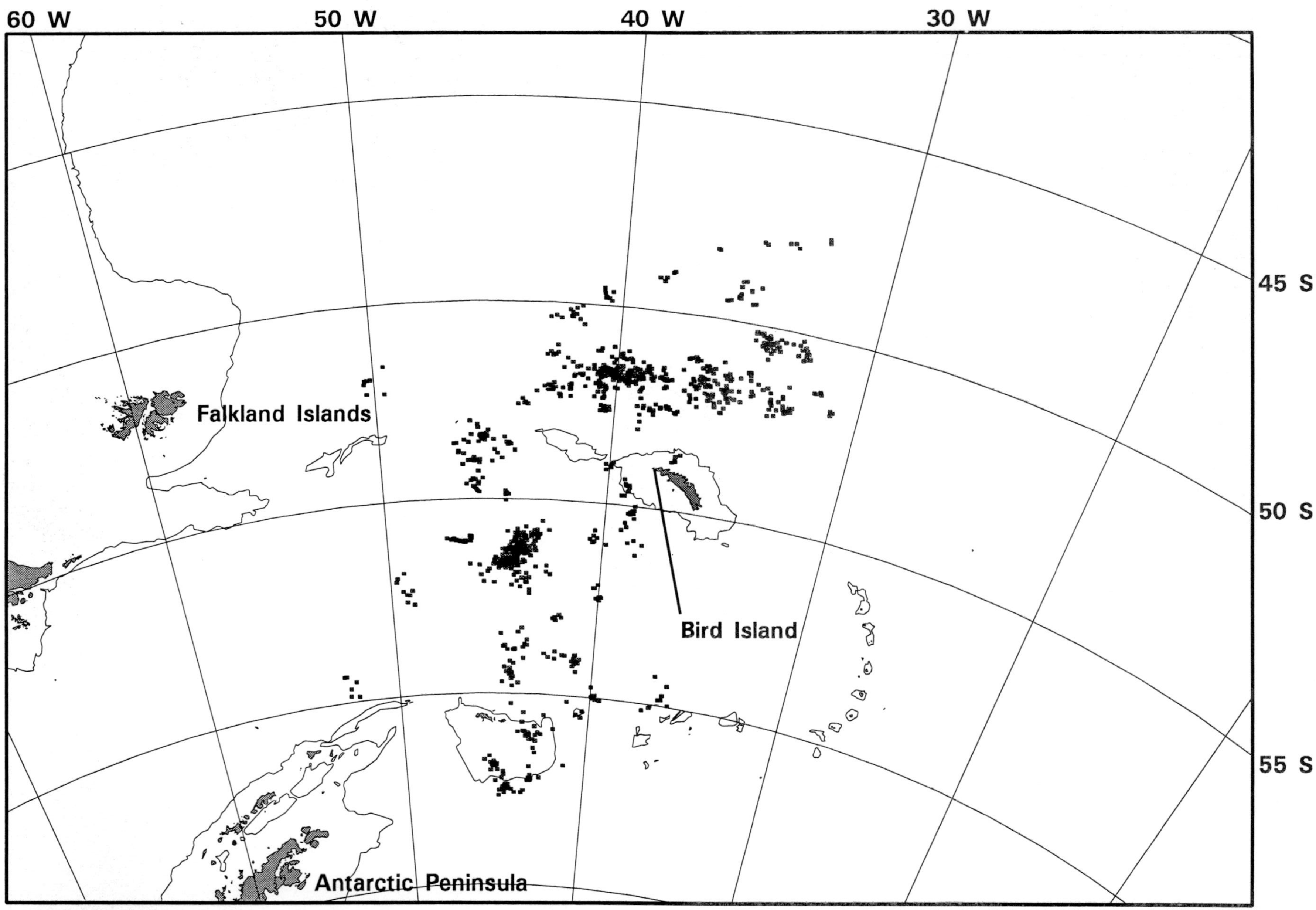

Fig. 6d. Distribution of uplinks from satellite-tracked Grey-headed Albatrosses breeding at Bird Island, South Georgia in January–March 1993. Shown are uplinks as for Figure 6c but with transit movements excluded.

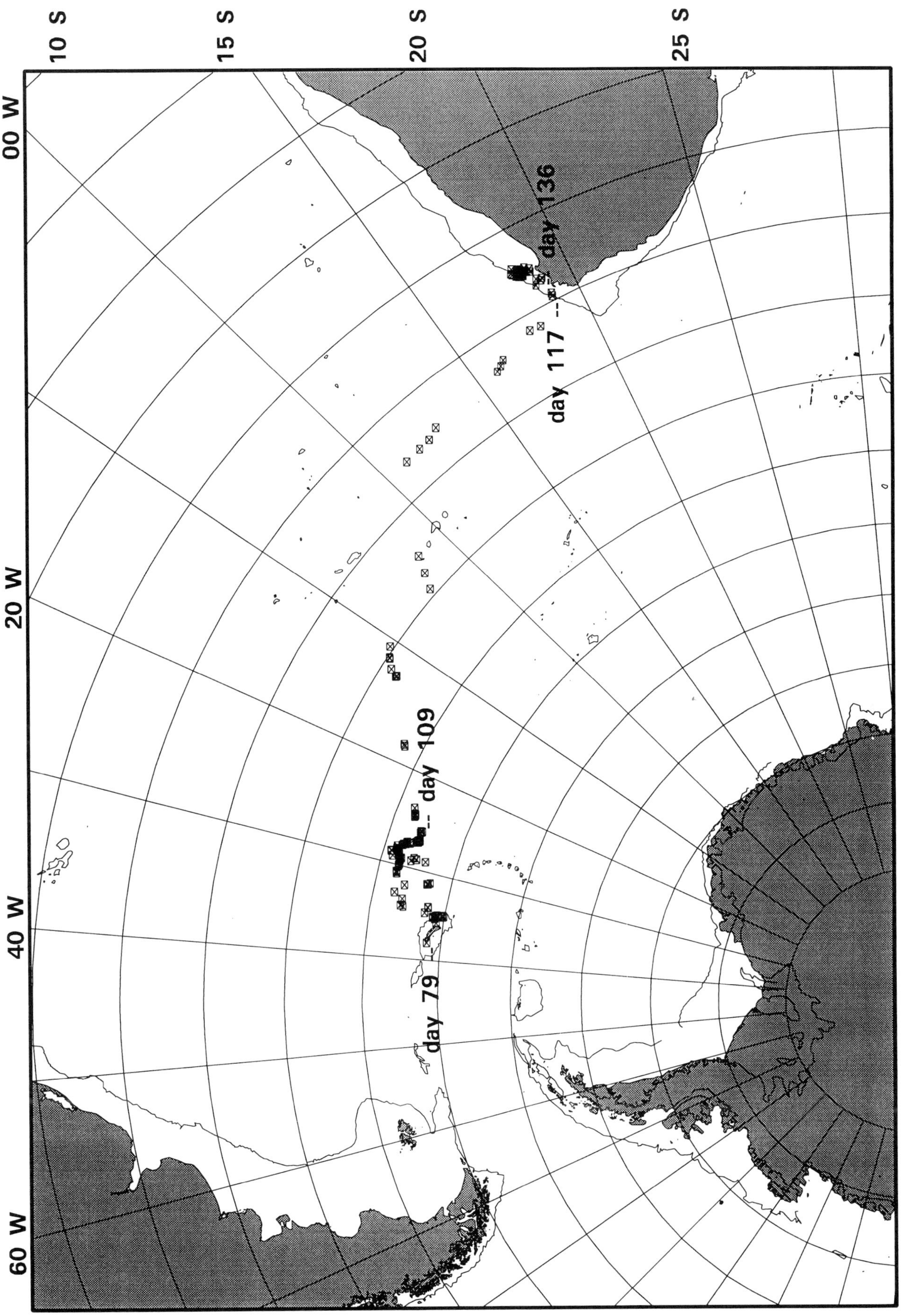

Fig. 7a. Track of Black-browed Albatross after concluding breeding activities at Bird Island, South Georgia. Shown is the track of male 1440 (130–1195); distance travelled 14 012 km in 58.5 days, March–April 1992. Numbers on map refer to Julian days (day 79 = 20 March, day 136 = 16 May). The position of the 1 000 m isobath is shown.

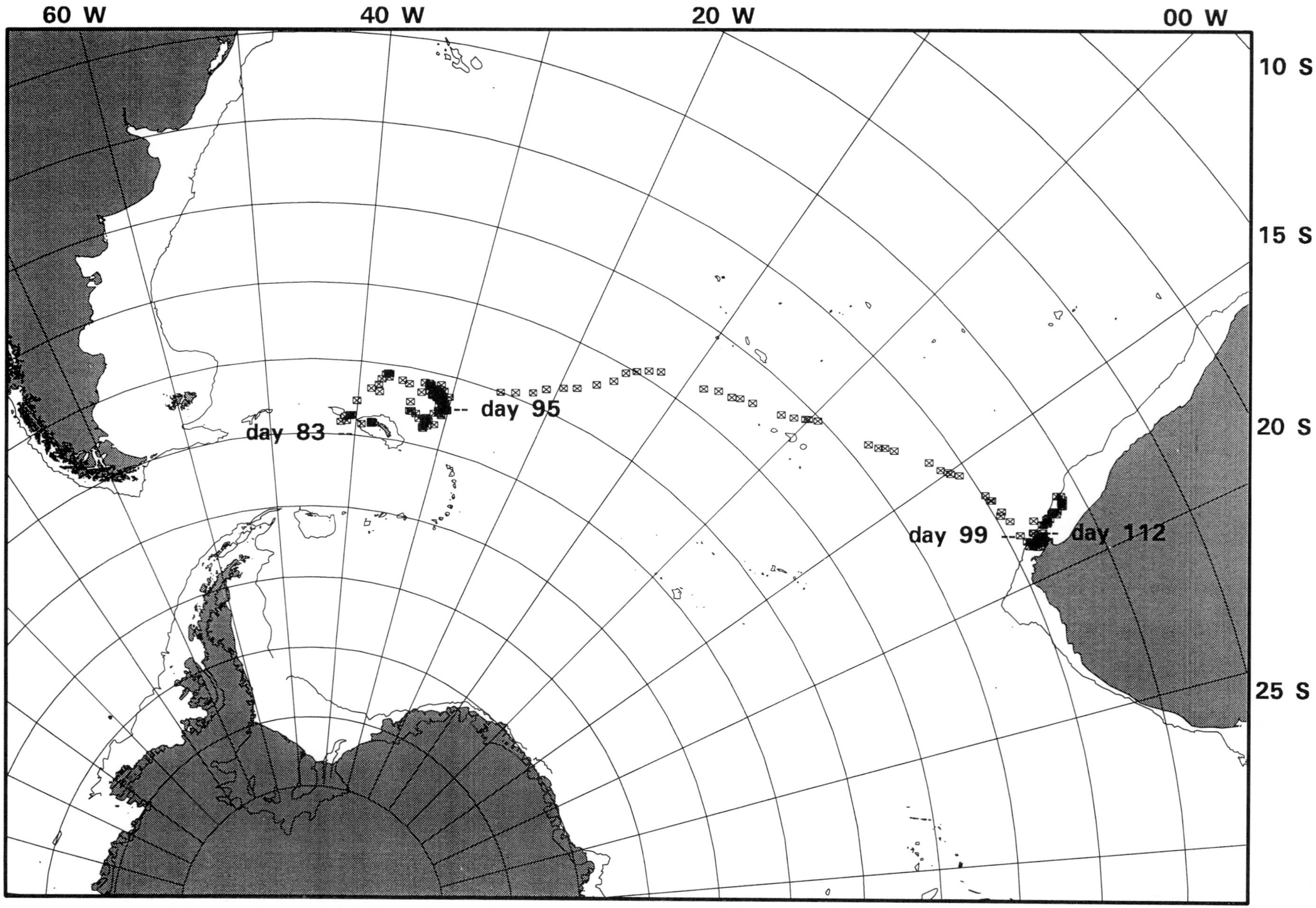

Fig. 7b. Track of Black-browed Albatross after concluding breeding activities at Bird Island, South Georgia. Shown is the track of unsexed bird 3247 (130–1102); distance travelled 16 090 km in 28.4 days, March–April 1993. Numbers on map refer to Julian days (day 79 = 20 March, day 136 = 16 May). The position of the 1 000 m isobath is shown.

greater use of the APFZ (especially to the west of South Georgia) and in the details of distribution in the central Scotia Sea.

We have only one track of a Grey-headed Albatross at the end of its chick-rearing activities with which to assess post-breeding movements (Fig. 8). This bird had been making regular, typical foraging trips (see Fig. 6) until it returned to the breeding colony on 16 March 1996 (day 76), to find its chick was dead. It then spent 8 h near Bird Island before travelling 794 km in 17 h (at 46.0 km h^{-1}) west to the shelf edge south of the South Orkney Islands at 60°S, continuing south-west along the shelf of the Antarctic Peninsula to 68°S and 75°W (west of Anvers Island), having covered 2 977 km at 23.0 km h^{-1}. It then headed north-west into the South Pacific Ocean before moving in a broad northerly arc to an area at 45°S, 80°W, 200 km south-east of Chiloe Island, off the west coast of Chile, having covered 3 415 km at 38.7 km h^{-1}. Five days were spent in this area (travelling 1 148 km at 9.1 km h^{-1}) before moving west for 1 150 km (at 16.1 km h^{-1}) reaching 41°S, 85°W before the transmitter failed on 5 April (day 95).

DISCUSSION

Distribution of South Georgia albatrosses and interactions with fisheries

Wandering Albatross

During their breeding season Wandering Albatrosses from South Georgia are confined to the south-west Atlantic Ocean where they range widely between 25–65°S and 20–80°W. They probably forage over this whole range with particular aggregations in certain shelf-slope areas. In their oceanic range they overlap extensively with the main area of operation in the south-west Atlantic (35–50°S, 30–60°W) of the Japanese longline fishery for Southern Bluefin Tuna (Brothers 1991; Croxall and Prince 1990; Croxall *et al.* 1990). Over shelf and shelf-slope waters, Wandering Albatrosses have the potential to interact with a wide variety of fisheries. In Patagonian (south of 42°S) and Falkland Islands waters these include large trawl fisheries for Hake *Merluccius hubbsi* and Southern Blue Whiting *Micromesistius australis* (Thompson and Riddy 1995; Schiavini *et al.* 1997). Although net-sonde monitor cables may have caused mortality in the past (Bartle 1991) few, if any, trawlers currently use such equipment (Schiavini *et al.* 1997). Indeed, offal or discarded material obtained by scavenging around trawlers may have been of net, albeit short-term, benefit to Wandering Albatrosses.

Of greater importance, however, is the development of longline fisheries in water depths of 50–1 000 m. The target species include Swordfish *Xiphius gladius*, Big-eye *Thunnus obesus*, Long-fin *T. alalunga* and Yellow-fin *T. albacares* Tuna — with a significant bycatch of Blue Shark *Prionace glauca* — in Uruguayan and Brazilian waters (Barea *et al.* 1994; Stagi *et al.* 1997; Neves and Olmos 1997), Hake and Kingclip *Genypterus blacodes* off Patagonia and also Patagonian Toothfish around southern Chile, Patagonia, the Falkland Islands and South Georgia (Schiavini *et al.* 1997). As the maps in Stagi *et al.* (1997) and Schiavini *et al.* (1997) show, these fisheries overlap extensively with the distribution of Wandering Albatrosses from South Georgia (Fig. 2). Indeed, in some areas the congruence of distribution might suggest that we are seeing the locations of albatrosses attracted by fisheries. However, it is clear that many, if not most, of these areas, were favoured by Wandering Albatrosses before longline (or even trawl) fisheries were developed. This reflects the existence of particular oceanographic features (frontal systems, upwellings, etc.) stimulating high productivity of nutrients and rich aggregations of macrozooplankton and attendant predatory fish. Similarly Murphy (1936) summarized several accounts of the abundance of Wandering Albatrosses in the region of the Brazilian and Falkland currents (which run parallel in opposite directions) along the coastal shelf of eastern South America.

We might conclude, therefore, that even today the broad pattern of Wandering Albatross distribution in the South Atlantic Ocean reflects their propensity for travelling between seasonally rich feeding grounds while searching vast areas of ocean for other suitable prey. However, the attraction of Wandering Albatrosses to fisheries undoubtedly focuses their distribution within these traditional areas, with all the attendant risks of incidental mortality. These risks and the resulting mortality are now becoming well documented, not just in the area of southern Brazil-northern Argentina (Stagi *et al.* 1997; Neves and Olmos 1997; Fig. 3) with the recognition that many beach-cast birds owe their origin to longline mortality offshore, but also in most shelf-slope areas between northern Patagonia and Tierra del Fuego (Schiavini *et al.* 1997). An acute concern is the development of fisheries at times of year and/or in places where Wandering Albatrosses have particularly restricted requirements and/or distributions. This is notably so during the brooding period when Wandering Albatrosses make only short trips and are disproportionately at risk to longline fishing for Patagonian Toothfish in the vicinity of South Georgia, their only breeding site in the whole region (Fig. 4; Croxall and Prince 1996).

Outside the breeding season, recoveries (and limited satellite-tracking data) combine to suggest that Wandering Albatrosses from South Georgia tend to head north before crossing the

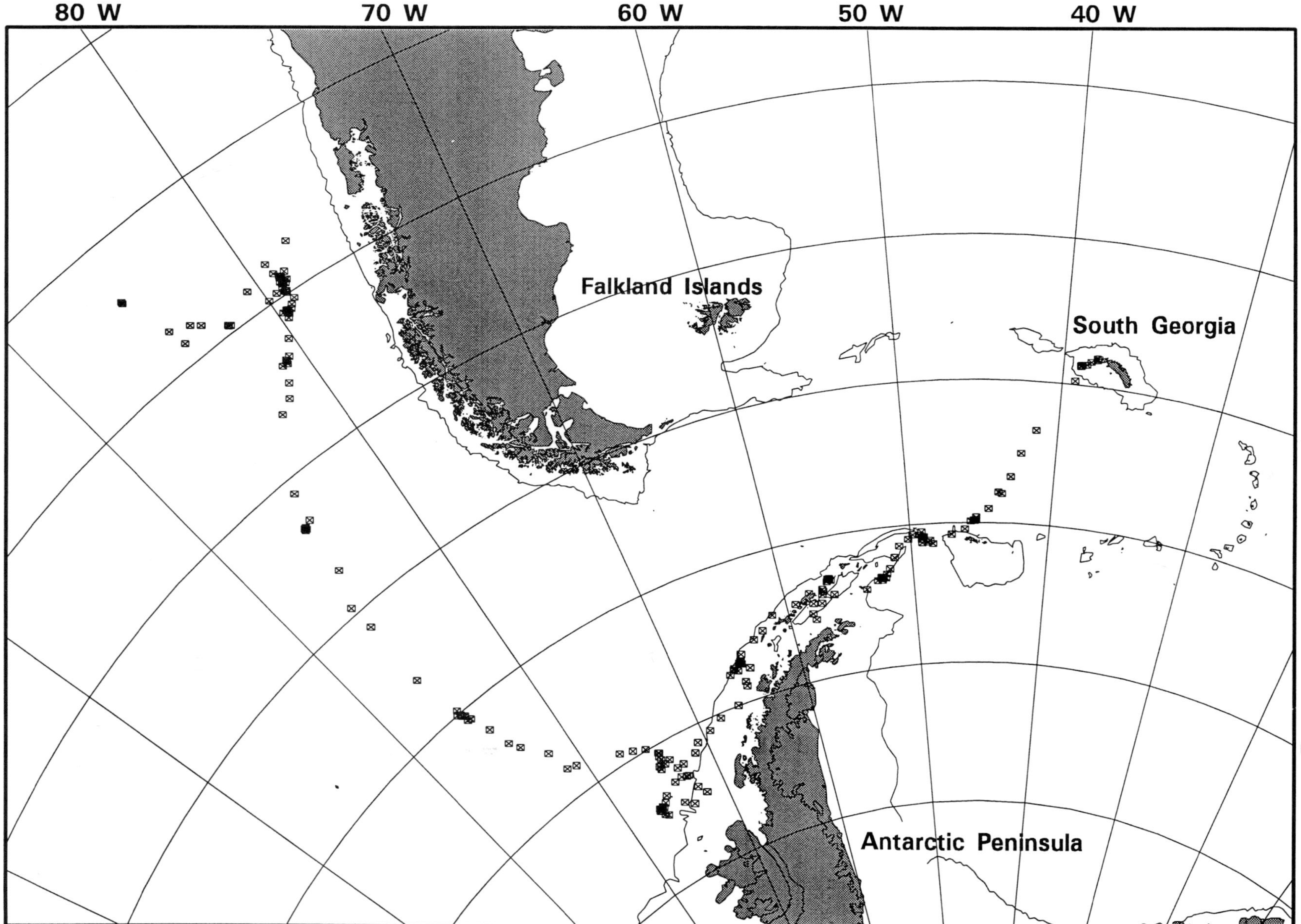

Fig. 8. Track of Grey-headed Albatross (unsexed 1570 132–0100) after concluding breeding activities at Bird Island, South Georgia. Distance travelled 9 745 in 18.5 days, March–April 1996. The position of the 1 000 m isobath is shown.

Atlantic Ocean at relatively low latitudes (30–40°S). Off South Africa they frequent shelf, shelf-slope and adjacent oceanic areas, the last of these being a principal area of longline fishing for Southern Bluefin Tuna (Brothers 1991; Weimerskirch *et al.* 1997). Recoveries and satellite tracks all indicate the importance of this region as a staging area for Wandering Albatrosses on their eastward journey towards Australia. We are less confident of the precise route they take towards Australia, but if travelling between 35–45°S (see Fig. 5a) they would, except between 55–90°E, coincide with the major regional fishing grounds for Southern Bluefin Tuna (Brothers 1991). Assuming that South Georgia Wandering Albatrosses follow the track to New South Wales of birds returning thence to their breeding grounds on Iles Crozet (Nichols *et al.* 1995), then they are transiting the areas within the Australian fishing zone which, historically and currently, are most heavily fished with longlines for Southern Bluefin Tuna (Brothers 1991; Klaer and Polacheck, in press). Although Iles Crozet birds return westwards from eastern Australia, recoveries of South Georgia Wandering Albatrosses in the Tasman Sea, off eastern and southern New Zealand (again important fishing grounds for Southern Bluefin Tuna (Murray *et al.* 1993)), in the Pacific Ocean and off western Chile suggest that some, if not most, may return to their breeding grounds by continuing to fly east. Almost nothing seems to be documented about fisheries in the South Pacific Ocean that have the potential to interact with albatrosses.

Hitherto it has been customary to emphasize similarities between the South Georgia and Iles Crozet populations of Wandering Albatrosses. Both populations were decreasing, comprise birds of similar size inhabiting isolated sub-Antarctic islands and show similar amplitude of, and seasonal changes in, foraging range (compare Weimerskirch *et al.* 1993 and Fig. 1). However, we can now recognize two important differences which may reflect the different population trajectories at the two sites. First, South Georgia birds spend appreciably more time over waters of the nearest (South American) continental shelf; in contrast Iles Crozet birds stay in more oceanic waters and barely reach the South African shelf-slope (Weimerskirch *et al.* 1993). Furthermore, for much of the 1970s and 1980s the Crozet population was near the centre of operations of the longline fishery for Southern Bluefin Tuna, whereas the similar fishery in the south-west Atlantic Ocean was much smaller and only partially overlapped with the at-sea distribution of South Georgia Wandering Albatrosses. At such times the Iles Crozet populations were in precipitous decrease (Weimerskirch and Jouventin 1987) whereas South Georgia ones showed a much lower rate of decrease (Croxall *et al.* 1990). Currently, however, the balance of advantage has changed substantially. Longline fisheries for Southern Bluefin Tuna have decreased generally and now operate further away from Iles Crozet (Weimerskirch *et al.* 1997; Weimerskirch 1997). The recent massive development of new longline fisheries throughout the length of the South American coastal shelf-slope has greatly increased interactions with Wandering Albatrosses from South Georgia. This may account for the continuing decrease in this population compared with the current relative stability of the Iles Crozet populations.

Second, South Georgia Wandering Albatrosses remain at much greater risk than do Iles Crozet birds by virtue of their longer migrations, including global circumnavigation. Within any two-year period South Georgia Wandering Albatrosses are at risk from interactions with fisheries throughout the Atlantic, Indian, Australasian and Pacific sectors of the Southern Ocean; Iles Crozet birds are apparently at risk only in the Indian and Australian sectors. The outlook for the South Georgia population, therefore, remains more serious, given the diversity of threats it faces. However, while we might expect a continuing recovery of the Iles Crozet populations, the development of new longline fisheries (e.g., for Hake) off South Africa could attract substantial numbers of Wandering Albatrosses from Iles Crozet, especially during incubation and late chick-rearing periods.

Black-browed Albatross

South Georgia Black-browed Albatrosses have a foraging lifestyle very different from that of Wandering Albatrosses. At all times of year they appear largely restricted to shelf and shelf-slope areas (and associated frontal zones) and this seems characteristic of the species in other areas (Cherel and Weimerskirch 1995). During the breeding season, therefore, Black-browed Albatrosses from South Georgia have been at low risk from fisheries. Indeed they probably benefited from the scavenging opportunities presented by the development of fin-fisheries in the vicinity of South Georgia in the mid-1960s (Kock 1992) in the same way as described for conspecifics at the Falkland Islands (Thompson 1992; Thompson and Riddy 1995). Thus the South Georgia Black-browed Albatross population may well have increased throughout the 1970s and 1980s. Subsequently, trawl fisheries around South Georgia have largely collapsed through over-fishing. The only current threat to South Georgia Black-browed Albatrosses during their breeding season is from the longline fishery for Patagonian Toothfish which has been operating in South Georgia shelf-slope waters throughout the last decade. Black-browed Albatrosses have been the commonest species in the seabird bycatch (Ashford *et al.* 1994, 1995) and the recent decrease in breeding populations and adult and juvenile survival rates are likely to reflect this (Croxall *et al.* 1997).

Outside the breeding season the restriction of most (*c.*85%) of the South Georgia Black-browed Albatross population to the coastal shelf of South Africa circumscribes potential interactions with fisheries. For many years the trawl fisheries in the Benguela region killed numerous albatrosses, but also provided scavengeable offal and discards. Present trawl fisheries are unlikely to kill many albatrosses. However, the recent development of longline fisheries (especially for Hake), now poses a considerable threat to adult and juvenile Black-browed Albatrosses alike. The remarkably small proportion (4%) of the South Georgia population that migrates to the Patagonian shelf restricts mortality from fisheries in this area which are having potentially very serious effects on Black-browed Albatrosses from the Falkland (and possibly also Chilean) populations (Stagi *et al.* 1997; Schiavini *et al.* 1997). The greater proportion of recent recoveries of Black-browed Albatrosses from longline fisheries for tuna to the south-east of South Africa, in the Indian Ocean and around Australia is of particular concern if it represents a tendency for Black-browed Albatrosses increasingly to visit these areas, with higher attendant risk of mortality. Although the South Georgia population of Black-browed Albatrosses (after the Falkland Islands numerically the most important in the world) has, until recently, been relatively well isolated from adverse interactions with fisheries, the development of substantial longline fisheries around its breeding grounds and also in its principal wintering areas do not augur well for the future.

Grey-headed Albatross

The overall geographical distribution of adult Grey-headed Albatrosses from South Georgia during their breeding season is not greatly different from that of Black-browed Albatrosses. However, Grey-headed Albatrosses are much less frequently associated with fishing vessels (Ashford *et al.* 1995) and, because they spend little time foraging over the shelf areas, are less at risk from local longline fisheries for Patagonian Toothfish. Although breeding birds are killed in this fishery, we assess adult Grey-headed Albatrosses as at lower risk from shelf-slope longline fisheries than either Black-browed or Wandering Albatrosses. The location and pattern of recoveries of Grey-headed Albatrosses indicate a very different lifestyle from Black-browed Albatrosses and suggest that the former species remains in oceanic (perhaps particularly frontal system) habitats at higher latitudes and is particularly abundant in the eastern Indian Ocean, the Tasman Sea and off northern and eastern New Zealand. Substantial longline fisheries for Southern Bluefin Tuna exists in all these areas (Murray *et al.* 1993) and it is likely that South Georgia Grey-headed Albatrosses are and have been much more affected by the fisheries in these areas than have Black-browed Albatrosses.

It is in these areas that we presume the high mortalities of juveniles — the main factor accounting for the steady decrease in South Georgia breeding populations over the last 20 years (Croxall *et al.* 1997) — occur. However, so little is known of the at-sea distributions of juveniles in the decade before they start breeding that we cannot exclude the possibility that events in other areas (especially those less frequented by adults, whose survival rates have been less affected) are also important.

Given the distribution of recoveries, the post-breeding migration of the single satellite-tracked Grey-headed Albatross into the eastern Pacific Ocean was unexpected. It is, therefore, possible not only that part of the population moves westward into these areas in the austral autumn but even that birds may continue across the Pacific Ocean to the New Zealand area — a movement similar to that of the return of Royal Albatrosses *D. epomophora* to their New Zealand breeding sites. Grey-headed Albatrosses could, therefore, have a similarly extensive (i.e., virtually worldwide in the Southern Ocean) vulnerability to pelagic longline fisheries as Wandering Albatrosses — and therefore be at much greater potential risk than Black-browed Albatrosses.

With at least 56% of the world population of Grey-headed Albatrosses breeding at South Georgia (and the other main population being in southern Chile), the prospects of improved survival rates and population recovery are not encouraging. More information on their distribution at sea outside the breeding season and on interactions with fisheries at this time is urgently needed.

Finally, any development of commercial squid fisheries in the Southern Ocean would have substantial potential implications for Grey-headed Albatrosses. At South Georgia their diet and foraging activities are principally directed at the ommastrephid squid *Martialia hyadesi* (Rodhouse *et al.* 1990, 1996), a species which has major potential for commercial exploitation (Rodhouse 1990; Rodhouse *et al.* 1993). Any development of squid fisheries, especially around South Georgia, will need to proceed with great care so as not to pose further threats to the survival of Grey-headed Albatrosses.

CONCLUSIONS

Despite the considerable differences in their distributions, movements and lifestyles, Wandering, Black-browed and Grey-headed Albatrosses are all immensely vulnerable to a similar suite of threats posed by the recent activities of longline fishing for tuna and the current development of numerous similar fisheries for a range of other target species. An important element in reducing current and potential mortality of albatrosses associated with fishing activities is knowing where albatrosses are at different times of year. Satellite tracking should ultimately be able to provide most of the necessary information, but this will be costly and time consuming. At present, however, while we may be able to describe adequately the distribution of albatrosses during their breeding season, knowledge of their movements at other times of year (potentially just as critical in terms of fisheries interactions and albatross survival) still depends principally on interpreting recoveries of banded birds.

The current plight of many albatross populations, especially those at South Georgia, suggests that we cannot delay in taking appropriate action to reduce mortality of albatrosses caused by longline fisheries: First, appropriate mitigating measures (e.g., setting lines at night, always using correctly rigged streamer lines, no discharge of offal) must be used by all longline fisheries that operate where albatrosses occur. Second, properly trained observers should be aboard a high proportion of vessels, helping fishing masters to reduce seabird bycatch and to maximize fishing efficiency. Third, new fishing techniques designed to eliminate seabird bycatch (e.g., underwater setting) should be developed, tested and made widely available. Fourth, new longline fisheries (in the sense of operating in new areas, targetting new fish species or using new methods) should be developed and monitored with particular care.

In terms of albatrosses from South Georgia, areas in which better management and regulation of fisheries is particularly critical are: a) on the coastal shelf and adjacent waters of South America (also protecting Black-browed Albatrosses from the Falkland Islands and Royal Albatrosses from New Zealand); b) similar areas around South Africa (also protecting Atlantic Yellow-nosed Albatrosses *Diomedea (Thalassarche) chlororhynchus* and populations of several species of albatrosses from the Indian Ocean); and c) oceanic habitats throughout the Indian Ocean and especially in Australasian waters (also protecting a variety of resident and migratory albatross species). If this can be achieved through national initiatives and international co-operation then populations of albatrosses at South Georgia (and elsewhere) may slowly stabilize and recover within the next decade or so. If not, the survival of some albatrosses could be in the balance — though they may well outlive many of the longline fisheries which are currently contributing to their downfall.

ACKNOWLEDGEMENTS

We thank all staff and colleagues who have contributed to the albatross research programme at Bird Island by banding or retrapping birds and by assisting with the satellite-tracking studies. We thank Nic Huin for assistance with the synthesis of banding data and Julie Leland for typing the manuscript.

APPENDIX 1

Summary of banding and recoveries of Wandering, Grey-headed and Black-browed Albatrosses banded at Bird Island, South Georgia 1959–1996.

	No. ringed	Recoveries 1959–64 No.	1959–64 Rate (%)	1965–74 No.	1965–74 Rate (%)	1975–84 No.	1975–84 Rate (%)	1985–94 No.	1985–94 Rate (%)	1995– No.	1995– Rate (%)	Total No.	Total Rate (%)
Wandering Albatross													
ADULT													
1959–64	4 383	8	0.183	16	0.365	3	0.068	3	0.068	1	0.023	31	0.707
1965–74	1 020			0	0.000	0	0.000	0	0.000	0	0.000	0	0.000
1975–84	1 435					2	0.139	4	0.279	0	0.000	6	0.418
1985–94	1 108							2	0.181	3	0.271	5	0.451
1995–	67									0	0.000	0	0.000
Total	8 013	8		16		5		9		4		42	0.524
Indian		1		5		1		1		0		8	0.100
Pacific		0		2		0		0		0		2	0.025
S. Atlantic		7		7		2		8		4		28	0.349
Tasmania		0		2		2		0		0		4	0.050
CHICK													
1959–64	1 400	5	0.357	4	0.286	1	0.071	0	0.000	0	0.000	10	0.714
1965–74	443			0	0.000	0	0.000	1	0.226	0	0.000	1	0.226
1975–84	7 837					13	0.166	29	0.370	2	0.026	44	0.561
1985–94	9 177							35	0.381	7	0.076	42	0.458
1995–	973									0	0.000	0	0.000
Total	19 830	5		4		14		65		9		97	0.489
Indian		1		1		2		12		2		18	0.091
Pacific		0		1		3		5		0		9	0.045
S. Atlantic		3		2		6		38		7		56	0.282
Tasmania		1		0		3		10		0		14	0.071
TOTAL													
1959–64	5 783	13	0.225	20	0.346	4	0.069	3	0.052	1	0.017	41	0.709
1965–74	1 463			0	0.000	0	0.000	1	0.068	0	0.000	1	0.068
1975–84	9 272					15	0.162	33	0.356	2	0.022	50	0.539
1985–94	10 285							37	0.360	10	0.097	47	0.457
1995–	1 040									0	0.000	0	0.000
Total	27 843	13		20		19		74		13		139	0.499
Indian		2		6		3		13		2		26	0.093
Pacific		0		3		3		5		0		11	0.040
S. Atlantic		10		9		8		46		11		84	0.302
Tasmania		1		2		5		10		0		18	0.065
Grey-headed Albatross													
ADULT													
1959–64	991	0	0.000	0	0.000	2	0.202	0	0.000	0	0.000	2	0.202
1965–74	141			0	0.000	0	0.000	0	0.000	0	0.000	0	0.000
1975–84	1 652					2	0.121	1	0.061	0	0.000	3	0.182
1985–94	658							0	0.000	0	0.000	0	0.000
1995–	9									0	0.000	0	0.000
Total	3 451	0		0		4		1		0		5	0.145
Aust./NZ		0		0		2		1		0		3	0.087
Indian		0		0		0		0		0		0	0.000
S. Africa		0		0		0		0		0		0	0.000
S. Atlantic		0		0		2		0		0		2	0.058
CHICK													
1959–64	13 518	8	0.059	12	0.089	3	0.022	1	0.007	0	0.000	24	0.178
1965–74	320			1	0.313	1	0.313	0	0.000	0	0.000	2	0.625
1975–84	3 118					2	0.064	0	0.000	0	0.000	2	0.064
1985–94	3 933							7	0.178	0	0.000	7	0.178
1995–	87									0	0.000	0	0.000
Total	20 976	8		13		6		8		0		35	0.167
Aust./NZ		4		8		4		3		0		19	0.091
Indian		0		1		0		3		0		4	0.019
S. Africa		4		4		0		1		0		9	0.043
S. Atlantic		0		0		2		1		0		3	0.014

Appendix 1 — *continued.*

	No. ringed	Recoveries 1959–64		1965–74		1975–84		1985–94		1995–		Total	
		No.	Rate (%)	No.	Rate (%)	No.	Rate (%)	No.	Rate (%)	No.	Rate (%)	No.	Rate (%)
Grey-headed Albatross — *continued*													
TOTAL													
1959–64	14 509	8	0.055	12	0.083	5	0.034	1	0.007	0	0.000	26	0.179
1965–74	461			1	0.217	1	0.217	0	0.000	1	0.021	2	0.434
1975–84	4 770					4	0.084	1	0.021	0	0.000	5	0.105
1985–94	4 591							7	0.152	0	0.000	7	0.152
1995–	96									0	0.000	0	0.000
Total	24 427	8		13		10		9		0		40	0.164
Aust/NZ		4		8		6		4		0		22	0.090
Indian		0		1		0		3		0		4	0.016
S. Africa		4		4		0		1		0		9	0.037
S. Atlantic		0		0		4		1		0		5	0.020
Black-browed Albatross													
ADULT													
1959–64	781	1	0.128	3	0.384	1	0.128	1	0.128	0	0.000	6	0.768
1965–74	0			0	0.000	0	0.000	0	0.000	0	0.000	0	0.000
1975–84	755					3	0.397	0	0.000	0	0.000	3	0.397
1985–94	414							0	0.000	0	0.000	0	0.000
1995–	0											0	0.000
Total	2 905	1		3		4		1		0		9	0.310
Aust/NZ		0		0		1		0		0		1	0.034
Indian		0		0		0		0		0		0	0.000
S. Africa		1		1		3		1		0		6	0.207
S. Atlantic		0		2		0		0		0		2	0.069
CHICK													
1959–64	20 464	145	0.709	89	0.435	21	0.103	1	0.005	0	0.000	256	1.251
1965–74	118			2	1.695	1	0.847	0	0.000	0	0.000	3	2.542
1975–84	1 528					8	0.524	1	0.065	0	0.000	9	0.589
1985–94	2 169							4	0.184	0	0.000	4	0.184
1995–	16									0	0.000	0	0.000
Total	24 295	145		91		30		6		0		272	1.120
Aust/NZ		8		7		6		3		0		24	0.099
Indian		1		3		0		0		0		4	0.016
S. Africa		132		75		20		3		0		230	0.947
S. Atlantic		4		6		4		0		0		14	0.058
TOTAL													
1959–64	21 245	146	0.687	92	0.433	22	0.104	2	0.009	0	0.000	262	1.233
1965–74	118			2	1.695	1	0.847	0	0.000	0	0.000	3	2.542
1975–84	2 283					11	0.482	1	0.044	0	0.000	12	0.526
1985–94	2 583							4	0.155	0	0.000	4	0.155
1995–	16									0	0.000	0	0.000
Total	26 245	146		94		34		7		0		281	1.071
Aust/NZ		8		7		7		3		0		25	0.095
Indian		1		3		0		0		0		4	0.015
S. Africa		133		76		23		4		0		236	0.899
S. Atlantic		4		8		4		0		0		16	0.061

REFERENCES

Ashford, J. R., Croxall, J. P., Rubilar, P. S. and Moreno, C. S., 1994. Seabird interactions with long-lining operations for *Dissostichus eleginoides* at the South Sandwich Islands and South Georgia. *CCAMLR Science* **1:** 143–53.

Ashford, J. R., Croxall, J. P., Rubilar, P. S. and Moreno, C. S., 1995. Seabird interactions with long-lining operations for *Dissostichus eleginoides* around South Georgia, April to May 1994. *CCAMLR Science* **2:** 111–22.

Barea, L., Loinaz, I., Marin, Y., Rios, C., Saralegui, A., Stagi, A., Vaz-Ferreira, R. and Wilson, N., 1994. Mortality of albatrosses and other seabirds produced by tuna longline fisheries in Uruguay. CCAMLR WG–IMALF–94/17, Hobart, Australia.

Bartle, J. A., 1991. Incidental capture of seabirds in the New Zealand subantarctic squid trawl fishery, 1990. *Bird Conserv. Internat.* **1:** 351–59.

Brothers, N. P., 1991. Albatross mortality and associated bait loss in the Japanese longline fishery in the Southern Ocean. *Biol. Conserv.* **55:** 255–68.

Brothers, N. and Foster, A., in press. Seabird catch rates: an assessment of causes and solutions in Australia's domestic tuna longline fishery. *Mar. Ornithol.*

Cherel, Y. and Weimerskirch, H., 1995. Seabirds as indicators of marine resources: Black-browed Albatrosses feeding on ommastrephid squids in Kerguelen waters. *Mar. Ecol. Prog. Ser.* **129:** 295–300.

Croxall, J. P., 1979. Distribution and population changes in the Wandering Albatross *Diomedea exulans* at South Georgia. *Ardea* **67:** 15–21.

Croxall, J. P. and Prince, P. A., 1990. Recoveries of Wandering Albatrosses *Diomedea exulans* ringed at South Georgia, 1956–86. *Ringing and Migration* **11:** 43–51.

Croxall, J. P. and Prince, P. A., 1996. Potential interactions between Wandering Albatrosses and longline fisheries for Dissostichus at South Georgia. *CCAMLR Science* **3:** 101–10.

Croxall, J. P., Prince, P. A., Hunter, I., McInnes, S. J. and Copestake, P. G., 1984. The seabirds of the Antarctic Peninsula, islands of the Scotia Sea and Antarctic Continent between 80°W and 20°W: their status and conservation. Pp. 637–66 *in* Status and Conservation of the World's Seabirds ed by J. P. Croxall, P. G. H. Evans and R. W. Schreiber. ICBP: Cambridge.

Croxall, J. P., Prince, P. A., Rothery, P. and Wood, A. G., 1997. Population changes in albatrosses at South Georgia. Pp. 69–83 *in* Albatross Biology and Conservation ed by G. Robertson and R. Gales. Surrey Beatty & Sons: Chipping Norton.

Croxall, J. P., Rothery, P., Pickering, S. P. C. and Prince, P. A., 1990. Reproductive performance, recruitment and survival of Wandering Albatrosses *Diomedea exulans* at Bird Island, South Georgia. *J. Anim. Ecol.* **59:** 773–94.

de la Mare, W. K. and Kerry, K. R., 1994. Population dynamics of the Wandering Albatross *(Diomedea exulans)* on Macquarie island and the effects of mortality from longline fishing. *Polar Biol.* **14:** 231–41.

Gales, R., 1993. Co-operative Mechanisms for the Conservation of Albatross. Australian Nature Conservation Agency and Australian Antarctic Foundation: Hobart, Australia.

Gales, R. and Brothers, N., 1995. Characteristics of seabirds caught in the Japanese tuna longline fishery in the Australian region. P. 29 *in* Abstracts of the First International Conference on the Biology and Conservation of Albatrosses. Hobart, Australia.

Gibson, J. D., 1963. Third report of the New South Wales Albatross Study Group. *Emu* **63:** 215–23.

Gibson, J. D., 1973. New South Wales Albatross Study Group, 1972 season. *Aust. Bird Bander* **11:** 5–7.

Klaer, N. and Polacheck, T., in press. Bycatch of albatrosses and other seabirds by Japanese longline fishing vessels in the Australian Fishing Zone from April 1992 to March 1995. *Emu.*

Kock, K.-H., 1992. Antarctic Fish and Fisheries. Cambridge University Press: Cambridge, England.

Morant, P. D., Brooke, R. K. and Abrams, R. W., 1983. Recoveries in southern Africa of birds breeding elsewhere. *Ringing and Migration* **4:** 257–68.

Murphy, R. C., 1936. Oceanic Birds of South America. Macmillan: New York.

Murray, T. E., Bartle, J. A., Kalish, S. R. and Taylor, P. R., 1993. Incidental capture of seabirds by Japanese Southern Bluefin Tuna longline vessels in New Zealand waters, 1988–1992. *Bird Conserv. Internat.* **3:** 181–210.

Neves, T. and Olmos, F., 1997. Albatross mortality in fisheries off the coast of Brazil. Pp. 214–19 *in* Albatross Biology and Conservation ed by G. Robertson and R. Gales. Surrey Beatty & Sons: Chipping Norton.

Nicholls, D. G., Murray, M. D., Battam, H., Robertson, G., Moors, P., Butcher, E. and Hildebrandt, M., 1995. Satellite tracking of the Wandering Albatross *Diomedea exulans* around Australia and in the Indian Ocean. *Emu* **95:** 223–30.

Prince, P. A., Rothery, P., Croxall, J. P. and Wood, A. G., 1994. Population dynamics of Black-browed and Grey-headed Albatrosses *Diomedea melanophris* and *D. chrysostoma* at Bird Island, South Georgia. *Ibis* **136:** 50–71.

Prince, P. A., Wood, A. G., Barton, T. and Croxall, J. P., 1992. Satellite tracking of Wandering Albatrosses *(Diomedea exulans)* in the South Atlantic. *Antarctic Sci.* **4:** 31–36.

Robertson, C. J. R., 1993. Survival and longevity of the northern Royal Albatross *Diomedea epomophora sanfordi* at Taiaroa Head 1937–1993. *Emu* **93:** 269–76.

Rodhouse, P. G., 1990. Cephalopod fauna of the Scotia Sea at South Georgia: potential for commercial exploitation and possible consequences. Pp. 289–98 *in* Antarctic ecosystems: ecological change and conservation ed by K. R. Kerry and G. Hempel. Springer-Verlag: Berlin.

Rodhouse, P. G., Croxall, J. P. and Prince, P. A., 1993. Towards an assessment of the stock of the ommastrephid squid *Martialia hyadesi* in the Scotia Sea: data from predators. Pp. 433–40 *in* Recent advances in cephalopod fisheries biology ed by T. Okutani, R. K. O'Dor and T. Kubodera. Tokai University Press: Tokyo.

Rodhouse, P. G., Prince, P. A., Clarke, M. R. and Murray, A. W. A., 1990. Cephalopod prey of the Grey-headed Albatross *Diomedea chrysostoma. Mar. Biol.* **104:** 353–62.

Rodhouse, P. G., Prince, P. A., Trathan, P. N., Hatfield, E. M. C., Watkins, J. L., Bone, D. G., Murphy, E. J. and White, M. G., 1996. Cephalopod community associated with mesoscale oceanographic features at the Antarctic Polar Front exploited by satellite tagged predators. *Mar. Ecol. Prog. Ser.* **136:** 37–50.

Schiavini, A., Frere, E., Gandini, P., García, N. and Crespo, E., 1997. Albatross–fisheries interactions in Patagonian shelf waters. Pp. 208–13 *in* Albatross Biology and Conservation ed by G. Robertson and R. Gales. Surrey Beatty & Sons: Chipping Norton.

Stagi, A., Vaz-Ferreira, R., Marín, Y. and Joseph, L., 1997. The conservation of albatrosses in Uruguayan waters. Pp. 220–24 *in* Albatross Biology and Conservation ed by G. Robertson and R. Gales. Surrey Beatty & Sons: Chipping Norton.

Thompson, K. R., 1992. Quantitative analysis of the use of discards from squid trawlers by Black-browed Albatrosses *Diomedea melanophris* in the vicinity of the Falkland Islands. *Ibis* **134:** 11–12.

Thompson, K. R. and Riddy, M. D., 1995. Utilization of offal and discards from "finfish" trawlers around the Falkland Islands by the Black-browed Albatross *Diomedea melanophris. Ibis* **137:** 198–206.

Tickell, W. L. N., 1967. Movements of Black-browed and Grey-headed Albatrosses in the South Atlantic. *Emu* **66:** 357–67.

Tickell, W. L. N., 1968. The biology of the great albatrosses *Diomedea exulans* and *D. epomophora. Antarctic Res. Ser.* **12:** 1–55.

Vaske, T., 1991. Seabirds mortality on longline fishing for tuna in southern Brazil. *Ciência e Cultura* **43:** 388–90.

Weimerskirch, H., 1997. Foraging strategies of Indian Ocean albatrosses and their relationships with fisheries. Pp. 168–79 *in* Albatross Biology and Conservation ed by G. Robertson and R. Gales. Surrey Beatty & Sons: Chipping Norton.

Weimerskirch, H. and Jouventin, P., 1987. Population dynamics of the Wandering Albatross *(Diomedea exulans)* of the Crozet Islands: causes and consequences of the population decrease. *Oikos* **49:** 315–22.

Weimerskirch, H. and Jouventin, P., 1997. Changes in population sizes and demographic parameters of six albatross species breeding on the French sub-Antarctic islands. Pp. 84–91 *in* Albatross Biology and Conservation ed by G. Robertson and R. Gales. Surrey Beatty & Sons: Chipping Norton.

Weimerskirch, H., Brothers, N. and Jouventin, P., 1997. Population dynamics of Wandering Albatrosses *Diomedea exulans* and Amsterdam Albatross *Diomedea amsterdamensis* in the Indian Ocean and their relationships with longline fisheries: conservation implications. *Biol. Conserv.* **79:** 257–70.

Weimerskirch, H., Salamolard, M., Sarrazin, F. and Jouventin, P., 1993. Foraging strategy of Wandering Albatrosses through the breeding season: a study using satellite telemetry. *Auk* **110:** 325–42.

CHAPTER TWELVE

Foraging strategies of Indian Ocean albatrosses and their relationships with fisheries

HENRI WEIMERSKIRCH[1]

Since 1990 the foraging strategies of Wandering, Amsterdam, Yellow-nosed, Black-browed, Sooty and Light-mantled Sooty Albatrosses have been studied at Amsterdam Island, Iles Crozet and Iles Kerguelen in the southern Indian Ocean. The major foraging strategies of these species vary extensively over the breeding season, both within and among species. Albatrosses either commute to specific areas located close to or far from colonies, or forage continuously in loops or on linear flight paths. Iles Crozet Wandering Albatrosses and Iles Kerguelen Black-browed Albatrosses foraging in the Kerguelen continental shelf area overlap with fisheries, the overlap resulting from the birds natural foraging ranges, rather than from changes in their foraging behaviour to scavenge from fishing vessels. If conservation measures are to be implemented to avoid contact between albatrosses and fishing vessels, a good knowledge of the foraging strategies of each species is necessary. Closure of areas to fishing within a radius of 200–300 km around the breeding grounds would be a useful first step in reducing albatross mortality during the breeding seasons.

Key words: Albatrosses, Foraging Strategies, Indian Ocean, Fisheries Interactions, Conservation Implications.

INTRODUCTION

THE life history attributes of albatrosses and other oceanic seabirds are assumed to be adapted to the marine environment they exploit, particularly the distances between nesting and feeding sites (Lack 1968) and the patchiness and unpredictability of prey resources (Ashmole 1971). Although the breeding biology and population demography of some albatross species have been extensively studied (e.g., Croxall *et al.* 1997; Weimerskirch and Jouventin 1997), little information is available on the ecology at sea of these seabirds, which is surprising, given that albatrosses spend more than 80% of their life at sea. Knowledge of the foraging behaviour and at-sea distributions of albatrosses is necessary to understand the relationships between albatrosses and the marine environment, and the relationships between albatrosses and the fisheries that are presently causing many albatross populations to decrease (Croxall and Gales 1997; Weimerskirch and Jouventin 1997; Prince *et al.* 1997).

Since 1989, when Wandering Albatrosses *Diomedea exulans* were satellite tracked for the first time (Jouventin and Weimerskirch 1990), satellite transmitters have become smaller and lighter and stomach temperature loggers (Wilson *et al.* 1992) and activity recorders (Wilson *et al.* 1995) have been developed. These instruments, when deployed together on the same albatross, allow information to be gathered on foraging ranges and locations, the timing and rates of food consumption, and behaviour at sea. Also, the application of these technologies provide inferences to be made about prey distribution and abundance, and enable estimates to be made of the profitability of albatross foraging trips (Weimerskirch and Wilson 1992; Weimerskirch *et al.* 1994a).

This chapter describes the types of foraging strategies used by six albatross species from Iles Crozet, Iles Kerguelen and Amsterdam Island, in the southern Indian Ocean, and the relationships between some species and fisheries operating in their foraging areas. The chapter also considers the usefulness of studies of foraging behaviour in attempts to minimize fisheries-related effects on albatrosses.

FORAGING BEHAVIOUR OF WANDERING ALBATROSSES

Studies since 1989 of Wandering Albatrosses at Iles Crozet (Weimerskirch *et al.* 1992; Salamolard and Weimerskirch 1993; Weimerskirch *et al.* 1993; Weimerskirch 1995; Weimerskirch *et al.* 1997a)

[1]Centre d'Etudes Biologiques de Chizé, Centre National de la Recherche Scientifique, F79360 Villiers-en-Bois, France.

Pages 168–79 *in* ALBATROSS BIOLOGY AND CONSERVATION, ed by G. Robertson and R. Gales. Surrey Beatty & Sons, Chipping Norton. 1997.

have shown that foraging trips during incubation are of long duration with a long, looping route (hereafter referred to as "loop searching") (Fig. 1). Birds forage at an average maximum distance of 1 280 km from Iles Crozet, with some birds reaching up to 4 000 km, and travel on average 5 990 km per foraging trip (maximum distance: 15 200 km). Males forage in Antarctic and sub-Antarctic waters, whereas females usually forage to the north of these waters. Birds reduce the duration and maximum range of their foraging trips at the end of the incubation period. However, during the brood/guard period Wandering Albatrosses change their foraging strategy, and commute repeatedly to a specific area on the Iles Crozet continental shelf edge (hereafter referred to as "short commuting", Fig. 2). Following the brood/guard period, when the chick is left alone on the nest, birds adopt a new foraging strategy, alternating short commuting trips to the shelf edge with long looping trips to more distant oceanic waters. During the long foraging trips, instead of flying in loops, about 20% of the birds (mostly males) commute to the Iles Kerguelen shelf, 1 200 km away (hereafter referred to as "long commuting", Fig. 1). Preliminary results from South Georgia (south Atlantic Ocean), Amsterdam Island and the Auckland Islands (south-west Pacific Ocean) suggest that individuals from other populations of closely related taxa behave in a similar manner (Prince *et al.* 1992; Walker *et al.* 1995). However, Auckland Island birds undertake long commuting trips more often than do Iles Crozet birds, since birds from the Auckland Islands frequent waters in the middle of the Tasman Sea, a great distance from their breeding site (Walker *et al.* 1995).

One consequence of the change in foraging tactics over the breeding season is that bird densities at sea vary as the breeding season progresses. During incubation Wandering Albatrosses from Iles Crozet fly long, looping courses between Antarctic and tropical waters. High densities of Wandering Albatrosses occur only around the breeding grounds which act as a bottle neck for outbound and inbound birds (Fig. 3). Peaks of high density also occur north of Iles Crozet. In winter, when adults are attending large chicks, Iles Crozet Wandering Albatrosses alternate long, looping trips with short commuting trips, and at any time about 65% of the birds occur within a 200 km radius of the island (Fig. 3).

On short commuting trips Wandering Albatrosses from Iles Crozet and capture prey, particularly the Squid *Moroteuthis ingens*, and fish, on continental shelf slopes (Weimerskirch *et al.* 1997a) where these prey are encountered irregularly. On long oceanic looping routes, birds feed on a range of patchily-distributed squid species that are encountered regularly throughout their trips. On these long flights the birds use the wind to cover great distances with low energy costs. The energy costs of catching prey by birds on these flights are probably lower than the energy costs by birds on short commuting trips when birds often fly in unfavourable wind conditions.

FORAGING BEHAVIOUR OF THE SMALLER ALBATROSS SPECIES

The foraging behaviours of Light-mantled Sooty Albatrosses *Phoebetria palpebrata* have been studied at Macquarie Island (Weimerskirch and Robertson 1994) and at Iles Crozet (Weimerskirch *et al.*, unpubl. data). Light-mantled Sooty Albatrosses commute quickly (up to 80 km/h) to specific areas where they spend most of their foraging time (Fig. 4). Birds from Macquarie Island foraged along the Antarctic continent 1 700 km south-west from the island, while with birds from the Iles Crozet average distances to commuting zones were shorter, although some birds forage up to 2 400 km from their nest. Long commuting by Light-mantled Sooty Albatrosses at Iles Crozet resembles that by Wandering Albatrosses from the same island. However, the foraging areas used by Light-mantled Sooty Albatrosses do not appear to be related to bathymetric features or known oceanographic features, such as frontal zones, and are located in oceanic sub-Antarctic and Antarctic waters.

The foraging behaviour of Sooty Albatrosses has been studied at Iles Crozet from 1993–1996 (Weimerskirch *et al.*, unpubl. data). Sooty Albatrosses undertake long commuting flights in the early incubation period, moving an average maximum distance of 600 km from their breeding sites (Fig. 4), but late in the incubation period they undertake loop searching flights. During the incubation period the distribution of Sooty Albatrosses in oceanic waters does not appear to be related to specific oceanographic features (Fig. 5). During brooding, Sooty Albatrosses radiate in all directions from the colony on, more-or-less, straight flight paths (hereafter referred to as "linear searching") (Fig. 6).

Foraging trips of Black-browed Albatrosses *D. (Thalassarche) melanophrys* at Iles Kerguelen during chick rearing last, on average, two days, and are typical short commuting flights. Black-browed Albatrosses commuted to specific places located mostly on the eastern Kerguelen continental shelf break, with some birds commuting to a small shelf north of Heard Island (Fig. 7). Birds tend to return to the same site on consecutive foraging trips. The north-eastern slopes of the Kerguelen continental shelf, where birds concentrate, is also used from one year to the next (Weimerskirch *et al.*, in press). Yellow-nosed Albatrosses *D. chlororhynchos*

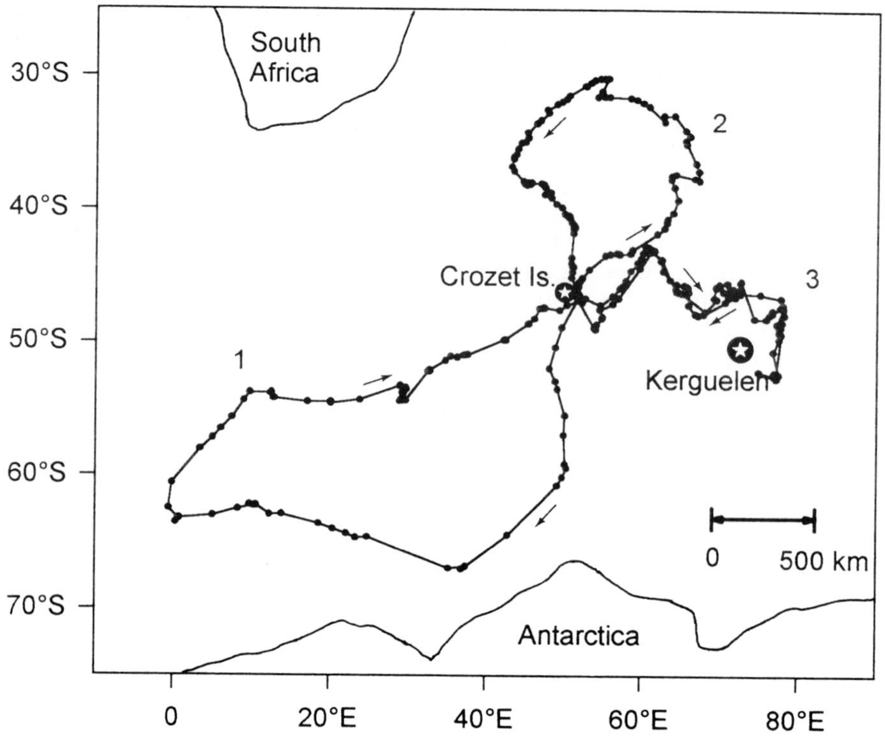

Fig. 1. Movements of Wandering Albatrosses from Possession Island, Iles Crozet Islands, during incubation. Searching loop in Antarctic waters by a male (1) and in tropical waters by a female (2), and long commuting trip to the Iles Kerguelen continental shelf area by a male (3). Location of Iles Crozet and Kerguelen indicated by stars.

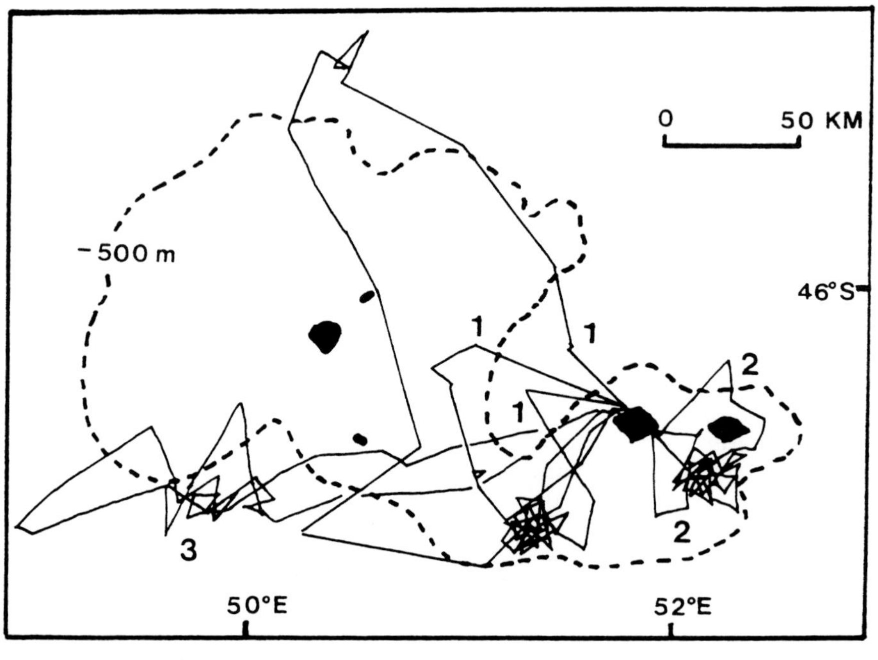

Fig. 2. Movements of Wandering Albatrosses from Possession Island during the chick brooding period. Shown are successive short commuting trips to the shelf break by two males (1 and 2), and to oceanic waters near the shelf break by one female (3).

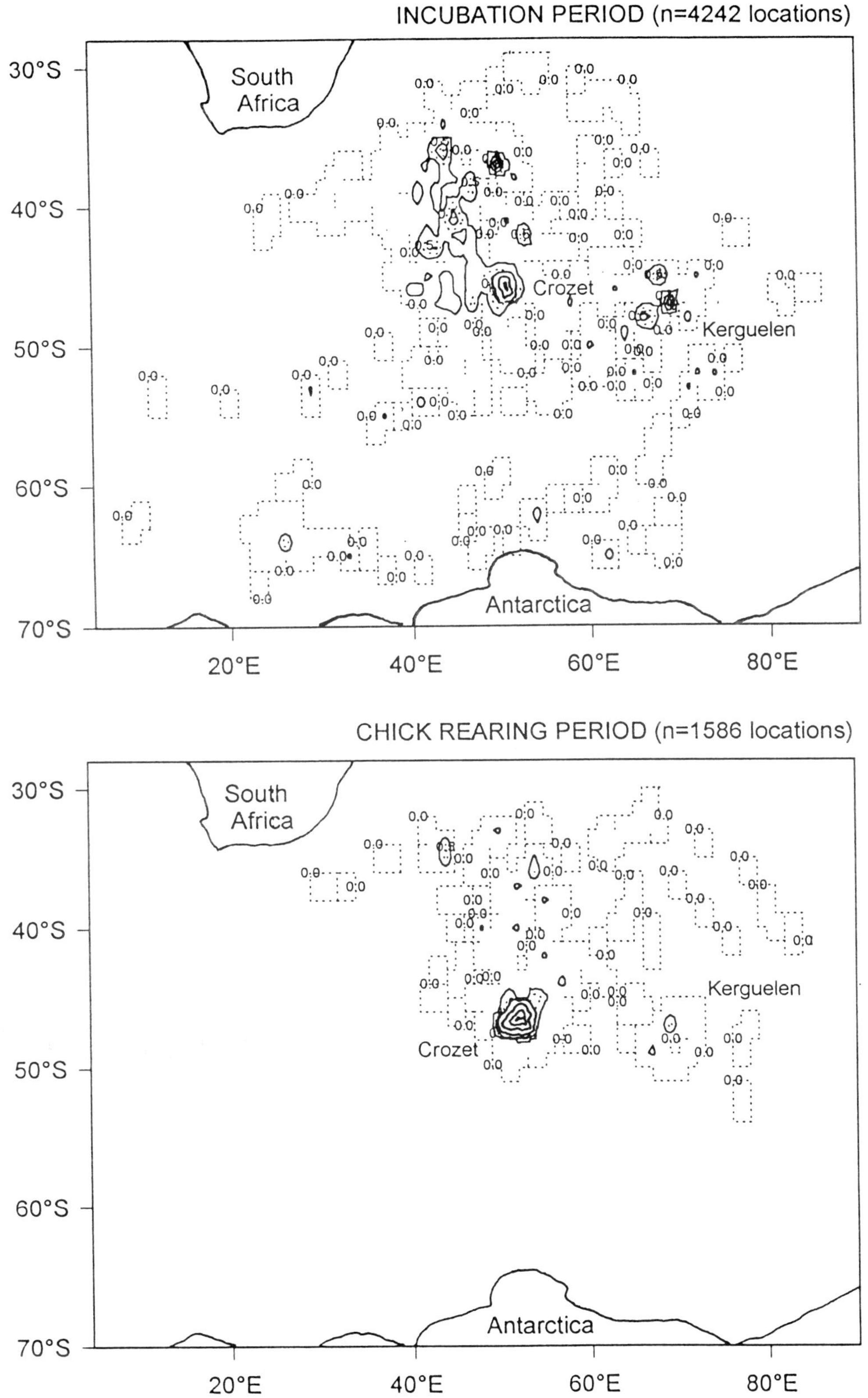

Fig. 3. Zones of concentration of Wandering Albatrosses during incubation (January–March) and chick rearing (May–October) indicated by core areas of satellite fixes. Sectors encompassed by dotted lines indicate 0–0.25% of locations. Isopleths ranging from 0.25 to 4.5 in full lines indicate centre of core where 0.25 to 4.5% of locations are concentrated.

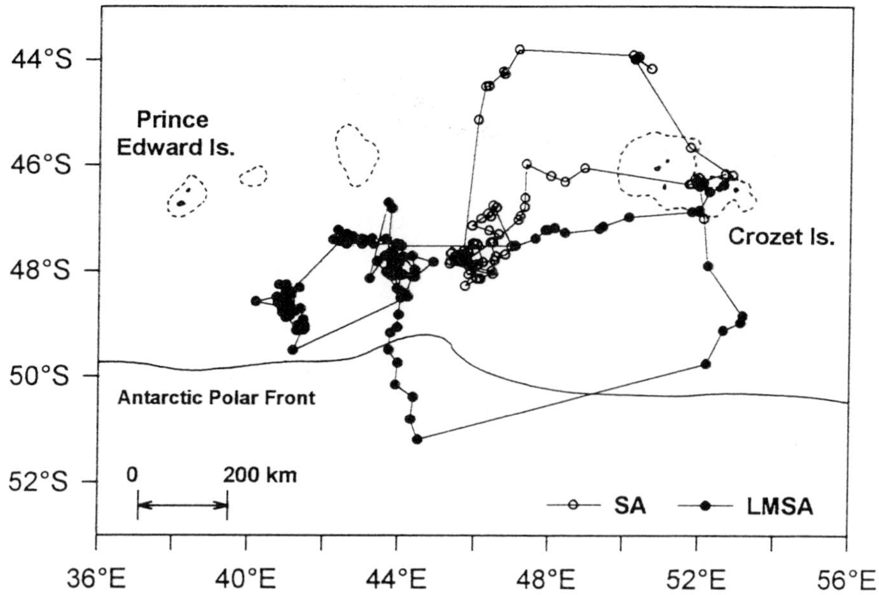

Fig. 4. Long commuting movements of a single Light-mantled Albatross and a single Sooty Albatross from Possession Island during the incubation period.

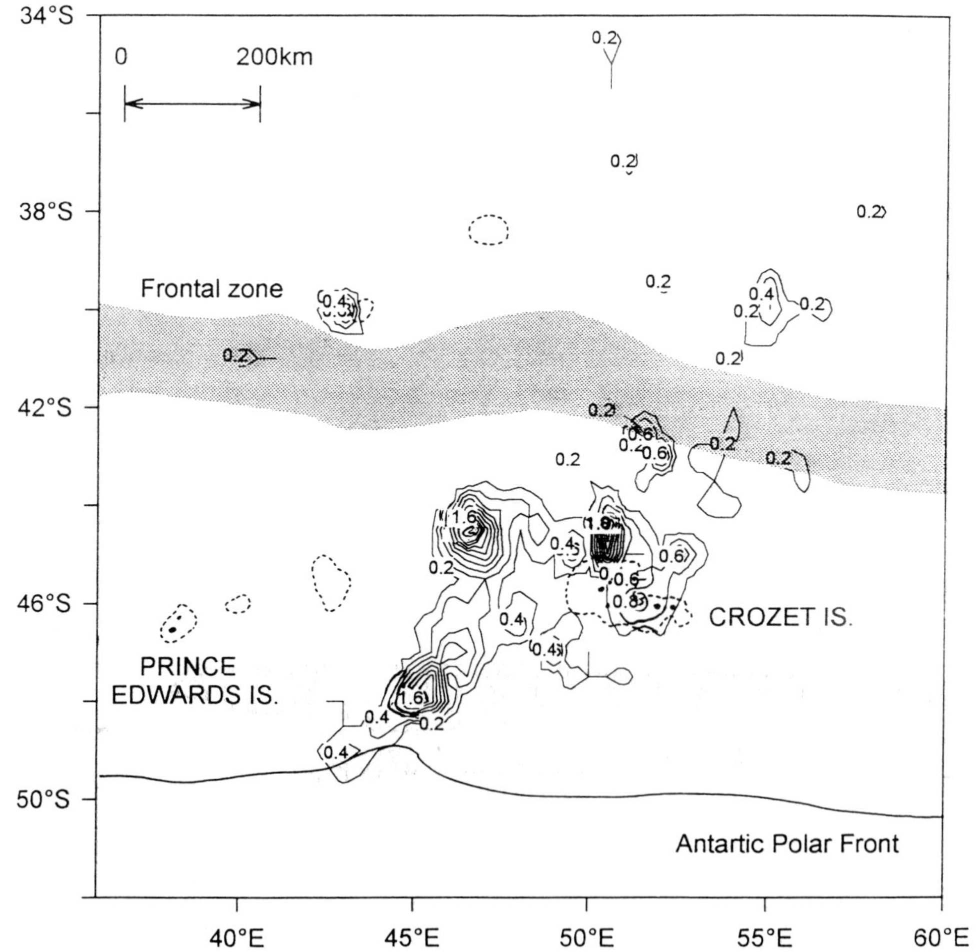

Fig. 5. Zones of concentration of Sooty Albatrosses during incubation (October–December) indicated by core areas of satellite fixes. Isopleths are at 0.2% intervals, ranging from 0.2 to 1.8%.

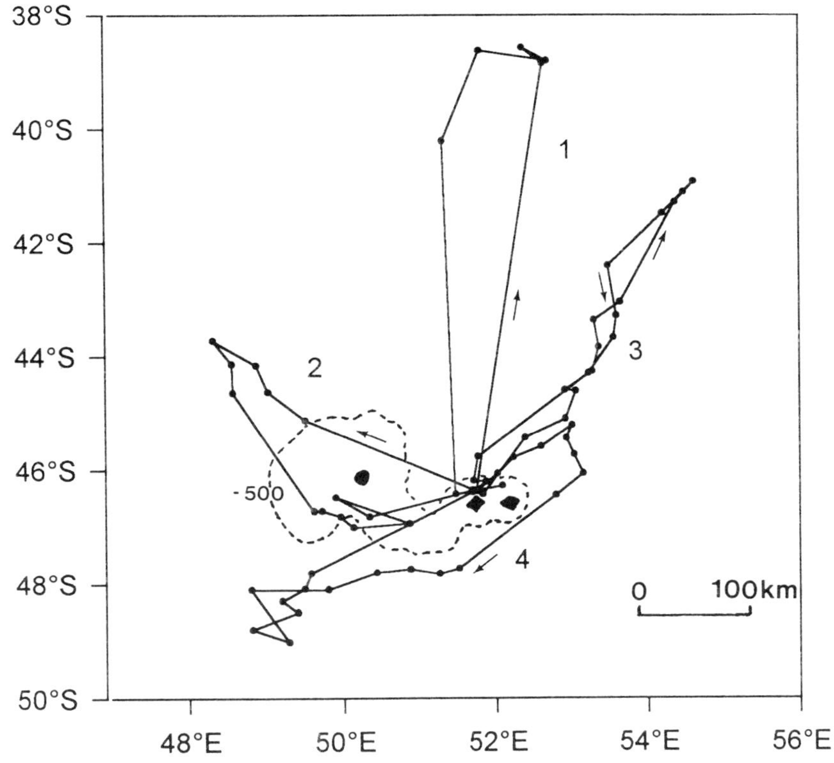

Fig. 6. Short linear search tracks during the brooding period of Sooty Albatrosses from Possession Island.

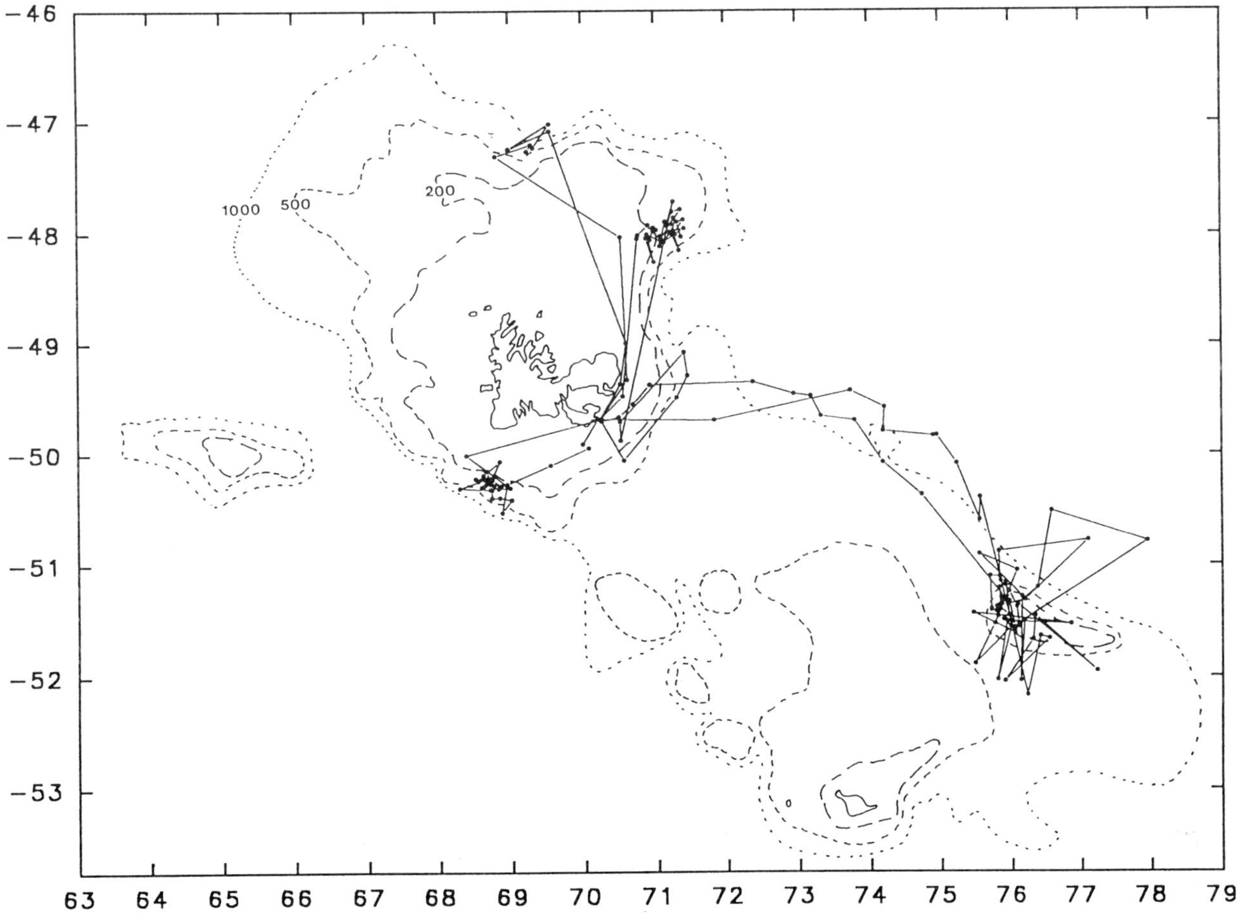

Fig. 7. Short commuting foraging trips by Black-browed Albatrosses to the Iles Kerguelen continental shelf break.

Table 1. Main four foraging strategies undertaken by six albatross species for different stages of the breeding season. Number of dots indicate occurrence of foraging strategy; four dots represent all birds using a particular strategy.

	Commuting		Searching	
	Long	Short	Loop	Linear
Maximum range (km)	500–2 300	50–300	500–4 200	100–1 000
Zone prospected	Oceanic (O) Shelf edge other than breeding grounds (S) Frontal zones	Shelf edge around breeding grounds (S)	Oceanic (O)	Oceanic (O) Shelf edge (S)
Wandering Albatross				
Incubation	● (S)		●●● (O)	
Brooding		●●●● (S)		
Chick rearing		●● (S)	●● (O)	
Amsterdam Albatross				
Incubation			●●●● (O)	
Sooty Albatross				
Incubation	●● (O)		●● (O)	
Brooding				●●●● (O)
Chick rearing			●● (O)	●● (O)
Light-mantled Sooty Albatross				
Incubation	●●● (O)		● (O)	
Brooding				●●●● (O)
Yellow-nosed Albatross				
Chick rearing	●●●(O)		● (O)	
Black-browed Albatross				
Chick rearing		●●●● (S)		

bassi (Thalassarche bassi) attending chicks at Amsterdam Island conducted long commuting flights into oceanic tropical waters north of the island, although some birds undertake long looping flights to these waters (Weimerskirch *et al.*, unpubl. data).

THE ROLE OF DIFFERENT FORAGING STRATEGIES IN ALBATROSSES

The foraging strategies described above for the six albatross species studied in the Indian Ocean can be summarized thus (Table 1):

- commuting flights, where birds fly rapidly to specific marine localities and spend most of the foraging time at these locations before returning to the nest. Commuting flights can be short or long, both in duration and range.
- searching flights, where birds fly continuously, usually over great distances. Searching flights can be conducted in looping or linear courses, for example when birds move along the edge of a continental shelf.

Different marine environments are used by the different albatross species, and in some species different environments are used on successive foraging trips within the same breeding season (Table 1). Sooty, Light-mantled Sooty and Yellow-nosed Albatrosses are oceanic foragers, Black-browed Albatrosses forage in neritic waters and Wandering and Amsterdam Albatrosses exploit both oceanic and neritic environments.

Linear searching, used by Sooty and Light-mantled Sooty Albatrosses at Iles Crozet during brood-stage foraging trips, have also been observed in Royal Albatrosses *D. epomophora* (Nicholls *et al.* 1994) and Buller's Albatrosses *D. (Thalassarche) bulleri* (Sagar and Weimerskirch 1996) near New Zealand. However, the linear searching trips made by these species differs from those of Sooty and Light-mantled Sooty Albatrosses in that the former species search along the New Zealand shelf break and the latter two forage far from their breeding sites in oceanic waters. Grey-headed Albatrosses *D. (Thalassarche) chrysostoma* at South Georgia also undertake long commuting flights to reach waters near the Antarctic Polar Front (Prince *et al.* 1997) as do Waved Albatrosses *D. (Phoebastria) irrorata* commuting from the Galapagos Islands to the coast of Peru (Anderson *et al.* 1997).

Commuting flights probably evolved to catch prey whose distribution is spatially predictable, whereas long looping flights are likely an adaptation to catch prey that is unpredictable to locate. Linear searching along continental shelf edges is probably related to prey being associated with well-defined bathymetry. The significance of linear searching in oceanic waters is unknown, but given that it occurs during the brooding stage, when foraging time is limited by the food demands of the chick, it could be expected to be an adaptation to unpredictable prey location.

Foraging strategies of albatrosses differ greatly between species, but also within a species between periods of the breeding season, and can differ between sexes. For example Wandering Albatrosses raising chicks at Iles Crozet undertook 80% looping route flights, and 20% long commuting flights, to the Kerguelen continental shelf (by males), whereas at Auckland Island Wandering Albatrosses at the same stage in the breeding cycle commuted to the middle of the Tasman Sea (Walker *et al.* 1995). These different strategies are probably explained by differences in the prey species exploited.

For a particular stage of the breeding season, individual albatrosses may alternate among these foraging strategies. Wandering, Light-mantled, probably Sooty, and Yellow-nosed Albatrosses employ the alternating long-short strategy (with a bimodal distribution of the duration of foraging trips), whereas Black-browed Albatrosses only undertake short neritic trips (Fig. 8). The alternating long-short strategy is probably common in oceanic species of seabird, as it has been documented in oceanic petrel species (Weimerskirch *et al.* 1994b). The selection of a particular foraging strategy is probably related to the location of prey and the predictability by which it can be located, its energy yield and to the stage of the breeding cycle, albatrosses having less time to forage when chicks are brooded. For example, during chick rearing, Wandering Albatrosses gather food at a faster rate on long looping trips than on short commuting trips, but birds still make short trips, presumably to maintain the delivery rate of food to the chick (Weimerskirch *et al.* 1997a). Hence albatross species foraging far from the breeding grounds during incubation are likely to use a two-fold foraging strategy during chick rearing, unless prey availability and distribution changes between these two stages of the breeding cycle.

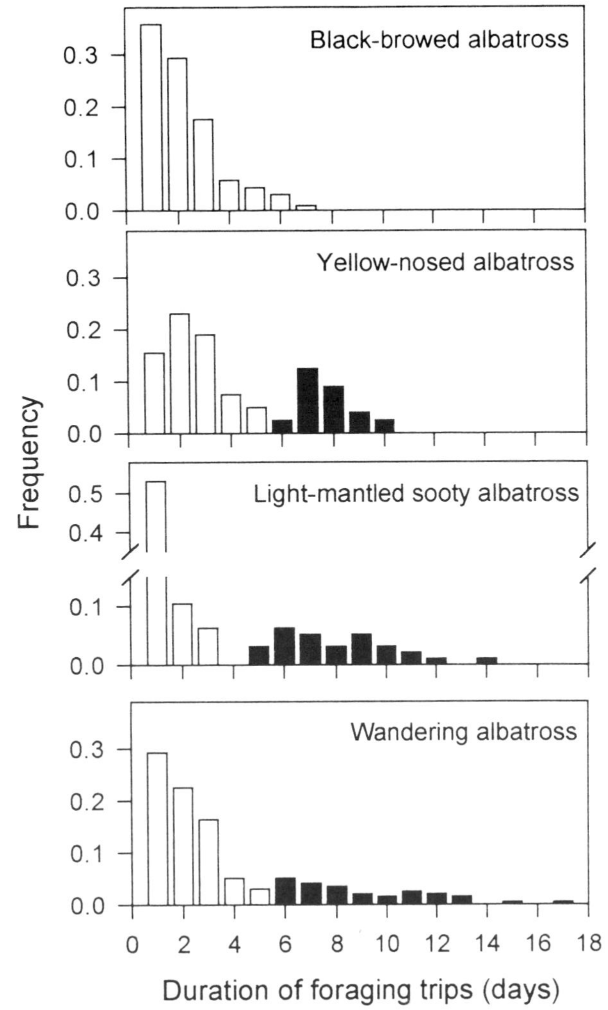

Fig. 8. Distribution of the duration of foraging trips during the chick rearing period in four species of albatrosses. For three species the distribution is bimodal, indicating a two-fold foraging strategy. White columns represent short trips, black columns long trips.

ALBATROSS FORAGING AND RELATIONSHIPS WITH FISHERIES

Most albatross species are habitual followers of ships, from which they scavenge fisheries discards (Thompson 1992). In addition to offal, albatrosses are also attracted by the baits of longlines and are at risk of being killed (Alexander *et al.* 1997). Trawl fisheries may also be a cause of mortality for albatrosses, especially those using net sonde cables (Bartle 1991). Three types of fisheries operate in the Economic Exclusion Zone around Iles Kerguelen and each is characterized by the amount of fisheries discards made available to birds: a trawl fishery, which discards large quantities of offal, a trawl fishery which discards small quantities of offal, and a longline fishery which provides moderate quantities of offal and baits. Between 1990 and 1995 these fisheries operated in three distinct zones on the Iles Kerguelen continental shelf and shelf beak (Figs 9 and 10). Trawlers targeting ice-fish *Champsocephalus gunnari* have been operating in the eastern and northern parts of the Iles Kerguelen shelf break since the early 1970s. A new fishery longlining for Patagonian Toothfish *Dissostichus eleginoïdes* started in 1992 and operates along the western part of the shelf break, and on the north edge of the shelf. Wandering Albatrosses from Iles Crozet and Black-browed Albatrosses from Iles Kerguelen forage over the Kerguelen shelf, and both species are known to associate with fishing vessels. Average numbers of Wandering and Black-browed Albatrosses attending vessels differ from one fishing type to the next, trawlers providing offal being the most attractive.

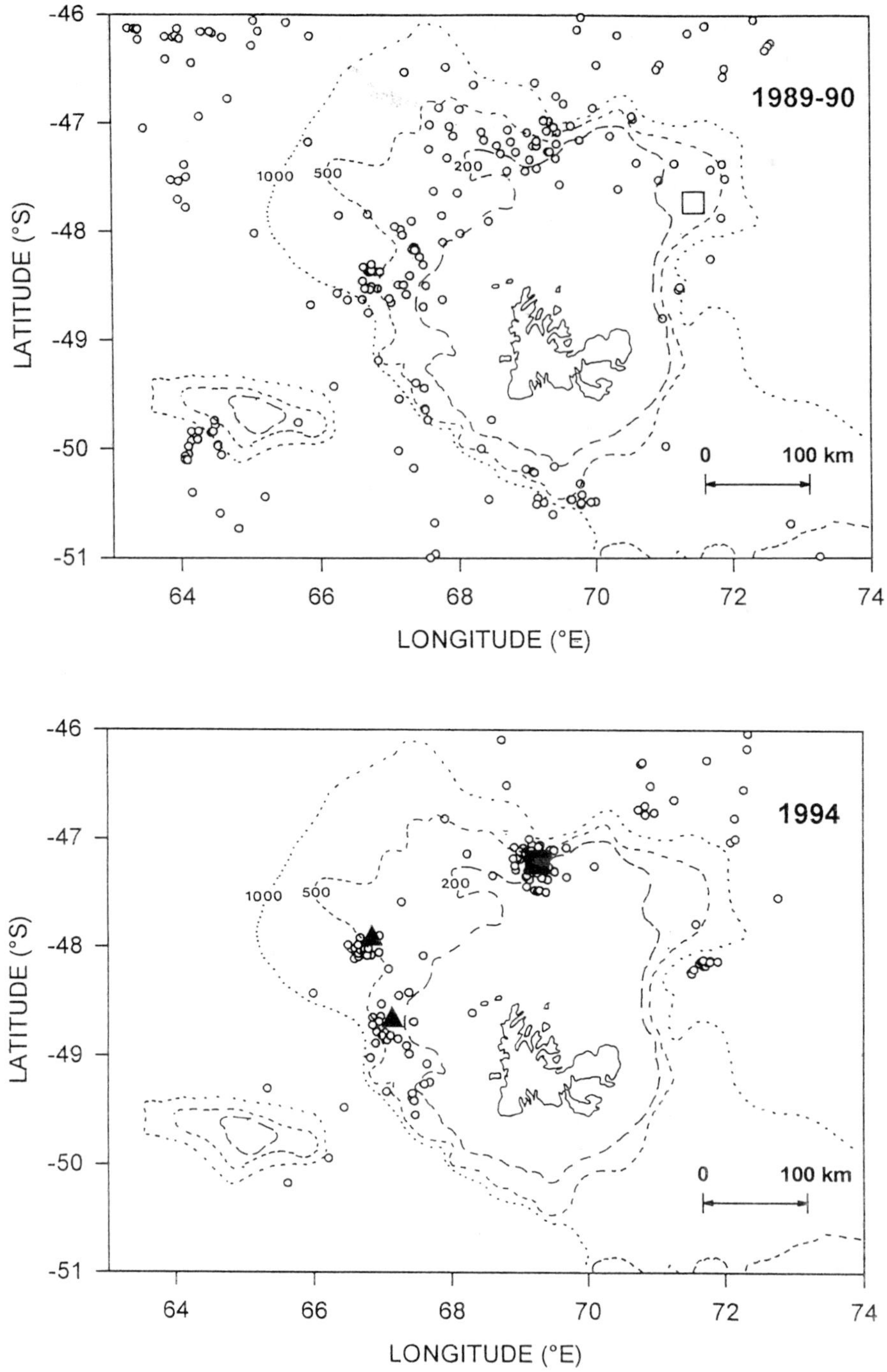

Fig. 9. Distribution of locations of Wandering Albatrosses (open circles) breeding at the Iles Crozet and foraging in Kerguelen waters, prior to 1989–90 and after 1994, the latter being the start of fisheries for Patagonian Toothfish in the western and northern shelf break area. Positions of longliners are indicated by black triangles, positions of trawlers with substantial offal discards by black squares, and positions of trawlers with limited quantities of offal by white square.

In 1994, while longline vessels operated in the western part of the Iles Kerguelen shelf break and a trawler operated in the northern part, Wandering Albatrosses were concentrated in the same areas, indicating that the albatrosses were attracted to the fishing vessels (Fig. 9). It is possible that the birds changed their normal foraging behaviours to take advantage of these new food sources. However, in 1989–90, when no fishing occurred in the western part of the shelf, Wandering Albatrosses were spread widely over the western area, and were absent from the eastern shelf where the trawlers operated. This suggest that the presence of Wandering Albatrosses behind longliners may be the result of an overlap between the natural range of this species and the fishing areas. (Fig. 9).

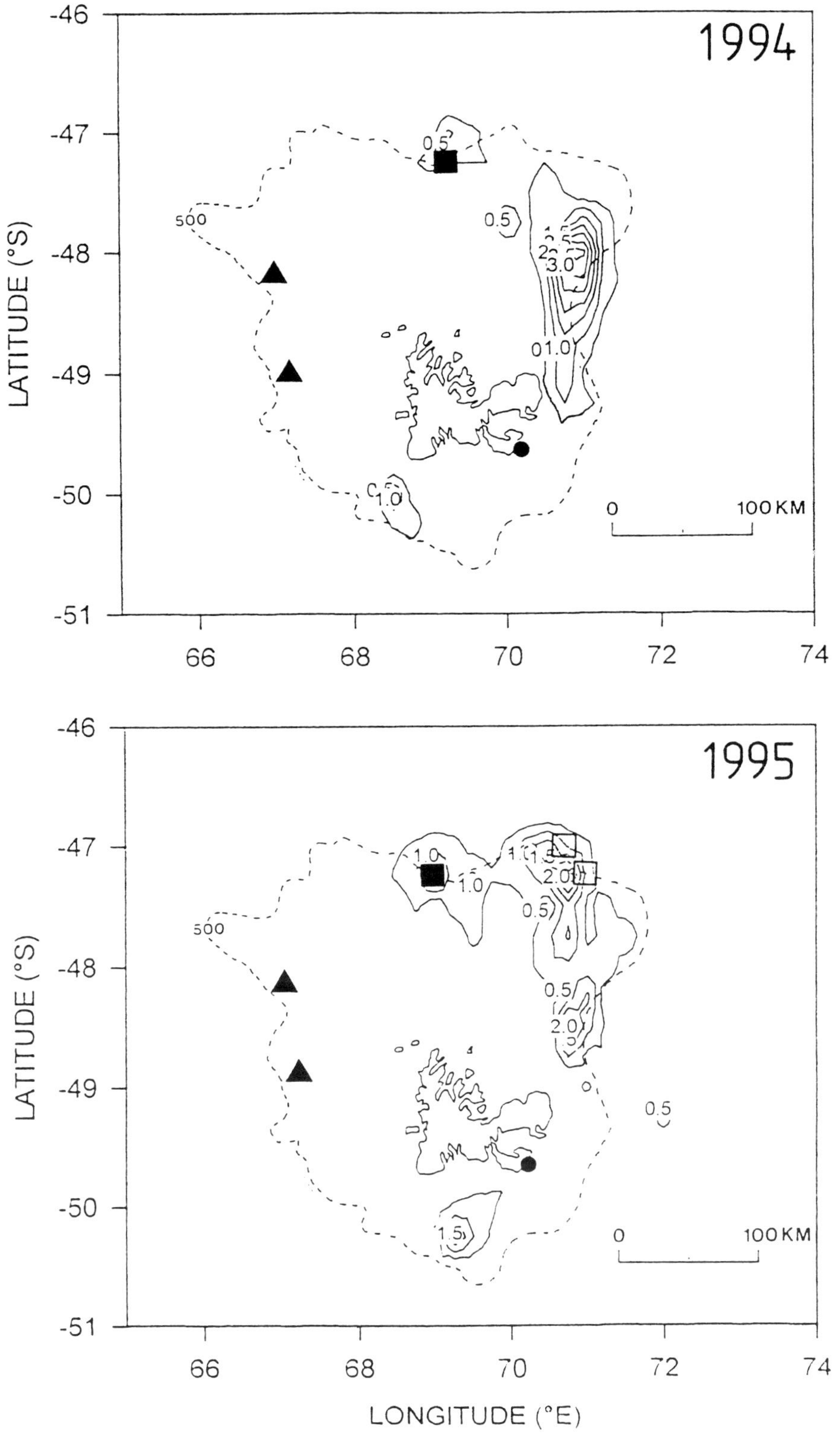

Fig. 10. Concentrations of Black-browed Albatrosses breeding in the southeastern end of the Iles Kerguelen mainland in 1994 and 1995 in relation to the presence of fishing areas. Position of longliners is shown by black triangles, trawlers with substantial offal discards by black squares and trawlers with limited quantities of offal by white squares.

In 1994, Black-browed Albatrosses from the southern study colony at Iles Kerguelen concentrated in an area at sea where no fishing occurred, and with only a few birds attending a trawler on the north shelf break (Fig. 10a). The albatrosses caught natural prey (squid and fish) on the northeastern slopes (Cherel and Weimerskirch 1995) and in that year fledging success was the highest on record. In 1995, fishing operations was concentrated in the three main fishing zones mentioned above (Fig. 10b) and Black-browed Albatrosses from the southern colony were able to attend these operations within hours of their commencement, but birds were still concentrated in the northeastern shelf break. Breeding success was lower in 1995 than in 1994. These results show almost no overlap between the longline fishery and the breeding birds studied, suggesting that the longline fishery had only a limited effect on this population with respect to altering the birds foraging behaviours. Black-browed Albatrosses probably rely mostly on natural prey, but if the trawl fishery overlaps with their natural feeding zones they may take advantage of this interaction. This suggestion is reinforced by the correlation between high catch rates by the fishery during the 1980s and high breeding successes of Black-browed Albatrosses. Interestingly, the total catches by the fishery — and thus amount of offal made available to the birds — was not statistically related to the breeding success (Weimerskirch and Jouventin 1989). Although Black-browed Albatrosses from the southern colony on Iles Kerguelen do not overlap with longline vessels the birds from the north-western colony are likely to be in contact with this fishery. Indeed, it has been shown that these two colonies have different foraging areas (Weimerskirch *et al.* 1988), suggesting that fisheries impacts are likely to differ between colonies.

CONSERVATION IMPLICATIONS

Several fisheries threaten albatross population throughout the world (Alexander *et al.* 1997; Gales 1997). In the Southern Ocean, the Japanese longline fishery for Southern Bluefin Tuna *Thunnus maccoyi* causes high mortality of albatrosses in some areas (Brothers 1991; Murray *et al.* 1992) and is likely to have been responsible for the decrease of the Indian Ocean Wandering Albatross populations (Weimerskirch *et al.* 1997b; Weimerskirch and Jouventin 1997). However, other fisheries are probably also involved in the decreases in albatross numbers (e.g., Bartle 1991). Limiting contact between foraging birds and fisheries, by restricting fishing times and locations, is one potential conservation measure, but for this to be successful the foraging strategies of albatrosses must be fully understood.

Breeding albatrosses are central place foragers: the closer the fishery is to their nesting place, the higher the risk that birds will encounter fishing vessels. Interactions between longline fishing vessels and Wandering Albatrosses can be reduced by prohibiting fishing within a 200–300 km radius of Iles Crozet and Iles Kerguelen, this radius delimiting the area in which highest concentrations of brood- and guard-stage Wandering Albatrosses occur. This would also protect Wandering Albatrosses at Iles Crozet during the incubation period when high densities of albatrosses occur immediately around the island and during the long and short foraging trips during chick rearing, when Wandering Albatrosses spend 65% of their foraging time within 200–300 km of the island. The 200–300 km radius would also minimize contact between Black-browed Albatrosses and longline fisheries (due to the short duration of foraging flights), and afford some protection to the long-commuting Sooty Albatross, since high densities of birds occur around the island during the commencement and completion of foraging trips.

Although fisheries exclusion zones around Iles Crozet and Iles Kerguelen would benefit the albatrosses breeding there, birds would still be vulnerable when they migrate to Australian waters, where they contact Japanese and Australian longline tuna vessels. Sooty and Light-mantled Sooty Albatrosses would also be affected elsewhere because they breed biennially and disperse throughout the Southern Ocean during their year off. These dispersals make it difficult to determine important areas for non-breeding albatrosses and to derive recommendations for managing fisheries to limit interactions with albatrosses. Clearly, studies of foraging behaviour, particularly of juvenile and immature birds, outside the breeding seasons are critical for future management strategies. Closure of areas to fishing, in concert with other conservation measures such as night setting and underwater setting, would greatly reduce the incidental bycatch of albatrosses in longline fisheries.

ACKNOWLEDGEMENTS

The studies received financial and logistical support by the Administration of "Terres Australes et Antarctiques Françaises", Institut Français pour la Recherche et la Technologie Polaire", CNRS–GDRE no. 1069, Ministère des Affaires Etangères. I wish to thank all involved in the studies in the field, and Nigel Brothers and Graham Robertson for their extensive help with this manuscript.

REFERENCES

Alexander, K., Robertson, G. and Gales, R. 1997. The incidental mortality of albatrosses in longline fisheries. Australian Antarctic Division. Tasmania. 44 pp.

Anderson, D. J., Schwandt, A. J. and Douglas, H., 1997. Foraging ranges of Waved Albatrosses in the eastern Tropical Pacific Ocean. Pp. 180–85 *in* Albatross Biology and Conservation ed by G. Robertson and R. Gales. Surrey Beatty & Sons: Chipping Norton.

Ishmael, N. P., 1971. Seabird ecology and the marine environment. Pp. 223–86 *in* Avian Biology ed by D. S. Farner and J. R. King. Academic Press: New York.

Bartle, J. A., 1991. Incidental capture of seabirds in the New Zeland subantarctic squid trawl fishery, 1990. *Bird Conserv. Int.* **1991:** 351–59.

Brothers, N., 1991. Albatross Mortality and Associated Bait Loss in the Japanese Longline Fishery in the Southern Ocean. *Biol. Conserv.* **55:** 255–68.

Cherel, Y. and Weimerskirch, H., 1995. Seabird indication of localization of unknown marine resources: Black-browed Albatrosses feeding on Ommastrephid squids in Kerguelen waters. *Mar. Ecol. Progr. Ser.* **129:** 295–300.

Cherel, Y., Weimerskirch, H. and Duhamel, G., 1995. Interactions between long-line vessels and seabirds in Kerguelen waters and a method to reduce seabird mortality. *Biol. Conserv.* **75:** 63–70.

Croxall, J. P., Prince, P. A., Rothery, P. and Wood, A. G., 1997. Population changes in albatrosses at South Georgia. Pp. 69–83 *in* Albatross Biology and Conservation ed by G. Robertson and R. Gales. Surrey Beatty & Sons: Chipping Norton.

Croxall, J. P. and Gales, R., 1997. An assessment of the conservation status of albatrosses. Pp. 46–65 *in* Albatross Biology and Conservation ed by G. Robertson and R. Gales. Surrey Beatty & Sons: Chipping Norton.

Gales, R., 1997. Albatross populations: status and threats. Pp. 20–45 *in* Albatross Biology and Conservation ed by G. Robertson and R. Gales. Surrey Beatty & Sons: Chipping Norton.

Jouventin, P. and Weimerskirch, H., 1990. Satellite tracking of Wandering Albatrosses. *Nature* **343:** 746–48.

Lack, D., 1968. Ecological adaptations for breeding in birds. Methuen: London.

Murray, T. E., Bartle, J. A., Kalish, S. R. and Taylor, P. R., 1993. Incidental capture of seabirds by Japanese southern bluefin longline vessels in New Zealand waters, 1988–1992. *Bird Conserv. Internat.* **3:** 181–210.

Nicholls, D., Murray, D. and Robertson, C. J. R., 1993. Oceanic flight of the northern royal albatross *Diomedea epomophora sandfordi* using satellite telemetry. *Corella* **18:** 50–52.

Prince, P. A., Wood, A. G., Barton, T. and Croxall, J. P., 1992. Satellite tracking of Wandering Albatrosses *(Diomedea exulans)* in the South Atlantic. *Antarctic Sci.* **4:** 31–36.

Prince, P. A., Croxall, J. P., Trathan, P. N. and Wood, A. G., 1997. The pelagic distribution of South Georgia albatrosses and their relationships with fisheries. Pp. 137–67 *in* Albatross Biology and Conservation ed by G. Robertson and R. Gales. Surrey Beatty & Sons: Chipping Norton.

Sagar, P. and Weimerskirch, H., 1996. Satelitte tracking of Southern Buller's Albatrosses from the Snares, New Zeland. *Condor* **98:** 649–52.

Salamolard, M. and Weimerskirch, H., 1993. Relationship between foraging effort and energy requirement throughout the breeding season in the Wandering albatross. *Funct. Ecol.* **7:** 643–52.

Thompson, K. R., 1992. Quantitative analysis of the use of discards from squid trawlers by Black-browed Albatrosses *Diomedea melanophris* in the vicinity of the Falklands Islands. *Ibis* **134:** 11–21.

Walker, K., Elliott, G., Nicholls, D., Murray, D. and Dilks, P., 1995. Satellite tracking of Wandering albatross *(Diomedea exulans)* from the Auckland Islands: preliminary results. *Notornis* **42:** 127–37.

Weimerskirch, H., 1995. Regulation of foraging trips and incubation routine in male and female Wandering Albatrosses. *Oecologia* **102:** 37–43.

Weimerskirch, H. and Robertson, G., 1994. Satellite tracking of Light-mantled Sooty Albatross. *Polar Biol.* **14:** 123–26.

Weimerskirch, H. and Wilson, R. P., 1992. When do Wandering Albatrosses *Diomedea exulans* forage? *Mar. Ecol. Prog. Ser.* **86:** 297–300.

Weimierskirch, H. and Jouventin, P., 1997. Changes in population sizes and demographic parameters of six albatross species breeding on the French sub-Antarctic Islands. Pp. 84–91 *in* Albatross Biology and Conservation ed by G. Robertson and R. Gales. Surrey Beatty & Sons: Chipping Norton.

Weimerskirch, H., Salamolard, M. and Jouventin, P., 1992. Satellite telemetry of foraging movements in the Wandering albatross. Pp. 185–98 *in* Wildlife Telemetry. Ellis Horwood: Chichester.

Weimerskirch, H., Salamolard, M., Sarrazin, F. and Jouventin, P., 1993. Foraging strategy of Wandering Albatrosses through the breeding season: a study using satellite telemetry. *Auk* **110:** 325–42.

Weimerskirch, H., Doncaster, P. and Cuenot Chaillet, F., 1994a. Pelagic seabirds and the marine environment: foraging of Wandering Albatrosses in relation to the availability and distribution of their prey. *Proc. R. Soc. London* **B255:** 91–97.

Weimerskirch, H., Mougey, T. and Hindermeyer, X., in press. Foraging and provisioning strategies of black-browed albatrosses in relation to the requirements of the chick: natural variation and experimental study. *Behav. Ecol.*

Weimerskirch, H., Chastel, O., Chaurand, T., Ackerman, L., Hindermeyer, X. and Judas, J., 1994b. Alternate long and short foraging trips in pelagic seabird parent. *Anim. Behav.* **47:** 472–76.

Weimerskirch, H., Cherel, Y., Cuenot Chaillet, F. and Ridoux, V., 1997a. Alternative strategies of food provisioning to chick by male and female Wandering Albatrosses. *Ecology* **78:** in press.

Weimerskirch, H., Brothers, N. and Jouventin, P., 1997b. Population dynamics of Wandering Albatrosses *Diomedea exulans* and Amsterdam albatross *D. amsterdamensis* in the Indian Ocean and their relationships with longline fisheries: conservation implications. *Biol. Conserv.* **79:** 257–70.

Wilson, R. P., Cooper, J. and Plotz, J., 1992. Can we determine when marine endotherms feed? A case study with seabirds. *J. Exp Biol.* **167:** 267–75.

Wilson, R. P., Weimerskirch, H. and Lys, P., 1995. A device for measuring seabird activity at sea. *J. Avian Biol.* **26:** 172–76.

CHAPTER THIRTEEN

Foraging ranges of Waved Albatrosses in the eastern Tropical Pacific Ocean

DAVID J. ANDERSON,[1] ANDREA J. SCHWANDT[1] and HECTOR D. DOUGLAS[1]

Satellite telemetry was used to determine the foraging ranges of five Waved Albatrosses during the incubation period from Punta Cevallos, Isla Española in the Galápagos Islands. All tracked birds visited a clearly defined area of cold, upwelled water of high productivity near the Peruvian coast approximately 1 200 km south-east of the breeding colony. Flight paths between the breeding colony and the foraging site were direct. These preliminary results suggest that incubating Waved Albatrosses do not forage within the Galápagos Archipelago, and indirect evidence suggests that they are not affected by longline fishing to the same extent as are other albatross species.

Key words: Waved Albatross, Galápagos Archipelago, Foraging Ranges, Peruvian Coast, Longline Fishing.

INTRODUCTION

MOST albatross species have shown a high susceptibility to incidental bycatch in longline fishing operations, primarily by becoming hooked during setting when the birds attack baited hooks (Brothers 1991; Alexander *et al.* 1997; Gales 1997). Some albatross populations are known to be decreasing, and the evidence points to longline bycatch as the principal cause. The Waved Albatross *Diomedea (Phoebastria) irrorata* is one of the two species for which documented mortality in longline fisheries is lacking (Gales 1993). This situation was correctly interpreted as a lack of data rather than a demonstrated absence of interaction.

Virtually the entire world population of Waved Albatrosses breed at one site: Isla Española in the Galápagos Islands, Ecuador (1°S, 90°W; see Fig. 1a) (Harris 1973; Anderson and Ricklefs 1987). The at-sea distributions, accumulated from ship-based observations, have recorded Waved Albatrosses around the Galápagos group, off the South American coast from the equator to 12°S, and on a flyway between these two areas (Murphy 1936; Harris 1973; Tickell, in press). During the non-breeding season (January–March) adults vacate the Galápagos Islands, and the entire population is thought to contract its range to the area off the South American coast.

In this study we used birds from the Punta Cevallos subcolony, Isla Española, on the eastern end of the island, which is one of the larger subcolonies on the island (Harris 1973). We used satellite telemetry to determine the feeding localities of breeding adults during the nine month breeding season. This chapter reports on the foraging ranges of satellite-tagged birds and describes the relationship between these foraging trips and longline fishing activity. We also report on the performance of the tracking system and the effects of the transmitters on the instrumented birds.

METHODS

Attachment and accuracy of satellite transmitters

We used the methods of Prince *et al.* (1992) to attach PTTs (Platform Transmitter Terminals) to each of five breeding adult Waved Albatrosses (three females, two males). Both members of two breeding pairs received PTTs, and we assumed that the smaller member of each pair was a female (Harris 1973). We presumed the fifth bird to also be female given its smaller body size than its mate. We mounted transmitters on birds within 25 hrs of their egg being laid, so tracking began during early incubation.

We attached each PTT to a flexible plastic sheet with cable ties, and we glued the sheet to the bird's dorsal contour feathers. Three PTTs were recovered by cutting the cable ties; two were lost at sea when they fell off the birds. Feathers anterior to the backpack covered the front end of the transmitters. The backpack weighed approximately 85 g and added 1–3%

[1]Department of Biology, Wake Forest University, Winston–Salem, NC USA 27109.

Pages 180–85 *in* ALBATROSS BIOLOGY AND CONSERVATION, ed by G. Robertson and R. Gales. Surrey Beatty & Sons, Chipping Norton. 1997.

to the bird's gross mass. The transmitters (PTT100, Microwave Telemetry, Columbia, MD USA) had a duty cycle of 8 h on: 24 h off. We mounted all five transmitters on 23 April, 1995 and used the Argos satellite system to track the birds for the following 32–65 days. We monitored the birds' nest sites twice per day during the tracking period, in the morning and late in the afternoon, and recorded the presence/absence of each instrumented bird. We evaluated the accuracy of the locations provided by the Argos system by comparing Argos fixes with the known locations of birds when in the colony.

Effects of instruments on birds

We used the computer program Bird Flight Performance (Program 1, Pennycuick 1989) to estimate the effects of the transmitters on flight speeds and power requirements of tagged birds. This allows the calculation of several relevant parameters given the bird's mass and wingspan. On 30 June 1995 we photographed 20 Waved Albatrosses held in the flight position according to Pennycuick's (1989) recommendation, measured bill length with callipers, and weighed each bird in a bag with a Pesola spring balance. We then projected each dorsal view onto a digitizing tablet and measured the wing span (wingtip to wingtip) with an Imaging Program (Scan, Jandel Scientific, San Rafael, CA USA). We assumed that these 20 birds included nearly equal numbers of males and females, based on the distribution of the culmen lengths.

We obtained preliminary data on the proportion of time spent in flapping flight by 16 Waved Albatrosses that came within sight of our boat as we travelled on the open ocean within the Galápagos on 2 July, 1995. The duration of the samples averaged 35.9 sec. (s.d. = 10.5). Two of these were excluded from the subsequent analyses because the birds appeared to be involved in kleptoparasitic interactions, and our intent was to sample typical flight.

Interactions with fisheries

Summaries of the spatial distribution of Japanese longline fishing vessels came from Miyabe and Bayliff 1987, Nakano and Bayliff 1992 and Caulkins *et al.* 1993. We could locate no historical data on fishery bycatch of Waved Albatrosses, although a recent co-operative agreement between Ecuador and Japan that places Ecuadorian observers on Japanese boats may yield some information in the future.

RESULTS

Performance of the tracking system

Equatorial locations provide fewer opportunities for transmitter fixes than higher latitude sites because the girth of the Earth at that point limits the number of satellite passes over a given spot to 6–8 per day (Service Argos 1988). Our 8 h on: 24 h off duty cycle prolonged battery life but was a further limitation on location frequency. On average we received 0.8 locations/day/bird, and at least one location per bird on only 60.1% of days at sea.

The Argos system currently provides six classes of location; the confidence in the locations accuracy varies between classes. Average errors ranged from 0.23 km (location class 3) to 17.8 km (location class B). There was no consistent bias in the bearing of the errors, although there may have been a tendency toward larger east-west errors than north-south errors. A total of 337 fixes were attempted from the five birds; of these 108 were unsuccessful (location class Z). Of the 229 (68%) successful attempts, 191 (83%) were in location classes A (56), B (77), or 0 (58), with associated errors that were minor compared to the overall distances travelled by the birds (see below).

Influence of transmitters on the birds

The model of Pennycuick (1989) covers effects on flapping flight only; the untelemetered birds we observed at sea spent approximately 30% of their time in flapping flight (Mean = 0.285; s.d. = 0.066; median = 0.311; range: 0.201–0.424; n = 14). We assumed that the backpack added no frontal area to the bird because the leading edges of the transmitters were covered by the contour feathers anterior to the transmitter. The results indicated that if birds flew so as to minimize the power used to fly, instrumented birds should increase flight speeds by 1.06% (range: 0.77–1.80%) and power requirements by 3.15% (range: 2.26–6.13%). If the birds flew to maximize the distance travelled per unit of energy expended (i.e., efficiency), the model predicts that speeds should increase 1.01% (range: 0.05–1.64%) and power requirements should increase 3.02% (range: 2.21–4.08%). These estimates indicate only minor effects of the PTTs on the flights of the albatrosses. We were not at the colony during fledging and were unable to determine if the instrumented birds fledged their chicks. However, four of the tracked birds exhibited colony attendance behaviour consistent with chick attendance and presumably the chicks of these birds fledged; the bird that was tracked for 65 days abandoned the colony.

Flight paths and foraging behaviour

We monitored one trip each from three birds, and two trips each from two birds. On six of the seven trips the bird flew directly to an area of cool, upwelled water and high productivity off the Peruvian coast (Fig. 1b–g) about 1 200 km away. On the remaining trip the bird first flew to

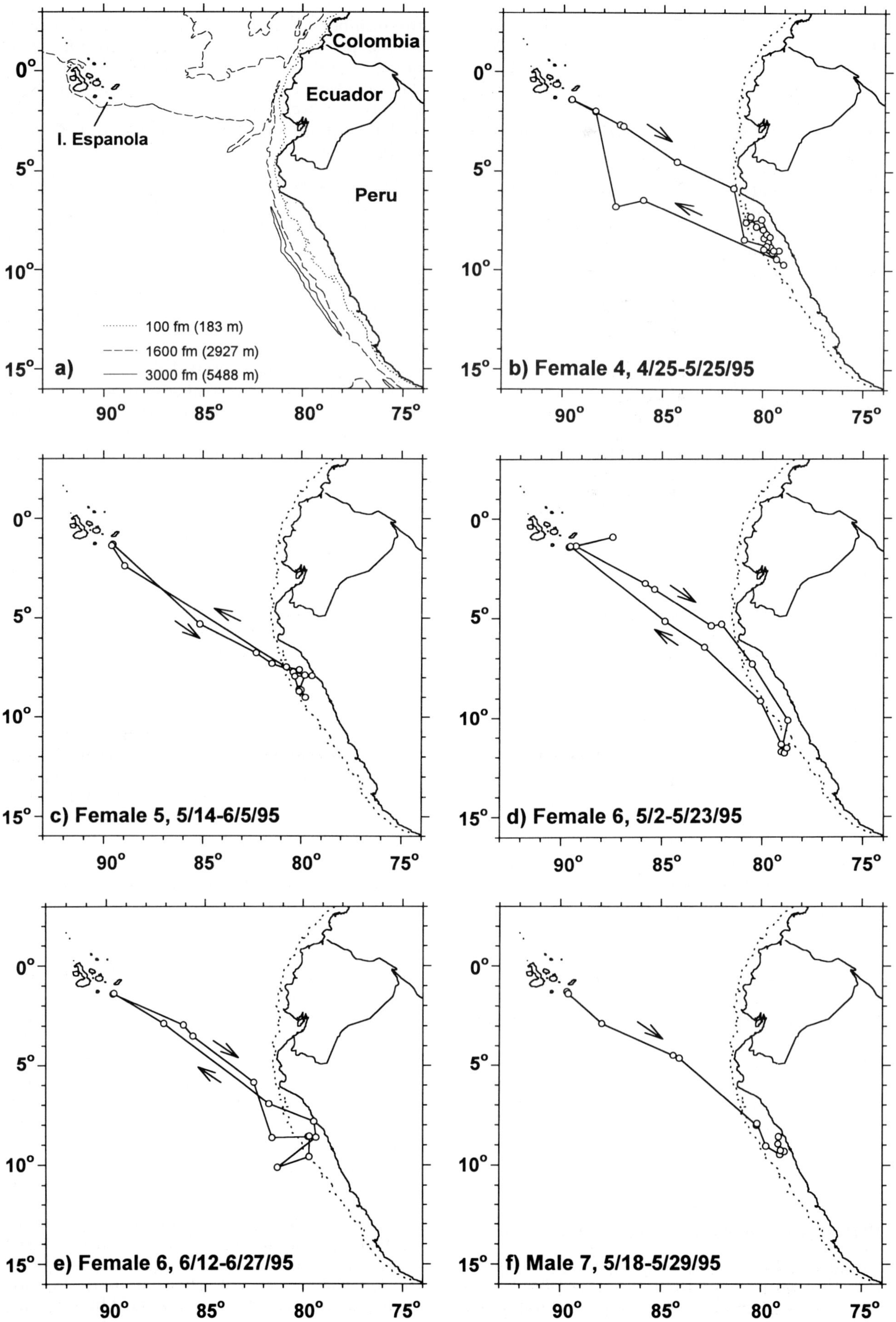
Colombia
Ecuador
Peru
I. Espanola
100 fm (183 m)
1600 fm (2927 m)
3000 fm (5488 m)
a)
b) Female 4, 4/25-5/25/95
c) Female 5, 5/14-6/5/95
d) Female 6, 5/2-5/23/95
e) Female 6, 6/12-6/27/95
f) Male 7, 5/18-5/29/95
0°
5°
10°
15°
90°
85°
80°
75°

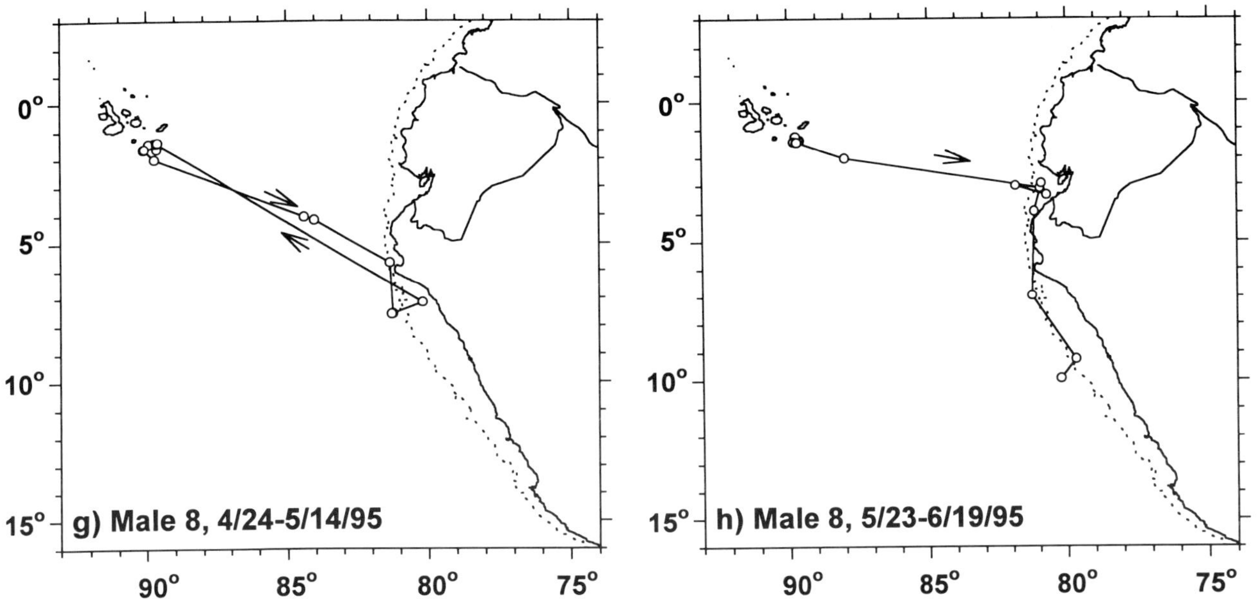

Fig. 1. (a) The Galápagos Islands and Isla Española, the main nesting site of the Waved Albatross; the South American continent; and submarine topographic features of the region in fathoms. Axes show degrees latitude S and longitude W. (b–h) Routes of incubating birds during foraging trips, departing from Isla Española. Each point is an Argos-determined location. The dotted line is the 1 000 m isobath.

the Gulf of Guayaquil and spent six days there before flying to the Peruvian upwelling region (Fig. 1h). Flight times between the breeding colony and the upwelling region, into the prevailing wind, ranged from 80.8–163.6 h. The straight-line ground speed for the fastest bird over the 1 200 km journey to the upwelling region was 14.9 km/h. Return trips from the region to the breeding colony, following the prevailing wind, ranged from 48.1–150.6 h, with a maximum straight-line speed of 24.9 km/h. These speeds are minimum estimates, since they do not account for tacking and other indirect flight behaviour. Although outbound ground speeds tended to exceed those inbound, the difference was not significant (Mann-Whitney $U = 1.22$, $N_1 = 7$, $N_2 = 5$, $p = 0.22$).

The birds' movements suggest that the upwelling region was the feeding site. Between the Galápagos Islands and the continental upwelling region (Region 1) their flight paths were direct, whereas along the South American shelf (Region 2) birds frequently reversed direction (Fig. 1). Additionally, straight-line ground speeds between consecutive satellite fixes were significantly higher in Region 1 (median = 13.55 km/h; range: 3.63–46.81; $n = 13$) than in Region 2 (median = 3.30; range: 0.40–115.83; $n = 39$) (Mann-Whitney $U = 420$; $N_1 = 13$; $N_2 = 39$; $p < 0.001$). Birds apparently travelled rapidly between the Galápagos Islands and Region 2, where they changed flight direction more frequently, flew at lower speeds, and flew less often than during the travel phase of their foraging trips. The change in the albatrosses' movements corresponds with broad-scale oceanic primary productivity: between the Galápagos Islands and the South American coast the surface water has low productivity typical of the eastern tropical Pacific, whereas the coastal area used by the birds is consistently one of the most highly productive marine habitats on the planet (Feldman *et al.* 1984; Feldman 1986).

DISCUSSION

In this study we have identified the Peruvian upwelling zone as a foraging area for incubating Waved Albatrosses. Incubating adults did not appear to forage to any significant degree within the Galápagos group. While adult Waved Albatrosses are present at sea within the Galápagos Islands during the breeding season, we suspect that these are non-breeding birds that are prospecting for mates and sites on Isla Española.

Waved Albatrosses differ in at least two respects from some other albatross species that have been studied by satellite tracking. First, male and female Wandering Albatrosses *D. exulans* from the Iles Crozet (Weimerskirch *et al.* 1993) and the Auckland Islands (Elliot *et al.* 1995) forage at different locations with respect to their breeding sites (see Weimerskirch 1997). We saw no such difference in foraging locations of male and female Waved Albatrosses, similar to the results of Nicholls *et al.* (1995) for non-breeding Wandering Albatrosses. Note, however, that our sample size (five birds) was small. Second, while breeding individuals of some albatross species follow looping courses at sea (Jouventin and Weimerskirch 1990; Prince *et al.* 1997; Weimerskirch 1997), Waved Albatrosses in our study made use of one feeding area to which they commuted in direct flights. While Wandering Albatrosses have been shown to adjust their flight paths to make energetically efficient use of prevailing winds (Jouventin and Weimerskirch 1990), Waved Albatrosses showed no such adjustment, flying into a headwind en route to the feeding area, and flying with a tailwind when returning to their colonies.

A longline fishery for tuna has operated in the eastern tropical Pacific for decades (Miyabe and Bayliff 1987; Nakano and Bayliff 1992; and Caulkins *et al.* 1993) (Fig. 2). There appears to be little longline activity in what we suspect is the albatrosses primary feeding area, but consistent activity on the flight path between the Galápagos Islands and the Peruvian coast. Similar overlap between longline activities and the flight paths obtained in this study is indicated by datasets covering the period between 1971 and 1991 (Miyabe and Bayliff 1987; Nakano and Bayliff 1992; Caulkins *et al.* 1993), so incidental bycatch of Waved Albatrosses is possible. We have not found distributional data from other nations' longline fisheries in this area, so it is possible (and perhaps likely) that Figure 2 underestimates the potential overlap between Waved Albatrosses and longline activities.

At present we cannot directly evaluate whether bycatch occurs. However, indirect data indicate that little or no bycatch presently occurs. First, in over 12 person-years of field studies in the Punta Cevallos albatross colony longline hooks or gear on land have never been found (although one hook did wash up on the beach). A total of 10 Waved Albatrosses with facial cuts, including inside the mouth, have been seen, and like Harris (1973), we have attributed these injuries to fights, although the mouth injuries could have been due to longline hooks. Second, Waved Albatrosses are wary of boats and do not follow or approach them (S. Chapman, in Harris 1973). Murphy (1936) was unable to approach them in a boat, and our experiences in the Galápagos Islands are similar. Local fishing boats discarding offal in the vicinity of the Punta Cevallos colony attract frigatebirds *Fregata* spp. and Brown Pelicans *Pelecanus occidentalis*, but not the much more abundant albatrosses. The albatrosses also avoid boats in the Peruvian upwelling zone (L. Spear, pers. comm.). We do not know why Waved Albatrosses are different in this respect from some other albatross species that crowd around boats for baits or offal. Thirdly, there is no evidence that the Isla Española breeding population has decreased in size during the period 1971–1994; indeed the population may even have increased during this period (see comments in Anderson and Cruz 1997). Nonetheless, we stress the indirect nature of these data. No satisfactory substitute exists for direct, on-site observations of Waved Albatross behaviour in the vicinity of longline fishing operations and we currently lack these observations.

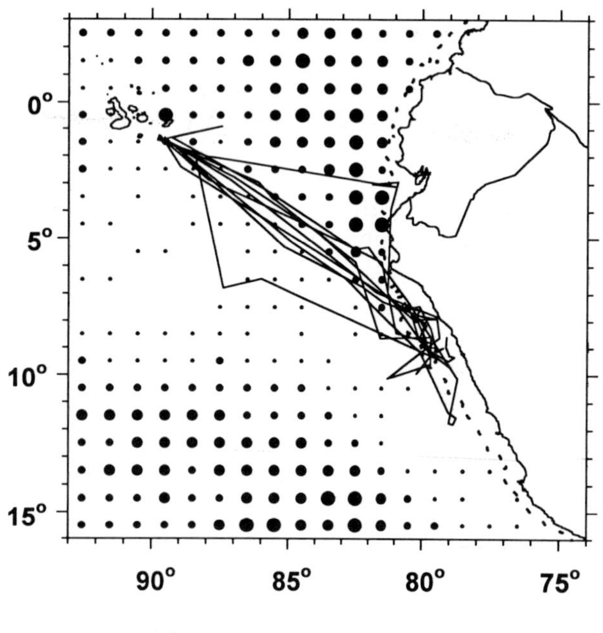

Fig. 2. Total longline hooks reported set in 1981–87 by the Japanese longline fleet, by 1° block, from Nakano and Bayliff (1992), and tracks of foraging Waved Albatross from our study.

In contrast to the bleak picture that is currently being painted for most albatross-fisheries interactions, our preliminary results suggest that Waved Albatrosses may not have been affected in the same way. Although their movements at sea appear to put them in proximity to Japanese longline fishing vessels, Waved Albatrosses not only do not follow boats in general but appear to avoid them. This behaviour requires further research, especially within the feeding area. We presently lack data on the reaction of Waved Albatrosses during offal discarding by fishing vessels and, especially, during longline hook-setting. Nonetheless, the current data presented here make us cautiously optimistic that the Waved Albatross population has not, and will not, be affected by longline operations as they are presently conducted.

ACKNOWLEDGEMENTS

This study was funded primarily by the National Geographic Society, and secondarily by U.S. National Science Foundation Grant DEB 93–04579, both to DJA. We thank the Charles Darwin Research Station, the Galápagos National Park Service, Doris Welsh, and the captains and crews of the M/V *Isabela II* and M/V *Corinthian* for facilitating logistical aspects of the study; H. Townsend for assistance in the field; P. Howie for technical advice on transmitters; and Graham Robertson and Phil Moors for comments on an earlier version of the chapter. The study was conducted under permit from INEFAN (Ecuador) and the Wake Forest University Animal Care and Use Committee.

REFERENCES

Alexander, K., Robertson, G. and Gales, R., 1997. The incidental mortality of Albatrosses in longline fisheries. Australian Antarctic Division, Tasmania. 44 pp.

Anderson, D. J. and Cruz, F., 1997. Biology and management of the Waved Albatross at the Galápagos Islands. Pp. 105–09 *in* Albatross Biology and Conservation ed by G. Robertson and R. Gales. Surrey Beatty & Sons: Chipping Norton.

Anderson, D. J. and Ricklefs, R. E., 1987. Radio-tracking masked and blue-footed boobies (*Sula* spp.) in the Galápagos Islands. *Nat. Geog. Res.* **3:** 152–63.

Brothers, N. P., 1991. Albatross mortality and associated bait loss in the Japanese longline fishery in the Southern Ocean. *Biol. Conserv.* **55:** 255–68.

Caulkins, T. P., Yamaguchi, M. and Miyabe, N., 1993. Some observations on bigeye tuna *Thunnus obesus* caught by the surface and longline fisheries for tunas in the eastern Pacific Ocean. *Bull. Inter-Amer. Trop. Tuna Comm.* **20:** 443–99.

Elliot, G. P., Walker, K. J., Nicholls, D. G. and Murray, M. D., 1995. Foraging patterns of wandering Albatross from the Auckland Islands. Abstracts of the First International Conference on the Biology and Conservation of Albatrosses. Hobart, Tasmania, Australia.

Feldman, G., 1986. Variability in the productive habitat in the eastern equatorial Pacific. *Eos* **67:** 106–08.

Feldman, G., Clark, D. and Halpern, D., 1984. Satellite color observations of the phytoplankton distribution in the eastern equatorial Pacific during the 1982–83 El Niño. *Science* **226:** 1069–71.

Gales, R., 1997. Albatross populations: status and threats. Pp. 20–45 *in* Albatross Biology and Conservation ed by G. Robertson and R. Gales. Surrey Beatty & Sons: Chipping Norton.

Harris, M. P., 1973. The biology of the Waved Albatross *Diomedea irrorata* of Hood Island, Galápagos. *Ibis* **115:** 483–510.

Jouventin, P. and Weimerskirch, H., 1990. Satellite tracking of wandering Albatrosses. *Nature* **343:** 746–48.

Miyabe, N. and Bayliff, W. H., 1987. A review of the Japanese longline fishery for tunas and billfishes in the eastern Pacific Ocean, 1971–1980. *Bull. Inter-Amer. Trop. Tuna Comm.* **19:** 3–163.

Murphy, R. C., 1936. Oceanic Birds of South America. Macmillan: New York.

Nakano, H. and Bayliff, W. H., 1992. A review of the Japanese longline fishery for tunas and billfishes in the eastern Pacific Ocean, 1981–1987. *Bull. Inter-Amer. Trop. Tuna Comm.* **20:** 187–355.

Nichols, D., Murray, D., Battan, H., Robertson, G., Moors, P., Butcher, E. and Hildebrandt, M., 1995. Satellite tracking of the wandering Albatross *Diomedea exulans* around Australia and in the Indian Ocean. *Emu* **95:** 223–30.

Pennycuick, C. J., 1989. Bird Flight Performance: A Practical Calculation Manual. Oxford University Press: Oxford.

Prince, P. A., Wood, A. G., Barton, T. and Croxall, J. P., 1992. Satellite tracking of wandering Albatrosses *Diomedea exulans* in the South Atlantic. *Antart. Sci.* **4:** 31–36.

Prince, P. A., Croxall, J. P., Trathan, P. N. and Wood, A. G., 1997. The pelagic distribution of South Georgia albatrosses and their relationships with fisheries. Pp. 137–67 *in* Albatross Biology and Conservation ed by G. Robertson and R. Gales. Surrey Beatty & Sons: Chipping Norton.

Service Argos Inc., 1988. User Manual, Ver. 1. Landover, MD USA.

Tickell, W. L., in press. Galápagos Albatrosses at sea. *Sea Swallow.*

Weimerskirch, H., Salamolard, M., Sarrazin, F. and Jouventin, P., 1993. Foraging strategy of wandering Albatrosses through the breeding season: a study using satellite telemetry. *Auk* **110:** 325–42.

Weimerskirch, H., 1997. Foraging strategies of Indian Ocean albatrosses and their relationships with fisheries. Pp. 168–79 *in* Albatross Biology and Conservation ed by G. Robertson and R. Gales. Surrey Beatty & Sons: Chipping Norton.

Shy Albatrosses preparing for the new nesting season on Albatross Island, Tasmania.
GARY NUNN

Chick feeding in albatrosses involves the adult regurgitating a slurry of partially digested squid and fish and delivering it to the chick in the cross-bill manner shown by these Yellow-nosed Albatrosses at Amsterdam Island, southern Indian Ocean.
HENRI WEIMERSKIRCH

Mortality due to fisheries and other factors

The fate of albatrosses has been linked to the actions of humans for hundreds of years. As early as the 1700s albatrosses were killed by seafarers travelling between England and Australasia on voyages of exploration and emigration (Chapter 14). Seafarers caught albatrosses on baited hooks and shot them in flight, exploiting the inclination of albatrosses to follow ships and scavenge food items from surface waters. These behaviours were to have major repercussions for albatrosses centuries later when large scale drift net fishing commenced in world oceans; this fishing practice claimed the lives of many thousands of albatrosses in the North Pacific Ocean alone (Chapter 15) before being banned in the early 1900s. Ship following and foraging from surface waters would also cause widespread mortality of albatrosses with the global expansion of longline fishing fleets from the 1950s onwards. This fishing practice is currently believed to be the most serious threat to the survival of albatross populations worldwide. Some of the highest catch rates of albatrosses have been reported from longline fisheries in South American waters (Chapters 16, 17 and 18) which are frequented by a wide range of albatross species from many parts of the world. There are also other human-induced threats to albatrosses. The effects of organochlorine contamination of the marine food chain (Chapter 19) and plastics ingestion (Chapter 20) by albatrosses now occurs on a scale that may compromise health and compound fisheries-related effects on albatross populations.

Wandering Albatrosses drowned in longline fishing operations. Determination of the sex, age class and provenance of albatrosses caught by fishing vessels allows an assessment to be made of the impact of longline fisheries on albatross populations.
TASMANIAN PARKS AND WILDLIFE SERVICE

CHAPTER FOURTEEN

Human-induced mortality of Southern Ocean Albatrosses at sea in the 19th century: a brief historical review

DAVID G. MEDWAY[1]

Evidence from a variety of sources indicates that human-induced mortality of albatrosses at sea in the Southern Ocean increased considerably from about the end of the 18th century. This occurred as a direct consequence of a major increase in the volume of shipping into the Pacific Ocean after the European voyages of exploration undertaken in the second half of the 18th century. Persons sailing on all types of ships were responsible for killing albatrosses. Those who sailed on emigrant vessels, which operated throughout most of the 19th century from England to Australia and New Zealand, played a major role in the mortality of albatrosses. It is not possible to determine how many albatrosses were killed at sea, but it is clear that the combination of mortality at sea and at many of the breeding localities throughout the 19th century must have reduced the total number of Southern Ocean Albatrosses. The evidence indicates that the Wandering Albatross may have been the species most severely affected. It seems unlikely that the Wandering Albatross could have recovered numerically from the depredations of the 19th century before its remaining populations became subjected in recent times to what appears to be an even higher level of human induced mortality at sea in the form of longline fishing.

Key words: 19th Century, Southern Ocean, Whaling Ships, Emigrant Ships, Human-induced Albatross Mortality.

INTRODUCTION

IN 1928 Alexander (1928) stated that from a becalmed ship or one travelling slowly, albatrosses could often be captured without injury by hook and line, with salted pork being used as bait. He claimed the number destroyed in this way was only small. That may have been so at the time, but this review of a subset of the available evidence shows that a large number of Southern Ocean Albatrosses of more than one species were killed at sea by humans in the 19th century, particularly by "fishing", but also by shooting. The advent of faster shipping made it progressively more difficult to capture albatrosses on hook and line. Nevertheless other human activities, particularly longline fishing, have commenced in more recent times and these activities continue to affect many species of Southern Ocean Albatrosses.

"FISHING" FOR ALBATROSSES AT SEA

William Bayly, an astronomer on Cook's second circumnavigation, described a procedure by which albatrosses were caught on hook and line. He detailed this in his journal (1772–74) as they sailed toward the Cape of Good Hope:

> saw . . . some of the Birds called Albatrosses, they are a large bird or near the size of a Goose, they are easily catched, by the following method: take a piece of deal board about a foot square more or less fix a line to it so that the hook when baited with any sort of meat may lay on the board and be thereby floted, put the board on the water with the hook baited laying on it let the board fall a stern by veering out your line perhaps 100 yards the whiteness of the board attracts the bird so that he pitcheth on the water by it and swims along side of it, and seeing the bait greedily swallows it hook and all and is therby catched.

Some 60 years later, Bennett (1834), after two voyages from England to Australia, recorded:

> One of my amusements on board was *fishing* for albatrosses, Cape petrels, or pigeons as they are commonly termed . . .

[1]PO Box 476, New Plymouth, New Zealand.
Pages 189–98 *in* ALBATROSS BIOLOGY AND CONSERVATION, ed by G. Robertson and R. Gales. Surrey Beatty & Sons, Chipping Norton. 1997.

with a hook and line, and a piece of fat pork as a bait, and a quantity of spare line to let out when the bird attacked my bait, so as to allow for the sailing of the ship, and give him time to hook himself as comfortably as possible, or as securely as desired. Many a laugh, and sapient gaze, and passing remark, have I had from my fellow-passengers, at the nonsense and folly of *"fishing"* for birds: I recollect having a crowd of eager, idle gazers looking on to see what success would attend this novel method of sporting among the feathered creation: the wind was light; and after some time spent in expectation, as is customary with that humane body, the fishers, fortune favoured me, for a huge Albatross greedily seized my bait, and hauling the ponderous bird for some distance through the water, at last I *landed* him safely on deck: the strain of the line was very great, (with the ship not going above two or three knots an hour through the water,) and I was momentarily expecting it to give way. On this gigantic bird arriving on deck, its enormous size, and elegant, sleek plumage, was exposed to the close gaze of all the before disbelieving passengers, who flocked from cabin and steerage to view the splendid prize, which excited both their surprise and admiration, as it lay on the deck, with expanding wings, and half drowned from having been dragged some distance through the water.

Some albatrosses took a baited hook more readily than others. John Gould (1840–48) considered of the Black-browed Albatross *D. (Thalassarche) melanophrys* that:

> Of all the species with which I am acquainted, this is the most fearless of man, it often approaches many yards nearer the vessel than any other; I have even observed it approach so near that the tips of its pinions were not more than two arms' length from the taffrail. It is very easily captured with a hook and line . . .

Green (1887) likewise considered the Black-browed Albatross to be the most sociable of all the family: "It is the most easily caught of all the Albatrosses, as it comes down quicker and takes a bait less leisurely than the others of the *Diomedea*".

Gould (1865) observed that the Wandering Albatross "readily seizes a hook baited with fat of any kind". Frederic Lucas (1887) considered the capture of a Wandering Albatross to be "the crowning triumph of the avian angler". Green (1887) said that the Wandering Albatross:

> is easily caught at sea with a baited hook, the only difficulty being to pay out sufficient line to keep the bait stationary while the ship is forging ahead through the water and the bird is making up its mind to seize the bait. The birds see it directly, as their sight is marvellous, but they have to settle on the water first, and they well know that rising again is a tedious job . . . To entice a bird down as quickly as possible it is a good plan to throw overboard a handful of scraps, making the bait keep amongst them. The line is floated by means of corks placed close to the hook, which should be a treble one, as there is no time to lose, and the bird must be hooked directly he seizes the bait. Both line and hook should be light, to float as long as possible, and yet very strong, as if well hooked the hauling of the bird on board is simply a question of your tackle holding, and from the fact that you are travelling all the time through the water, no give and take, or "playing" your bird, can be indulged in. Farlow of the Strand, Holroyd of Gracechurch Street, and others, make waterproof cable-laid sea-line in lengths of sixty yards, at about five shillings each, and three such lengths will answer the purpose admirably, while for hooks the largest size treble jack-hooks, such as are used on a spoon-bait, are most suitable. For all the other species of Albatross ("Mollies"), and also the Giant Petrel, exactly the same tackle is necessary, only everything on a slightly smaller scale.

Lucas (1887) recorded the procedure to be followed should a Wandering Albatross be hooked:

> the spirit of opposition is aroused within him, and sitting upright, with big webbed feet thrust firmly forward, the Albatross beats the water vigorously with his wings, or holds them stiffly extended in an effort to free himself from the hook by mere dead weight. The strain of such a bird pulling at the end of a hundred and fifty feet of line is considerable; but so long as he pursues these tactics his capture is pretty well assured. For the Albatross is not "hooked" in the ordinary sense of the word, but is simply held by the hook catching in the curved bill. So long, therefore, as the strain on the line is steady, so long will the hook hold; once slackened, it drops by its own weight, and if, as sometimes happens, the bird flies towards the ship he is soon free.

Green (1887) also mentioned the difficulties that could be experienced in bringing in a Wandering Albatross caught on hook and line:

> In common with all other fishing, but more particularly here, as the bird is only hooked in the beak, a tight line must be kept.

> When you commence hauling or winding him in (I always wound up on a winch like a huge log-reel) the bird generally throws back his head, pushes out both his feet, and flaps "hard astern" with his powerful wings. Care must then be taken to prevent his being dragged bodily under instead of along the top; as, should this happen the bird gets filled with water, and the weight to pull in is immensely increased. Sometimes, however, he rises in the air, especially in certain latitudes with the wind right aft . . . It is then a difficult job to bring him on board, as trying to keep a tight line often results in pulling him right over, and he falls on his back in the water, the hook comes out, and — worst of all — the poor bird rarely succeeds in righting himself again.

Bennett (1860) provides a fitting conclusion for this section. He recorded that in seas near the Cape of Good Hope they captured the "unusual" number of seven specimens of the Wandering Albatross in one day. He gave a detailed account of the capture of those birds which is worth repeating here for it graphically describes an event which likely occurred at sea on a great number of occasions in the 19th century:

> Our sporting after these noble birds commenced by observing two specimens sweep close to the ship in majestic and graceful flight, sailing round the vessel as if eager for food. A line and hook, to which some fat pork was attached, was dropped astern, and kept as slack as the ship's way rendered feasible. As soon as the line reached the water, a number of Pintados or Cape Petrels (the Albatros's jackals) fluttered about it, pecking at the bait, screaming, tumbling one against the other, diving and jostling together in their eagerness to get at the fatty morsel. Whilst these screaming, fluttering, quarrelsome birds were pecking at the large piece of flesh (which they were, fortunately for themselves, incapable of swallowing), their noise attracted our nobler game, that, swooping down upon the tempting lure with an independent dignity of manner, swept away all the small birds assembled around it with its huge wings, and quietly seized the dainty meal: he was soon hooked, and his capture effected. Although at this time we observed only two Albatroses about the ship, yet their capture, or some other cause, appeared to attract others, until we secured the seven specimens before mentioned (being all that visited us) in one day. At first these Albatroses alighted with upraised wings, and then settled down to the bait, which was kept stationary by letting out the line as rapidly as possible. When the bait was dragged along the water, they would swim rapidly after it; but this exertion they cannot continue long: then they would again elevate themselves in the air, finding they could make more rapid progress by flying than swimming, and alight close to, and make another attack upon the bait. Should two or more alight near it at the same time, they will endeavour to dislodge one another by fighting, and a loud tumult of their hoarse trumpeting voices ensues, expressing great anger, until one of them is hooked; and then the others follow the captured bird close to the ship until he is hauled on board.

SHOOTING ALBATROSSES AT SEA

Reinhold Forster (1785), who sailed as naturalist on Cook's second circumnavigation (see Medway 1997), appears to have been the first person to record the ease with which some albatrosses could be shot at sea:

> As they must cross the sea from one end to the other in order to find food, we found them to be extremely curious; for when a longboat had been put to sea to check the currents or the water temperature at a certain depth, or even to find fresh provisions that we lacked on occasions for four months, the Albatrosses came at once to explore it; but they paid with their lives for this curiosity.

Gould (1865) also observed that if a boat be lowered the attention of the Wandering Albatross "is immediately attracted, and while flying round it is easily shot". The accompanying illustration of Wandering Albatrosses in the Indian Ocean in July 1842, by E. A. Porcher of HMS *Fly*, aptly demonstrates the attraction of that species to any strange object floating on the ocean (Fig. 1).

It is clear that many albatrosses and other seabirds were shot at sea for no better reasons than amusement or target practice. In January 1801, Baudin (in Cornell 1974) noted that they amused themselves firing at several albatrosses and petrels "but did not hit one, although they were well within range"; and, a few days later, they spent time "amusing ourselves watching numbers of Cape gannets and albatrosses, as well as various kinds of petrel . . . Several were fired at, but none killed, although each of the marksmen claimed that his shot had completely done for the bird he had aimed at".

Gould (1865) said that *"Diomedea fuliginosa" (Phoebetria palpebrata)* was the "only species that passes directly over the ship, which it frequently

Fig. 1. Wandering Albatrosses congregating around a bottle floating in the Indian Ocean, July 1842. From a painting by Edwin Porcher, 1842. Courtesy of the National Library of Australia.

does in blowing weather, often poising itself for several seconds over the mast-head, as if inquisitively viewing the scene below; it is at this moment it offers so inviting a mark for the gunner, that it often forfeits its life, and if shot when a little to windward of the vessel, it is almost certain to fall on board". Green (1887) considered that the great object of "*Diomedea fuliginosa*" at sea

> appears to be to study minutely the trucks of the masts, especially if vanes are carried, round which it sails in the most marvellous manner, keeping beautifully even pace with the vessel the while. This over-curiosity on its part is tempting to passengers anxious to try the merits of their newly-bought fire-arms. This, however, as might be imagined, would be likely to lead to serious consequences to the rigging; so in the ship on which the specimen was shot only rifles were allowed, and that at stated times, whilst only one person was permitted to fire at a time. The birds themselves took absolutely no notice of the proceedings unless hard hit, and it is a curious fact that no amount of noise or close whirr of bullets appears to have the slightest effect on an Albatross.

In the South Atlantic in December 1841, in the course of an emigrant voyage from England to New Zealand, Fell (1973) recorded that they were

> Shooting on the poop. A beautiful bird called the Cape hen, in great abundance, flying about the vessel, as well as some albatross, a noble looking bird measuring 12 and 14 feet from tip to tip of the wings. We had an interesting discussion at lunch, whether it was justifiable to shoot those birds merely for wantonness and amusement, when it is impossible to reach them after they are dead. I had only Mr. Otterson and Mr. Barnicoat on my side of the question, so it was agreed by the sportsmen that they were justified to continue the sport if only for practice.

Bennett (1834) did not share this opinion. Although he was by no means averse to the capture of albatrosses on hook and line, he did not think likewise when it came to shooting them:

> Sometimes the passengers on board will amuse themselves by firing with ball at the poor bird as it follows the bait; one I saw struck by the deadly shot, turned on its back, dead, and, as the ship passed it, we could see its companions fall upon it eager to make it their prey. We were inclined to consider this wanton act a useless destruction of birdlife . . . I cannot refrain

Fig. 2. "Warm reception that ye first Albatross met with". From a sketch by Edward Moore, 1854, depicting the widespread practice of shooting albatrosses for amusement during voyages from England to Australasia. Courtesy of the National Library of Australia.

> from making some remarks on the cruelty of those who pass away a leisure hour on board, by firing at them as they fly about the ship. These little "indefatigables", as some are pleased to term them, are too often doomed to become the subjects of this murderous sport. Often and often, with broken wings, they are left to linger on the wide expanse of waters, unable to procure any food but that which may accidentally pass them; buffetted about by the waves, and helpless in themselves, they linger out a miserable existence, until death puts a period to their misery. It is revolting to the feelings to see these beautiful and perfectly innocent birds destroyed, solely to gratify the inclination to destroy.

The widespread practice of shooting seabirds for amusement or target practice is graphically illustrated in the accompanying sketch by E. C. Moore done on an emigrant voyage from England to Australia in 1854 (Fig. 2).

USES FOR DEAD ALBATROSSES

Green (1887) described a method of preparing an albatross skin, and some of the uses to which various parts of an albatross might be put:

> an extra strong dose of prussic acid should be poured down its throat, and, having tied up its mouth, you may prepare for skinning . . . In skinning an Albatross, — always a long and tedious job, — too much time cannot be spent in scraping the layers of fat off the skin, and then thoroughly curing with arsenical soap. This being satisfactorily accomplished, at every available opportunity bring the skin up on deck, and leave it in the wind and out of the sun. If these directions are not fully carried out a most disagreeable smell will always remain, and the grease will eventually show through the feathers and spoil the specimen . . . Everyone knows the final uses the specimen may then be put to. Its webbed feet make capital tobacco-pouches by drawing out all the bone and leaving on the claws as ornaments. The skin of the

> feet should be well stuffed with tow, and then pinned out on a board to dry. The wing-bones make excellent pipe-stems; the breast, if carefully cured, a warm though somewhat conspicuous muff; and the beak, in the hands of a skilled artificer, a handsome paper-clip.

Bennett (1860) noted that under the feathers of the albatross

> there is an enormous quantity of fine down, covering the body, which is in great requisition for tippets and boas. Nor does the admiration the bird excites when captured prevent its being destroyed, on account of the profusion of this valuable material obtained from its plumage.

James Harry was master of the merchant barque *Templar* on a voyage from Glasgow to Melbourne in the mid 19th century. He left Glasgow in September 1853 and sailed by way of the vicinity of the Cape of Good Hope and then east across the south Indian Ocean, arriving at Melbourne in early January 1854. It was of his voyage to Australia in 1853 that Harry (1844–63) described the reason for his destruction of albatrosses. It is reasonable to assume he was not the only seafarer of the time who killed albatrosses for the purpose described:

> During our passage from the vicinity of the Cape of Good Hope we were accompanied by any number of Cape Hens, Cape Pigeons, and Albatross, as if it was a calm, and not many of those birds about, the grease bucket capsized overboard, will invariably in a short time, attract them in flocks so that anyone need not want wild fowl if fond of shooting; whether their scent is so strong as to smell the grease when thrown on the water or whether they have a telegraphic mode of communication I dont know, but certain it is, that shortly after the grease is on the water one will see birds arriving from every quarter; when it happened to be a calm I used to adopt this plan to congregate albatross, and shoot them for the eggs contained in them, as the other kinds are no use and it would only be cruelty to kill them . . . The way I used to manage was to bait a hook with fat pork and entice them gently up to within shooting distance then aim at its head and blind it, as I defy any one to get hold of it, as where they catch with their bill if on a man's body they take the peice out, I have seen them bring the upper and lower beak to meet on a deal board of an inch thickness, the eggs found in the Albatross are about the size of those found in the common fowl and are quite a treat at sea. . .

THE EXTENT OF HUMAN-INDUCED MORTALITY OF ALBATROSSES

The early years of the 19th century brought significant changes in the volume and types of shipping which plied the waters of the Southern Ocean. It is obvious that persons travelling on all types of shipping "fished" for albatrosses and other seabirds, or shot them, as opportunity offered. Two types of shipping have been selected for mention here.

Whaling ships

By 1775 whalers were sailing as far south as the Falkland Islands. The first whalers to operate out of Sydney, Australia, did so in 1791 and it was from there that the New Zealand whaling grounds were opened. By 1801 whalers were using bays and inlets on the coast of New Zealand as bases (Spence 1980). A "Whale Chart" reproduced by Whipple (1982) shows that whales occurred in large areas of the Southern Ocean. A significant portion of those areas lie below about 30°S. A considerable number of the whaling vessels which operated in the southern oceans during the 19th century must have been present in areas where albatrosses were abundant.

Whaling cruises in the southern oceans could last up to five years. On the whaling grounds much time was spent waiting for whales to appear. Fresh food lasted no longer than a few days out of port; after that it was salt beef and salt pork, often of poor quality, smelling unpleasant and badly cooked. It is no wonder that seabirds should be killed whenever possible, not only for amusement but also as a source of fresh meat. There can be little doubt that albatrosses were among the birds taken from whaling ships whenever the opportunity offered. As an example, Bennett (1840) was a passenger on the *Tuscan* during its whaling voyage round the world from 1833 to 1836. When off the south-east coast of South America in December 1833, he noted that

> Many ocean birds of the high south latitudes were now visible around us, as nellies (*Procellaria gigantea*); blue-petrels, or sperm-birds (*Prion pachyptila*); pintados, or cape-pigeons; and pios, or cape-hens (*Diomedia fuliginosa);* with several other species of the albatross family. Among the number of these birds we captured by hook and line, baited with fat meat, was a kind of petrel, which has not been described by our ornithologists.

Emigrant ships

For a substantial part of the 19th century sailing ships transported settlers from England to Australia and New Zealand. Between 1788 and

1880 about 1.3 million free emigrants arrived in Australia from the United Kingdom (Hassam 1995). Emigration reached its peak in 1874–75 when 30 000 people were transported to New Zealand in 93 ships (Church 1985). The usual route for those vessels was into the south-west Atlantic, and then east across the south Indian Ocean in the "roaring 40's and 50's", latitudes inhabited by large numbers of Southern Ocean Albatrosses. It is clear that albatrosses were killed as opportunity offered, for one reason or another, by persons sailing on these ships. Hassam (1995) indicates that there are about 850 extant shipboard diaries kept by persons who emigrated to Australia from the United Kingdom last century, and considers that this was only a small proportion of the diaries written. He also noted that emigrants considered that diaries could hardly be authentic without a record of the wing-span of an albatross.

The following examples of the capture of albatrosses at sea provide ample evidence of universal practices on emigrant ships which cumulatively must have resulted in the death of a large number of albatrosses. The preface in Green (1887) was apparently intended primarily to enable "every passenger who embarks on board a ship bound for Australasia" to identify oceanic birds as they flew near the ship, and to give "hints as to the best method of capture to be adopted by those desirous of adding specimens to a collection".

By 1834 Bennett (1834) had made two voyages from England to Australia. At least one of them, in 1832, was on an emigrant ship:

> When the wind falls light, the Albatross fishers are on the alert; it is then that all the hooks and lines that can be procured on board are put into requisition, and the steward receives frequent solicitations for pieces of fat pork for bait; some of these birds when hooked, before they can be brought on board either manage to detach themselves from the hook, or escape by its breaking. It is laughable to see, when one is hooked, with what eagerness he is followed by the others, (while dragged through the water by his capturers,) all eager to seize the bait, they consider he is monopolising; they pounce down upon it, and look surprised to find themselves quietly perched on the waves while the other poor hooked bird, bait and all, is before them in rapid progress towards being taken on board. Numerous eager gazers are all ready assembled on the poop of the vessel to receive the stranger; where, on its arrival, it remains the patient, stupid-looking subject of a hundred passing remarks: the ladies give their due share of praise to its large bright eyes, fine sleek and clean plumage; the expansion of wings is duly measured, the bird, during the time remaining sitting in a quiet manner, appearing very sulky, and apparently unmindful of all around, until a few practical jokes keep its bright eyes on the alert against all personal attacks. After having been admired and pitied (as much as its situation may be considered to require) by the fair sex, and having satisfied the curiosity of both sexes, it is consigned either to the anatomist to be dissected and stuffed for the benefit of science, or to form a skeleton to grace some of the museums in England; or sometimes regains its liberty, with a piece of ribbon tied around the neck, by which it may be readily recognized should it continue to follow the ship, which it usually does for a long time after.

In November 1840, Weekes (1840–54) left Plymouth in England on the *William Bryan* bound for New Zealand. His journal contains the following entries:

> West of the Cape of Good Hope on 26 January 1841 "Birds are becoming numerous — 5 Albatrosses have been flying about the ship. We have been firing at them and I think killed one for he was obliged to sit on the water". During the passage across the south Indian Ocean on 6 February "Caught an Albatross with a hook and line. We have skinned him to send home. The measurement from wing tip to do was 10ft — from tail to beak 6ft (sic) 4in."; on 14 February "Caught an Albatross but the poor wretch broke away with about 30 yds or more of line"; on 19 February "We caught four birds ysty of the Albatros kind and have been skinning them to-day — They were very ravenous"; on 21 February "The people (Sailors) have made a good dinner off the Albatrosses cooked in the form of a sea-pie. They say that by soaking them in water over night they get rid of any fishy taste they may be supposed to have. Mr McClarty, the 3rd mate, told me they were 'very good indeed'".

In November 1841, Flight (1841–52) left Plymouth in England on the *Timandra* bound for New Zealand. When in the South Atlantic west of the Cape, he recorded that there were "Several albatrosses about the ship. Caught one measuring 10 feet from the tip of one wing to that of the other. One of the sailors, with Wm Allen, skinned the bird. The skin was rubbed with mercurial ointment, then with pepper and salt; the brain taken out and the head filled with pepper and salt. The flesh was eaten by some of the emigrants and said to be good".

Fig. 3. "In the albatross seas — a morning's catch". Photograph by F. D. Green taken on the voyage of the *Sobraon* between England and Australia, 1890–91. Courtesy of the Mitchell Library, Sydney.

In February 1850, George Hepburn left Gravesend in England on the *Poictiers* bound for New Zealand. His journal (in Stewart 1934) contains the following entries:

> Near the Cape of Good Hope on 30 April 1850, "Yesterday morning the third mate caught a live albatross with a line and hook. It is a large bird about the size and very like a swan, but short in the neck, long beak like a duck, pointed like a hawk, wings measuring about nine feet from tips, yet it cannot fly from off the deck. It walked about for some time until all who liked saw it. It is almost pure white except the wings which are mostly black, and is really a beautiful bird. However, they killed him and plucked him to-day for the feathers, giving the carcass to the dogs, being of too strong a nature to eat. Several others were shot a day or two before which fell into the sea"; and on 14 May 1850 in the south Indian Ocean "Ever since rounding the Cape we have a goodly company of flying fowls attending the ship. The largest of these is the albatross, about the size of a good swan or goose, with wings extended from nine to 12 feet, a beautiful bird, of which two have been caught with a simple bait and thread line by some of the young gents. Mrs H. would very fain have got the entire skin of one or two of them to send home to certain gentlemen in the Lang Toon for being stuffed, but the birds were foolishly dissected".

Couchman (1879) left Plymouth in England for New Zealand on board the *Arethusa.* She noted while crossing the south Indian Ocean:

> On 29 October 1879 "The albatross are very large when they throw their wings wide open they measure 10 and 12 feet across their body and wings, and if they once get on the deck of the ship they cannot fly up again, so they are caught"; on 5 November in very stormy weather "the men are catching the albatross for they are tired out and keep falling on the deck. They are splendid birds. They skin them and throw the bodies away while the plumage they keep for sale, and the quills in the wings they make pipe stems of. I have got a fine large skin given to me to make Jane a set of furs. The first mate had it dressed and nailed out on a board to dry it"; on 17 November "Our constables are very busy catching albatrosses for their skins. They have caught 16 today"; and on 2 December 1879 "The men have caught 14 more albatrosses today".

Fig. 4. "Taxidermists at work". Photograph by F. D. Green taken on the voyage of the *Sobraon* between England and Australia, 1890–91. Courtesy of the Mitchell Library, Sydney.

F. D. Green took photographs on the clipper ship *Sobraon* during its final emigrant voyage from England to Australia in 1890–91. An album of his photographs (Doyle 1890–91) is in the Mitchell Library in Sydney. Among them are photographs which depict captured albatrosses. They illustrate practices which also occurred on other emigrant voyages. Two of those photographs are reproduced here (Figs 3 and 4).

EFFECTS OF HUMAN-INDUCED MORTALITY OF ALBATROSSES

It is not possible to obtain any idea of the number of Southern Ocean Albatrosses killed at sea by humans in the 19th century. It is clear, however, that the mortality must have increased greatly in the 19th century with the expansion of shipping into those parts of the Southern Ocean inhabited by albatrosses. Some seafarers took considerable numbers of seabirds. When speaking of the Wandering Albatross, Green (1887) said that he "once made the voyage round the world with a captain who had studied Albatrosses for forty years. During that period he had caught an incredible number, and I myself have frequently seen him capture twelve or so in a day". Green himself caught and measured over 100 albatrosses.

Gould (1844) said that the Wandering Albatross was "very numerous between the 30th and 60th degrees of S. lat., and is to be met with in every part of the circle encompassing the world bounded by those degrees". Gould (1840–48) also wrote that "So frequently does the boldness of this species cost it its life, that hundreds are annually killed without, however, its numbers being apparently in any degree lessened". Gould's figure of "hundreds" was probably an underestimate given what must have been at the time a large volume of shipping in the Southern Ocean. A few years later, Bennett (1860) said "it has been remarked by many voyagers that the large Wandering Albatros has been during the last few years more rarely seen, either about the shores of Australia, or in its still more general range in the latitudes of Cape Horn and the Cape of Good Hope". At the same time Bennett noted that "The smaller species of these birds are still observed roaming about the ocean, apparently in undiminished numbers". Both Gould (1844) and Green (1887) considered the Black-browed Albatross to be the most abundant albatross of the Southern Ocean.

CONCLUSIONS

The slaughter of large numbers of nesting seabirds on Palmerston Atoll, Cook Islands, in April 1777 reminded Clerke (in Beaglehole 1967) of "the general destruction that ever accompanies Man". At virtually every breeding location, and when at sea, Southern Ocean Albatrosses must to all intents and purposes have been free from human molestation until people from the Northern Hemisphere began arriving in the Southern Ocean from about the end of the 18th century. There can be no doubt that a very considerable number of Southern Ocean Albatrosses of more than one species must have been killed at sea by humans during the 19th century. There is no way of knowing how many were killed, but it is clear that a combination of mortality at sea and at many breeding locations throughout the 19th century must have reduced the number of albatrosses inhabiting the Southern Ocean. The evidence indicates that the Wandering Albatross may have been the species most severely affected and the level of mortality inflicted, both at sea and at its breeding locations, must have substantially reduced its numbers. It seems certain that the Wandering Albatross could not have recovered numerically from the depredations of the 19th century before its remaining populations became subjected in recent times to what appears to be an even higher level of human induced mortality at sea in the form of longline fishing.

ACKNOWLEDGEMENTS

I am grateful for assistance readily given at the Mitchell Library in Sydney and for the permission of that library, and of the National Library of Australia in Canberra, to reproduce the illustrations which accompany this chapter. I am particularly grateful to Sue Superville of the Hector Library at the Museum of New Zealand in Wellington for her prompt and interested attention to my continuing requests. The co-operation of both the Director and the Archivist of the Taranaki Museum in New Plymouth has enabled me to study and to quote from unpublished journals in their care. I am also grateful to Joy Chong of New Plymouth for the careful attention she has given to the typing of the manuscript, and to Lance Tickell and John Warham whose constructive comments have improved it.

REFERENCES

Alexander, W. B., 1928. Birds of the Ocean. A Handbook for Voyagers. G. P. Putnam's Sons: New York and London.

Bayly, W., 1772–74. Private journal kept on the *Adventure*. Manuscript in Alexander Turnbull Library: Wellington.

Beaglehole, J. C. (ed), 1967. The Journals of Captain James Cook on his voyages of discovery. The voyage of the *Resolution* and *Discovery* 1776–1780. 2 vols. Hakluyt Society: Cambridge.

Bennett, F. D., 1840. Narrative of a whaling voyage round the globe from the year 1833 to 1836 . . . 2 vols. Richard Bentley: London. Bibliotheca Australiana facsmile reprint 46 (1970).

Bennett, G., 1834. Wanderings in New South Wales, Batavia, Pedir Coast, Singapore, and China; being the journal of a naturalist in those countries during 1832, 1833 and 1834. Richard Bentley: London. Australiana facsimile editions No. 115 (1967).

Bennett, G., 1860. Gatherings of a Naturalist in Australasia . . .John van Voorst: London. Facsimile edition 1982. Currawong Press: Milson's Point.

Church, I., 1985. Survival on the Crozet Islands. The wreck of the *Strathmore* in 1875. Heritage Press: Waikanae.

Cornell, C. (Trans.), 1974. The Journal of Post Captain Nicolas Baudin Commander-in-Chief of the Corvettes *Geographe* and *Naturaliste* assigned by Order of the Government to a Voyage of Discovery. Libraries Board of South Australia: Adelaide.

Couchman, M., 1879. Letter in diary form describing a voyage on the *Arethusa* from Plymouth to New Zealand August–December 1879. Typescript copy Ms 275 in Taranaki Museum: New Plymouth.

Doyle, H. M., 1890–91. Voyage of the *Sobraon* 1890–1891. Album of photographs at Q656.5D in Mitchell Library: Sydney.

Fell, A., 1973. A Colonist's Voyage to New Zealand under sail in the "Early Forties" . . . James Townsend and Sons: Exeter. First published 1926. Reprint by Capper Press: Christchurch.

Flight, J., 1841–52. Diary, including a journal of his voyage from Plymouth to New Zealand on the *Timandra* November 1841–February 1842. Original Ms 002 and typescript copy in Taranaki Museum: New Plymouth.

Forster, J. R., 1785. Memoire sur les Albatros. *Memoires Math. Phys. Paris. (Acad. Sci.)* **10:** 563–72, pls. xiii–xv.

Gould, J., 1840–48. The birds of Australia. 7 vols. Published by the author: London.

Gould, J., 1844. On the Family *Procellaridae*, with descriptions of Ten new Species. *Annals Mag. Nat. Hist.* **13:** 360–68.

Gould, J., 1865. Handbook to the birds of Australia. 2 vols. Published by the author: London.

Green, J. F., 1887. Ocean Birds. R. H. Porter: London.

Harry, J., 1844–63. Journal of Life at Sea. Ms A2047 in Mitchell Library: Sydney.

Hassam, A., 1995. Sailing to Australia. Shipboard diaries by nineteenth-century British emigrants. Melbourne University Press: Carlton.

Lucas, F. A., 1887. Notes of a bird catcher. *Auk* **4**(1): 1–6.

Medway, D. G., 1997. Type specimens of albatrosses collected on Cook's second voyage. Pp. 3–12 *in* Albatross Biology and Conservation ed by G. Robertson and R. Gales. Surrey Beatty & Sons: Chipping Norton.

Moore, E. C., 1854. Sketchbook of scenes on board the *Frances Henty*. NK4270/1 to 18 in National Library of Australia: Canberra.

Spence, B., 1980. Harpooned. The Story of Whaling. Conway Maritime Press: Greenwich.

Stewart, W. D. (ed.), 1934. The Journal of George Hepburn On His Voyage from Scotland to Otago in 1850 . . . with extracts from his letters written from Otago. Coulls Somerville Wilkie Ltd and A. H. Reed: Dunedin.

Weekes, H., 1840–54. Journal of Common Things, including his voyage from Plymouth to New Zealand on the *William Bryan* November 1840–March 1841. Ms 022 in Taranaki Museum: New Plymouth.

Whipple, A. B. C., 1982. The Seafarers. The Whalers. Time-Life Books: Amsterdam.

CHAPTER FIFTEEN

Laysan and Black-footed Albatrosses: trophic relationships and driftnet fisheries associations of non-breeding birds

PATRICK GOULD,[1] PEGGY OSTROM,[2] WILLIAM WALKER[3] and KEVIN PILICHOWSKI[2]

Laysan and Black-footed Albatrosses are sympatric throughout the North Pacific Ocean and Bering Sea. In our study of the trophic relationships of marine birds in the central North Pacific Ocean, non-breeding Black-footed Albatrosses occupied the top trophic level. On average, non-breeding Laysan Albatrosses obtained food at approximately one trophic level below non-breeding Black-footed Albatrosses. Prior to the cessation of the high-seas driftnet fishery in 1992, non-breeding individuals of both albatross species supplemented their diets with prey made available by the fisheries (e.g., squid entangled in nets and offal discards), but Black-footed Albatrosses appeared to spend more time scavenging than Laysan Albatrosses. Driftnet fisheries appeared to exert a great influence on the behaviors, diets and distribution patterns of non-breeding Laysan and Black-footed Albatrosses.

Key words: Laysan Albatross, Black-footed Albatross, Trophic Level, Diets, Driftnet Fisheries, Longline Fisheries, North Pacific Ocean.

INTRODUCTION

ALBATROSSES are among the most frequent marine birds to interact with commercial fisheries. Impacts of fisheries on albatrosses and other seabird species include capture by fishing vessels (Alexander *et al.* 1997; Johnson *et al.* 1993), the provision of food trapped in fishing nets, and supplementation of diets through the disposal of offal (Thompson 1992). Variability in food utilization within and between albatross species has been documented in the Southern Ocean albatrosses (Prince 1980; Weimerskirch *et al.* 1988; Cherel and Klages 1997). Resource partitioning among albatross species is manifested by differences in their foraging ranges, feeding zones, and diets. Resource partitioning may occur both in the breeding and non-breeding seasons, and competition among albatrosses may be intensified in times of food scarcity. Commercial fishing operations may allow the overlap of otherwise disjunct albatross feeding zones by providing a new, relatively abundant, and easily obtained food supply (Weimerskirch *et al.* 1988).

In this chapter we review the findings of Gould *et al.* 1997, Gould and Hobbs 1993 and Johnson *et al.* 1993 on the food habits and driftnet fisheries associations of non-breeding Laysan *Diomedea (Phoebastria) immutabilis* and Black-footed *D. (Phoebastria) nigripes* Albatrosses prior to the cessation of the fisheries in 1992. We also present new data on the trophic relationships of non-breeding albatrosses, on their at-sea distributions and their interactions with driftnet fisheries.

ALBATROSS DISTRIBUTION AT SEA

Laysan and Black-footed Albatrosses co-occur in the temperate North Pacific Ocean. Both species concentrate around sub-tropical breeding grounds in the boreal winter (October–May) and disperse north across the Pacific Ocean into transitional and sub-arctic waters in the boreal summer (June–September). Transition waters occur between 30°N–45°N in an area between sub-arctic and sub-tropic waters with a west-to-east flowing current. Rapid changes in water temperature and salinity form frontal zones at the north and south borders of this region (Roden 1980; Roden and Robinson 1988; Roden 1991). Laysan and Black-footed Albatrosses may spend the first one-to-three years of their lives at sea (Shuntov 1974; Fisher and Fisher 1969;

[1]Alaska Science Center, US Geological Survey, 1011 E Tudor Road, Anchorage, AK, 99503 USA.
[2]Michigan State University, Department of Geological Sciences, 206 East Lansing, MI, 48824 USA.
[3]National Marine Mammal Laboratory, Alaska Fisheries Science Center, National Marine Fisheries Service, NOAA, 7600 Sand Point Way, NE, Seattle, WA 98115 USA.
Pages 199–207 *in* ALBATROSS BIOLOGY AND CONSERVATION, ed by G. Robertson and R. Gales. Surrey Beatty & Sons, Chipping Norton. 1997.

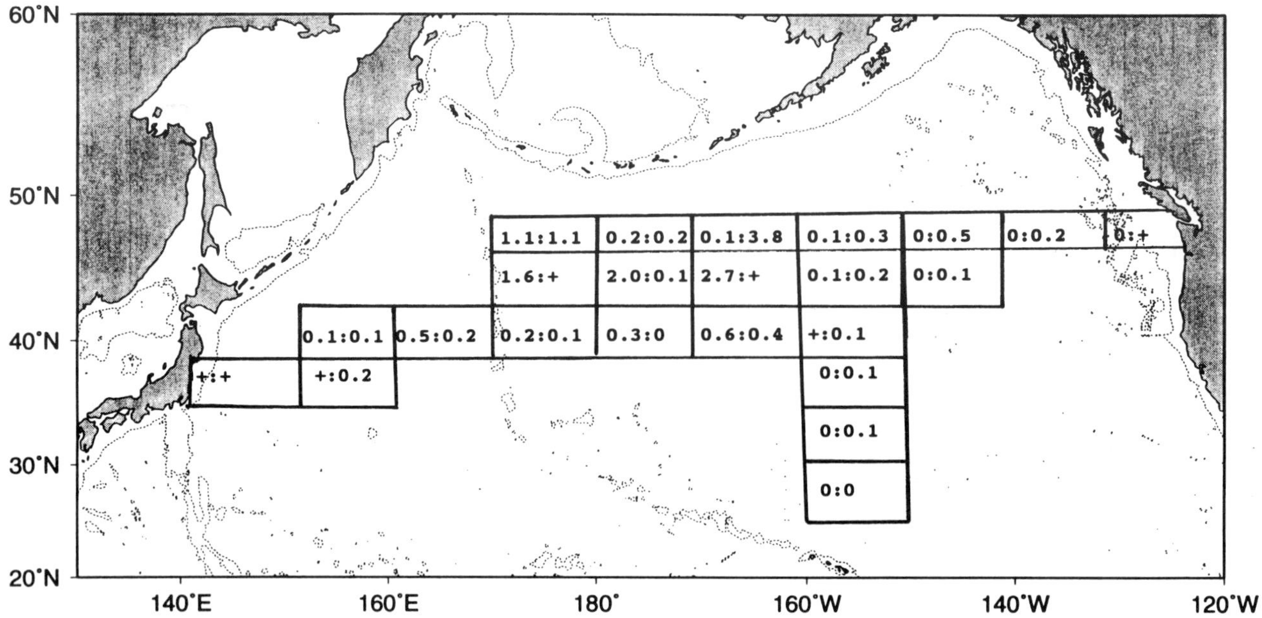

Fig. 1. Laysan and Black-footed Albatross density indices (Laysan Albatrosses/km^2:Black-footed Albatrosses/km^2) from marine surveys between 30 June and 15 September 1990, from Tokyo to Vancouver, Vancouver to Honolulu, Honolulu to Tokyo. Sample sizes per latitude-longitude block ranged from 3.1 km^2 to 221.0 km^2 (average = 54.9 ± 51.4 km^2). Dotted lines represent 1 500 fathom contours. The principal study area was 35°N–46°N and 160°E–145°W. Figure adapted from Gould and Piatt (1993).

Rice and Kenyon 1962a, 1962b). Non-breeding Laysan Albatrosses and Black-footed Albatrosses occur throughout the sub-arctic North Pacific Ocean from Japan to the west coast of North America (Sanger 1974a, 1974b; Brazil 1991; McDermond and Morgan 1993). Some non-breeding birds move into the Gulf of Alaska to about 60°N between March and November, with peak numbers occurring in September. In the Bering Sea Black-footed Albatrosses frequent waters near the Aleutian Islands while Laysan Albatrosses disperse north to about 58°N (Gould *et al.* 1982). Densities of both species appear to be highest over continental shelf slopes or ocean basins (Gould *et al.* 1982; Briggs *et al.* 1987). Other factors identified as affecting at-sea distributions include oceanic conditions (Kuroda 1991; Gould and Piatt 1993), age and breeding status (Fisher and Fisher 1972; Robbins and Rice 1974), food availability and commercial fishing operations (Wahl and Heineman 1979; Briggs *et al.* 1987).

World populations of Laysan Albatrosses outnumber Black-footed Albatrosses by more than 12:1 (McDermond and Morgan 1993; see Croxall and Gales 1997). At sea, Laysan Albatrosses outnumber Black-footed Albatrosses in the western Pacific Ocean and Black-footed Albatrosses outnumber Laysan Albatrosses in the eastern Pacific Ocean (Rice and Kenyon 1962a; Shuntov 1974; Briggs *et al.* 1987). In the Bering Sea and Gulf of Alaska, ratios of Laysan to Black-footed Albatrosses are 16:1 and 1:12, respectively (Gould *et al.* 1982). Ratios of Laysan to Black-footed Albatrosses of 2.3:1 in the far western Pacific Ocean and 0.2:1 in the far eastern Pacific Ocean are clinal (Kuroda 1988).

As part of this study marine bird surveys were also conducted across the transitional North Pacific Ocean aboard the research vessel *Shoyo maru* from 30 June to 15 September 1990, (see Gould and Forsell 1989 for survey methodology). Both Laysan and Black-footed Albatrosses were most abundant in the central region, an area intensively fished by the squid driftnet fleets. Black-footed Albatrosses were found throughout the survey area while Laysan Albatrosses were not seen in the south and east, although they probably occurred there in small numbers (Fig. 1). Johnson *et al.* (1993) found the ratio of Laysan Albatrosses to Black-footed Albatrosses of 4:1 for albatrosses caught in high seas squid and large-mesh driftnets in the central North Pacific. However, Black-footed Albatrosses are more likely to follow ships than Laysan Albatrosses (Gould *et al.* 1997), so the ratios reported above are probably biased in favour of Black-footed Albatrosses.

FORAGING AND DIETS

There are few observations published of Laysan Albatrosses and Black-footed Albatrosses feeding in the wild other than by scavenging near vessels. Gould and Piatt (1993) reported two instances of Laysan Albatrosses feeding in association with dolphins, and another observation of a Laysan Albatross and eight Black-footed Albatrosses feeding on a dead squid. Both species have been reported to take food in the upper one-metre of the ocean by surface seizing, contact dipping, and scavenging (Miller 1940; Ainley and Sanger 1979; Harrison *et al.* 1983; Johnson *et al.* 1993).

Although both Laysan Albatrosses and Black-footed Albatrosses have access to the same food resources, differences in diet exist. Laysan Albatrosses have better night vision, longer bills, and longer but lighter bodies than Black-footed Albatrosses (Harrison *et al.* 1983; Fefer *et al.* 1984; McDermond and Morgan 1993). They may thus be more capable than Black-footed Albatrosses of rapid retrieval of small prey that are active in surface waters at night. The heavier, and more compact Black-footed Albatrosses, with their shorter, stockier bills, may be better adapted to scavenging naturally occurring large carrion or refuse from ships.

Stomach samples

Laysan Albatrosses and Black-footed Albatrosses feed opportunistically on almost any potential food item, living or dead, that becomes available including waste items discarded from ships and indigestible pieces of plastic (Cottam and Knappen 1939; Fisher 1945; Anderson 1954; Kenyon and Kridler 1969; Ainley and Sanger 1979; Harrison *et al.* 1983; Sileo *et al.* 1990; Gould *et al.* 1997). Black-footed albatrosses, in particular, have been identified as scavengers (Miller 1940; Harrison 1990). In the study by Harrison *et al.* (1983) squid was the main component of the diets of Laysan Albatross chicks while fish (primarily fish eggs) was the main component of the diet of Black-footed Albatross chicks. Non-cephalopod invertebrates accounted for 13% and 5% of Laysan Albatrosses and Black-footed Albatrosses nestling diets, respectively. In a large area (between 35°N–46°N and 160°E–145°W) of the transitional region of the North Pacific Ocean, Gould *et al.* (1997) found more than 40 species from 25 families of marine organisms in the digestive tracts of non-breeding Laysan Albatrosses and Black-footed Albatrosses. The diets of non-breeding albatrosses in the transitional region differed from those of the nestlings studied by Harrison *et al.* (1983). Among non-breeding albatrosses within the transition zone, fishes were more numerous than squids in the diets of non-breeding Laysan Albatrosses while squids were more important than fishes in the diets of non-breeding Black-footed Albatrosses.

Although the sample sizes in Gould *et al.* (1997) were small (99 Laysan Albatrosses and 65 Black-footed Albatrosses), there appeared to be several geographical trends in the composition of albatross diets (Figs 2 and 3). Cephalopods dominated the diets of Black-footed Albatrosses throughout the study area, perhaps being more common in the east than the west. For Laysan Albatrosses, the occurrence of squids varied from west to east, but no clear pattern was evident (Fig. 2). The occurrence of squids decreased in the diets from 46°N to about 41°N (the sub-arctic boundary area) and then increased south of this area (Fig. 3). The occurrence of fishes in the diets of both Laysan Albatrosses and Black-footed Albatrosses increased from 45°N to about 41°N, and then decreased farther south. This pattern was the opposite of that for squids in Laysan Albatrosses. Fish percentages decreased from west to east in Laysan Albatrosses, while the pattern was unclear for Black-footed Albatrosses. Non-cephalopod invertebrates were more common in diets of both albatross species in eastern waters than in western waters of the Pacific Ocean.

Stable isotope analysis

As a second method of diet analysis, the digestive tract data of Gould *et al.* (1997) were supplemented with stable nitrogen isotope ($\delta^{15}N$) data. The nitrogen isotopic composition of a consumer differs by a constant amount from that of its food source (Deniro and Epstein 1981; Harrigan *et al.* 1989), the average difference for marine organisms being +3.2‰ (per mil.) (Michener and Schell 1994). The $\delta^{15}N$ values increase with trophic position, the increase being associated with preferential excretion of (^{14}N by the consumer (Minigawa and Wada 1984). Nitrogen isotope values increase with trophic position because the 3.2‰ enrichment occurs between each tropic level within a food web (Harrigan *et al.* 1989; Wada *et al.* 1987; Ostrom and Fry 1993). Consequently, $\delta^{15}N$ values are robust indicators of trophic position. The increase in $\delta^{15}N$ values allows organisms to be placed along a trophic continuum rather than at discrete trophic positions. For example, two organisms whose $\delta^{15}N$ values separated by 1.6‰ differ by 0.5 trophic levels. Another characteristic of the isotope approach is that $\delta^{15}N$ values represent material that has been assimilated by the birds and are, therefore, indicative of longer-term diet (the half-life for turnover of $\delta^{15}N$ in muscle tissue is 12.4–27 days: Tiezen *et al.* 1983; Hobson and Clark 1992; Hobson *et al.* 1994). Given that our samples of Black-footed Albatrosses and Laysan Albatrosses were widely distributed across the north Pacific Ocean, and that the standard deviation of the average $\delta^{15}N$ value for each of these albatross species was small, we assumed that if spatial variation in $\delta^{15}N$ did occur the influence on both species would be similar.

The nitrogen isotope data suggested that all age classes of Laysan Albatrosses, including nestlings, fed at lower average trophic levels than corresponding age classes of Black-footed Albatrosses (Gould *et al.* 1997). In the non-breeding season, breeding-age Laysan Albatrosses

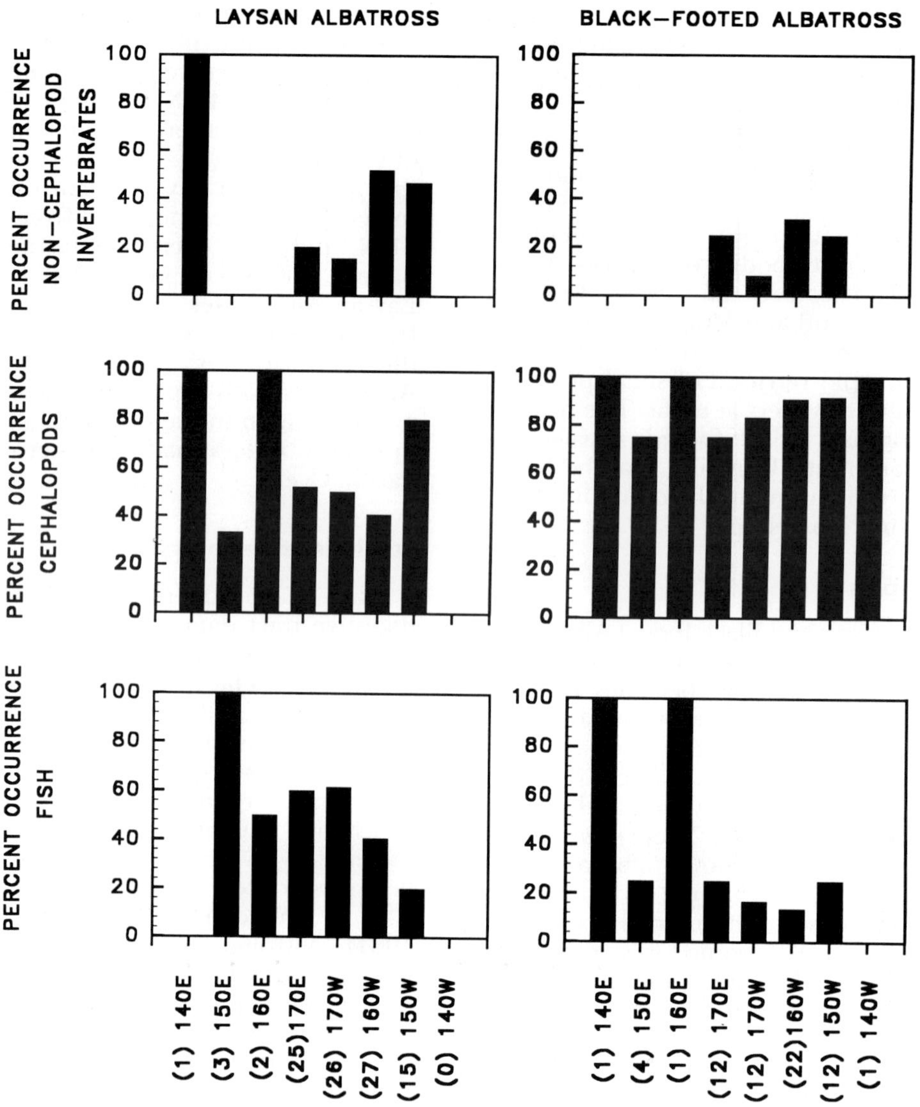

Fig. 2. North-south variability in per cent occurrence of major food types in diets of Laysan and Black-footed Albatrosses. Sample sizes are in parentheses.

fed, on average, one trophic level below breeding-age Black-footed Albatrosses (differences in $\delta^{15}N$ between Laysan Albatrosses and Black-footed Albatrosses = 3.1‰). In contrast, the average trophic level of the food of nestling Laysan Albatrosses is about one-third lower than that of nestling Black-footed Albatrosses (the difference in mean $\delta^{15}N$ between Laysan Albatrosses and Black-footed Albatrosses nestlings was 1.3‰).

FISHERIES INTERACTIONS

An estimated 17 500 Laysan Albatrosses and 4 400 Black-footed Albatrosses were killed in high seas squid and large-mesh driftnet fisheries in 1990 (Johnson *et al.* 1993; Gould and Hobbs 1993). The principal target species for these fisheries were Neon Flying Squid *Ommastrephes bartrami*, Tunas (e.g., *Thunnus* sp. and *Katsuwonus* sp.) and Billfishes (e.g., *Tetrapturus* sp.). The estimated incidental catch of non-target species was 10s of 1 000s for mammals, 100s of 1 000s for birds, and 10s of millions for fish (Yatsu *et al.* 1993).

At driftnet fishing operations, Laysan Albatrosses and Black-footed Albatrosses scavenged animals immobilized by the nets, and offal discarded from vessels during processing of the catch (P. Gould, pers. obs.). Albatrosses shredding large animals caught in the nets attracted other marine birds which fed on the scraps dropped by the albatrosses. Since hard parts (e.g., otoliths) may last for many days in the digestive tracts of birds, while flesh and cartilage are digested more quickly, examination of the mass of material in the digestive tracts provided an index of an albatrosses last meal. The digestive tracts of Laysan Albatrosses and Black-footed Albatrosses killed while feeding from driftnets indicated that undigested cephalopods accounted for 75% by mass of the total food taken from driftnets (Gould *et al.* 1997). The Neon Flying Squid comprised 67.9% and 67.2% by mass of Laysan Albatrosses and Black-footed Albatrosses digestive tract contents, respectively. The only other prey species to

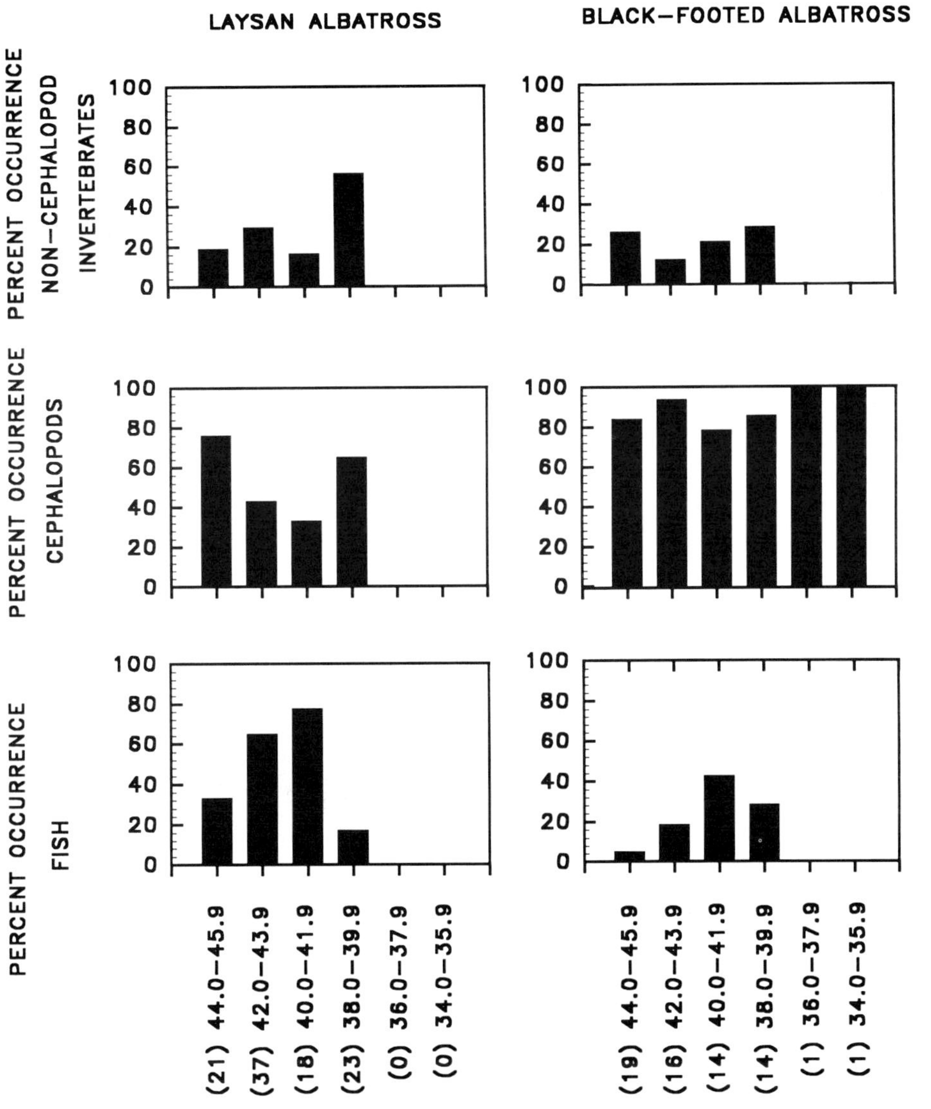

Fig. 3. West-east variability in per cent occurrence of major food types in diets of Laysan and Black-footed Albatrosses. Sample sizes are in parentheses.

comprise more than 10% by mass was Pacific Pomfret *Brama japonica* (Laysan Albatrosses = 14.3%, Black-footed Albatrosses = 18.1%). These results indicate that Laysan and Black-footed Albatrosses preyed heavily on food made available by driftnet fishing operations. However, it is not known how frequently albatrosses made use of the fishing operations. Stable nitrogen isotope data assist in clarifying the impact of driftnet fisheries on albatross food habits in diets assimilated over a period of at least 2–4 weeks.

To evaluate the effects of driftnet fisheries on albatross diets we assessed the difference between the $\delta^{15}N$ value of the diet of albatrosses feeding from driftnets, or offal, and that of albatrosses that were not influenced by fisheries. Our first step was to calculate the predicted $\delta^{15}N$ values of the diets of the albatrosses in out study. To do this we subtracted the estimated trophic level fractionation (3.2‰) from the average $\delta^{15}N$ of each albatross species. We then used mass balance equations to estimate the composition of a driftnet-related diet (DRD) based on digestive tract contents of albatosses killed while feeding from driftnets. To estimate the diet of the birds in the absence of fisheries we made three assumptions concerning Neon Flying Squid and Pacific Pomfret: that they were the primary components of the last meal of the albatrosses, that they were the most common species captured by driftnets, and that free swimming individuals were too large to be captured by albatrosses. Thus, we excluded Neon Flying Squid and Pacific Pomfret from our data and reanalysed the results to estimate a non-driftnet related diet (NDRD). The composition of these two diets are approximations only, because some large squid and fish are likely to be taken by albatrosses when found in a moribund condition away from driftnets, and because smaller organisms may sometimes be captured in driftnets and scavenged by albatrosses.

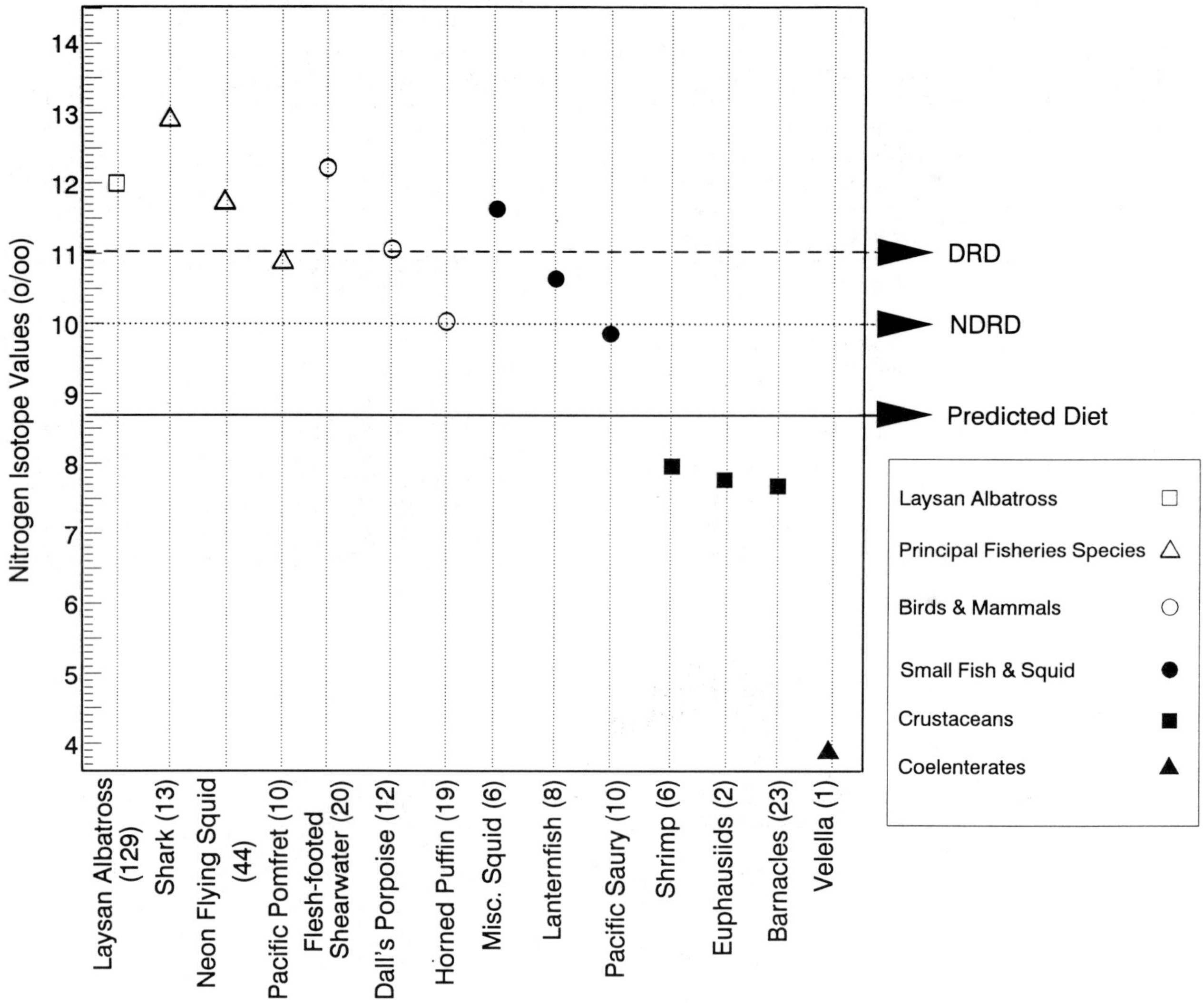

Fig. 4. Trophic relationship between the Laysan Albatross, other consumers, and driftnet fisheries in the central North Pacific (adapted in part from Table 3 in Gould *et al.* 1997). Nitrogen isotope values are positively correlated with trophic level. DRD = mean value for driftnet related diet; NDRD = mean value for non-driftnet related diet; Predicted Diet = $\delta^{15}N_{albatross} - 3.2$. Shark *Prionace* sp., Flesh-footed Shearwater *Puffinus carneipes;* Dall's Porpoise *Phocoenoides dalli;* Horned Puffin *Fratercula corniculata;* Lanternfish (Myctophidae); Pacific Saury *Cololabis saira;* Shrimp (Decapoda); Euphausiids (Euphausiacea); Barnacles *Lepas fascicularis.* Miscellaneous squid = 1 *Berryteuthis anonychus,* 2 *Histioteuthis dofleini,* 1 *Taonius pavo,* and 1 *Octopoteuthis deletron.*

The average $\delta^{15}N$ of the DRDs and NDRDs were determined from mass balance equations of Harrigan *et al.* (1989) and Ostrom and Fry (1993). The mass balance equations relate the isotopic composition of different food sources to that of the consumer:

$$\delta_{diet} = \sum_{i=1}^{j} f_j \delta_j$$

where $\delta_{diet} = \delta_{consumer}$ — estimate of shift in $\delta^{15}N$ between the consumer and its diet, f_j = the fractional contribution of the food item j, and δ_j = the isotopic composition of the food item

j, and $\sum_{i=1}^{j} f_j = 1$ (Harrigan *et al.* 1989; Ostrom and Fry 1993).

The contributions of prey to the diets were derived using the Index of Relative Importance of Pinkas *et al.* (1971) of the contents of digestive tracts of net-killed birds. To evaluate the implication of driftnet fisheries to Black-footed Albatrosses and Laysan Albatrosses, we compared the $\delta^{15}N$ data for the DRD, the NDRD and the prey items from the fishery (Figs 4 and 5). This showed that Laysan Albatrosses occupied the same trophic level as the main driftnet fisheries species (Fig. 4). The predicted $\delta^{15}N$ of an average Laysan Albatross diet (8.8‰) is intermediate between the $\delta^{15}N$ of small fish and squid (9.9–11.6‰) and crustaceans (7.8‰). The mean $\delta^{15}N$ of the NDRD is higher than that of the predicted diet (10.0‰), possibly because gelatinous invertebrates with low $\delta^{15}N$ values (e.g., *Velella* sp.) were under-represented in digestive tracts due to rapid digestion. The DRD is higher still (11.0‰), because it emphasized the high trophic level fisheries-related prey in Laysan Albatrosses diets. Since the DRD also under-represented invertebrates, the influence of

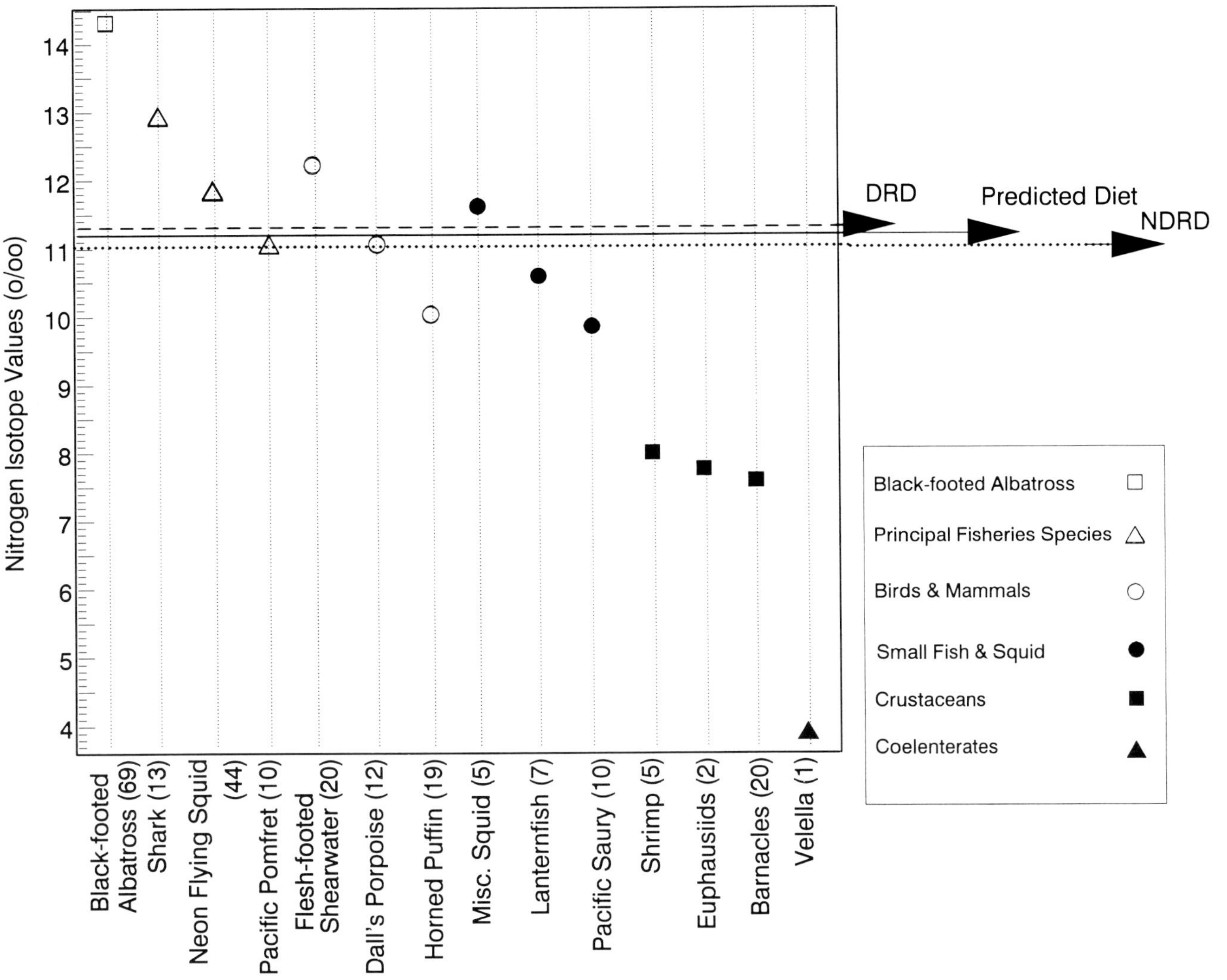

Fig. 5. Trophic relationship between the Black-footed Albatross, other consumers, and driftnet fisheries in the central North Pacific (adapted in part from Table 3 in Gould *et al.* 1997). Nitrogen isotope values are positively correlated with trophic level. DRD = mean value for driftnet related diet; NDRD = mean value for non-driftnet related diet; Predicted Diet = $\delta^{15}N_{albatross}$ − 3.2. Shark *Prionace* sp., Flesh-footed Shearwater; Dall's Porpoise; Horned Puffin; Lanternfish; Pacific Saury; Shrimp; Euphausiids; Barnacles. Miscellaneous squid = 1 *Berryteuthis anonychus*, 2 *Histioteuthis dofleini*, 1 *Taonius pavo*, and 1 *Octopoteuthis deletron*.

driftnets on albatross diets was best represented by the difference between the mean DRD and mean NDRD values rather than between the DRD and predicted values. This difference was about 1‰ and suggests that driftnets effectively raised the mean trophic level of Laysan Albatrosses diet by about one-third.

Of the taxa analysed, Black-footed Albatrosses occupied the highest trophic level within the central Pacific Ocean (Fig. 5). The $\delta^{15}N$ values for the estimated diets of the Black-footed Albatrosses we analysed, the predicted NDRD and DRD (11.2, 11.1, and 11.5, respectively), indicated a diet of large organisms such as those that may be caught and processed, or discarded, by driftnet fisheries (e.g., Neon Flying Squid, Pacific Pomfret, sharks, and even birds and mammals). This suggests that unlike Laysan Albatrosses, the trophic level of the diet of Black-footed Albatrosses is not elevated as a consequence of a fisheries interactions.

DISCUSSION

Laysan and Black-footed Albatrosses have evolved different feeding strategies that result in a separation of one full trophic level for non-breeding individuals within the central North Pacific Ocean. Small fish species are major constituents of non-breeding Laysan Albatrosses diets, especially in western parts of the transitional North Pacific Ocean. Squids are the main components of non-breeding Black-footed Albatrosses diets throughout the transitional area. These diets reflect Laysan Albatrosses and Black-footed Albatrosses distribution patterns. The spatial coincidence in fish consumption and abundance of Laysan Albatrosses (both highest abundance in the western sector of the Pacific Ocean) suggest that preferred food resources may be a mechanism for concentrating these birds in a specific location. This possibility would be supported if the preferred food of

Black-footed albatrosses (medium and large squids) are more available in the eastern Pacific Ocean.

The advent of high seas driftnet fisheries in the North Pacific Ocean had large impacts on Laysan and Black-footed Albatrosses through entanglement of birds in fishing gear and supplementation of food resources. Although the mortality problem has been well documented (Johnson *et al.* 1993; Gould and Hobbs 1993), the food supplementation issue has remained speculative and unquantified until now. We have shown that non-breeding Laysan Albatrosses and Black-footed Albatrosses scavenged considerable amounts of squid and fish directly from driftnets and from offal discards. Although our data lack precision due to small sample sizes, they suggest that Black-footed Albatrosses and Laysan Albatrosses scavenged similar percentages of net-caught squid and fish. Assessments of trophic relationships based on stable nitrogen isotope values indicate that food scavenged from driftnets was more typical of Black-footed Albatrosses diets than of Laysan Albatrosses diets. Food supplements from driftnets may thus have had a greater impact on Laysan Albatrosses than on Black-footed Albatrosses. Black-footed Albatrosses appeared to be more prone to scavenging than Laysan Albatrosses, and are top predators/scavengers in the local ecosystem. Over the long term, Black-footed Albatrosses probably used the fisheries more frequently than Laysan Albatrosses.

Food supplementation by commercial fisheries may mitigate incidental mortality caused by those fisheries, especially for birds inexperienced at foraging, and for all birds during periods of low natural food abundance. With the cessation of driftnetting in 1992, and the refocusing of much of that effort into longline fishing, a new problem is developing. Longline fisheries kill many albatrosses (Alexander *et al.* 1997) but, due to differences in fishing methods, may not be providing the large quantities of supplemental food that were once available from driftnet fisheries. It is also possible that the attraction of Laysan Albatrosses and Black-footed Albatrosses to these fisheries has altered pre-fishery distribution patterns of the birds. The balance of positive and negative impacts of these fisheries on Laysan Albatrosses and Black-footed Albatrosses populations is unknown, complex and probably in a continual state of flux.

ACKNOWLEDGEMENTS

Funding for this work was provided by the National Marine Fisheries Service and US Fish and Wildlife Service. We thank Linda Jones and Kenton Wohl of those organizations for their support. Larry Spear sorted and weighed all of the digestive tracts and Michael Newcomer assisted with the *Shoyo Maru* bird survey work. We thank Shannon Fitzgerald, Keith Hobson, Vivian Mendenhall, Nathaniel Ostrom, Graham Robertson, and Barbara Wienecke for their reviews and constructive criticisms of the manuscript.

REFERENCES

Alexander, K., Robertson, G. and Gales, R., 1997. The incidental mortality of albatrosses in longline fisheries. Australian Antarctic Division, Tasmnia. 44 pp.

Ainley, D. G. and Sanger, G. A., 1979. Trophic relations of seabirds in the northern Pacific Ocean and Bering Sea. Pp. 95–122 *in* Conservation of Marine Birds of Northern North America ed by J. C. Bartonek and D. N. Nettleship. U.S. Dept. Int., U.S. F.W.S., Wildlife Research Report 11.

Anderson, W. G., 1954. Notes on food habits of sea birds of the Pacific. *Elepaio* **14:** 80–84.

Brazil, M. A., 1991. The birds of Japan. Smithsonian Institute Press: Washington, D.C.

Briggs, K. T., Tyler, W. B., Lewis, D. B. and Carlson, D. R., 1987. Bird communities at sea off California: 1975 to 1983. *Studies Avian Biol.* **11:** 1–74.

Cherel, Y. and Klages, N., 1997. A review of the food of albatrosses. Pp. 113–36 *in* Albatross Biology and Conservation ed by G. Robertson and R. Gales. Surrey Beatty & Sons: Chipping Norton.

Cottam, C. and Knappen, P., 1939. Food of some uncommon North American birds. *Auk* **56:** 138–69.

Croxall, J. P. and Gales, R., 1997. An assessment of the conservation status of albatrosses. Pp. 46–65 *in* Albatross Biology and Conservation ed by G. Robertson and R. Gales. Surrey Beatty & Sons: Chipping Norton.

DeNiro, M. J. and Epstein, S., 1981. Influence of diet on the distribution of nitrogen isotopes in animals. *Geochim. Cosmochim. Acta* **45:** 341–51.

Fefer, S. I., Harrison, C. S., Naughton, B. N. and Shallenberger, J. S., 1984. Synopsis of results of recent seabird observations in the Northwestern Hawaiian Islands. Pp. 9–76 *in* Resource Investigations in the Northwestern Hawaiian Islands, Vol. I ed by R. W. Grigg and K. Y. Tanoue. Univ. Hawaii Sea Grant College Program: Honolulu, Hawaii.

Fisher, H. I., 1945. Black-footed Albatrosses eating flyingfish. *Condor* **47:** 128–29.

Fisher, H. I. and Fisher, L. F., 1969. The visits of Laysan albatross to the breeding colony. *Micronesica* **5:** 173–221.

Fisher, H. I. and Fisher J. R., 1972. The oceanic distribution of the Laysan albatross, *Diomedea immutabilis. Wilson Bull.* **84:** 7–27.

Gould, P. J. and Forsell, D. J., 1989. Techniques for shipboard surveys of marine birds. U.S. Dept. Inter., Fish and Wildl. Serv. Tech. Rept. 25.

Gould, P. J., Forsell, D. J. and Lensink, C. J., 1982. Pelagic distribution and abundance of seabirds in the Gulf of Alaska and eastern Bering Sea. U.S. Fish and Wildl. Serv. Off. Biol. Serv. FWS/OBS–82/48: Washington D.C.

Gould, P. J. and Hobbs, R., 1993. Population dynamics of the Laysan and other albatrosses in the North Pacific. Pp. 485–97 *in* Symposium on Biology, Distribution and Stock Assessment of Species Caught in the High Seas Driftnet Fisheries in the North Pacific Ocean ed by J. Itô, W. Shaw and R. L. Burgner. Int'l. No. Pac. Fish. Comm., Bull. No. 53, Vol. 3, Vancouver, Canada.

Gould, P., Ostrom, P. and Walker, W., 1997. Trophic relationships of albatrosses associated with squid and large-mesh drift-net fisheries in the North Pacific Ocean. *Can. J. Zool.* **75:** 549–62.

Gould, P. J. and Piatt, J. F., 1993. Seabirds of the central North Pacific. Pp. 27–38 *in* The Status, Ecology, and Conservation of Marine Birds of the North Pacific ed by K. Vermeer, K. T. Briggs, K. H. Morgan and D. Siegel-Causey. Can. Wildl. Serv. Spec. Publ., Ottawa.

Harrigan, P., Zieman, J. C. and Macko, S. A., 1989. The base of nutritional support for the gray snapper *Lutjanus griseus*: an evaluation based on a combined stomach content and stable isotope analysis. *Bull. Mar. Sci.* **44:** 65–77.

Harrison, C. S., 1990. Seabirds of Hawaii: natural history and conservation. Cornell Univ. Press: New York.

Harrison, C. S., Hida, T. S. and Seki, M. P., 1983. Hawaiian seabird feeding ecology. *Wildl. Monogr. No.* **85** 1–71.

Hobson, K. A. and Clark, R. C., 1992. Assessing avian diets using stable isotopes II: factors influencing diet-tissue fractionation. *Condor* **94:** 189–97.

Hobson, K. A., Piatt, J. F. and Pitocchelli, J., 1994. Using stable isotopes to determine seabird trophic relationships. *J. Anim. Ecol.* **63:** 786–98.

Johnson, D. H., Shaffer, T. L. and Gould, P. J., 1993. Incidental catch of marine birds in the North Pacific high seas driftnet fisheries in 1990. Pp. 473–83 *in* Symposium on Biology, Distribution and Stock Assessment of Species Caught in the High Seas Driftnet Fisheries in the North Pacific Ocean, Int'l. No. Pac. Fish. Comm., Bull. ed by J. Itô, W. Shaw, R. L. Burgner. No. 53, Vol. 3, Vancouver, Canada.

Kenyon, K. W. and Kridler, E., 1969. Laysan albatross swallow indigestible matter. *Auk* **86:** 339–43.

Kuroda, N., 1988. A distributional analysis of *Diomedea immutabilis* and *D. nigripes* in the north Pacific. *J. Yamashina Inst. Ornithol.* **20:** 1–20.

Kuroda, N., 1991. Distributional patterns and seasonal movements of procellariiformes in the North Pacific. *J. Yamashina Inst. Ornithol.* **23:** 23–84.

McDermond, D. K. and Morgan, K. H., 1993. Status and conservation of North Pacific albatrosses. Pp. 70–81 *in* The Status, Ecology, and Conservation of marine birds of the North Pacific, Can. Wildl. Serv. Spec. Publ., Ottawa, ed by K. Vermeer, K. T. Briggs, K. H. Morgan, D. Siegel-Causey.

Michener, R. H. and Schell, D. M., 1994. Stable isotope ratios as tracers in marine aquatic food webs. Pp. 138–57 *in* Stable Isotopes in Ecology and Environmental Science ed by K. Lajtha and R. H. Michener. Blackwell Scientific: Oxford.

Miller, L., 1940. Observations of the black-footed albatross. *Condor* **42:** 229–38.

Minagawa, M. and Wada, E., 1984. Stepwise enrichment of δ^{15}N along food chains: further evidence and the relation between δ^{15}N and animal age. *Geochim. Cosmochim Acta* **48:** 1135–140.

Ostrom, P. H. and Fry, B., 1993. Sources and cycling of organic matter within modern and prehistoric food webs. Pp. 785–98 *in* Organic Geochemistry. Plenum Press: New York.

Pinkas, L., Oliphant, M. S. and Iverson, I. L. K., 1971. Food habits of albacore, bluefin tuna, and bonito in California waters. Pp. 5–10 *in* Food habits of albacore, bluefin tuna and bonito in Californian waters ed by L. Pinkas, M. S. Oliphant and I. L. K. Iverson. Calif. Fish and Game, Fish Bull. 152.

Prince, P. A., 1980. The food and feeding ecology of grey-headed albatross *Diomedea chrysostoma* and black-browed albatross *D. melanophris*. *Ibis* **122:** 476–88.

Rice, D. W. and Kenyon, K. W., 1962a. Breeding distribution, history, and populations of North Pacific albatrosses. *Auk* **79:** 365–86.

Rice, D. W. and Kenyon, K. W., 1962b. Breeding cycles and behavior of Laysan and black-footed albatrosses. *Auk* **79:** 5117–567.

Robbins, C. S. and Rice, D. W., 1974. Recoveries of banded Laysan albatrosses *Diomedea immutabilis* and Black-footed Albatrosses *D. nigripes*. Pp. 232–77 *in* Pelagic Studies of Seabirds in the Central and Eastern Pacific Ocean ed by W. B. King. Smithsonian Contrib. to Zool. No. 158, Smithsonian Institute Press: Washington, D.C.

Roden, G. I., 1980. On the subtropical frontal zone north of Hawaii during winter. *J. Phys. Oceanogr.* **10:** 342–62.

Roden, G. I., 1991. Subarctic-subtropical transition zone of the North Pacific: large-scale Aspects and mesoscale structure. Pp. 1–38 *in* Biology, Oceanography, and Fisheries of the North Pacific Transition Zone and Subarctic Frontal Zone ed by J. A. Wetherall. NOAA Tech. Rep. NMFS 105, U.S. Dept. Commerce.

Roden, G. I. and Robinson, A. R., 1988. Subarctic frontal zone in the northeastern Pacific: mesoscale structure and synoptic description. Harvard Open Ocean Model Reports, *Meteorol. Oceanogr.* **31:** 1–27.

Sanger, G. A., 1974a. Black-footed Albatross *Diomedea nigripes*. Pp. 96–128 *in* Pelagic Studies of Seabirds in the Central and Eastern Pacific Ocean ed by W. B. King. Smithsonian Contrib. to Zool. No. 158. Smithsonian Institute Press: Washington, D.C.

Sanger, G. A., 1974b. Laysan Albatross *Diomedea immutabilis*. Pp. 129–53 *in* Pelagic Studies of Seabirds in the Central and Eastern Pacific Ocean ed by W. B. King. Smithsonian Contrib. to Zool. No. 158. Smithsonian Institute Press: Washington, D.C.

Shuntov, V. P., 1974. Sea birds and the biological structure of the ocean. From: Dalnevostrochnoe Knizhnoe Izdatelstvo, Vladivostok, 1972. Translated for Bureau of Sport, Fish. and Wildl., U.S. Dept. Inter., and Nat'l. Sci. Found., Washington, D.C., by the Agence Tunisienne de Public-Relations, Tunis, Tunisia, 1974, Nat'l. Tech. Info. Serv. No. TT-74-55032.

Sileo, L., Sievert, P. R., Samuel, M. D. and Fefer, S. I., 1990. Prevalence and characteristics of plastic ingested by Hawaiian seabirds. Pp. 665–81 *in* Proceedings of the Second International Conference on Marine Debris, 2–7 April 1989, Honolulu, Hawaii ed by R. S. Shomura and M. L. Godfrey. U.S. Dept. Commer., NOAA Tech. Memo. NMFS, NOAA-TM-NMFS-SWFSC-1154.

Tieszen, L. L., Boutton, T. W., Tesdahl, K. G. and Slade, N. A., 1983. Fractionation and turnover of stable carbon isotopes in animal tissues: Implications for δ^{13}C analysis of diet. *Oecologia (Berlin)* **57:** 32–37.

Thompson, K. R., 1992. Quantitative analysis of the use of discards from squid trawlers by black-browed albatrosses *Diomedea melanophris* in the vicinity of the Falkland Islands. *Ibis* **134:** 11–21.

Wada, E., Miagawa, M., Mizutani, H., Tsuji, T., Imaizumi, R. and Karasawa, K., 1987. Biogeochemical studies on the transport of organic matter along the Otsuchi River Watershed, Japan. *Estuarine Coastal Shelf Sci.* **25:** 321–36.

Wahl, T. R. and Heinemann, J., 1979. Seabirds and fishing vessels: occurrence and attractions. *Condor* **81:** 390–96.

Weimerskirch, H., Bartle, J. A., Jouventin, P. and Stahl, J. C., 1988. Foraging ranges and partitioning of feeding zones in three species of southern albatrosses. *Condor* **90:** 214–19.

Yatsu, A., Hiramatsu, K. and Hayase, S., 1993. Outline of the Japanese squid driftnet fishery with notes on the by-catch. Pp. 5–24 *in* Symposium on Biology, Distribution and Stock Assessment of Species Caught in the High Seas Driftnet Fisheries in the North Pacific Ocean, Int'l. No. Pac. Fish. Comm., Bull. No. 53, Vol. 3, Vancouver, Canada ed by J. Itô, W. Shaw, R. L. Burgner.

CHAPTER SIXTEEN

Albatross–fisheries interactions in Patagonian shelf waters

ADRIÁN SCHIAVINI,[1,2,3] ESTEBAN FRERE,[3,4] PATRICIA GANDINI,[3,4] NESTOR GARCÍA[2,5] and ENRIQUE CRESPO[2,3,5]

Patagonian waters attract several species of albatrosses during their breeding and post-breeding periods: Wandering, Royal, Black-browed and Grey-headed Albatrosses are the most important of them. The main fisheries that take albatrosses as bycatch in Patagonian waters are longline fisheries for Patagonian Toothfish and trawl fisheries for Hake and shrimp. Twenty-four trawlers are reported to use net monitor cables which are known to kill albatrosses during collisions. Of the 19 longline vessels currently operating from Patagonian harbours, 17 vessels set lines manually and two set lines automatically (autoline); only the autoline vessels use seabird bycatch mitigation measures and bird scaring lines and streamers are the only measure currently being adopted. During the autumn of 1995 manual-set longline vessels commenced fishing in the CCAMLR area, which commences 650 nm east of the Argentine coast. For the 18 month period between 1994 and mid 1995 a total of 3 972 longline fishing operations (sets and hauls) took place in Patagonian shelf waters and the CCAMLR area combined. Of the 2 052 sets that took place in Patagonian waters 1 921 were by autoline vessels and 131 were by manual-set vessels; an estimated 25 386 000 hooks were deployed for the period reported for both areas combined. Seabird catch rates reported for vessels fishing in CCAMLR waters (0.19–0.67 birds/1 000 hooks) applied to vessels in Patagonian waters suggest that about 3 832–13 514 and 261–922 birds would have been killed by manual-set longline fisheries in Patagonian waters and in the CCAMLR area, respectively, during 1994–mid 1995.

Key words: Wandering Albatross, Black-browed Albatross, Grey-headed Albatross, Longline Fishing, Trawl Fishing, Argentina, Patagonian Shelf Waters, Patagonian Toothfish.

INTRODUCTION

CONCERN has emerged in recent years about the effect of longline fisheries on albatrosses and other seabird species (Brothers 1991; Alexander *et al.* 1997). Recently this concern has extended to the longline fisheries in Brazilian, Uruguayan and Patagonian waters (collectively, the south-western Atlantic Ocean) where large numbers of longline vessels operate. There is a dearth of information on seabird mortality rates in longline fisheries in Patagonian waters, which lie south of 40°S, and expected catch rates must be extrapolated from rates reported for vessels fishing in the area regulated by the Convention for the Conservation of Antarctic Marine Living Resources (CCAMLR), which commences about 650 nautical miles east of the Argentine coast. Potentially a greater number of birds may be killed by longline fisheries in waters outside the CCAMLR area than inside the area because of the greater number of hooks being set outside the area (CCAMLR 1995).

In this chapter we report briefly on the albatrosses species that frequent Patagonian shelf waters, and the spatial distribution of trawl and longline fishing effort in these waters. We use the seabird catch rates reported for longline fisheries in CCAMLR waters to derive an approximation of the seabird bycatch rates that could be expected for Patagonian waters. We also discuss the relationships between fisheries and the seabird communities of the southwestern Atlantic Ocean and the conservation needs of seabirds in this area.

ALBATROSS DISTRIBUTION IN PATAGONIAN SHELF WATERS

The seabird community has been described previously (Murphy 1936; Cooke and Mills 1972; Tickell and Woods 1972; Brown *et al.* 1975; Veit 1995). In the present study, we deal only with Wandering *Diomedea exulans exulans,* Royal *Diomedea* spp., Black-browed *D. (Thalassarche) melanophrys melanophrys* and

[1]Centro Austral de Investigaciones Científicas (CONICET), CC 92, (9410) Ushuaia, Argentina.
[2]Universidad Nacional de la Patagonia San Juan Bosco, Argentina.
[3]Patagonian Coastal Zone Management Plan, Fundación Patagonia Natural. Marcos Zar 760, (9120) Puerto Madryn, Argentina.
[4]Universidad Federal de la Patagonia Austral, Argentina.
[5]Centro Nacional Patagónico (CONICET), Boul. Brown s/nro. (9120) Puerto Madryn, Argentina.

Pages 208–13 *in* ALBATROSS BIOLOGY AND CONSERVATION, ed by G. Robertson and R. Gales. Surrey Beatty & Sons, Chipping Norton. 1997.

Grey-headed *D. (Thalassarche) chrysostoma* Albatrosses which occur commonly in Patagonian shelf waters and have been reported as bycatch in fishing operations in other parts of the Southern Ocean (Dalziell and de Porter 1993; Jones and Parkes 1994; Cherel *et al.* 1994; Ashford *et al.* 1994a,b).

For Wandering and Royal Albatrosses, the waters of the southwestern Atlantic Ocean are an important dispersion area. Band recoveries from South Georgia Wandering Albatrosses indicate the birds disperse to the southwestern Atlantic Ocean off Patagonia (Croxall and Prince 1990; Prince *et al.* 1997). Satellite telemetry has supported this finding and suggests possible sexual segregation, males frequenting the southern part of the south-west Atlantic Ocean, while females move to the mouth of Río de la Plata at about 36°S (Prince *et al.* 1992; Prince *et al.* 1997). Royal Albatrosses are distributed widely over the Patagonian shelf. Post-breeding dipersal from their breeding grounds in New Zealand includes the southwestern Atlantic Ocean, and juveniles may remain in Patagonian waters for several years before returning to their breeding grounds (Gales 1993). Royal Albatrosses may move up the east coast of South America as far as the mouth of the Río de la Plata (Watson *et al.* 1971).

Black-browed Albatrosses are the commonest albatross species in Patagonian waters. The Islas Malvinas (Falkland Islands) (260 nautical miles east of the Argentina coast) breeding populations represents 86% of the world population, or about 585 000 pairs (deduced from Croxall and Gales 1997). Black-browed Albatrosses are commonly seen in the waters adjacent to Isla de los Estados (about 55°S; 64°W) and Isla Grande de Tierra del Fuego (52–55°S; 65–72°W) (Schiavini *et al.* 1994). This species has been declining since 1990 at both Iles Kerguelen (Weimerskirch and Jouventin 1997) and South Georgia (Croxall *et al.* 1997). At Islas Malvinas (Falkland Islands), the Black-browed Albatross population may be increasing (Croxall and Gales 1997) possibly due to the abundance of offal and discards from trawl fisheries (Thompson and Riddy 1995).

Several colonies of Black-browed Albatrosses are distributed near fishing grounds in the south-west Atlantic Ocean (Olrog 1950; Croxall *et al.* 1984; Schlatter 1984). The Islas Malvinas (Falkland Islands) populations are thought to remain close to the islands during the breeding season, and to disperse to waters off the east coast of South America after the breeding season (Tickell, pers. comm. in Gales 1993). Other colonies in the southwestern South Atlantic Ocean occur in Chilean waters near the Fuegian Archipelago (52–56°S; 64–73°W): Islas Diego Ramírez, Islas Ildefonso and Islas Evout (Olrog 1950; Watson *et al.* 1971; Schlatter 1984). About 20 000 pairs of Black-browed Albatrosses breed at Islas Diego Ramírez (Schlatter 1984); population sizes at Islas Ildefonso and Islas Evout are unknown.

Grey-headed Albatrosses have two important breeding grounds near the Patagonian shelf: Islas Diego Ramírez (about 10 000 pairs, Schlatter 1984) and South Georgia (about 100 000 pairs [deduced from Croxall and Gales 1997]). The origin of Grey-headed Albatrosses that visit Patagonian waters has not been determined, but most likely includes birds from both Islas Diego Ramírez and South Georgia.

FISHING TYPE AND FISHING EFFORT

General

The south-west Atlantic Ocean off Patagonia has traditionally been an important fishing ground for Argentine Hake *Merluccius hubbsii*, Argentine Short-fin Squid *Illex argentinus*, Shrimp *Pleoticus muelleri* and Southern Blue Whiting *Micromesistius australis*. In the last 10 years the fishing effort in this area has increased through joint ventures between Argentina and foreign companies. About 70% of the ships in the Argentine fishing fleet trawl in this area (Crespo *et al.* 1995; E. Crespo, unpubl. data) for Argentine Hake, shrimp and Southern Blue Whiting. Longline fishing for Hake and Kingclip *Genypterus blacodes* commenced in these waters in 1992 and since 1994 the fishery has shifted to Patagonian Toothfish *Dissostichus eleginoides* which is caught over the shelf break and in deeper waters between 54°–57°S. Currently 19 longline ships (9% of the national fishing fleet) operate in these waters: two use the autoline method of deploying lines and the remainder set lines manually, adopting the Spanish system of demersal fishing as described by Ashford *et al.* (1994a).

Trawling

In northern and central Patagonian waters between 1992 to 1994 the trawling fleet comprised 149 vessels and the fishery operated for a total of 33 103 days/year (Crespo *et al.* 1995) (Fig. 1). Twenty-four of these vessels used net monitor cables which are known to kill albatrosses (Bartle 1992). In southern Patagonian waters 24 trawlers targetting Southern Blue Whiting for surimi (fish paste) in the 18 month period between 1994 and mid 1995 made 4 370 trawlings operations involving 25 179 hours of trawling. None of these ships used net monitor cables.

Longlining

The fishing declarations of 12 of the 19 demersal (bottom-set) longline ships based in Patagonian harbours were surveyed for

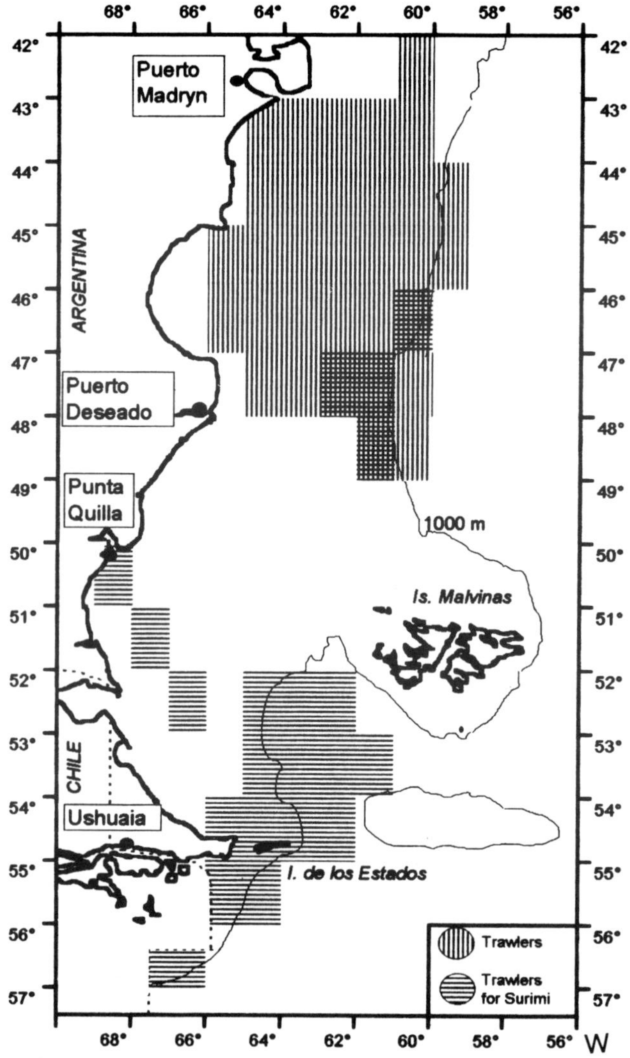

Fig. 1. Spatial distribution of the trawl fishing effort in Patagonian waters by vessels based in Patagonian harbours.

information on fishing locations, fishing methods and number of lines set, the latter being to gain a measure of potential seabird bycatch rates. The survey period spanned the 18 months between 8 December 1993 and 17 July 1995 and included fishing activities in all calendar seasons. No information was available on times of line setting (day or night). The 12 ships made a total of 74 fishing trips of up to 50 days/trip; of the 12 ships 10 were manual-set vessels (61 trips) and two were autoline vessels (13 trips). We assumed the number of hooks set was the same as reported by the CCAMLR for these types of vessels: 10 500 hooks/set for manual-set vessels and 2 000 hooks/set for autoline vessels. A total of 3 972 fishing operations (settings and haulings) was performed from the 12 ships based in Patagonian harbours in the 18 months, involving the deployment of about 25 386 000 hooks. The manual operations totalled 2 052 (1 921 in the southwestern South Atlantic and 131 in the CCAMLR area) and the autoliner operations totalled 1 920 sets in the south-west Atlantic Ocean only. The total number of hooks set was more-or-less constant through 1994 until the summer 1994/1995, when there was an increase in the number of manual-set operations. In Patagonian waters the fishing effort covered two main areas of the shelf and shelf break: from 43°–50°S and 53°–57°S; Fig. 2). The fishery in the southern portion of the fishing area comprised 64% of the total number of hooks set.

SEABIRD MORTALITY IN FISHING OPERATIONS

Trawling

No data are available on seabird mortality by trawlers operating in Patagonian waters. However, albatrosses and other seabird species have been killed in collisions with net sonde cables used by trawlers in New Zealand waters and around Iles Kerguelen (Bartle 1991; Duhamel 1991) and seabirds in Patagonian waters are likely to be similarly affected. Interviews with fishing captains, who have identified seabirds killed from identification booklets, reveal that several seabird species have been reported colliding with net sonde cables, including Black-browed Albatrosses, gulls *Larus* spp., terns *Sterna* spp., White-chinned Petrels *Procellaria aequinoctialis* and Giant Petrels *Macronectes* spp. The continued use of net sonde cables can be expected to result in continued collisions, and deaths, of these and other species of seabird.

Longlining

Anecdotal reports of bycatch are available for five of the 12 longline ships surveyed. The reports of catch rates are highly variable, and include estimates of up to "50 birds/day" and "one bird/100 hooks". Two captains from the same vessel reported that birds "learned" to avoid being hooking, because the number of birds caught has decreased in recent years (sic).

In the absence of bycatch data acquired by dedicated bird observers on vessels, we can only provide a crude estimation of the potential incidental mortality of seabirds in longline fisheries, based catch rates in longline fisheries reported for the CCAMLR area (CCAMLR 1995). We took as minimum and maximum values the rates of 0.19 birds/1 000 hooks (Jones and Parkes 1994) and 0.67 birds/1 000 hooks (Dalziell and de Porter 1993). Only catch rates for manual-set vessels were used, because we did not have access to catch rates from autoline vessels. We derived a range of implied total mortality of 3 832–13 514 birds for the Patagonian shelf and 261–922 birds for the CCAMLR area, for the 18 month period considered, which is equivalent to an annual rate of about 2 387–8 418 and 1 096–3 866 for the

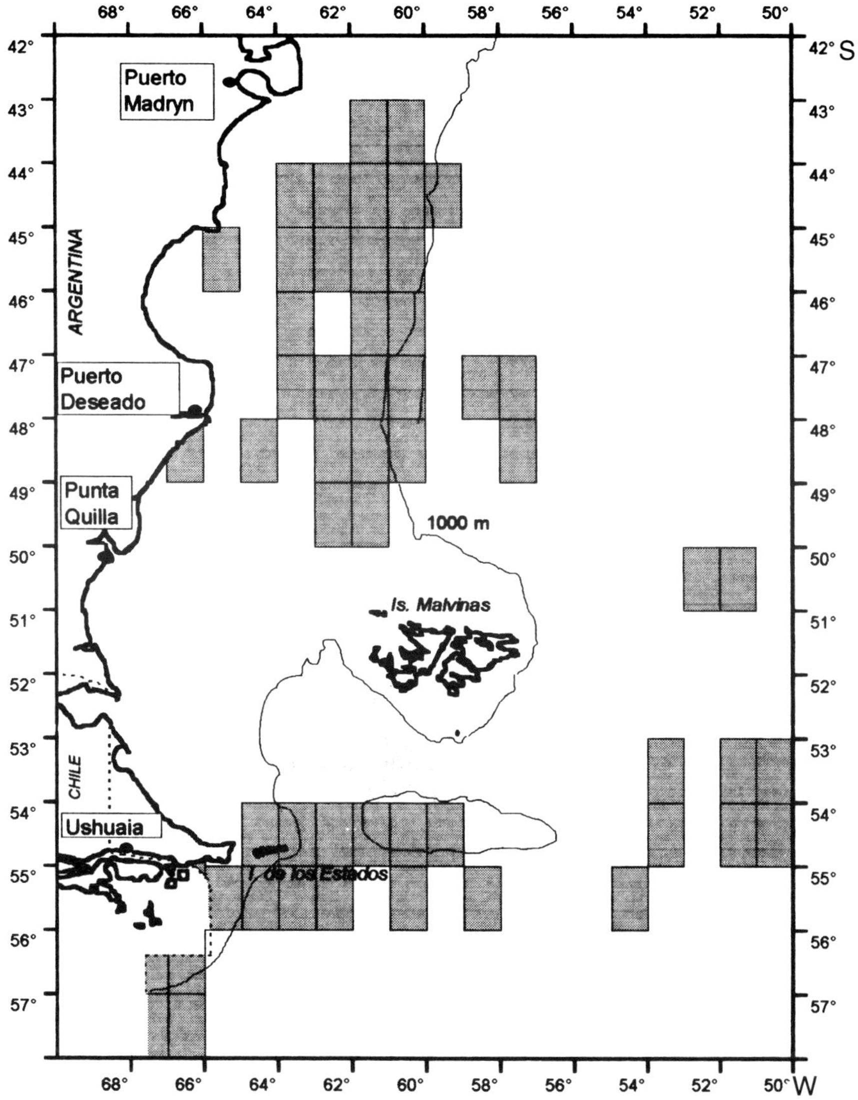

Fig. 2. Spatial distribution in longline fishing effort in Patagonian waters by vessels based in Patagonian harbours.

two areas, respectively. These are preliminary estimates because we lack precise information on the differences in fishing effort, fishing method, seabird abundance and species composition of bycatch between CCAMLR waters and waters of the south-west Atlantic Ocean. Based on its abundance in the area the Black-browed Albatross is probably the commonest species taken by longline vessels fishing for Patagonian Toothfish over the Patagonian shelf and shelf break (M. Bingham, pers. comm.).

CONSERVATION OF ALBATROSSES IN ARGENTINE WATERS

Legal aspects

Each province of Argentina is responsible for the management of their fisheries and wildlife resources up to 3 nautical miles offshore. Offshore fisheries, as well as their environmental related issues, are managed by the national government (Secretaría de Agricultura, Alimentación Pescă). National fisheries legislation is in the process of changing, but at present existing laws do not include issues relating to seabird bycatch. Only the new fishing law of Tierra del Fuego (provincial law 244) gives authority to minimize bycatch of seabirds and marine mammals, but this law does not apply to fisheries outside the 3 nautical mile limit which is managed by the national legislation.

Monitoring seabird-fisheries interactions and mitigation measures

Although dedicated seabird observers are being placed only on fishing vessels in the CCAMLR area the longlining Toothfish fishery in Patagonian waters remains unmonitored. At present the manual set longline vessels in Patagonian waters do not use methods to reduce seabird bycatch and autoline vessels use a single measure only — a simplified, and probably less effective, version of the bird scaring lines and streamers recommended by CCAMLR. The captains of ships using manual operations assume that bycatch is part of the "natural" process of fishing and most do not think bycatch reduction is necessary. Recently, the national fisheries administration (Secretaría de Agricultura Ganadería y Pesca) recommended the use of larger hooks (size code 08 of the CCAMLR code) in order to reduce bycatch of seabirds (CCAMLR Resolution 377/95). Clearly much more need to be done to minimize seabird/ fisheries interactions in Patagonian waters.

Conservation needs

The information presented in this chapter highlights the importance of the Patagonian shelf as an area of potentially high mortality of albatrosses due to interactions with trawl and longline fisheries. This calls attention to the need for co-ordinated management of the relationship between fisheries and seabirds, which is complex. Conservation needs of relevance to fisheries in Patagonian waters that should be adopted immediately are listed in Alexander *et al.* (1997) and in Conservation Measures 29/XV and 30/X of the CCAMLR (1996) and include:

1. The establishment of a programme of dedicated seabird observers on vessels fishing in Patagonian waters.
2. The establishment of a legal framework for the reduction of seabird bycatch, by means of mitigation measures already available for longline fisheries, such as bird scaring lines and streamers, night setting, discharge of offal to avoid attract birds to the line hauling site, weighted lines and the use of thawed bait. Net monitoring cables should be prohibited in trawl fisheries.
3. Provision of educational material to fishermen, such as seabird identification guides and manuals to educate fishermen about seabirds, the impact of seabirds on their fishing efficiency (via bait stealing) and methods to minimize incidental bycatch of seabirds.

ACKNOWLEDGEMENTS

We are grateful to the organizers of the inaugural albatross conference for supporting our attendance at the conference and for inviting us to present this chapter. Fundación Patagonia Natural funded this research through the Patagonian Coastal Zone Management Plan (project GEF/UNDP). Dirección de Recursos Naturales from Tierra del Fuego provided statistical data on fishing effort. Natalie Goodall reviewed part of this manuscript.

REFERENCES

Alexander, K., Robertson, G. and Gales, R., 1997. The incidental mortality of albatrosses in longline fisheries. Australian Antarctic Division, Tasmania. 44 pp.

Ashford, J. R., Croxall, J. P., Rubilar, P. S. and Moreno, C. A., 1994a. Seabird interactions with longlining fishing operations for *Dissostichus eleginoides* at the South Sandwich Islands and South Georgia. *CCAMLR Science* **1:** 143–53.

Ashford, J. R., Croxall, J. P., Rubilar, P. S. and Moreno, C. A., 1994b. Seabird interactions with longlining operations for *Dissostichus eleginoides* around South Georgia. Paper WG-IMALF-94/15

Bartle, J. A., 1991. Incidental capture of seabirds in the New Zealand and sub-Antarctic squid trawl fishery, 1990. *Bird. Conserv. Internat.* **1:** 351–59.

Brothers, N. P., 1991. Albatross mortality and associated bait loss in the Japanese longline fishery in the Southern Ocean. *Biol. Conserv.* **55:** 255–68.

Brown, R. G. B., Cooke, F., Kinnear, P. K. and Mills, E. L., 1975. Summer seabird distributions in Drake Passage, the Chilean fjords and off southern South America. *Ibis* **117:** 339–56.

CCAMLR, 1995. Information on CCAMLR measures to prevent incidental mortality of seabirds during longline fishing operations. CCAMLR Secretariat, Hobart, Tasmania, Australia.

CCAMLR, 1996. Schedule of conservation measures in force 1996/97. CCAMLR Secretariat, Hobart, Tasmania, Australia.

Cherel, Y., Weimerkirsch, H. and Duhamel, G., 1994. Interactions between longline vessels and seabirds in Kerguelen waters and a method to reduce seabird mortality. Paper WG-IMALF-94/12.

Cooke, F. and Mills, E. L., 1972. Summer distribution of pelagic birds off the coast of Argentina. *Ibis* **114:** 245–51.

Crespo, E. A., Pedraza, S. N., Dans, S. L., Alonso, M. K., Reyes, L. M., García, N. A., Coscarella, M. and Schiavini, A. C. M., 1995. Direct and indirect effects of the high seas fisheries on the marine mammal populations in the northern and central Patagonian coast. *In* Evaluation in Symposium "The role of Marine Mammals in the Ecosystem". Dartmouth, Nova Scotia, Canada, 6–8 September 1995. Special Issue of North Atlantic Fishery Organization (NAFO) Scientific Journal.

Croxall, J. P., McInnes, S. J. and Prince, P. A., 1984. The status and conservation of seabirds at the Falkland Islands. Pp. 271–91 *in* Status and Conservation of the World's Seabirds ed by J. P. Croxall, P. G. H. Evans and R. W. Schreiber: International Council of Birds Preservation, Technical Publication No. 2.

Croxall, J. P. and Gales, R., 1997. An assessment of the conservation status of albatrosses. Pp. 46–65 *in* Albatross Biology and Conservation ed by G. Robertson and R. Gales. Surrey Beatty & Sons: Chipping Norton.

Croxall, J. P., Prince, P. A., Rothery, P. and Wood, A. G., 1997. Population changes in albatrosses at South Georgia. Pp. 69–83 *in* Albatross Biology and Conservation ed by G. Robertson and R. Gales. Surrey Beatty & Sons: Chipping Norton.

Dalziel, J. and de Porter, M., 1993. Seabird mortality in longline fisheries around South Georgia. *Polar Record* **29** (169): 143–45.

Duhamel, G., 1991. Incidental mortality rising from fisheries activities around Kerguelen island (Division 58.5.1). Scientific Commission for the Conservation of Antarctic marine Living Resources. SC-CCAMLR-X/BG/14.

Gales, R., 1993. Co-operative mechanisms for the conservation of Albatross. Australian Nature Conservation Agency. 132 pp.

Jehl, J. R., 1974. The distribution and ecology of marine birds over the continental shelf of Argentina in winter. *Trans. San Diego Soc. Nat. Hist.* **17:** 217–34.

Jones, C. and Parkes, G., 1994. Report on incidental mortality and effectiveness of mitigation measures during demersal longlining by IHN Sung 66 in Subarea 48.3–December 1993 to February 1994. Paper WG-IMALF-94/15.

Murphy, R. C., 1936. Oceanic Birds of South America. American Museum of Natural History: New York.

Olrog, C. C., 1950. Mamíferos y aves del Archipiélago de Cabo de Hornos. *Acta Zoológica Lilloana* **9:** 505–32.

Prince, P. A., Wood, A. G., Barton, T. and Croxall, J. P., 1992. Satellite tracking of wandering albatrosses *Diomedea exulans* in the South Atlantic. *Antarc. Sci.* **4**(1)**:** 31–36.

Prince, P. A., Rothery, P., Croxall, J. P. and Wood, A. G., 1994. Population dynamics of Black-browed and Grey-headed Albatross *Diomedea melanophrys* and *D. chrysostoma* at Bird Island, South Georgia. *Ibis* **136:** 50–71.

Prince, P. A. Croxall, J. P., Trathan, P. N. and Wood, A. G., 1997. The pelagic distribution of South Georgia Albatrosses and their relationships with fisheries. Pp. 137–67 *in* Albatross Biology and Conservation ed by G. Robertson and R. Gales. Surrey Beatty & Sons: Chipping Norton.

Schiavini, A., Lizurume, M. E., Frere, E., Bertelotti, M., García, N. and Chizzini, A., 1994. Relevamiento aéreo otoñal de la fauna superior costera de Tierra del Fuego. Unpublished report prepared for Fundación Patagonia Natural.

Schlatter, R. P., 1984. The status and conservation of seabirds in Chile. Pp. 261–69 *in* Status and Conservation of the World's Seabirds ed by J. P. Croxall, P. G. H. Evans and R. W. Schreiber: International Council of Birds Preservation, Technical Publication No. 2.

Thompson, K. R. and Riddy, 1995. Utilization of offal discards from "finfish" trawlers around the Falkland Islands by the Black-browed Albatross *Diomedea melanophys* in the vicinity of the Falkland Islands. *Ibis* **137:** 198–206.

Tickell, W. L. N. and Woods, R. W., 1972. Ornithological observations at sea in the South Atlantic Ocean, 1954–1964. *British Antarctic Survey Bull.* **31:** 63–84.

Veit, R. R., 1995. Pelagic communities of seabirds in the South Atlantic Ocean. *Ibis* **137:** 1–10.

Watson, G. E., Angle, J. P., Harper, P. C., Brodge, M. A., Schlatter, R. P., Tickell, W. L. N., Boyd, J. C. and Boyd, M. M., 1971. Birds of the Antarctic and Subantarctic. Antarctic Map Folio Series, Folio 14. American Geographical Society.

Weimerkirsch, H., Zotier, R. and Jouventin, P., 1989. The avifauna of the Kerguelen Islands. *Emu* **89:** 15–29.

Weimerskirch, H. and Jouventin, P., 1997. Changes in population sizes and demographic parameters of six albatross species breeding on the French sub-Antarctic islands. Pp. 84–91 *in* Albatross Biology and Conservation ed by G. Robertson and R. Gales. Surrey Beatty & Sons: Chipping Norton.

CHAPTER SEVENTEEN

Albatross mortality in fisheries off the coast of Brazil

TATIANA NEVES[1] and FABIO OLMOS[2]

Albatrosses off the Brazilian coast are known to interact with two main fisheries: mid-water longline fishing and demersal (bottom) longline fishing. Monitoring of 14 mid-water vessels and one demersal longline vessel in 1994–95 showed that although fishing occurs year-round, seabirds are incidentally captured mostly during the winter (70% for the demersal vessel and 94% for the mid-water vessels). Mid-water longline vessels captured 0.12 birds/1 000 hooks, mostly albatrosses (71% of all captures). The main captured species were Black-browed and Yellow-nosed Albatrosses and White-chinned Petrels; over 80% of the albatrosses caught were immature females. In 19 cruises the demersal longline vessel caught 0.3 birds/1 000 hooks, mostly Great Shearwaters but also Black-browed and Yellow-nosed Albatrosses, White-chinned petrels, Great Shearwaters and Spectacled Petrels. For both fishing types two cruises in winter and were responsible for most birds killed, suggesting there are spatial and temporal differences in the vulnerability of seabirds to longline fishing. Mid-water longline vessels fishing in Brazilian waters have low capture rates of albatrosses compared to other countries, apparently due to the widespread use of setting lines at night. Petrels and shearwaters, which are proficient at diving, seem more vulnerable to capture by demersal longline vessels than albatrosses. The size of the Brazilian longline fishing fleet and the importance of the Brazilian coast for seabirds, especially during the winter, suggest Brazil is a country where incidental mortality may have a significant impact on migratory seabird populations.

Key words: Brazilian Waters, Longline Fishing, Black-browed Albatross, Yellow-nosed Albatross, White-chinned Petrel, Great Shearwater.

INTRODUCTION

ALTHOUGH no albatrosses breed near Brazil, Brazilian waters are an important foraging area for adult and immature birds dispersing from breeding colonies during the austral winter. Seven species of albatrosses are known to occur in Brazilian waters (Sick 1993; Vooren and Fernandes 1989), three of which are regularly observed interacting with fishing vessels: Black-browed Albatross *Diomedea (Thalassarche) melanophrys*, Yellow-nosed Albatross *D. (Thalassarche) chlororhynchos* and Wandering Albatross *D. exulans* (Vooren and Fernandes 1989; Vaske Jr 1991). The greatest concentrations of albatrosses are found along the southern Brazilian coast, close to Uruguay, where the influence of the Sub-tropical Convergence is strongest (Thomsen 1962; Lino and Mascarenhas Jr 1985). Albatrosses are also commonly found northward up to 23–24°S following the winter penetration of the Falklands Current over the South American continental shelf (Lino and Macarenhas Jr 1985; Stech *et al.* 1994), and occasionally even to north-east Brazil (13°S), especially in waters of high productivity (Coelho *et al.* 1990; Lima *et al.* 1994).

Attention has recently been drawn to the impact of fisheries on seabirds in Brazilian waters (Vaske Jr 1991). Albatrosses have been found to interact with two kinds of fisheries: mid-water (tuna) longline fisheries and demersal (bottom) longline fisheries, but there is also potential for interaction with trawl and driftnet fisheries.

In this chapter we summarize the available information on albatrosses-fisheries interactions and present the first data on research to assess the incidental mortality of seabirds in fisheries, the seasonal distribution of fishing effort in Brazilian waters, and studies of fishing methods aimed at developing seabird bycatch mitigation measures to be employed by the Brazilian fishing fleet.

METHODS

Mid-water (tuna) longline fisheries

Mid-water longline fisheries in Brazilian waters have traditionally targeted tuna *Thunnus*

[1]FURG, Universidade do Rio Grande, Lab. de Aves Marinhas, C.P. 474, CEP., 96.201–900 Rio Grande, RS Brazil.
[2]Instituto Florestal, C.P. 1322, CEP. 01059–970 São Paulo, SP Brazil.
Pages 214–19 *in* ALBATROSS BIOLOGY AND CONSERVATION, ed by G. Robertson and R. Gales. Surrey Beatty & Sons, Chipping Norton. 1997.

spp., however, an increase in the market prices of swordfish *Xiphias gladius* and shark has resulted in a switch in fishing effort to these species. To assess fishing methods adopted and bird byctatch rates, data were collected from the captains of 14 Brazilian tuna vessels from November 1994 to October 1995. The vessels surveyed are based at the Santos Fishing Terminal (23°59′S, 46°21′W), south-east Brazil, and represent 38% of the Brazilian tuna fishing fleet (ICCAT 1995). Data was collected from fishing log books and included information on fishing seasons, fishing locations, fishing methods and fishing times (duration of settings and haulings). Data were collected in both summer (November to April) and winter (May to October) months.

Half of the vessels surveyed used monofilament nylon main lines, and half used main lines constructed from steel and twisted nylon. The monofilament main lines were 80 km long and contained, on average, 802 hooks (range: 300–1 200 hooks, n = 382 sets). Hooks were set at 45–80 meters depth and mainly targeted swordfish. Most sets (69%, n = 863) were made at night, between 18:00 h and 21:00 h local time (see Fig. 2). The bait used was the Squid *Illex argentinus*, which was occasionally tinted (green, yellow, orange, red) to attract swordfish. Light sticks were attached to the line in an attempt to improve fish catch rates. This contrasts with the traditional Japanese method which employs steel main lines with an average of 1 672 hooks (n = 380 settings) ranging from 900 to 1 700 hooks/set. Hooks are set to a depth of 70–120 m and baited with Brazilian Sardines *Sardinella brasiliensis*, Chub Mackarels *Scomber japonicus* or Short-finned Argentinean squids.

Demersal (bottom) longline fisheries

The information on demersal longline fishing comes from one vessel only, the RV *Orion*, of the Fisheries Institute of São Paulo. Log books from 19 fishing cruises conducted between April 1994 and May 1995 were examined for fishing locations, fishing effort, time of setting and hauling and fish species caught. Observations on seabird abundance and behaviour were made onboard during the cruises. Dead birds captured were held by the crew for later identification, but due to operational problems not all birds caught could be studied. The fishing grounds were 160–320 km from the Brazilian coast between 23°40′S and about 28°00′S, in 150–500 metre deep water.

The fishing gear consisted of an 8–13 km long steel mainline with over 1 000 hooks, baited with either squid or fish, attached to the mainline by nylon monofilament branch lines. The weight of the mainline caused it to sink rapidly during the set. Line setting took place during the day, lasted between one and two hours, and hauling took about twice that time. Line hauling usually commenced 30 minutes to one hour after the completion of the set. Fishing was conducted mostly by day, because fishing at night results in an increases in the number of fish damaged, or completely eaten, by carnivorous isopod crustaceans.

The RV *Orion* fishing cruises occurred in all months except December 1994 and January 1995, and lasted from 5 to 16 days each. Usually the longline to be set twice/day. The mean number of hooks/set was 1 107 (n = 253) but ranged from 147–1 887 hooks/set. A total of 280 197 hooks were set in 157 fishing days. This fishery targeted Tilefish *Lopholatilus villarii*, Namorado *Pseudopercis numida* and Groupers *Epinephelus* spp.

RESULTS

The Santos fleet

The Santos tuna-fishing fleet fishes mostly between 15°S and 35°S and 25°W and 50°W. We observed noticeable seasonal differences in the distribution of the fishing effort between summer and winter (Fig. 1). In the southern sector (25° to 35°S), during winter 4 720 hooks/day were set compared to 3 700 hooks/day during summer. In 91 fishing cruises 118 birds were captured, or 0.12 birds/1 000 hooks. We consider this capture rate to be an underestimate because there are many unmonitored foreign vessels fishing in Brazilian territorial waters, and because the fishermen sometimes do not retrieve or record all birds captured. The time of day of bird capture was closely linked to setting time, most of which occurred between 1800 and 2100 hrs local time (Fig. 3).

The captured species were Wandering Albatrosses (1), Black Browed Albatrosses (33), Yellow-nosed Albatrosses (17), White-chinned Petrels *Procellaria a. aequinoctialis* (14) and *Procellaria aequinoctialis* subsp. (6), Spectacled Petrels *P. a. conspicillata* (6) and Great Shearwaters *Puffinus gravis* (1). A further 40 birds were captured but not identified. Twenty-eight birds (31% of the total number) were captured during a single cruise on which tinted bait was used. The pigment used was luminescent and had a strong smell, which probably made the bait more attractive to birds and fish.

Of the 118 birds caught most (94.4%) were caught during the winter months (Fig. 3), this being especially true for the albatrosses (71% of all birds caught). During winter the Sub-tropical Front reaches the southern coast of Brazil bringing nutrients from Antarctic waters and an increase in biological productivity. For this

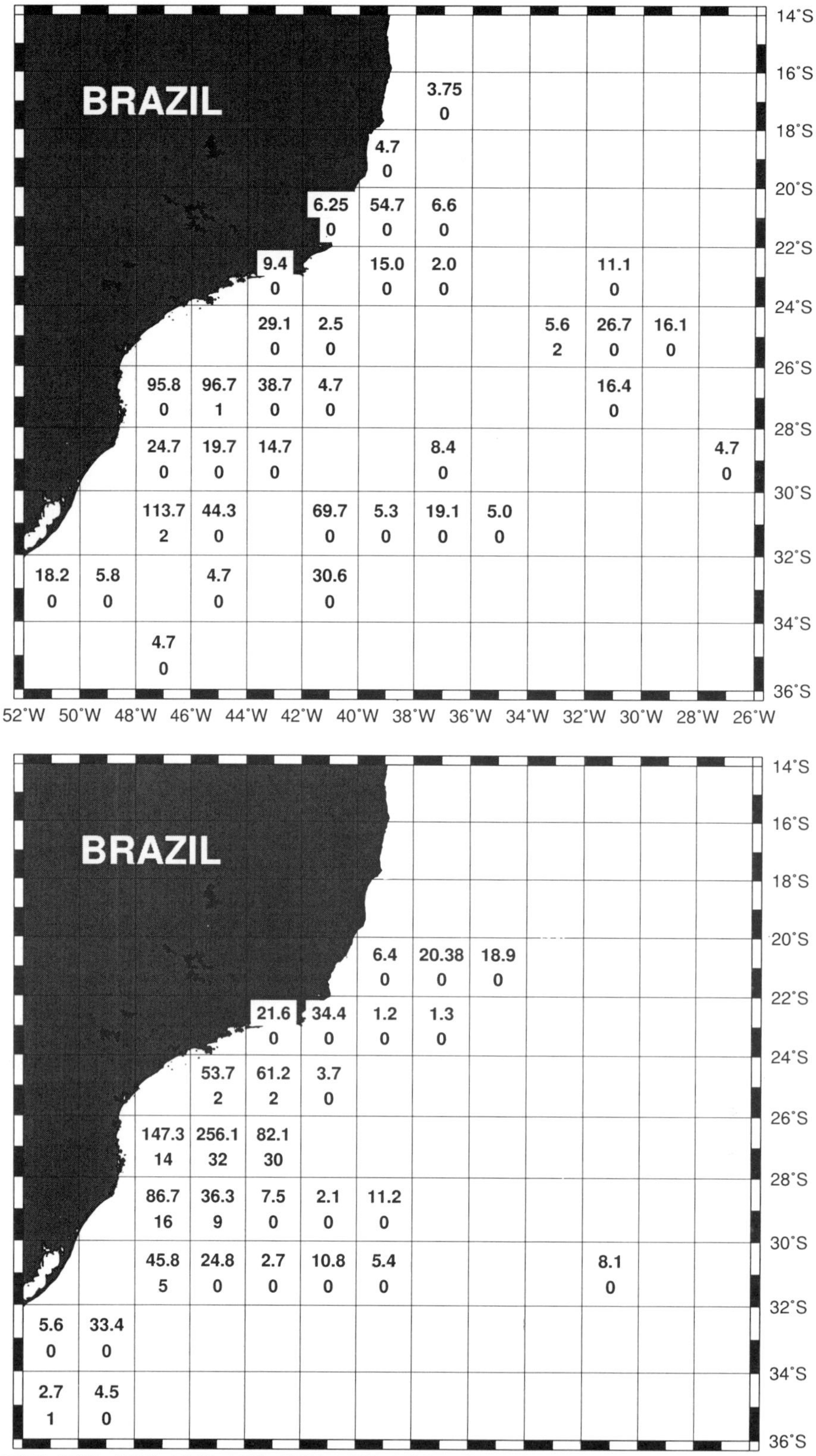

Fig. 1. Fishing effort and number of seabirds caught by mid-water (tuna) longline vessels in Brazilian waters between (a) November 1994 and April 1995 and (b) May and October 1995. The upper figures in squares indicate numbers (in thousands) of hooks deployed, and the lower figures indicate numbers of birds caught.

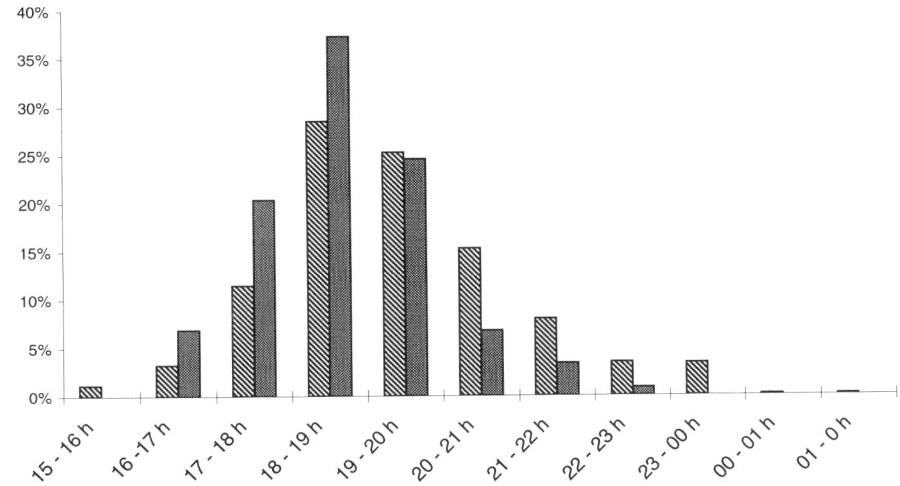

Fig. 2. Per cent of line settings (grey bar, n = 863) and seabird captures (black bar, n = 118) as a function of time of day by mid-water longline fishing vessels in Brazilian waters between November 1994 and October 1995.

reason the greatest number of birds can be observed in winter in southern Brazilian waters, and this coincides with the southward shift of the fishing effort.

All but one of the 33 Black Browed Albatrosses were immature birds (97%), with an overall sex ratio of 1:2.9. Of the 17 Yellow-nosed Albatrosses caught, 89% were immature birds, with a sex ratio of 1:3. The only captured Wandering Albatross was an immature (stage 3) (Harrison 1983) male.

The RV *Orion*

A total of 81 seabirds was incidentally captured by the RV *Orion* (almost 0.3 birds/1 000 hooks), of which 49 were retrieved for later identification. Birds were captured in cruises on 21–28 April (4 birds), 6–15 May (31 birds), 26 May–5 June (15 birds), 15 June–1 July (3 birds), 11–19 July (1 bird), 5–15 August (7 birds), 6–15 September 1994 (14 birds), and 30 April–9 May 1995 (6 birds). The 49 birds identified included six Yellow-nosed Albatrosses, one Black-browed Albatross, six White-chinned Petrel, two Spectacled Petrel and 34 Great Shearwaters.

The greatest number of birds (56.7%, n = 46) was captured during two cruises to the Cabo Frio upwelling region (around 23°40′S) during May and June, and during one cruise to 26°S in late winter (17.3%, n = 14). Albatrosses were captured in May–June (two Yellow-nosed Albatrosses), June–July (two Yellow-nosed

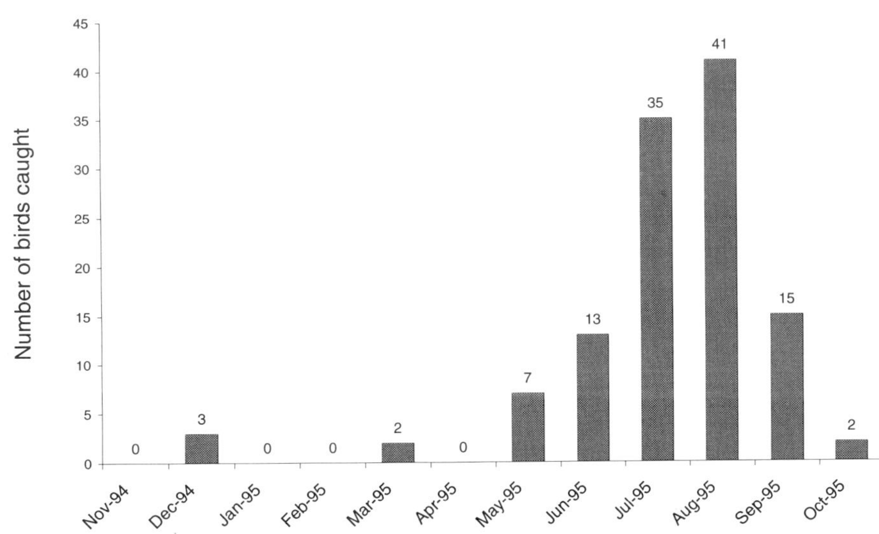

Fig. 3. Number of seabirds caught by mid-water and demersal longline vessels fishing off the Brazilian coast between November 1994 and October 1995. The total number of birds caught was 118, 94.4% of which were caught from May–October 1995. Data for October comes from two vessels only. Figures above the bars indicate the actual number of birds caught.

Albatrosses), August (one Yellow-nosed Albatross) and September (one Yellow-nosed and one Black-browed Albatrosses), with all except the September birds being captured in the Cabo Frio upwelling region. The only Black-browed Albatross captured was an immature bird, while all but two of the Yellow-nosed Albatrosses captured were adults.

DISCUSSION

The main species captured by the demersal longline was the Great Shearwater which is seasonally common off the Brazilian coast and a proficient diver for baited hooks. It is interesting that fishing in the same areas during other times of the year yielded either none or far fewer birds. The predominance of immature Black-browed Albatross and adult Yellow-nosed Albatrosses in southeastern Brazil has been reported by Olmos *et al.* (1995).

Although there is common agreement over the great abundance of birds during the winter, when Giant Petrels *Macronectes* sp., Antarctic Fulmas *Fulmarus glacialoides*, Cape Petrels *Daption capense*, White-chinned Petrels and others are more likely to be observed (Vooren 1997), there are some species which stay in the area all through the year. Good examples are Yellow-nosed Albatross, Great Shearwater and Spectacled Petrel from Tristan da Cunha Island (Vooren 1997). Apparently during the summer months there is sufficient food from discarded bait and viscera for the birds not to bother diving after baited hooks. Increased competition for food due to an increase in the number of birds during the winter months is probably the reason for most captures occurring in this season.

Assuming the species composition of retrieved birds was representative of all birds captured by the RV *Orion*, about 12 albatrosses would have been captured in 1994, most of them Yellow-nosed Albatrosses, or about 0.04 albatrosses/1 000 hooks deployed. Albatrosses seem to suffer a lower mortality than other seabird species as they preferred to scavenge discarded bait and fish, and seem reluctant to dive for baited hooks, especially Yellow-nosed Albatrosses. Also, they are less capable at diving after the rapidly sinking bait than petrels and shearwaters. Great Shearwaters are more abundant and are better divers than albatrosses, and more vulnerable to being hooked and drowned. Note, however, that the species composition of birds caught could change if demersal fishing commences closer to the Subtropical Convergence, where there are greater numbers of albatrosses and other seabird species. If such a development occurs heavy losses of petrels and albatrosses could be expected, as is the case in Uruguayan waters (A. Stagi, pers. comm.; Stagi *et al.* 1997).

The results show that incidental mortality of albatrosses in Brazilian waters by tuna vessels mainly involves immature female Yellow-nosed Albatrosses. It is not known if this reflects a greater vulnerability of this age and sex class, or merely their potentially greater abundance off the Brazilian coast during certain seasons of the year. In northern waters (25°–26°S) at least, immature Yellow-nosed Albatrosses comprise only about one-third of all albatrosses observed, suggesting that longline fishing has a proportionally greater effect on young birds of this species. Another interesting point is that only a few cruises by tuna vessels accounted for most of the seabird mortality, presumably because of more efficient capture techniques by particular vessels and/or a greater density of birds were the vessels fished.

We have no measure of the significance of the mortality of albatrosses in the demersal longline fishery. Being introduced into Brazil barely one year ago this fishing technique will probably expand in the near future and seabird mortality will probably increase. As the species composition and abundance of pelagic seabird species off the southeastern coast of Brazil are largely unknown there is no way of assessing the impact of the development of the demersal longline fishery on seabirds in southern Brazilian waters.

Adapting fishing gear or the setting procedure to make the bait sink faster, and the use of bird scaring lines, would probably reduce bird mortality in Brazilian waters; these techniques are being tested by the RV *Atlantico Sul* during fishing prospection cruises. Because of the increased emphasis on catching swordfish and sharks, night setting, usually considered to be beneficial to albatrosses, is now the standard procedure for many tuna fishing vessels in Brazilian waters and probably accounts for the low observed seabird capture rate when compared to other parts of the world. Uruguay, which borders Brazil to the south, has reported very high seabird capture rates in tuna fisheries (CCAMLR 1994; Stagi *et al.* 1997). Similarly, Vaske Jr (1991), working in southern Brazil, observed very high capture rates (3.82 birds/1 000 hooks) in five winter cruises, but the vessels he monitored set lines during the day, and the observations were made only during the season of greatest bird abundance.

Management possibilities exist, one of them being closure of areas for fishing vessels when large numbers of migratory seabirds, such as Great Shearwaters, are present in fishing areas. Seasonal closures of fishing areas has been suggested for fisheries around South Georgia (Croxall and Prince 1996) and Illes Kerguelen (Weimerskirch 1997) as a method to minimize fisheries impacts on breeding albatrosses. The fact that most birds in this study were caught during a brief period of time and in a small area suggests that area closures would be a useful conservation measure. However, the lack of necessary knowledge on the seasonal distribution of seabirds in Brazilian waters, added to legislative and financial difficulties in ensuring area closures are effective, make the implementation of this management option impractical at present.

Alternative management options do, however, exist. Preliminary observations of bird scaring streamer lines, the prevention of line tensioning during the set (which pulls hooks to the surface and exposes them to bird attack) and night setting have shown that these measures are effective in keeping seabirds away from the danger zone behind vessels. These measures should be adopted as a matter of urgency by longline vessels operating in Brazilian waters.

ACKNOWLEDGEMENTS

Thanks are due to Mr Sebastião Medeiros and Mr Luis Alonso of the *Museu do Mar* of Santos for supporting the researches in their laboratory, to Dr. Carolus Maria Voorem, Prof. Fabio Giordano, Prof. Fabio Enir Costa, for their important suggestions and advice. We thank Gastão Cesar Cyrino Bastos for allowing access to the RV *Orion* data and for his friendship before, during, and after the research cruises. Those data were gathered as part of a co-operative project between the São Paulo Fisheries Institute (IP) and *Kawai-Suissan Com. e Ind. Pescados Ltda.* We thank Mr Katutoshi Ono of *Olinda Ind. and Com de Pescados Ltda* who allowed the researchers to accompany the cruises, which opened an important door to the mutual co-operation between the fishing companies and Brazilian marine researchers. We would like to thank the skippers and crews for their collaboration and the trainees who, with voluntary work, effort and dedication were the master key of the study. Finally we thank Msc. Theodoro Vaske Jr and Oc. Antonio Greig for reviewing the manuscript.

REFERENCES

CCAMLR, 1994. Report of the Ad Hoc working group o incidental mortality arising from longline fishing (Hobart, Australia, 21 and 22 October 1994). SC-CCAMLR-XIII, Annex 8.

Coelho, E. P., Alves, V. S., Soneghet, M. L. L. and Carvalho, F. S., 1990. Levantamento das aves marinhas no percurso Rio de Janeiro — Bahia (Brasil).Bolm.Inst. *Oceanogr., S. Paulo* **38**(2): 161–67.

Harrison, P., 1983. Seabirds: an identification guide. Christopher Helm, London.

ICCAT, 1995. Statistical Bulletin (Madrid, Spain). Vol. **26**.

Lima, P. C., Castro, J. O., Santos, S. S., Sampaio, C. L. S., Neto, F. P., Campos, S. S. and Lima, R., 1994. Ocorrência e mortandade de aves oceânicas no litoral baiano. Anais IV Congresso Brasileiro de Ornitologia p. 135. Sociedade Brasileira de Ornitologia.

Lino, G. J. and Mascarenhas, A. S., Jr, 1985. Sobre a propagação da Corrente das Malvinas ao longo da Costa Sul do Brasil no período de 1980–1983. Anais do IV Congresso Brasileiro de Engenharia de Pesca. Curitiba, PR.

Olmos, F. P., Martuscelli, R. and Silva e Silva e Neves, T. S., 1995. The seabirds of São Paulo, southeastern Brazil. *Bull. British Ornith. Club* **115**(2): 117–28.

Sick, H., 1993. Birds in Brazil: a Natural History. Princeton University Press.

Stagi, A., Vaz-Ferreira, R., Marin, Y. and Joseph, L., 1997. The conservation of albatrosses in Uruguayan waters. Pp. 220–24 *in* Albatross Biology and Conservation ed by G. Robertson and R. Gales. Surrey Beatty & Sons, Chipping Norton.

Stech, J. L., Lorenzzetti, J. A., Silva, C. L., Jr, da Rosa, S. M. and Campos, E. J. D., 1994. Some wintertime Brazil Current and shelf circulation features as observed from AVHRR/NOAA images. Symposium: The South Atlantic: Present and Past Circulation. Bremen, Germany, 15–19 August 1994.

Thomsen, H., 1962. Massas de água características del Oceano Atlântico — Parte Sudoeste (Reporte H. 632) serviço de Hidrografia Naval — secretaria de Marina, Buenos Aires — República Argentina. 28 Pages.

Vaske, T. Jr, 1991. Seabirds mortality on longline fishing for tuna in southern Brazil. *Ciência e Cultura* **43**(5): 388–90.

Vooren, C. M. and Fernandes, A. C., 1989. Guia de Albatrozes e Petréis do Sul do Brazil. Sagra Editora, Porto Alegre.

Vooren, C. M., 1997. Sea and shore birds. *In* Subtropical Convergence Environments: the coast and the sea in the southwestern Atlantic ed by U. Seeliger, C. Odebrecht and J. P. Castello. Springer-Verlag, Berlin.

CHAPTER EIGHTEEN

The conservation of albatrosses in Uruguayan waters

A. STAGI,[1] R. VAZ-FERREIRA,[1] Y. MARÍN[2] and L. JOSEPH[3]

At least five species of albatrosses visit Uruguayan waters from breeding sites in the Southern Ocean. The threats to albatrosses in Uruguayan waters include intentional killing by fishermen, oil spills and mid-water (tuna) and demersal longline fishing activities. Of these longline fishing poses by far the most serious threat to albatross survival. On 14 tuna fishing trips between 1993 and 1994 catch rates of 4.7 albatrosses/1 000 hook deployed were recorded. The Black-browed Albatross was the main species caught. Catch rates by demersal vessels were lower but there is potential for the absolute number of albatrosses caught by demersal vessels to be higher, due to the greater number of hooks deployed in each set. An educational programme for fishermen and the implementation of seabird bycatch mitigation measures on fishing vessels are urgently required.

Key words: Albatrosses, Uruguayan Waters, Longline Fishing, Seabird Bycatch.

INTRODUCTION

THE Uruguayan coast runs along the southwestern Atlantic Ocean and includes one of the world's largest estuaries. The Rio de la Plata, which is an estuary formed by the Rio Paraná and the Rio Uruguay, has a surface area of 30 212 km² and a coast that extends for some 280 km between the mouth of the Rio Uruguay and Punta del Este. The southwestern Atlantic Ocean extends for approximately 200 km between Punta del Este and Chuy between 34°S and 38°S. Between the coast and the continental shelf is an ocean surface of about 133 000 km.² In this area the water masses are highly dynamic owing to their diverse origins either from the Malvinas current, the Brazil current, coastal waters, or the Rio de la Plata. In summer tropical waters from the Brazil current predominate and during winter cold water from the Malvinas current advances over the continental shelf. The two currents meet and form the Sub-tropical Convergence, which is a zone of upwelling and high productivity. This upwelling area is favoured by a rich variety of marine life including sea mammals and seabirds; it is also favoured by commercial fisheries.

In this chapter we provide an overview of the occurrence of albatrosses in Uruguayan waters and discuss the conservation-related problems affecting them. We also describe the impacts of fishing activities on albatrosses and our attempts to minimize fisheries-induced mortality of albatrosses in Uruguayan waters. The information presented complements the unpublished report by Vaz-Ferreira *et al.* (1994).

ALBATROSSES IN URUGUAYAN WATERS

Although no albatross species breed near Uruguay five species have been recorded in Uruguayan waters. The most common of these are the Wandering *Diomedea exulans*, Black-browed *D. (Thalassarche) melanophrys* and Yellow-nosed *D. (Thalassarche) chlororhynchos* Albatrosses. Royal Albatrosses *D. epomophora* are an occasional visitor and there is one record of a Grey-headed Albatross *D. (Thalassarche) chrysostoma* visiting Uruguayan waters.

Wandering Albatrosses

This species is most abundant in winter but there are some summer records (Escalante 1970; Gore and Gepp 1978; A. Stagi and R. Vaz-Ferreira, pers. obs.). Wandering Albatrosses are usually seen as single birds or as pairs; only rarely are groups of several birds observed. Data from banding recoveries (Croxall and Prince 1990) and from satellite-tracked birds (Prince *et al.* 1992; Prince *et al.* 1997) have shown that Wandering Albatrosses in Uruguayan waters originate from South Georgia and are most

[1]Seccion de Vertebrados, Facultad de Ciencias, Universidad de la Republica, Tristan Narvaja 1674, Montevideo 11200, Uruguay.
[2]Instituto Nacional de Pesca (INAPE), Constituyente 1497, Montevideo 11200, Uruguay.
[3]Laboratorio de Evolucion, Facultad de Ciencias, Universidad de la Republica, Tristan Narvaja 1674, Montevideo 11200, Uruguay.
Pages 220–24 *in* ALBATROSS BIOLOGY AND CONSERVATION, ed by G. Robertson and R. Gales. Surrey Beatty & Sons, Chipping Norton. 1997.

probably females, the males foraging further south. A Wandering Albatross caught off Uruguay had been banded twice, once off eastern Australia and once at South Georgia; this albatross holds the record for the longest time (32 years) between banding and recovery in Uruguayan waters. Five band recoveries have been of birds banded as nestlings at South Georgia.

Black-browed Albatrosses

This species is most abundant in Uruguayan waters during the winter and only occasionally are birds seen in summer (Escalante 1970; Gore and Gepp 1978; A. Stagi and R. Vaz-Ferreira, pers. obs.). Black-browed Albatrosses follow fishing boats and commonly form groups of 20–90 birds and occasionally more than 500. Tickell (1967) showed that 88% of recoveries of Black-browed Albatrosses banded at the Islas Malvinas (Falkland Islands) came from the Atlantic coast of South America, including Uruguayan waters, whereas about 95% of recoveries of South Georgia birds came from South Africa. Only 1.6% of the latter had been recorded off the South American coast.

Yellow-nosed Albatrosses

This species accompanies fishing boats during the day and occasionally is found dead on beaches of the Uruguayan coast; in August 1967 10 individuals were found dead on beaches near Punta del Este (35°S; 55°W, F. Achaval, pers. comm.). Although at-sea observations have revealed groups of up to 250 Yellow-nosed Albatrosses in the first half of April, most frequently seen have been small groups of albatrosses or single, immature birds.

ALBATROSS MORTALITY IN URUGUAYAN WATERS

The factors that affect the survival of albatrosses in Uruguayan waters include deliberate capture and killing by fishermen, shooting by fishermen (up until about 1990), oil spills and commercial fisheries.

Oil spills

The potential for mortality through oiling is of special concern in Uruguay and occasionally oiled Black-browed and Yellow-nosed Albatrosses are washed up on beaches, suggesting mortality through the direct and indirect consequences of oil spills. The Uruguayan oil refining company ANCAP has an oil processing plant on the Atlantic coast and oil tankers regularly anchor a few kilometres offshore while oil is pumped to a nearby plant. Oil spills are not uncommon and there is concern from the general public about the effect of oil spills on the sea, marine life, the coastal environment and the tourism industry.

Intentional killing by fishermen

From interviews with fishermen and casual observations on fishing boats it seems that an unknown number of albatrosses has died through wanton destruction by members of boat crews (the reaction of fishermen to hooked and drowned birds is generally one of antipathy). For example, up until about 1990 members of some crews used to entertain themselves by shooting albatrosses that flew near fishing boats. Albatrosses that did not die outright may have been killed with blows, or by being hooked. In one instance an albatross had its wings tied together before being thrown into the water simply so the crew could see what would happen. In another instance two baited hooks on a line were cast into the water so that the crew could watch two albatrosses fighting for the bait; the birds became hooked and died trying to free themselves. Other cases involve painting captured albatrosses, placing bow ties around their necks and removing the feathers from their heads. In all cases the observers were unable to prevent these actions from taking place.

Fisheries-induced mortality

Several fishing methods are practiced in Uruguayan waters (Fig. 1). These include traditional net and hook fishing and coastal trawling, neither of which leads to significant mortality of albatrosses. Deep-sea trawling attracts many birds but is also a minor cause of mortality. The greatest cause of albatross mortality in Uruguayan waters is from longline fishing for Tuna *Thunnus* spp., Merluza (Hake) *Merluccius hubbsi* and rays (Vaske 1991). The most important of these for the conservation of seabirds is longline fishing for tuna. Tuna (mid-water) fishing began in 1959 and in 1995 five tuna vessels were fishing in Uruguayan waters (owing to the depletion of tuna stocks Swordfish *Xiphias gladius* is the main target species of these vessels). The number of hooks in each set varies from 300–1 000 and the hooks are baited with whole squid. The lines usually are set at night in an operation lasting between three and six hours. The fishing seasons are March–May, July–August and September–October. We have observed the deployment of 61 545 baited hooks during 139 sets and hauls in 14 tuna fishing trips. The demersal (bottom-set) long lines have involved 723 950 baited hooks set in two trips and 46 nets set on one trip fishing for Merluza. Collectively, we have spent a total of 324 days at sea comprised of 247 days on tuna vessels, 65 on deep sea longline boats and 12 days on Merluza fishing vessels.

We have most detailed data from a total of nine trips on tuna vessels between 1993 and 1994 in which 26 364 hooks were set. A total of

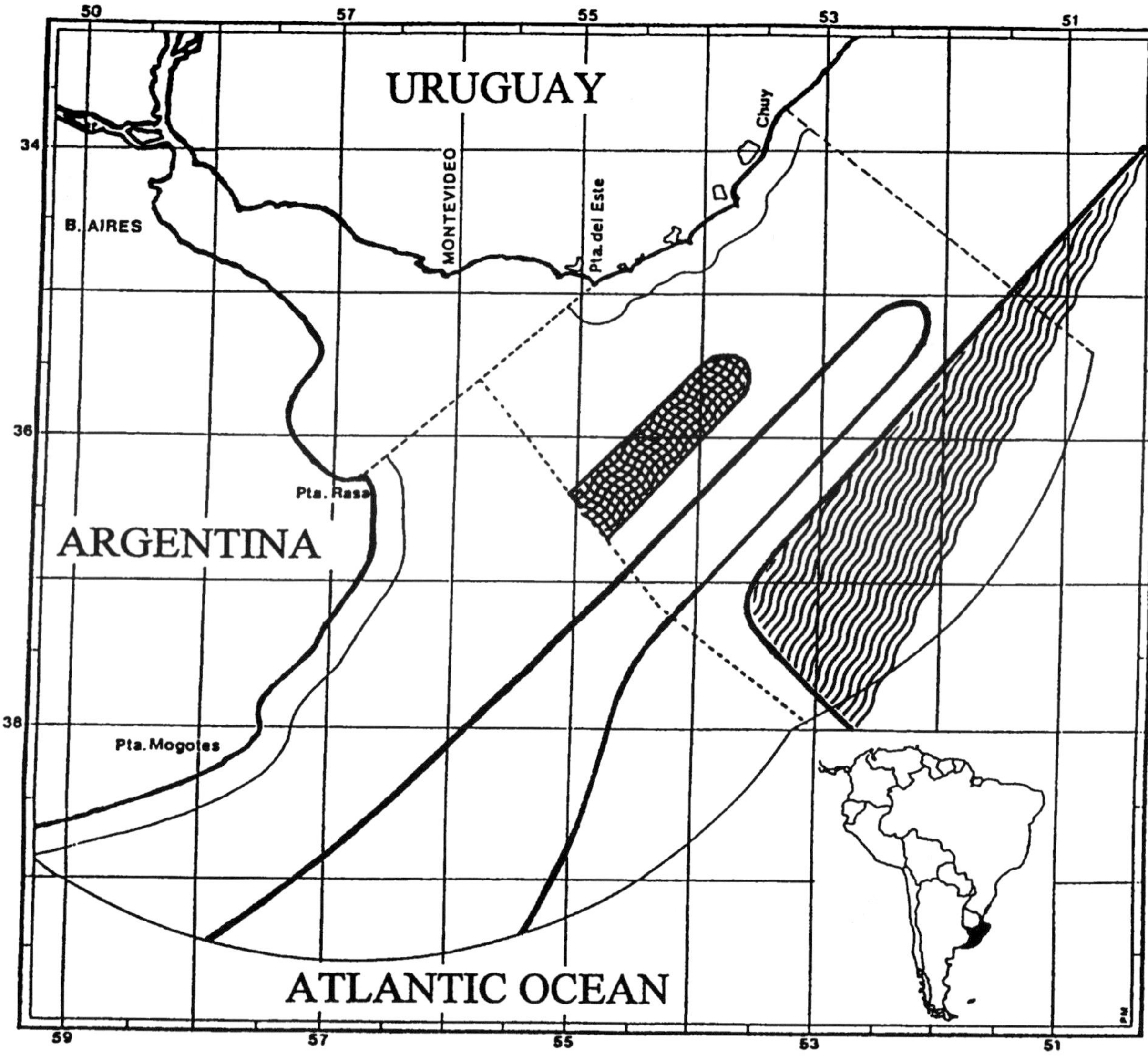

Fig. 1. Map of Uruguayan territorial waters showing the locations of longline fishing for tuna (wavy shading), demersal (bottom-set) longline fishing for rays (checkered shading) and trawl fishing for Merluza (Hake) (open area delineated by emboldened line).

277 seabirds were hooked and drowned and of these 265 were albatrosses, mostly Black-browed Albatrosses. This represents 10.5 birds/1 000 hooks deployed, though the numbers are driven upwards by the absence of weighted swivels on the first trip. On this first trip 154 albatrosses were killed, so it was decided to place an 80 gm swivel 3.6 m from the hook to accelerate the sink rate of baits. Omission of the catch rates for the first trip gives a mean of 4.7 albatrosses/ 1 000 hooks deployed (Table 1). As well as using weighted lines crews have been advised to set lines after sunset to minimize attracting the birds. When birds are seen at night, viscera and remains of fish are occasionally thrown overboard to distract their attention away from where the lines are being set. Nonetheless, lines are sometimes set by day to guarantee that lines are set before those of a competing vessel. We believe that fishermen do not wish to hook

Table 1. Numbers of albatrosses hooked on longline tuna fishing trips in Uruguayan waters in 1994. The 123 albatrosses caught comprised 120 Black-browed Albatrosses and three Wandering Albatrosses. The first trip used lines with unweighted swivels (see text for explanation). Totals in parentheses exclude estimates for the first trip.

Trip number	Number of hook set	Number of albatrosses caught	Number of albatrosses caught/ 1 000 hooks
1	320	154	481.3
2	1 740	70	40.2
3	1 624	5	3.1
4	4 068	3	0.7
5	3 008	6	2.0
6	3 112	0	0.0
7	1 984	3	1.5
8	3 278	22	6.7
9	7 230	14	1.9
Totals	26 364	277	10.5
	(26 044)	(123)	(4.7)

albatrosses on purpose because of the potential loss of a hook and a fish (a 60 kg Swordfish can bring up to US $360).

The catch rate of albatrosses caught by tuna vessels in Uruguayan waters varies for males and females. For example, in 1994 a sample of 37 Black-browed Albatrosses killed by longline fishing activities included 35 females, one male and one bird of unknown sex. We are aware that more female than male Wandering Albatrosses frequent Uruguayan waters (Prince *et al.* 1997) and that most Yellow-nosed Albatrosses caught by longline vessels in Brazilian waters are female (Neves and Olmos 1997). A biased sex ratio of albatrosses killed has implications for pair formation, and if consistent in time can be expected to have more serious consequences for albatross conservation than if mortality was spread more evenly between the sexes. In 1995 a single demersal longline fishing trip resulted in 83 dead Black-browed Albatrosses, but for logistical reasons we were unable to determine the sex of these birds.

The demersal longline fishery for rays and other deep sea species commenced in Uruguayan waters in 1994 with one Uruguayan boat, and in 1995 a Korean boat was involved. This fishing practice potentially poses the most serious threat to albatrosses. Although the hooking rate of albatrosses in this fishery is lower than in the tuna longline fishery many more hooks are used and a great number of birds can be killed (Table 2). For example, in two trips in 1995 78 Black-browed Albatrosses were killed on the first trip and five on the second trip; these numbers are known to be underestimates of the actual numbers of birds killed because not all captured birds were landed on deck and our observations did not include the entire duration of the fishing trips.

Table 2. Numbers of Black-browed Albatrosses killed in two dermersal (bottom-set) longline fishing trips for rays in Uruguayan waters in 1995.

Trip	Number of times line set	Number of hooks set	Number of hooks sampled	Number of birds killed
1	64	423 700	135 550	78
2	42	300 250	67 100	5
Total	106	723 950	202 650	83

FUTURE PROSPECTS

Regulations currently in force in Uruguay prohibit driftnet fishing and allow observers on foreign tuna fishing boats controlled by the National Institute of Fisheries (INAPE). Foreign tuna boats normally fish freely outside Uruguayan waters even if they use Uruguay as a base. Uruguayan boats, of which there are presently four, are free to fish within or outside territorial waters and are not required to have observers on board. Clearly there are aspects of these regulations that we would like to change, and it is pleasing that biologists from both INAPE and the Universidad de la Republica, Montevideo, are working together in an attempt to improve regulations and their enforcement.

Uruguay is a small country of some three million people. While there is much concern for environmental issues outside the University and government departments, relatively few people are equipped to provide environmental education, particularly away from the capital, Montevideo. Uruguayans are aware that much of their countries national resources lie in the sea and we hope that the coming years will see an increase in Uruguayan's concern for ensuring the future of these resources and for birds such as albatrosses. If Uruguay is to contribute substantially to the conservation of albatrosses an improvement in the efforts to work with and educate those involved in fisheries and help them appreciate the international concern about these birds is needed. Though more resources are badly required, both from within and outside Uruguay to accomplish these tasks, it is pleasing to note some progress has been made.

We would like to initiate a campaign to inform crews of tuna and demersal fishing boats about the conservation of albatrosses. Informative posters and leaflets could be produced that educate fishermen about the biology of albatrosses, the economic costs of losing baits to albatrosses and albatross-friendly fishing methods. Other measures such as using weighted swivels, using fewer and less intense lights on the boats and only setting lines at night will also reduce the bycatch of albatrosses. In Uruguay we have limited resources, but a joint effort between INAPE and our University could produce such a campaign. Visits to Uruguay by international ornithologists will also help in raising the awareness of the public and fishing crews to the problem of albatross mortality in longline fisheries.

ACKNOWLEDGEMENTS

Our presence on fishing boats in Uruguayan waters was possible through the kind co-operation and interest shown by C. Rios and the authorities of INAPE, Montevideo. We thank P. A. Prince for assistance with the literature and for sharing unpublished satellite tracking data of Wandering Albatrosses. A.S. and R.V-F especially thank Graham Robertson of the Australian Antarctic Division, and the Australian Antarctic Foundation, for making it possible for us to attend the First International Conference on the Biology and Conservation of Albatrosses.

REFERENCES

Croxall, J. P. and Prince, P. A., 1990. Recoveries of wandering albatrosses *Diomedea exulans* ringed at South Georgia 1958–1986. *Ringing and Migration* **11:** 43–51.

Escalante, R., 1970. Aves Marinas del Rio de la Plata y Aguas Vecinas del Oceano Atlantico. Barreiro y Ramos, Montevideo.

Gore, M. E. and Gepp, A. R. M., 1978. Las Aves del Uruguay. Mosca Hermanos, Montevideo.

Neves, T. and Olmos, F., 1997. Albatross mortality in fisheries off the coast of Brazil. Pp. 214–19 *in* Albatross Biology and Conservation ed by G. Robertson and R. Gales. Surrey Beatty & Sons: Chipping Norton.

Prince, P. A., Wood, A. G., Barton, T. and Croxall, J. P., 1992. Satellite tracking of wandering albatrosses *Diomedea exulans* in the South Atlantic. *Antarc. Sci.* **4:** 31–36.

Prince, P. A., Croxall, J. P., Trathan, P. N. and Wood, A. G., 1997. The pelagic distribution of South Georgia albatrosses and their relationships with fisheries. Pp. 137–67 *in* Albatross Biology and Conservation ed by G. Robertson and R. Gales. Surrey Beatty & Sons: Chipping Norton.

Tickell, W. L. N., 1967. Movements of black-browed and grey-headed albatrosses in the South Atlantic. *Emu* **66:** 357.

Vaske, T., 1991. Seabirds mortality on longline fishing for tuna in southern Brazil. *Ciencia e Cultura* **43:** 388–90.

Vaz-Ferreira, R., (co-ordinating author), 1994. Mortality of albatrosses and other seabirds produced by tuna long-line fisheries in Uruguay. United Nations Environment Programme Report CMSScC 5.6.

CHAPTER NINETEEN

The roles of organochlorine contaminants and fisheries bycatch in recent population changes of Black-footed and Laysan Albatrosses in the North Pacific Ocean

JAMES P. LUDWIG,[1] CHERYL L. SUMMER,[1] HEIDI J. AUMAN,[1] VANESSA GAUGER,[1] DARCY BROMLEY,[1] JOHN P. GIESY,[2] ROSALIND ROLLAND[3] and THEO COLBORN[3]

Organochlorine contaminants were measured in Laysan and Black-footed Albatrosses nesting at Midway Atoll, Hawaiian Leeward Islands. Samples from Black-footed Albatrosses were two to three times more contaminated with polychlorinated biphenyls (PCBs), polychlorinated dibenzo-*p*-dioxins (PCDDs), polychlorinated dibenzofurans (PCDFs), 2, 3, 7, 8-tetrachlorodibenzo-*p*-dioxin equivalents (TCDD-EQs) and the DDT group chemicals than were samples from Laysan Albatrosses. Black-footed Albatrosses were sufficiently contaminated to be at slight risk from eggshell thinning and subtle embryonic effects that increase risks of terata and decrease egg viability; observed hatching success was lower and egg crushing greater in Black-footed Albatrosses than Laysan Albatrosses. Organochlorine contamination probably reduced productivity by 2–3% in the Black-footed Albatross population. No contaminant effects were found in Laysan Albatrosses. Estimates of the effect of adult loss and widowing from fisheries bycatch revealed productivity reductions of 17.6% in Laysan Albatrosses and 27.15% in Black-footed Albatrosses, confirming the serious impact of fisheries on these species of albatrosses. Black-footed Albatross census data indicate a recent decrease in populations since the 1980s. Bycatch accounts for almost all (90%) of the recent productivity decrease in this population and contaminants were believed to cause about 10%. The Black-footed Albatross population is expected to decrease slowly owing to these mortalities. The Laysan Albatross population demonstrated no contaminant-related effects, and is increasing slowly in spite of increased bycatch in the last 20 years.

Key words: Black-footed Albatross, Laysan Albatross, Organochlorines, PCBs, Dioxins and Furans, TCDD-EQs, DDT, Bycatch, Fisheries.

INTRODUCTION

THE unintentional capture of albatross during fishing, or bycatch, kills thousands of albatrosses world-wide every year (Ainley *et al.* 1981; Brothers 1991; DeGange and Day 1991; Ogi *et al.* 1993). Johnstone *et al.* (1993) estimated 4 426 Black-footed Albatrosses *Diomedea (Phoebastria) nigripes* and 17 548 Laysan Albatrosses *D. (Phoebastria) immutabilis* were killed in 1990 in the now banned North Pacific high seas driftnet fisheries; these losses were estimated to be 2.5% and 1.5% of the world-wide populations of Black-footed and Laysan Albatrosses, respectively. Longline fisheries have recently replaced the banned driftnet fisheries in the North Pacific Oceans. Because longline fisheries are less well monitored than were driftnet fisheries, there are few estimates of current longline albatross bycatch. The few data available suggest the longline fisheries and the driftnet fisheries catch rates of Black-footed Albatrosses (about 4 500/year) are similar (derived from Johnstone *et al.* 1993; Wohl *et al.* 1995; Skillman and Flint 1997). Bycatch in unmonitored commercial fisheries is presently unknown.

Midway hosts the world's largest Laysan Albatross colony (~428 000 pairs), and the second largest Black-footed Albatross colony (~19 700 pairs in 1994–95) (Seto 1995; Flint 1995), but populations are probably still well below the historic maxima of the 1870s. In the 1960s the Laysan Albatross population grew by an estimated 7–8% per year (Fisher 1975a,b). However, over the period 1975–1995 the Laysan Albatross population has increased at some breeding sites and decreased at others

[1]The SERE Group Ltd, 138 Road 2 West, Kingsville, Ontario N9Y 2Z6 Canada.
[2]Department of Fisheries and Wildlife, Michigan State University, East Lansing, Michigan 48224 USA.
[3]World Wildlife Fund, 1250 24th Street NW, Washington DC 20037 USA.

Pp. 225–38 *in* ALBATROSS BIOLOGY AND CONSERVATION ed by G. Robinson and R. Gales. Surrey Beatty & Sons, Chipping Norton. 1997.

(Croxall and Gales 1997). The Black-footed Albatross population increased slowly until the 1960s, but has since stabilized or slightly decreased (Rice and Kenyon 1962; Johnstone *et al.* 1993; Flint 1995). Although the breeding Black-footed Albatross population at Midway fell from 19 757 nests to 18 731 nesting pairs between 1991 and 1994, this apparent decrease was caused by base demolition work on the U.S. naval facility at Midway undertaken in November 1994 (Seto 1995). A 50% decrease of Black-footed Albatross nests was recorded at French Frigate Shoals in the last decade (Croxall and Gales 1997). Smaller decreases were recorded at other Hawaiian Islands' Black-footed Albatross nesting colonies, especially those nearer to Honolulu. Driftnet and longline fishing bycatch associated with the Honolulu-based fishing fleet have suspected roles in the Black-footed Albatross population decreases.

The albatrosses at Midway were selected by United States Environmental Protection Agency in 1992 for a pilot study of the global dispersion and possible impacts of organochlorine chemicals on oceanic organisms. Albatrosses are ideal indicators of oceanic contaminants owing to their size, pelagic distributions and feeding habits. Their intakes of contaminants were believed to be from food and other items ingested at great distances from coastal point sources of organochlorine pollutants. Albatrosses at the top of the food web were expected to be minimally contaminated because the Pacific Ocean is an immense contaminant sink. One of the objectives of this study was to employ new techniques that measure dioxin-like toxicity as toxic equivalents (TCDD-EQs or TEQs) to 2,3,7,8 tetrachlorodibenzo-*p*-dioxin (Safe 1990; Tillitt *et al.* 1991a, 1992), and to use these estimates to establish cause-effect links using the Great Lakes Embryo Mortality Edema and Deformities Syndrome (GLEMEDS) paradigm (Gilbertson *et al.* 1991). GLEMEDS links subtle effects of dioxin-like contaminants to egg and embryo viability, the frequency of specific terata in embryos and hatchlings, and other structural and functional impairments to reproductive processes that appear in individuals exposed as embryos only after sexual maturation (Giesy *et al.* 1994a; Devito and Birnbaum 1994; Colborn and Clement 1993; Gilbertson *et al.* 1991). This subtle damage is not detected easily in nesting adults except as reproductive impairments in adults exposed *in ovo*. At low exposure concentrations, very large sample sizes are needed to detect subtle changes to reproductive processes and subsequent population effects.

This chapter reports on the breeding success and organochlorine contaminants of Laysan and Black-footed Albatrosses nesting at Midway Atoll between November 1992 and July 1995. Our findings are evaluated with available population data (e.g., Rice and Kenyon 1962; Fisher 1976; Flint 1995) to assess the impacts of human-induced mortality. Our goal was to discriminate between the effects of fisheries bycatch and organochlorine contaminants, and to evaluate their relative importance to albatross populations.

METHODS

Salient features of breeding cycles

Albatrosses are among the longest-lived of birds, with low reproductive rates and an extended prebreeding period. Mean age at first nesting varies from 8 to 13 years. Albatrosses are incapable of laying a replacement if their single egg is lost. Chicks require 6–13 months to fledge. Breeding occurs either annually or every two or three years, and all albatross species have many immature age classes. Natural annual adult death rates are very low, ranging from 4 to 9% or less (Fisher 1976; Weimerskirch and Jouventin 1987). Pair-bonds are life-long. Mate loss will cause the loss of one to five subsequent breeding seasons by the widowed bird (Fisher 1976). These are properties of populations with slow responses to unnatural mortality. The large numbers of immatures and long time to attain maturity can mask negative population changes for a decade or longer if population status is assessed by colony census data alone.

Pilot study

During the 1992–93 breeding season (hereafter referred to as 1993) we performed a pilot study to test methods of collecting samples and field data used to assess biological markers, contaminant burdens and reproductive output of Midway's albatrosses. The data quality met the objectives of the U.S. Environmental Protection Agency. In early January 1993 when incubation was 60–75% complete, 683 Laysan Albatross and 239 Black-footed Albatross nests and eggs were marked to measure hatching rates (Table 1). A few eggs were observed crushed when laid in November and December and an unknown number of eggs were lost or abandoned during incubation. We did not quantify these egg losses in 1993. Hatch rates were 90.5% in Laysan Albatross and 88.7% in Black-footed Albatross of those eggs that survived intact through the incubation period. A few eggshells fragmented near to the date of expected hatch, killing late stage embryos. Eggshell competence problems were slightly more frequent among Black-footed Albatrosses.

Studies in 1994 and 1995

Preliminary results from 1992/93 season (DDT group chemicals: 1.8 and 0.3 mg/kg for Black-footed and Laysan Albatrosses,

Table 1. Hatch rates of albatross eggs incubated in mid-January 1993.

		Eggs				Contaminant totals*		
Species	Albatrosses (n)	Infertile (n)	Unknown fate (n)	Hatched (n)	Hatched (%)	PCBs	DDT	TCDD-EQs
Laysan Albatross	683	30	43	618	90.5	1.1	0.3	17.5
Black-footed Albatross	239	13	22	212	88.7	3.8	1.8	37.5

*Units of total PCBs and total DDT-group chemicals are for all congeners and are reported as mg/kg wet weight. Units of TCDD-EQs are in ng/g wet weight equivalents to 2,3,7,8, tetrachlorodibenzo-*p*-dioxin.

Table 2. Summary of albatross productivity studies at Sand Island, Midway Atoll in 1994 and 1995.

		Eggs			Chicks				
Year	Albatrosses (n)	Abandoned or lost (n)	Crushed/ cracked (n)	Died (n)	Hatched (n)	Hatched (%)	Died (n)	Fledged (n)	Fledged (%)
Laysan Albatross									
1994	634	55	32	75	472	74.5	100	372	58.7
1995	1 117	22	39	113	943	84.4	125	818	73.2
Sum	1 751	77	71	188	1 415		225	1 190	
%	100	4.4	4.1	10.7		80.8	12.9		69.4
Black-footed Albatross									
1994	1 029	69	82	142	736	71.5	187	549	53.4
1995	1 578	49	71	147	1 311	83.1	207	1 104	70.0
Sum	2 607	118	153	289	2 047		394	1 653	
%	100	4.5	5.9	11.1		78.5	15.1		63.4

respectively; TCDD-Eqs: 37.5 and 17.5 pg/g ww in Black-footed and Laysan Albatross eggs, respectively) (Tillitt *et al.* 1991b; Jones *et al.* 1995; Auman 1994; Table 1) led us to design two years (1994 and 1995) of intensive reproductive monitoring using very large sample sizes of nests of Laysan and Black-footed Albatrosses. Field standard operating procedures were updated and new standard operating procedures prepared for *beta*-backscatter non-destructive in-field assessment of eggshells' quality after the methods of Fox *et al.* (1975). We monitored study plots for each species on Sand Island, Midway Atoll, which were representative of habitats, local human disturbance patterns and nesting densities. Each nest was marked within 0–7 days of egg laying. In December, 1993 we marked 634 Laysan Albatross and 1 029 Black-footed Albatross nests in 15 and 16 plots, respectively. In late November and early December 1994 we marked 1 117 Laysan Albatross and 1 587 Black-footed Albatross nests in 16 and 10 plots, respectively (Table 2). Each marked nest was monitored once per week for 34 weeks. Egg and chick fates were recorded as lost, crushed, cracked, embryo died, died with flaking shell, pipped and died, hatched, or chick died. Chicks were banded at ~2 months of age, just after the guard stage of adult care was completed. Fledging was recorded when a chick abandoned its territory and could not be found for two consecutive weeks of monitoring after June 15 each year. All nests were monitored weekly from December through late July in each year.

Chemical analyses and hazard indices

Sample collection, extraction and total contaminant quantification

Eggs used for analyses each year were supplied by U.S. Fish and Wildlife Service staff from nests destroyed as a part the closure of the naval base at Midway Atoll. Ten whole eggs were pooled from each species in November and early December each year into 4 L triple-cleaned USEPA grade 3 glass jars (Eagle-Picher, Miami, OK) and frozen at –20° C until analysed. Total polychlorinated biphenyls (PCBs) and TCDD-EQs were determined by extraction of a homogenized 10 gram sub-sample of each pool of egg contents, followed by drying with 40 grams of anhydrous sodium sulfate and column extraction with 200 ml methylene chloride. The sample extract was concentrated to 8 ml of which 5 ml was subjected to gel permeation chromatography on a column of Biobeads SX3 (Bio-Rad, Richmond CA). The sample was then cleaned on a column of H_2SO_4 silica gel to remove certain pesticides and residual lipids. IUPAC PCB 30 (2,4,6-trichlorobiphenyl) was added as an internal standard. Extraction efficiency for total PCBs was >92%, >90% for coplanar PCBs and >95% for polychlorinated dibenzo-*p*-dioxins (PCDDs) and polychlorinated dibenzofurans (PCDFs). Coplanar dioxin-like PCB congeners contribute most to TCDD-EQs and TEQs (Ahlborg *et al.* 1994). Gas Chromatography operating conditions and quantification of total PCBs were those of Tillitt *et al.* (1991a) where they are described in detail.

Solvents used in extract preparation were pesticide grade. Glassware and equipment used in extract preparation were rinsed with acetone and hexane before use, following strict QA/QC laboratory protocols at the Michigan State University Pesticide Research Center. H4IIE bioassay procedures which measure TCDD-EQs were the standard procedures and methods of Tillitt *et al.* (1991a,b, 1992).

Detailed descriptions of the methods used to quantify total concentrations of PCBs and TCDD-EQs are given in Tillitt *et al.* (1991a,b, 1992), Jones *et al.* (1995), Auman (1994) and Auman *et al.* (1997a). Total PCBs and DDTs are reported as mg/kg (parts per million) concentrations wet weight; TCDD-EQs are reported as ng/kg (parts per trillion) concentrations wet weight, uncorrected for extraction efficiencies. Hazard indices were calculated as the quotient of the concentration in eggs divided by concentrations in eggs of other avian species that were confirmed to have no effect (NOAECs). By convention, NOAECs are assumed to be 0.1 the concentration of the lowest concentration with a confirmed adverse effect (LOAECs). Therefore, a HI = 1 predicts no effect, and an HI = 10 is the minimal level where effects were observed in another avian species.

TEQ quantification from the concentration of coplanar congeners, and calculation by toxic equivalency factors

Fat and liver tissue extracts were shipped under chain-of-custody protocols to the ESR laboratories in New Zealand where concentrations of coplanar congeners of polychlorinated biphenyls (PCBs), polychlorinated dibenzodioxins (PCDDs), polychlorinated dibenzofurans (PCDFs), and some polychlorinated naphathalenes (PCNs) were measured instrumentally by advanced GC/MS techniques (Jones *et al.* 1994, 1996; Auman 1997a).

RESULTS

Breeding success

Black-footed Albatrosses hatched about 2.5% fewer eggs than Laysan Albatrosses in the three years of the study ($X^2 = 7.74$. df = 1; P = <0.002; Tables 1 and 2). The difference in egg mortalities between these two species was due to egg crushing shortly after the eggs were laid. Eggshell-related problems occurred at the start and end of incubation. During the first few days to one week of incubation eggs that were crushed typically collapsed from the pointed end. More than twice as many Black-footed Albatross eggs as Laysan Albatross eggs were crushed during this period (Fig. 1). Further, it is possible that we undercounted the number of crushed eggs laid in study plots, especially for Black-footed

Fig. 1. Contaminants-related eggshell thinning and breakage of Black-footed Albatross eggs from Midway Atoll, Hawaiian Leward Islands.

Albatrosses. We recorded two Laysan Albatross and 19 Black-footed Albatross nest cups that only had eggshell fragments present when we moved through plots to mark eggs initially. If all of these shell fragments were from recently crushed eggs, as is likely, then the rates of early egg crushing were 1.65% of Laysan Albatross nests and 4.02% of Black-footed Albatross nests of the eggs laid. Eggs that cracked in the last two weeks of incubation often fractured, and then flaked off large pieces of shell exposing the inner shell membrane which then dried to a hard leathery surface leading to embryo desiccation and death. The rates of late incubation shell fragmentation were similar in Black-footed Albatross and Laysan Albatross eggs. Overall, Laysan Albatrosses lost 4.07% and Black-footed Albatrosses 5.95% of their marked eggs to these problems during 1994 and 1995. Black-footed Albatrosses also had a slightly higher rate of embryo death (11.1% versus 10.7% of eggs laid). More Black-footed Albatross chicks died after hatching than did Laysan Albatross chicks. Although consistent over the three years, this difference was not statistically significant ($X^2 = 1.74$, 1 df, 0.20>P>0.10) and over both years, Black-footed Albatrosses fledged 6% fewer chicks than Laysan Albatrosses based on 1 751 Laysan Albatross and 2 607 Black-footed Albatross nests (Table 2).

Contaminant concentrations and hazard indices

The bioassay-derived TCDD-EQs account for agonistic: antagonistic interactions among the Ah-receptor active and inactive congeners in contrast to the calculated TEQs from the additive model which do not (Ahlborg *et al.* 1994; Giesy *et al.* 1994b). Jones *et al.* (1994, 1996) reported TEQs of 52.5 and 124 for Laysan Albatross and Black-footed Albatross eggs laid in November 1993, respectively. They also analysed eggs by H4IIE rat hepatoma cell bioassay for TCDD-EQs and found 12.5 and 52.5 TCDD-EQs in the same egg pools, respectively. These levels compared to 17.5 TCDD-EQs in Laysan Albatross eggs from December, 1992 (two year mean of 20 Laysan Albatross eggs 15.0 TCDD-EQs). Black-footed Albatross eggs in December 1992 held 37.5 TCDD-EQs (two-year mean of 20 Black-footed Albatross eggs of 45.0 TCDD-EQs).

Hazard Indices calculated from these data ranged from 5 to 33 for Laysan Albatross eggs, and 12 to 83 for Black-footed Albatross eggs based on TCDD-EQs, TEQs and the reported LOAECs for unrelated avian species (Auman 1994). When an average HI was calculated from the mean of all known avian lowest adverse effects concentrations (LOAECs) and the lower bioassay-derived TCDD-EQs were used as the contaminant concentrations, the Laysan Albatross eggs' HI = 7.4, and no dioxin-like effects were predicted. Black-footed Albatross eggs with a HI = 17.7 were projected to be well above the concentration where dioxin-like effects were expected for the average avian species (Auman *et al.* 1997a). Average HIs calculated from means of all residue concentration data confirmed our suspicion from the pilot study that Laysan Albatross were just below the threshold for dioxin-like chemical residue effects, but Black-footed Albatrosses were well above the threshold, assuming these are species with average susceptibilities to these toxicants.

DISCUSSION

Interspecific egg losses and shell qualities

Increased adult mortality has the most immediate impact on albatross populations because of their low natural death rates and delayed maturity (Fisher 1971, 1976). Laysan and Black-footed Albatrosses must share incubation and provisioning chores in order to fledge young (Fisher 1971, 1976). Bycatch could account for some of the egg loss or abandonment by increasing the rate of mate loss, and will cause some chick mortality. Incubation occupies two months of the year. For Black-footed Albatrosses losing mates at the reported bycatch rate of 2.5%/year for the population (Flint 1995; Gould *et al.* 1993), the probability of mate loss owing to bycatch during a two-month long incubation is 2.5% × 0.16 = 0.4%. For Laysan Albatrosses the probability of mate loss from bycatch is lower (1.5% × 0.16 = 0.24%). The net difference between the species of bycatch-induced egg loss is only 0.16%, consistent with different egg loss rates observed (4.4% in Laysan Albatross versus 4.5% in Black-footed Albatrosses, Table 2). Most eggs were lost when kicked out of nests during exchange of incubating birds or territorial squabbles. Once an egg was outside the nest, it was never returned to the nest.

Of greater interest is the difference in egg crushing/cracking rates (Fig. 1). Black-footed Albatrosses lost 3.34–4.02 % of their eggs to crushing, almost always on the pointed end of the egg, while Laysan Albatrosses lost 1.54–1.65% of their eggs to crushing; a further 2.53% of Black-footed Albatross and 2.51% of Laysan Albatross eggs had shells flake off just before hatching (Table 2). LOAECs in the egg for shell thinning by DDT-group compounds approximate 3 mg/kg ww (Bowerman *et al.* 1995). Jones *et al.* (1994) found that samples of Black-footed Albatross eggs had 60% (1.8 mg/kg) but Laysan Albatross eggs only 10% (0.3 mg/kg) of this critical threshold concentration which Auman (1994) calculated were below NOAECs for both species. However, contaminant concentrations vary widely between individuals in wildlife populations. Auman *et al.* (1997a) reported a 19-fold variance of mean DDT group chemical levels in plasma samples for 40 Laysan Albatross and 37-fold for 36 Black-footed Albatross for from Midway albatrosses in 1993. Thus we suggest that a few Black-footed Albatross females, perhaps 2–3% of the population, are likely to be more highly contaminated with concentrations well above the 1.8 ppm average concentration; this could cause direct eggshell thinning effects.

It is also possible that we found a different contaminant-related effect on eggshells at the low end of the dose:response curve in the large number of samples (2 607) of Black-footed Albatross eggs we monitored. Endocrine disrupting effects of DDT and PCB group compounds or their metabolic breakdown products from embryonic exposures of decades ago are also possible. Some of these compounds produce powerful hormone-like effects that could have altered the number and sensitivity of hormone receptors in the reproductive tracts of females exposed *in ovo* that are now egg-laying adults. If so, then these birds may produce eggs with defective shells as adults without any adult exposure, and the current *in ovo* concentrations of the DDT-group contaminants may have no association with the defective eggshells we found. Such inter-generational effects are known in wild birds (Fry *et al.* 1987) and American alligators *Alligator mississippiensis* (Vonier *et al.* 1996). These subtle contaminant effects are expressed by altered differentiation of gonads and the distribution and number hormone receptors in the reproductive tract.

Previous studies of eggshell thinning in wild bird populations were generally undertaken with much smaller sample sizes, typically in the range of 10–50 eggs (Klass *et al.* 1974). It is likely that our large sample sizes of (n = 1 751 Laysan Albatross and 2 607 Black-footed Albatrosses, Table 2) allowed us to detect eggshell thinning effects at markedly lower mean concentrations than other investigators reported as NOAECs. We examined the shells of some of the eggs that show the pointed end thinning by *beta*-backscatter (Fox *et al.* 1975) and conventional measurement techniques. We found shell thickness on the pointed ends to be up to 40% lower in eggs that were crushed compared to uncrushed eggs (Fig. 1). Black-footed Albatross whole eggshells in 1995 were 3–4% thinner than museum collections of shells from pre-WWII. We conclude that modest eggshell thinning is occurring in Black-footed Albatross but not in Laysan Albatross eggs. The most likely cause of this effect is organochlorine contaminants.

Hazard indices

The H4IIE bioassay and TEQ contaminant concentrations predict little or no hazard to Laysan Albatrosses and only a slight hazard to Black-footed Albatross embryos from the contaminant concentrations in Black-footed Albatross eggs. The Black-footed Albatross concentrations are cause for concern since the lethal dose for 50% (LD50s) of the very sensitive white leghorn chickens' eggs are 115–147 TCDD-EQs (Brunstrom and Andersson 1988; Henschel 1993). Other species' LD50 egg concentrations are 50–100 TEQs (for PCDFs only, other chemicals unknown) in wood ducks (White and Hoffman 1995), between 300 and 400 TCDD-EQs for great blue herons (Henshel *et al.* 1995), and about ~460 TCDD-EQs for Double-crested Cormorants (Tillitt *et al.* 1992; Ludwig *et al.* 1993; Giesy *et al.* 1994a,b). Black-footed Albatross eggs have concentrations of TCDD-EQs (52.5) and TEQs (124) that can produce elevated embryo mortality and the wide-ranging effects of GLEMEDS in sensitive avian species. Considering that we cannot determine when mate loss was a cause of embryo death, and that we expect a slight differential of 0.15% owing to greater bycatch of breeding Black-footed Albatrosses, then only about 0.21% of the differential mortality observed is likely attributable to another cause, such as dioxin-like toxicity and GLEMEDs. The slightly greater embryo death rate of Black-footed Albatross eggs we found is consistent with the greater TCDD-EQ and TEQ concentrations present in their eggs. There is no reasonable evidence in these data of contaminant effects to Laysan Albatross reproductive success or that organochlorine contaminants caused any chick death after hatching in either species.

All bird species thus far tested have been found to have very steep dose:response curves for dioxin-like chemicals in the embryo stage once a concentration is present associated with an effects threshold (Tillitt *et al.* 1992; Giesy *et al.* 1994a,b; Devito and Birnbaum 1994). The threshold for dioxin-like effects in these two albatross species is presently unknown since no controlled experiments have been done. Wild Double-Crested Cormorants *Phalacrocorax auritus* with 35 bioassay-derived TCDD-EQs hatched 91.5% of their eggs, 82% of their eggs at 161 TCDD-EQs, and only 63% of eggs with 337 TCDD-EQs. Tillitt *et al.* (1992) and Ludwig *et al.* (1996b) reported exceptionally robust co-efficient of determination values of 0.703 ($P < 0.003$) and 0.91 ($P < 0.001$), respectively in their studies correlating bioassay-derived TCDD-EQs with egg hatchability. Given the general effective metabolism of scavengers to detoxify all types of contaminants, the base of the albatross dose:response curve is not likely to be in the range of 7–10 TCDD-EQs/TEQs reported for very sensitive white leghorn chickens (Cheung *et al.* 1981; Brunstrom and Andersson 1988; Henschel *et al.* 1993). It is likely Black-footed and Laysan Albatrosses are relatively insensitive to TCDD-EQs/TEQs.

Other much more subtle effects may be present in the more highly exposed Black-footed Albatross chicks without a measurable difference in hatching rates. For example, asymmetries in brain development of TCDD-exposed Great Blue Herons *Ardea herodias* occurred in the ranges of 30–60 TCDD-EQs with no increase in egg death or other overt signs of GLEMEDS (Henshel *et al.* 1995). Other delayed reproductive effects associated with endocrine disruption of developmental processes may occur when the more exposed Black-footed Albatross offspring mature. Black-footed Albatross egg concentrations are sufficient that such increases in TCDD-EQs or TEQs will cause increases in egg death rates, and other wide-ranging effects characteristic of GLEMEDS (Fig. 1; Ludwig *et al.* 1996b; Giesy *et al.* 1994b; Jones *et al.* 1996). There is no dilution capacity left in the North Pacific Ocean ecosystem for either dioxin-like chemicals or total DDT chemicals without increasing damage to the Black-footed Albatross (Auman 1997a).

Contaminant trends

We know of only one previous measurement of organochlorine concentrations of total PCBs and DDT-group chemicals in North Pacific Ocean albatrosses, that of Fisher (1973) who reported 15.3 mg/kg total PCBs and 16.7 mg.kg total DDT (lipid nomalized, extraction corrected basis) in Black-footed Albatross visceral fat from samples collected in 1970. The Black-footed Albatross eggs analysed by Jones *et al.* (1994, 1995) and Auman *et al.* (1997a) averaged

10.6% lipid. Thus, on a lipid normalized basis (without correction for recoveries), the 1992 Black-footed Albatross egg concentration of 3.8 mg/kg of total PCBs (PCBs 3.8 × 100/10 = 38 mg/kg), and the 1.8 mg of DDE (1.8 × 100/10 = 18 mg/kg) suggest a near steady state in totals of both these generic chemical groups for Black-footed Albatrosses in the past two decades. Furthermore, Auman *et al.* (1997a) reported that the mean ratio of parent DDT:DDE in blood samples ranged from 1.1 to 19.7 ng/ml in Black-footed Albatrosses and from 1 to 23 ng/ml in Laysan Albatrosses, whereas the 1970 fat samples reported by Fisher held a DDT/DDE ratio of just 1:11.3. These data are strong evidence of recent DDT use since the environmental half-life of parent DDT is 1–3 years. Iwata *et al.* (1995) reported prominent DDT plumes in air and Pacific Ocean surface waters moving eastward in 1993 from south-east Asia where DDTs are still widely used.

The chemical analytical methods used by Fisher (1973) were very primitive by modern standards. In most contaminant reports before 1975, many compounds were routinely lumped together rather than separated by modern individual peak summing GC/MS techniques. Furthermore, no measurements of PCDDs or PCDFs were possible, and the pattern of food chain enrichment owing to either differential weathering (Oliver *et al.* 1989) or metabolism leading to accumulation of the most toxic congeners (Yamashita *et al.* 1993) were unknown to ecologists. The Fisher data are not comparable to the reports of Jones *et al.* (1994, 1996), nor suitable to calculate the historic toxicity hazard (the data are not sufficiently precise for congener analysis). However, the data suggest the hazards to Black-footed Albatrosses from the total PCB and DDT contaminant concentrations could have been as great or greater in 1992–1994 as they were in 1970.

Greater concentrations of the dioxin-like congeners may not have reached pelagic areas from continental point sources until recently. The PCDD and PCDFs in the contaminant mixtures contributed more than a third of the total TEQ toxicity found in the 1994 egg samples. These chemicals may be acquired through the consumption of partly burned plastic pieces (Jones *et al.* 1996). Plastics may also deliver PCBs to albatross by passive adsorption of PCBs from the sea surface microlayer (Ryan *et al.* 1988). Plastic consumption appears to have increased about 20 fold between the 1960s and 1980s (Kenyon and Kreidler 1969; Auman 1994). However, plastics are not confirmed to be a direct mechanical cause of death in seabirds (Sileo *et al.* 1990; Sievert and Sileo 1993; Auman 1994, 1997a). Albatross are surface feeders, exploiting food from the contaminant-enriched sea-surface (Hardy 1982). PCBs are reported to be enriched in the biogenic lipid surface foams by 10 fold above dissolved water concentrations (Napoliatano and Richmond 1995). The North Pacific Ocean albatrosses consume floating fish and invertebrate eggs, especially Black-footed Albatrosses (Harrison 1990). Floating fish and invertebrate eggs are essentially bags of lipid-rich yolk inside a semi-permeable membrane in direct contact with an enriched sea-surface microlayer source of contaminants. It is probable that these fish eggs passively accumulate the very dilute concentrations of contaminants from the sea-surface (Lebo *et al.* 1995), assuring a direct pathway into the fish eggs and thence to surface-feeding albatrosses. The dietary preference of Black-footed Albatrosses for fish eggs, compared to the Laysan Albatross preference for squid, may be largely responsible for the differing contaminant concentrations in these species.

Relative effects of contaminants and bycatch

Annual mortality in the Laysan Albatross population at Midway Atoll was about 6% of breeding adults during the 1960s when contaminants and bycatch were not thought to be important mortality factors (Fisher 1971, 1976). Fisher studied the same marked pairs in plots for 14 consecutive nesting seasons. We can find no reports of contaminant or bycatch effects until the 1970s. Because few Laysan Albatrosses breed successfully until age nine, the youngest regularly nesting birds Fisher studied in 1974 (at the end of his study) hatched in 1965. Most adult Laysan Albatrosses he studied between 1960 and 1974 were chicks hatched in 1925–1965, before or early in the world-wide release of organochlorine compounds. Thus, Fisher's estimate of 6% annual adult mortality is probably close to the natural mortality rate in the Laysan Albatross without contaminant effects.

However, the North Pacific Ocean albatross population responses are probably now driven by a complex interaction of natural, contaminant-related, and bycatch mortality factors. Reproductive opportunities are lost owing to adult and immature bycatch mortalities, broken pair bonds, lowered embryo survival from eggshell thinning, possibly a low incidence of GLEMEDS in Black-footed Albatrosses, natural mortality and senescence. Appendix 1 provides a set of assumptions and age-specific survivorship rates for a cohort of females (both species combined) entering these albatross populations at nine years of age, which is the age of first successful breeding. Population structure is estimated with four annual death rate scenarios in fisheries (the estimates are preliminary because it is assumed catch rates in driftnet and longline fisheries are similar [see above]). The estimates are 6%

natural loss rate (Fisher 1976), 6% + 1.5% bycatch for Laysan Albatrosses, 6% + 2.5% bycatch for Black-footed Albatrosses, and an excessive 10% rate (Appendix 2). Ten to 30 year-old Laysan Albatrosses were observed to reproduce 84% of their adult lives (Fisher 1976). Older birds from other albatross species have reduced reproductive rates (Tickell 1968; Carrick *et al.* 1960). Therefore, estimates of reproductive capability were adjusted for age-specific fecundity by reducing the breeding frequency to 75% among 30–34 year old females, 50% in 35–44 year olds, 25% in 45–50 year olds, and no reproduction after age 50. Under all scenarios, annual death rates were accelerated over the natural 6% rate by 3% in the 35–44 year age group, by 6% in the 45–50 year age group, and by 9% in the birds older than 50 years to account for increasing mortality with advancing age.

Based on the estimates above the albatrosses have 9 632–6 121 reproductive attempts over the life span of each cohort of 1 000 females (Appendix 3). The cohorts have different characteristics. The median age of breeding females (defined as that time when half the cohort has died) is 20.21 years with 94% survivorship (6% mortality), but drops to only 15.61 years at 90% survivorship. As survivorship decreases, the relative contribution of the youngest, least experienced age classes increases markedly. With 94% survivorship, the birds in the 9–19 year age classes provide 60% of the breeding effort while these youngest adults provide 73% of the breeding effort at 90% survivorship. Albatrosses engage in extensive courtship to acquire a replacement mate and always lose at least one breeding season after mate loss. The average Laysan Albatross female loses 1.9 breeding seasons after mate loss, and a few fail to nest for up to five years (Fisher 1976). Bycatch affects future breeding of the surviving mate causing the loss of 235 reproductive years for the Laysan Albatross cohort with an estimated 1.5% bycatch mortality rate (1.5% × 1.9 years × 8 260 = 235 years lost). The Black-footed Albatross cohort with an estimated 2.5% bycatch mortality rate (see above) will lose 358 reproductive years, while 504 years are lost in the 90% survivorship case from mate loss. The net effect of 1.5% increase in mortality rate from bycatch is a 16.7% (i.e., 1 607 ÷ 9 632, Appendix 3) loss of reproductive effort, 25.5% is lost with 2.5% bycatch mortality, and 36.5% is lost if the bycatch annual mortality rate reaches 4% (Appendix 3).

If the Black-footed Albatross population is currently experiencing a bycatch mortality of ~2.5%, then adding the 1.65% loss from consequential egg and chick abandonment (see above) to the 25.5% loss gives a total estimated loss of 27.15% of female reproductive effort, an order of magnitude greater than the reproductive losses we have attributed to contaminants (~2.5%). We therefore conclude that ~90% of the human-caused mortalities in the Black-footed Albatross population at Midway Atoll are caused by fishing, and ~10% by contaminants. The total reduction estimated from the combined effect of both factors (29.65%) is almost a third of the normal reproductive output for Black-footed Albatrosses. The total reduction in reproductive output owing to bycatch for Laysan Albatrosses is markedly less. Bycatch (there are no known contaminant effects) reduces Laysan Albatross reproductive output by 16.7% (Appendix 3) with a further 1% loss in the egg and chick stages from mate loss; hence the total loss in Laysan Albatrosses is 17.7%. These estimates suggest Black-footed Albatrosses are in far greater jeopardy than Laysan Albatrosses at Midway Atoll.

In addition to the lack of accurate estimates of bycatch in fisheries we lack of a good estimate from band returns of the survivorship rates from fledging to age of first recruitment to the breeding populations. Most data that could be used for this purpose are from bandings made with aluminium bands that are known to wear excessively in albatrosses and other long-lived seabird species (Ludwig 1967, 1981; Ludwig *et al.* 1996); band loss causes serious overestimation of preadult mortality rates (Ludwig 1967). In the absence of estimates for albatrosses we have assumed fledgling survival in the first year is similar to that (about one-third) for gulls *Larus* spp. and terns *Sterna* spp. (Lack 1954; Hickey 1952; Ludwig 1965, 1967, 1974, 1981) and that the second-year survival is about 82% (Hickey 1952; Ludwig 1981); these rates are generally in accordance with the 28% survivorship to age 10 years in Black-browed Albatrosses *D. (Thalassarche) melanophrys* (Prince *et al.* 1994). Thereafter, survivorship rates of immature age classes are usually indistinguishable from the adult rates (about 94%). We use these suvivorship estimates to estimate the potential for growth (Appendix 4) and the effect of bycatch and contaminants (Appendix 5) on the status of albatrosses at Midway Atoll.

Black-footed Albatrosses averaged 634 and Laysan Albatrosses 694 fledged young/1 000 nests/year in 1994 and 1995 (Table 2). However, these were two unusually good years for chick survival. The average fledging rate is most certainly lower for both species (Fisher 1971, 1976; SERE Group 1993, unpubl. data). Fisher (1976) estimated average productivity over 13 seasons at 640 fledglings/1 000 eggs laid by Laysan Albatrosses. Rice (1959) reported 49% of Midway Black-footed Albatross eggs produced fledglings. Woodward (1972) and Wittow (1993) reported Black-footed Albatross fledging rates at nearby Kure Atoll between 42.6 and 45.5% of eggs laid over three years. The mean of all published productivity data and this study is a fledging rate of 49.9% of Black-footed Albatross and 64% of Laysan Albatross eggs laid.

Appendix 4 estimates the survivorship of fledglings to maturity based on the assumptions about juvenile death rates listed above. Interestingly, the output of new recruits at age nine matches Fisher's (1976) Laysan Albatross data very well. The long-term productivity averaged for 13 seasons by Fisher (1976) of 640 fledglings/1 000 nests/ year, based on first and second year mortality rates (Appendix 4), and adult survivorship rates of 0.333, 0.82, and 0.94, respectively, produces 121 recruits at age nine to replace the 60 birds lost. Based on these estimates, the Laysan Albatross population should increase by 6.1% per year. Fisher measured annual population growth at 8%/year for 1960–1972. However, he included all birds that he caught between ages seven and nine years as first time breeders in his estimates, whereas we project recruitment in year nine. Our estimates provide for replacement plus 7.9% growth at age seven years, an excellent match to the actual population growth rate of 8% Fisher measured before bycatch or contaminants are believed to have affected this population. In the absence of bycatch and contaminant-driven mortalities, the Black-footed Albatross population should sustain modest growth in the range of 3.4–5.9%/year, and just 318 fledglings/year will maintain these populations (Appendix 4).

When these estimates of natural mortality in Laysan Albatrosses are augmented by a 1.5% bycatch mortality rate, recruitment to age nine is reduced to 112 birds, which is balanced against the annual loss of 75 adults for a net growth rate of 3.7%, which also matches the performance of this population after 1975 (Appendix 5). Hence the Laysan Albatross population would remain stable with an average productivity of 469 fledglings/year (694 × 75/111 = 469) and a 7.5% mortality rate. When the field productivity value we measured for Black-footed Albatrosses in 1994–95 of 634 fledglings/ 1 000 nests/year and the 8.5% mortality figure which includes the bycatch effect are estimated, then recruitment to age nine drops to 91 birds that enter the breeding population to replace the 85 lost annually, leaving very little margin for growth (0.6%). Under this scenario, the Black-footed Albatross population would remain stable with an average annual productivity of 592 fledglings/thousand (634 × 85/91 = 592). However, the average productivity in published studies is only 499 fledglings/1 000 eggs, nearly 20% below that needed to sustain this population long term (Woodward 1972; Rice and Kenyon 1962; this study). At the 0.499 average productivity, the population cannot sustain itself with a total annual mortality of 8.5%, and will decrease slowly at a rate of ~1.3% annually. An increase in annual mortality of Black-footed Albatrosses to 10% would drop recruitment to 77 birds/1 000 eggs, far below the replacement requirement for population stability. At 10% annual mortality, Black-footed Albatrosses would have to fledge 823 chicks/1 000 nests/year just to remain stable, which is well above the hatch rate we found of 785 eggs/1 000 eggs laid (Table 2). At 10% mortality, the Black-footed Albatross nesting population will decrease by ~3%/year.

Broken pair bonds further reduce output. At the 8.5% annual mortality rate, 4.75% of the reproductive effort is lost each year when widowed birds bond to new mates. Lost mates has the effect of reducing fledging rates for the population by an additional 24 fledglings/year (499 × 0.0475 = 24). When a 2.5% loss of eggs and embryos owing to contaminants (499 × 0.025 = 13) is deducted, then the long-term mean productivity in the population is reduced to 462 fledglings/1 000 eggs/year which leaves a shortfall of 1.9%/year. The very slight contaminants-related effect is confined to lowered chick productivity, while bycatch mortality causes losses among all age classes.

Future prospects

If these estimates are correct, then the Black-footed Albatross population at Midway Atoll cannot absorb any increase in mortality rates. The Laysan Albatross population at Midway Atoll is much less threatened and could sustain somewhat higher mortalities. These results are especially troubling for Black-footed Albatrosses since there has been little recovery of nesting numbers at Midway and a marked decrease at French Frigate Shoals since 1980. Rice and Kenyon (1962) estimated the Leeward Hawaiian Islands Black-footed Albatross breeding population in 1960 at 110 000 birds; Fefer *et al.* (1984) believed that the breeding population was 98 820 in 1980, based on more accurate Midway Atoll nest counts but with few survey data from other Leeward Hawaiian Islands' colonies. The 1994 U.S. Fish and Wildlife Service estimate from colony counts are of only 58 600 pairs, or 117 200 adults in all the Leeward Hawaiian Island colonies (E. Flint, pers. comm.). Thus, the population survey data provide evidence the Black-footed Albatross breeding population has been virtually static since 1959, while the Laysan Albatross population recovered markedly and continues to increase. One implication of the population counts and our estimates is that fisheries bycatch mortality increased markedly in the late 1960s to early-1970s. A seven to nine year delay in lowered recruitment occurred because the young adults require almost a decade to recruit. The estimate of a 1.9% annual decrease in the Black-footed Albatross breeding population with high bycatch is a good match for the field data obtained over the last 15 years at French Frigate Shoals and other Leeward Hawaiian Islands' colonies nearer to Honolulu (the home port of most of the North Pacific Ocean longline fishing fleet) than to Midway Atoll.

Driftnet and longline fishery bycatch data are fragmentary (Ogi *et al.* 1993; Johnstone *et al.* 1993; Flint 1995). There are no monitors of many nations' vessels fishing in the North Pacific Ocean outside of the U.S. 200 nm territorial limits. Further, rogue fishing vessels continue to use illegal driftnet gear, with no regard for their bycatch. One illegal driftnet vessel was seized a few hundred miles north of Midway Atoll in July 1995; crew members had fashioned ear-rings from five Black-footed Albatross bands. Our estimates and the precipitous decrease in Black-footed Albatrosses breeding at the French Frigate Shoals colony suggest that bycatch-related mortalities of those birds from the colonies nearer Honolulu could be above the current 2.5% rate. Bycatch has probably been a very serious threat to the Black-footed Albatross population for three decades, enough to prevent restoration of their historic numbers, and sufficient to slow the recovery of the Laysan Albatross population. Furthermore, bycatch will influence these populations for at least another decade because of the long time required to recruit breeding birds. Thus even if albatross bycatch were eliminated instantaneously, the effects of improved conservation on breeding numbers would not occur for almost a decade. In the near term, the prospect for breeding Black-footed Albatrosses in the Leeward Hawaiian Islands is stability or slow decrease in their numbers, and population stability or slow growth in the Laysan Albatross population.

ACKNOWLEDGEMENTS

The personnel of the SERE Group, Ltd. and World Wildlife Fund are grateful for the help of many U.S. Fish and Wildlife Refuge personnel including Ken McDermond, Ken Neithammer, Nanette Seto, Don Williamson, and Mike Nishimoto; their support and encouragement during field studies at Midway was invaluable. Three U.S. Navy Officers-in-Charge at Midway, LCDR Mike Driggers, LCDR James Ray, and Lt. David Black, granted permission for, and provided much help to, our field studies at Midway. For three years, CWO Gregg Diefenderfer loaned us the use of his barracks and much personal gear at Midway which we used as a field laboratory: his help was consistent and invaluable. Many personnel of the Piquiniq Management Corporation at Midway donated time and effort to our studies. This research was funded by grant # 8200227010 (A Baseline Assessment of Global Marine Contamination) from the U.S. Environmental Protection Agency Global Change Program to the Wildlife Contaminants Program of World Wildlife Fund, Washington, DC. Theo Colborn served as technical project officer. Michael Gilbertson critiqued early versions of the manuscript.

APPENDIX 1

Population dynamics estimates for Black-footed and Laysan Albatross populations at Midway Atoll. Assumptions are based of Fisher (1971, 1975, 1975).

1. Assumptions for female breeding age and age-specific death rates:
   - No effective reproductive contribution until nine years of age.
   - All breeders have an equal probability of success.
   - For ages 9–34 years, the natural age-specific survivorship is 0.94.
   - For ages 35–44 years, age specific survivorship drops to 0.91.
   - For ages 45–50 years, age-specific survivorship drops to 0.88.
   - In age classes >50 years, age-specific survivorship drops to 0.85.
   - No females older than 50 years contribute young to the population.
2. The probability of a female nesting is:
   - 0.84 for ages 9–29 years,
   - 0.75 for ages 30–34 years,
   - 0.50 for ages 35–44 years,
   - 0.25 for ages 45–50 years,
   - No females nest after age 50.

APPENDIX 2

Effects of different mortality rates in modelled Black-footed and Laysan Albatross populations. Shown are various mortality rates and survival in a cohort of females.

Age (years)	6%	7.5%	8.5%	10%
9	1 000	1 000	1 000	1 000
10	940	925	915	900
11	884	856	837	810
12	831	791	766	729
13	781	732	701	656
14	734	677	641	590
15	690	626	586	531
16	648	597	537	478
17	610	536	491	430
18	573	495	450	387
19	539	459	411	347
20	507	424	376	312
21	476	392	344	282
22	448	363	315	254
23	421	336	288	228
24	395	311	264	206
25	372	287	241	185
26	349	266	221	167
27	332	246	202	150
28	312	227	185	135
29	293	210	169	122
30	275	194	155	109
31	259	180	142	98
32	243	166	130	89
33	229	154	119	80
34	215	142	109	72
35	195	127	96	63
36	178	114	85	54
37	162	102	75	47
38	148	91	67	41
39	134	82	59	36
40	122	73	52	31
41	111	66	46	27
42	101	57	41	23
43	92	52	36	21
44	84	47	32	18
45	74	41	27	15
46	65	35	23	13
47	57	30	20	11
48	50	26	17	9
49	44	23	14	8
50	39	20	12	6
Last bird dies at age	77 years	71 years	66 years	61 years

APPENDIX 3

Effective reproductive effort by age intervals at various mortality rates of 1 000 females breeding from 9 to 50 years of age.

		Mean female survivorship rates			
		94%	92.5%	91.5%	90.0%
Age (years)	Proportion breeding	Total years of female reproductive effort			
9–19	0.84	5 808	5 416	5 175	4 839
20–29	0.84	2 755	2 160	1 838	1 440
30–34	0.75	746	470	368	252
35–44	0.50	302	203	148	90
45–50	0.25	21	11	7	4
Total for the cohort		9 632	8 260	7 536	6 625
Less pair-bond-related losses		0	235	358	504
Net reproductive effort		9 632	8 025	7 178	6 121
Net lost reproductive effort		0	1 607	2 454	3 511
Loss of reproductive effort (%)		0	16.7[1]	25.5[2]	36.5[3]

[1]at 1.5% byctach mortality; [2]at 2.5% bycatch mortality; [3]at 4% bycatch mortality.

APPENDIX 4

Reconcilation of population growth and natural mortality rates with published productivity estimates for Black-footed and Laysan Albatrosses at Midway Atoll. Assumptions (see text for justification): 0.333% first year survivorship; 0.82% second year survivorship; 0.94% annual survivorship at age three years and onward.

		Published estimates of productivity (#/1 000)			
		Laysan Albatross		Black-footed Albatross	
		694	640	634	499
Age years	Age specific survivorship	Number survivors at the end of this age interval			
0–1	0.333	231	213	211	166
2	0.82	189	175	173	136
3	0.94	178	164	163	129
4	0.94	167	154	153	120
5	0.94	157	145	144	113
6	0.94	148	136	135	106
7	0.94	139	128	127	100
8	0.94	131	121	119	94
Less the natural 6% loss of adults		60	60	60	60
Net excess for growth		71	61	59	34
Growth rate (%/year)		7.1	6.1	5.9	3.4

Under these assumptions, a long-term mean fledging rate of 318 fledglings/1 000 eggs laid will maintain a stable population.

APPENDIX 5

Reconciliation of accelerated death rates from longline fisheries and contaminants with published productivity rates. Assumptions: • Adult mortality rate for Laysan Albatrosses accelerated to 7.5%/year. • Adult mortality rate for Black-footed Albatrosses accelerated to 8.5%/year. • Post fledging mortality rates for each species immature age classes accelerated by 1.5% for Laysan Albatrosses and 2.5% for Black-footed Albatrosses. • Accelerated mortality rates are additive to the natural 6% rate. • Contaminant effects limited to eggshell thinning, and depressed hatch rates, in Black-footed Albatrosses.

			Published estimates of productivity/1 000 eggs laid					
			Laysan Albatross		Black-footed Albatross			
Age (years)	Age-specific survivoship: Laysan	Age-specific survivoship: Black-footed	Study mean 694	Grand mean 640	Study mean 634	Grand mean 499	Less broken pairs 475	*Less cont'nt effects 462
0–1	0.318	0.308	221	204	195	153	146	142
2	0.805	0.795	178	164	155	122	116	113
3	0.925	0.915	164	152	142	112	106	104
4	0.925	0.915	152	140	130	102	97	95
5	0.925	0.915	141	130	119	94	89	87
6	0.925	0.915	130	120	109	86	81	79
7	0.925	0.915	121	111	99	78	75	73
8	0.925	0.915	112	103	91	72	68	66
Less the annual loss of adults			75	85	85	85	85	85
Excess recruitment for population growth			37	28	6	−13	−17	19
Growth (%/year)			3.7	2.8	0.6	−1.3	−1.7	−1.9

*Less contaminant effects.

REFERENCES

Ahlborg, U. G., Becking, G. C., Birnbaum, L. S., Brouwer, A., Derks, H. J. G. M., Feeley, M., Golor, G., Hanberg, A., Larsen, J. C., Liem, A. K. D., Safe, S. H., Schlatter, C., Waern, F., Younes, M. and Yrjanheikki, E., 1994. Toxic equivalency factors for dioxin-like PCBs. Report on a WHO-ECEH and IPCS consultation, December, 1993. *Chemosphere* **28:** 1049–067.

Ainley, D. G., DeGange, A. R., Jones, L. L. and Beach, P. J., 1981. Mortality of seabirds in highseas salmon gill nets. *Fisheries Bull.* **79:** 800–06.

Auman, H. J., 1994. Plastic ingestion, biomarkers of health, PCBs and DDE in two species of albatrosses on Sand Island, Midway Atoll. MS Thesis, Fisheries and Wildlife, Michigan State University, East Lansing. 73 pp.

Auman, H. J., Ludwig, J. P., Summer, C. L., Verbrugge, D. A., Froese, K. L., Colborn, T. and Giesy, J. P., 1997a. PCBs, DDE, DDT, and TCDD-EQ in two species of albatross on Sand Island, Midway Atoll, North Pacific Ocean. *Environ. Toxicol. Chem.* **16:** 498–504.

Auman, H. J., Ludwig, J. P., Giesy, J. P., and Colborn, T., 1997b. Plastic ingestion by Laysan Albatross chicks on Sand Island, Midway Atoll in 1994 and 1995. Pp. 239–44 *in* Albatross Biology and Conservation ed by G. Robertson and R. Gales. Surrey Beatty & Sons: Chipping Norton.

Bowerman, W. W., Giesy, J. P., Best, D. A. and Kramer, V. J., 1995. A review of factors affecting productivity of bald eagles in the Great Lakes region: Implications for recovery. *Environ. Health Perspect.* **103:** 51–59.

Brothers, N., 1991. Albatross mortality and associated bait loss in the japanese longline fishery in the Southern Ocean. *Biol. Conserv.* **55:** 255–68.

Brunstrom, B. and Andersson, L., 1988. Toxicity and 7-ethoxyresorufin-O-deethtlase inducing potency of coplanar biphenyls (PCBs) in chick embryos. *Arch. Toxicol.* **62:** 263–66.

Carrick, R., Keith, K. and Gwynn, A., 1960. Fact and fiction on the breeding of the wandering albatross. *Nature* **188:** 112–14.

Colborn, T. and Clement, C. eds., 1993. Chemically-induced alterations in sexual and functional development: The wildlife-human connection. Princeton Scientific Publishing Co., Princeton, New Jersey. 394 pp.

Cheung, M. O., Gilbert, E. F. and Peterson, R. E., 1981. Cardiovascular teratogenicity of 2,3,7,8 tetrachlorodibenzo-*p*-dioxin in the chick embryo. *Toxicol. Appl. Pharmacol.* **61:** 197–204.

Croxall, J. P. and Gales, R., 1997. An assessment of the conservation status of albatrosses. Pp. 46–65 *in* Albatross Biology and Conservation ed by G. Robertson and R. Gales. Surrey Beatty & Sons: Chipping Norton.

DeGange, A. R. and Day, R. H., 1991. Mortality of seabirds in the Japanese landbased gillnet fishery for salmon. *Condor* **93:** 251–58.

Devito, M. J. and Birnbaum, L., 1994. Toxicology of dioxins and related chemicals. Pp. 139–62 *in* Dioxins and Health ed by A. Schecter. Plenum Press: New York.

Fefer, S. I., Harrison, C. S. and Naughton, B. N., 1984. Synopsis of the results of recent seabird observations in the Northwestern Hawaiian Islands. Pp. 9–76 *in* Resource investigations in the northwestern Hawaiian Islands ed by R. W. Grigg and K. Y. Tanoue. University of Hawaii Sea Grant Program.

Fisher, H. A., 1971. The Laysan albatross: its incubation, hatching, and associated behaviors. *Living Bird* **10:** 19–78.

Fisher, H. A., 1973. Pollutants in north Pacific albatrosses. *Pac. Sci.* **27:** 220–25.

Fisher, H. A., 1975a. Mortality and survival in the Laysan albatross, *Diomedea immutabilis*. *Pac. Sci.* **29:** 279–300.

Fisher, H. A., 1975b The relationship between deferred breeding and delayed mortality in the Laysan albatross. *Auk* **92:** 433–41.

Fisher, H. A., 1976. Some dynamics of a breeding colony of Laysan Albatrosses. *Wilson Bull.* **88:** 121–42.

Flint, E., 1995. Albatross populations and conservation issues in the northern hemisphere. Abstracts from the First International Conference on the Biology and Conservation of Albatrosses: Hobart, Tasmania, Australia.

Fox, G. A., Anderka, F. W., Lewin, V. and MacKay, W. C., 1975. Field assessment of eggshell quality by beta-backscatter. *J. Wildl. Manage.* **39:** 528–34.

Fry, D. M., Toone, C. K., Speich, S. M. and Peard, R. J., 1987. Sex ratio skew and breeding patterns in gulls: demographic and toxicological considerations. *Studies Avian Biol.* **10:** 26–43.

Giesy, J. P., Ludwig, J. P. and Tillitt, D. E, 1994a. Deformities in birds of the Great Lakes region: assigning causality. *Environ. Sci. Technol.* **28:** 128–35.

Giesy, J. P., Ludwig, J. P. and Tillitt, D. E., 1994b. Dioxins, dibenzofurans PCBs and similar chlorinated diaromatic hydrocarbons in, and their effects on, birds: Wildlife biomonitoring for hazards of complex environmental mixtures in the Laurentian Great Lakes. Pp. 254–307 *in* Dioxins and Health ed by A. Schecter. Plenum Press: New York.

Gilbertson, M. T., Kubiak, J., Ludwig, J. P. and Fox, G. A., 1991. Great Lakes Embryo Mortality, Edema, and Deformities Syndrome (GLEMEDS) in colonial fish-eating birds: Similarity to chick edema disease. *J. Toxicol. Environ. Health* **33:** 455–520.

Gould, P. J. and Hobbs, R., 1993. Population dynamics of the Laysan and other albatrosses in the north Pacific. *Nth Pac. Commission Bull.* **53:** 485–97.

Hardy, J. T., 1982. The sea-surface microlayer: Biology, chemistry and anthropogenic enrichment. *Progress Oceanog.* **11:** 307–28.

Harrison, C. L., 1990. Seabirds of Hawaii. Cornell University Press: Ithaca, NY.

Henshel, D. S., 1993. LD50 and teratogenecity studies of the effects of TCDD on chicken embryos. Abstracts of the 14th annual meeting of the Society of Environmental Toxicology and Chemistry, Houston, Texas, November 2–6, 1993.

Henshel, D. S., Martin, J. W., Norstrom, R., Whitehead, P., Steeves, J. D. and Cheng, K. M., 1995. Morphometric abnormalities in brains of great blue heron hatchlings exposed in the wild to PCDDs. *Environ. Health Perspect.* **103:** (Supplement 4) 61–66.

Hickey, J. J., 1952. Survival Studies of Banded Birds. Special Scientific Publication 15. U.S. Fish and Wildlife Service, USDI, 177 pp.

Iwata, H., Tanabe, S. and Tatsukawa, R., 1995. Global contamination by persistent organochlorines in the oceans. Abstract 80 of the Second World SETAC Congress, November 5–9, 1995. Vancouver, BC, Canada.

Jones, P. D., Hannah, D. J., Buckland, S. J., Day, P. J., Leathem, S. V., Porter, L. J., Giesy, J. P., Auman, H. J., Ludwig, J. P. and Summer, C., 1994. PCBs and 2,3,7,8-substituted dibenzo-*p*-dioxins and dibenzofurans in albatross tissues from Midway Island in the central Pacific Ocean. Pp. 137–41 *in* Short papers from DIOXIN '94, Kobe, Japan Vol. 20. Ecoinforma Press: Bayreuth, Germany.

Jones, P. D., Hannah, D. J., Buckland, S. J., Day, P. J., Leatham, S. V., Porter, L. J., Auman, H. J., Summer, C. L., Ludwig, J. P., Colborn, T. L. and Gisey, J. P., 1996. Persistent synthetic chlorinated hydrocarbons in albatross tissue samples from Midway Atoll. *Environ. Toxicol. Chem.* **15:** 1793–800.

Johnstone, D. H., Shaffer, T. L. and Gould, P. J., 1993. Incidental catch of marine birds in the north Pacific high seas drift net fisheries in 1990. *Nth Pac. Comm. Bull.* **53:** 473–83.

Kenyon, K. W. and Kreidler, E., 1969. Laysan albatrosses swallow indigestible matter. *Auk* **86:** 339–43.

Klass, E. E., Ohlendorf, H. M. and Heath, R. G., 1974. Avian eggshell thickness: variability and sampling. *Wilson Bull.* **86:** 156–64.

Lack, D., 1954. The natural regulation of animal numbers. Clarendon Press: Oxford.

Lebo, J. A., Gale, R. A., Petty, J. D., Tillitt, D. E., Huckins, J. N., Meadows, J. C., Orazio, C. E., Echols, K. R., Schroeder, D. J. and Inmon, L. E., 1995. Use of the semipermeable membrane device as an *in situ* sampler of waterborne bioavailable PCDD and PCDF residues at sub-parts-per-quadrillion concentrations. *Environ. Sci. Technol.* **29:** 2886–892.

Ludwig, J. P., 1965. Biology and structure of the Caspian tern population of the Great Lakes 1986–1964. *Bird Banding* **36:** 217–33.

Ludwig, J. P., 1967. Band loss-Its effect on banding data and apparent survivorship in the Great Lakes ring-billed gull population. *Bird Banding* **38:** 309–23.

Ludwig, J. P., 1974. Recent changes in the ring-billed gull population and biology in the laurentian Great Lakes. *Auk* **91:** 575–94.

Ludwig, J. P., 1981. Band wear and band loss in the Great Lakes Caspian tern population and a generalized model of band loss. *Colonial Waterbirds* **4:** 174–86.

Ludwig, J. P., Giesy, J. P., Summer, C. L., Bowerman, W., Aulerich, R., Bursian, S., Auman, H. J., Jones, P. D., Williams, L. L., Tillitt, D. E. and Gilbertson, M., 1993. A comparison of water quality criteria for the Great Lakes based on human and wildlife health. *J. Great Lakes Res.* **19:** 789–807.

Ludwig, J. P., Summer, C. L., Auman, H. J., Colborn, T. L., Ludwig, F. E. and Diefenderfer, G, 1996. Band loss in the north Pacific populations of Laysan albatross *Diomedea immutabilis* and Black-footed Albatross *D. nigripes*. *Nth Amer. Bird Bander* **21:** 1–12.

Ludwig, J. P., Kurita-Matsuba, H., Auman, H. J., Ludwig, M. E., Summer, C. L., Giesy, J. P., Tillitt, D. E. and Jones, P. D., 1996b. Deformities, PCBs, and TCDD-Equivalents in double-crested cormorants *Phalacrocorax auritus* and Caspian terns *Hydroprogne caspia* of the upper Great Lakes: Testing a cause-effects hypothesis. *J. Great Lakes Res.* **22:** 172–97.

Napoliatano, G. E. and Richmond, J. E., 1995. Enrichment of biogenic lipids, hydrocarbons, and PCBs in stream surface foams. *Environ. Toxicol. Chem.* **14:** 197–201.

Oliver, B. G., Charlton, M. N. and Durham, R. W., 1989. Distribution, redistribution and geochemistry of polychlorinated biphenyl congeners and other chlorinated hydrocarbons in Lake Ontario sediments. *Environ. Sci. Technol.* **23:** 200–08.

Ogi, H., Yatsu, A., Hatanaka, H. and Nitta, A., 1993. The mortality of seabirds by drift net fisheries in the north Pacific. *Nth Pac. Comm. Bull.* **53:** 499–518.

Prince, P. A., Rothery, P., Croxall, J. P. and Wood, A. G., 1994. Population dynamics of Black-browed Albatrosses *Diomedea melanophris* and *D. chrysostoma* at Bird Island, South Georgia. *Ibis* **136:** 50–71.

Rice, D. W., 1959. Birds and aircraft on Midway Islands, 1957–58 investigations. U.S. Fish and Wildlife Service. Special Scientific Report 44.

Rice, D. W. and Kenyon, K. W., 1962. Breeding distribution, history and populations of north Pacific albatrosses. *Auk* **79:** 365–86.

Ryan, P. G., Connell, A. D. and Gardner, B. D., 1988. Plastic ingestion and PCBs in seabirds: Is there a relationship? *Mar. Poll. Bull.* **19:** 174–76.

Safe, S., 1990. Polychlorinated biphenyls (PCBs), dibenzodioxins (PCDDs), dibenzofurans (PCDFs) and related compounds: environmental and mechanistic considerations which support the development of toxic equivalency factors. *Critical Rev. Toxicol.* **21:** 51–88.

Seto, N. W. H., 1995. Black-footed Albatross population census on Midway Atoll, 1994–1995 breeding season. Midway Islands National Wildlife Refuge. Unpublished Paper Report, 6pp.

Sileo, L., Sievert, P. R. and Samuel, M. D., 1990. Causes of mortality of albatross chicks at Midway Atoll. *J. Wildl. Diseases* **26:** 329–38.

Sievert, P. R. and Sileo, L., 1993. The effects of ingested plastic on growth and survival of albatross chicks. *In* The status, ecology and conservation of marine birds of the north Pacific. Canadian Wildlife Service Special Publication.

Skillman, R. and Flint, E. N., 1997. Mortality of Laysan and Black-footed Albatrosses in the Hawaii pelagic longline fishery. *Pac. Seabirds* **24:** 23.

Tickell, W. L. N., 1968. The biology of the great albatrosses *Diomedea exulans* and *Diomedea epomophora*. *Antarc. Res. Serv.* **12:** 1–55.

Tillitt, D. E., Ankley, G. T. and Giesy, J. P., 1991b. Characterization of the H4IIE rat hepatoma cell bioassay as a tool for assessing toxic potency of planar halogenated hydrocarbons in environmental samples. *Environ. Sci. Technol.* **25:** 87–92.

Tillitt, D. E., Ankley, G. T., Verbrugge, D. A., Giesy, J. P., Ludwig, James P. and Kubiak, T. J., 1991a. H4IIE rat hepatoma cell bioassay-derived 2,3,7,8-tetrachlorodibenzo-*p*-dioxin equivalents in colonial fish-eating waterbird eggs from the Great Lakes. *Arch. Environ. Contamin. Toxicol.* **21:** 91–101.

Tillitt, D. E., Ankley, G. T., Giesy, J. P., Ludwig, James, P., Kurita-Matsuba, H., Weseloh, D. V., Ross, P. S., Bishop, C. A., Sileo, L., Stromborg, K. L., Larson, J. and Kubiak, T. J., 1992. Polychlorinated biphenyl residues and egg mortality in double-crested cormorants from the Great Lakes. *Environ. Toxicol. Chem.* **11:** 1281–288.

Vonier, P. M., Crain, D. A., McLachlan, J. A., Guillette, L. J. and Arnold, S. F., 1996. Interaction of environmental chemicals with the oestrogen and progesterone receptors from the oviduct of the American alligator. *Environ. Health Perspect.* **104:** 1318–323.

White, D. H. and Hoffman, D. J., 1995. Effects of polychlorinated dibenzo-*p*-dioxins and dibenzofurans on nesting wood ducks *Aix sponsa* at Bayou Meto, Arkansas. *Environ. Health Perspect.* **103**(4): 37–39.

Wiemerskirch, H. and Jouventin, P., 1987. Population dynamics of the wandering albatross, *Diomedea exulans*, of the Crozet Islands: Causes and consequences of the population decline. *Oikos* **49:** 315–22.

Wittow, G. C., 1993. Black-footed Albatross *Diomedea nigripes*. *In* The birds of North America, No. 65 ed by A. Poole and F. Gill. Philadelphia: The Academy of Natural Sciences; Washington, D. C. The American Ornithologists Union.

Wohl, K. D., Gould, P. J. and Fitzgerald, S. M., 1995. Incidental mortality of seabirds in selected commercial fisheries in Alaska. Report to the Circumpolar Seabird Working Group, GAFF International Secretariate: Ottawa, Canada.

Woodward, P. W., 1972. The natural history of Kure Atoll, northwestern Hawaiian Islands. *Atoll Res. Bull.* **164:** 1–175.

Yamashita, N., Tanabe, S., Ludwig, J. P., Kurita, H., Ludwig, M. E. and Tatsukawa, R., 1993. Embryonic abnormalities and organochlorine contamination in double-crested cormorants *Phalacrocorax auritus* and Caspian terns *Hydroprogne caspia* from the upper Great Lakes. *Environ. Pollut.* **79:** 163–73.

CHAPTER TWENTY

Plastic ingestion by Laysan Albatross chicks on Sand Island, Midway Atoll, in 1994 and 1995

HEIDI J. AUMAN,[1] JAMES P. LUDWIG,[2] JOHN P. GIESY[3] and THEO COLBORN[4]

Laysan Albatross chicks from Midway Atoll, North Pacific Ocean, were assessed in 1994 and 1995 for impacts of plastic ingestion. Masses and incidence of plastic in chicks were compared between birds found dead of "natural causes" and those injured by motor vehicles. Naturally killed Laysan Albatross chicks had significantly greater masses of plastic in their proventriculi and gizzards and had significantly lighter body masses and lower fat indices than injured but otherwise healthy chicks. Ingested plastic probably does not cause significant direct mortality in Laysan Albatross chicks, but likely causes physiological stress as a result of satiation and mechanical blockages.

Key words: Laysan Albatross, Midway Atoll, Pacific Ocean, Plastic Ingestion, Physiological Stress

INTRODUCTION

SEABIRDS ingest floating plastic and other foreign matter while feeding on the surface of the ocean. Laysan Albatrosses *Diomedea (Phoebastria) immutabilis* accumulate plastic materials in their proventriculi and gizzards and are reported to have a greater incidence, a wider variety, and larger volume of ingested plastic than any other seabird (Kenyon and Kridler 1969; Sileo and Fefer 1987; Sileo *et al.* 1989). Fledglings normally regurgitate boli or castings of squid beaks, fish bones, and other indigestible matter, but large pieces of plastic may complicate or prohibit this process (Petit *et al.* 1981). At Midway Atoll (28°11′N, 177°22′W), Hawaiian Leeward Islands, Laysan Albatross chicks die by the thousands from disease, abandonment and starvation; chicks may also die of starvation or dehydration even if properly fed due to a build up of plastics in their stomachs (Sileo *et al.* 1990; Sievert and Sileo 1993).

This chapter reports on research to determine if ingested plastic adversely affected Laysan Albatross chicks on Midway Atoll and whether the rates of plastics ingestion have changed since studies conducted in the 1960s. Specifically, the research aimed to compare masses of plastic in Laysan Albatross chicks found dead with plastic loads in chicks accidentally injured or killed by vehicles, the latter group being assumed to be healthy before collision with vehicles. Although this comparison does not establish that greater plastic loads causes increased death rates (Ryan 1987), collecting plastic from dead chicks alone would have been insufficient to determine the cause of mortality, and since collection of uninjured chicks was considered unethical chicks accidentally injured by vehicles were used in the comparison.

METHODS

Collection of chicks

Dead and injured Laysan Albatross chicks were collected from 25 May to 21 July 1994, and 24 May to 30 June 1995 on Sand Island, Midway Atoll. Dead and injured birds were collected at random from populations within the areas of Sand Island populated by humans, since injured birds in uninhabited areas were rarely found. Throughout June of both years at least 10 fresh albatross chick carcasses were collected daily. Only chicks with glossy eyes were selected for necropsy, and only the most recently dead birds were necropsied in order to minimize the effects of desiccation and decomposition. Once the carcass was opened, a visual inspection of

[1]Midway Atoll No. 2, P.O. Box 660099 Limue HI 96766.
[2]138 Road 2 West, Kingsville, Ontario, Canada N9Y 2Z6.
[3]Department of Fisheries and Wildlife, Natural Resources, Michigan State University, East Lansing, MI 48824 USA.
[4]World Wildlife Fund, 1250 24th Street NW, Washington DC 20037 USA.

Pp. 239–44 *in* ALBATROSS BIOLOGY AND CONSERVATION ed by G. Robinson and R. Gales. Surrey Beatty & Sons, Chipping Norton. 1997.

the internal organs determined the decomposition state of the birds found dead. Carcasses were judged to be too decomposed to assess critically if the proventriculus was thin, peeling and transparent.

The majority of injured birds had compound fractures of their wings or legs. We assumed that ingested plastics did not contribute to chicks' injuries from collisions with vehicles. We assumed that no chicks had regurgitated their stomach contents as a reaction to their injury as no regurgitated boluses were observed near the site of injury. Chicks that had not collided with vehicles but had drooping wings, thought to be caused by lead poisoning, were excluded from the study since they were not considered healthy, although plastic load has not been correlated with either lead poisoning or droop wing (Sileo *et al.* 1990). Injured chicks were processed within 24 h of collection, but it was not possible to determine when a bird had been injured. Injured birds were restrained and 1–24 ml of blood was drawn from the brachial vein and stored; the birds were then killed by cervical dislocation and a necropsy performed within 10 minutes.

Sampling protocols

All collected chicks were weighed to the nearest 50 g using a Pesola spring scale. Bill length was measured to the nearest 1 mm to estimate age. The entire digestive tract was examined and subcutaneous fat were estimated visually using the following fat index: 0 = no fat, 1 = low fat, 2 = moderate fat, and 3 = high fat. The proventriculus and gizzard were removed by cutting anterior to the cardiac sphincter and posterior to the pyloric sphincter. Since there is no demarcation between the end of the oesophagus and the proventriculus, to sample consistently between carcasses each cut was made at a point immediately ventral to the caudal tip of the heart, about halfway between the taper of the oesophagus and the proventriculus. The entire proventriculus and gizzard were removed, any fat was peeled off, and both organs were weighed together on an electronic balance to 0.1 g. Contents of the proventriculus and gizzard were categorized as food, plastic, other human-made materials, and rock (mainly volcanic pumice). Plastic and other man-made materials were removed, washed in fresh water, air dried, and weighed to 0.1 g. (Although some authors used volume as the measure of plastics, we chose to measure mass because of the potential connection to chemical contaminants which are measured on a mass basis). The lining of the emptied proventriculus was examined for lacerations, ulcerations and punctures.

Statistical considerations

To determine a statistically viable sample size for the chicks found dead 30 birds were necropsied and the variance of the mean mass values of plastic in the proventriculus and gizzard was determined. Additional birds were necropsied until the estimate of the parametric variance did not change when plotted against the number of necropsied birds. This indicated that a statistically significant sample size had been reached. The sample size of birds injured by vehicles was determined by the number of chicks injured by vehicles. The non-parametric Wilcoxon tests (Wilcoxon 2-way sample test) were used to compare means of the injured and dead chicks. Samples collected in different years were tested separately.

RESULTS

Ninety-five dead and 39 injured Laysan Albatross chicks were necropsied in 1994 and 76 dead and 41 injured Laysan Albatross chicks were necropsied in 1995. Of these 251 chicks, only six (2.4%) did not contain plastic. Plastic items comprised chips and shards of unidentified plastic, Styrofoam, beads, fishing line, buttons, chequers, disposable cigarette lighters, toys, PVC pipe and other PVC fragments, golf tees, dish-washing gloves, magic markers and caylume light sticks. Non-plastic items included neoprene o-rings, pieces of rubber and a light-bulb (Fig. 1; see also Fig. 9 in Croxall 1997). Natural objects such as Ironwood *Casuarina* spp. cones, peach pits, walnuts, twigs, wood chips and pumice also were found. Squid beaks occurred in all chicks except those with completely empty proventriculi.

Only two chicks found dead had punctures or tears in the lining of their proventriculus. These two birds also had leaking gall bladders and puncture wounds in the livers and breast tissue, probably caused by a pitchfork, which is often used by people to pick up carcasses at Midway Atoll. Between December 1992 and April 1994 62 adult Laysan Albatrosses found injured or dying on Midway included four adults with scars on the proventricular lining (H. Auman, unpubl. data). One bird had an 11.3 cm long piece of plastic wedged crosswise in the proventriculus that projected through all but the last connective tissue layer on the outside of the proventriculus; it also had a conspicuous ulcer.

Chicks found dead in 1994 contained, on average, 23.8 g of plastic, while injured chicks averaged 11.3 g of plastic. In 1995 dead chicks averaged 18.1 g and the injured chicks 9.5 g; between-year differences in masses of plastics in dead and injured chicks were not statistically significant so the data for both years were pooled.

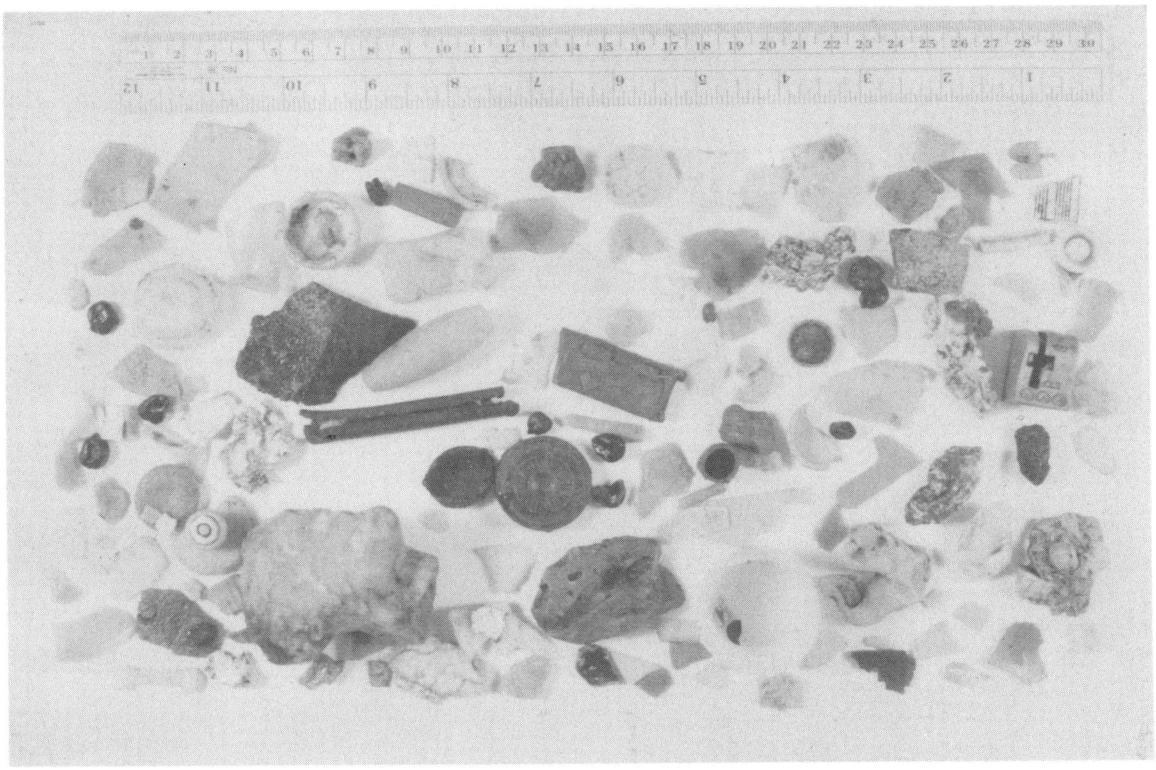

Fig. 1. Plastic items found in the proventricular bolus of a Laysan Albatross chick.

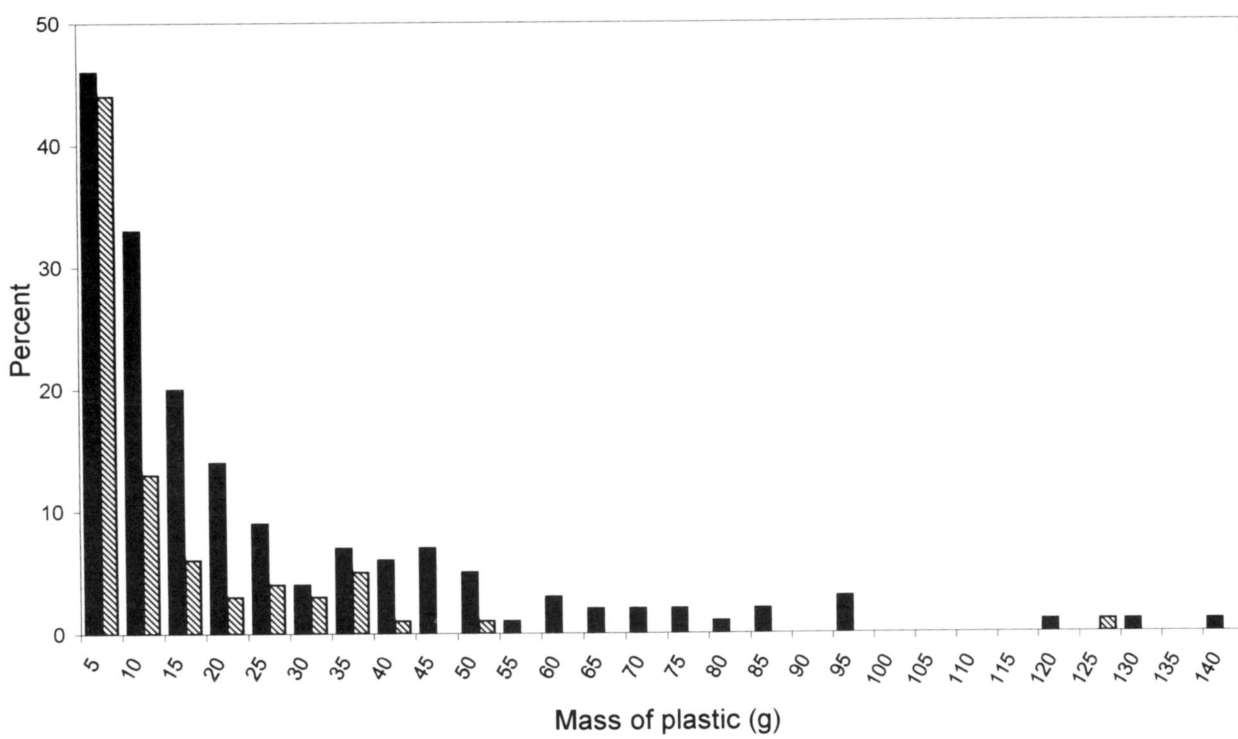

Fig. 2. Frequency of occurrence of plastic masses in 171 Laysan Albatross chicks found dead (black bars) and 80 chicks found injured by motor vehicles (diagonal bars) on Sand Island, Midway Atoll. Numbers for plastic masses represent 5 g increments up to and including the numbers shown.

Table 1. Parameters measured in dead and injured Laysan Albatross chicks at Sand Island, Midway Atoll, in 1994 and 1995. Figures in parenthesis indicate sample sizes.

	Mean				Range	
	Dead		Injured		Dead	Injured
Parameter	1994	1995	1994	1995	1994–1995	1994–1995
Body mass (g)	1 938 (95)	1 554 (73)	2 231 (39)	1 966 (40)	600–2 900	1 350–2 900
Beak length (cm)	10.6 (95)	9.8 (74)	11 (39)	10.0 (42)	7.9–11.7	9.2–11.9
Proventriculus length (cm)	13.3 (93)	12.4 (71)	12 (38)	13.2 (37)	6.5–21.0	8.5–23.0
Proventriculus width (cm)	5.9 (93)	4.9 (71)	3.8 (38)	4.2 (37)	2.1–9.4	1.6–7.8
Proventriculus height (cm)	4.7 (93)	3.2 (71)	2.9 (38)	3.0 (37)	0.3–8.3	0.6–6.9
Proventriculus mass (g)	140.4 (93)	103.5 (73)	85 (38)	104.0 (42)	15.2–393.7	18.5–313.9
Plastic mass (g)	23.8 (95)	18.1 (73)	11 (39)	9.5 (42)	0.0–136.3	0.0–122.7
Largest piece (g)	3.7 (95)	4.2 (50)	1.6 (39)	3.3 (27)	0.1–34.9	0.0–12.5
Fat index	2.0 (95)	1.3 (73)	2.5 (39)	2.2 (42)	1–3	1–3
Pumice mass (g)	30.0 (95)	NA (NA)	10 (39)	NA (NA)	0.0–194.8	0.0–121.1

However, the mean masses of plastic were significantly greater in the proventriculi and gizzards of birds found dead than injured in both 1994 (p = 0.0001) and 1995 (p = 0.01). The largest pieces of plastic in the dead chicks were significantly heavier than in injured chicks (p = 0.0099) (Table 1). In the sample of chicks injured in 1994 only one bolus of plastic greater than 40.0 g was found. If this 122.7 g sample is removed as an outlier, the mean plastic mass drops from 11.3 g to 8.4 g in the injured birds. The mass of plastic in injured and dead Laysan Albatross chicks is shown in Figure 2.

For both groups in both years there was no significant difference in the number of males and females sampled (p = 0.50). Additionally, there were no differences in chick masses (p = 0.11), proventricular width (p = 0.65), proventricular height (p = 0.21), mass of plastic boluses (p = 0.83), heaviest piece (p = 0.87) or fat index (p = 0.36) between male and female chicks. Males had significantly longer proventriculi (p = 0.03) and slightly heavier loads of proventricular pumice (p = 0.08) than females. Chicks found dead were significantly lighter than chicks found injured, and had greater proventricular lengths, widths and heights, lower fat indices, and greater masses of proventricular pumice (Table 1).

DISCUSSION

Effects of plastic ingestion

The differences in plastic masses between the dead and injured chicks in 1994 and 1995 were remarkable. However, the results are not evidence of a cause-and-effect link between plastic ingestion with chick death. While it is possible that the death of healthy chicks may result from ingested plastic, it is also possible that unhealthy chicks eat greater amounts of plastic from the ground as a result of their poor condition. It is also possible that unhealthy chicks have difficulty regurgitating indigestible matter. The appearance of glass shards, copper wire and other man-made items that do not float confirm that chicks ingest foreign items off the ground at Sand Island, where plastic brought to the island by previous albatross generations is abundant.

The possible effects of ingestion of plastics include starvation (Dickerman and Goelet 1987), suppressed appetite and reduced growth (in domestic chickens) (Ryan 1988), lower fledging masses (Sievert and Sileo 1993), decreased fat deposition (Connors and Smith 1982), increased PCB and other organochlorine assimilation (Ryan *et al.* 1988; Carpenter *et al.* 1972; Colton *et al.* 1974) and obstruction in the gut (Fry *et al.* 1987). In a trial involving White-chinned Petrels *Procellaria aequinoctialis* birds fed plastic had similar assimilation efficiencies and rates of plastic mass loss as did control birds, suggesting that plastic does not hamper digestive efficiency (Ryan and Jackson 1987). In a study on Sand Island, Midway Atoll, in 1987, no Laysan Albatross chick deaths, impactions or ulcerations in proventricular linings were attributed to ingested plastic (Sileo *et al.* 1990). Further, no conclusive evidence of negative effects from plastic consumption was reported by either Day (1980) or Furness (1985) in several seabird species; however, these authors considered that Procellariiformes may be more vulnerable than other seabirds due to their surface-seizing feeding methods, poor regurgitation capacity and smaller gizzards.

Thus it appears that plastics may add considerable stress to individuals, but probably have little or no direct impact at the population level. Satiation and reduced resistance to the effects of lead poisoning and avian pox virus may eventually contribute to death in birds with heavy plastic loads.

Satiation, or appetite suppression, is potentially caused by distention of the proventriculus, gizzard and intestines, dehydration, warm temperatures, and physical activity (Day 1980). Laysan Albatross chicks with plastic in their stomachs may experience all of these sensations in the hot summer months prior to fledging. Dehydration was considered to be the most common cause of death in Laysan Albatross chicks at Midway Atoll in 1987 (Sileo *et al.* 1990). The displacement of ingested food by plastic may be enough to starve or dehydrate chicks that are already in poor condition. Displacement of food may also reduce the mass and size of fat reserves of fledglings. The uninjured dead chicks in the present study were significantly lighter than the injured chicks. The mean fat indices for dead and injured chicks were 2.0 and 2.5, respectively, in 1994, and 1.4 and 2.2, respectively, in 1995; this difference may account for some of the mass difference between years. Plastic ingestion may also cause blockages in the digestive tract (Fry *et al.* 1987). Three (1.2%) of the chicks found dead contained a solid piece of plastic that completely blocked the junction of the oesophagus and proventriculus; however, it could not be determined if the blockage caused the death of the bird.

Plastic particles in the digestive system of seabirds may last for at least two years (Ryan and Jackson 1987) and radio transmitters inadvertently fed by Laysan Albatross adults to chicks lasted for over 40 d before being regurgitated (Petit *et al.* 1981). Only the smallest (<0.1 g) plastic pieces can enter the narrow gizzard opening. Interestingly, no plastic has been observed in seabird faeces (Day 1980; Petit *et al.* 1981), yet plastic in the intestine was noted in 39% of Laysan Albatrosses in 1986 and 1987 (Sileo *et al.* 1990). Presumably plastic that is not regurgitated is broken down in the digestive tract of Laysan Albatrosses.

Plastic ingestion may also lead to contamination by organochlorines. Polychlorinated biphenyls (PCBs) accumulate on the surfaces of plastic in the ocean and may reach five parts per million on polystyrene spherules (Carpenter *et al.* 1972). Phospholipid foam also forms on the sea surface and this can hyperconcentrate air-borne contaminants such as PCBs (Hardy 1982; Napoliatano and Richmond 1995). Plastics may serve as adsorbent surfaces for these toxicants. Additionally, the colorants, softeners, and anti-oxidants used in the conversion to user-friendly plastic may be harmful toxicants (van Franeker 1985). It is possible that large masses of plastic ingested by albatrosses may increase their PCB concentrations in addition to PCBs derived from dietary sources (Day 1980; Bourne and Imber 1982; van Franeker 1985). Positive correlations were reported between PCBs concentrations and masses of plastic in Great Shearwaters *Puffinus gravis* (Ryan *et al.* 1988) and PCBs, Polychlorinated dibenzo-*p*-dioxins, furans and naphthalenes have been found in adult Laysan Albatrosses from Midway Atoll (Jones *et al.* 1995). The dioxin-furan congener pattern Jones *et al.* (1995) found in albatross eggs and fat was unusual, but resembled the patterns found in incinerator ash and airborne emissions. It is quite likely that burned and melted plastics serve as a transport vehicle for these toxicants to albatrosses. Presumably, albatrosses extract dioxin and furan residues when the partially burned and melted particles are consumed at sea.

Trends in plastics ingestion

The incidence and mass of plastic ingested by Laysan Albatross chicks appears to be increasing. In 1966 74% of 91 Laysan Albatrosses sampled at Pearl and Hermes Reefs (north-west Hawaiian Islands) contained, on average, 1.87 g of plastic, with eight pieces being the greatest number found in any individual (Kenyon and Kridler 1969). In 1983 on Sand Island, Midway Atoll, plastic contents in live and dead Laysan Albatross chicks averaged 35.7 g (39.3 cc) and 76.7 g (85.0 cc), respectively (Fry *et al.* 1987). An average of 46 cc of plastic was collected from Laysan Albatross chicks in 1986 and 5 cc in 1987, and the frequency of occurrence in both these years was 90% (Sileo *et al.* 1990). In the current study plastics were found in 97.6% of chicks sampled, and the masses recorded (23.8 g in dead chicks in 1994 and 18.1 g in 1995) are in the same order as those for the 1980s, which suggests an upward trend in concentrations of plastic debris on the surface of the north central Pacific Ocean since the mid-1960s.

Sources of plastic

The majority of plastic refuse is generated from municipal solid waste disposal at sea, coastal landfills and runoff, and from sea-going vessels (Colton *et al.* 1974). Ocean currents, winds and the location of disposal influence the abundance and distribution of plastic in the North Pacific Ocean. The warm Kurishio current, which flows north-east near Japan, and the North Pacific current, which flows across the north central Pacific Ocean, move through the main feeding areas of North Pacific Ocean albatrosses and are probably responsible for transporting most plastic material from sources further to the west (Fry *et al.* 1987; Harrison 1990). In 1979–1980, 108 of

109 items found in Laysan Albatross regurgitates and carcasses at Midway Atoll were of Japanese origin (Petit *et al.* 1981). Increasing knowledge of the atmosphere as a transport medium for organochlorine toxicants from third world users to the oceans (Iwata *et al.* 1995), and the concentration processes in the phospholipid-enriched sea surface (Hardy 1982; Napolitano and Richmond 1995) lend a sense of urgency to the continuing task of monitoring plastics in seabirds since plastics may be another surface to which organochlorines may adhere. The effects of plastics may be far more subtle than mechanical blockage or reduced food consumption.

ACKNOWLEDGEMENTS

This research was made possible by a grant from the U.S. Environmental Protection Agency through the World Wildlife Fund (#CR 820227–01–0). We thank C.P.O. Gregg Diefenderfer, Darcy Bromley, Cheryl Summer and Katie Swift for their assistance with necropsies, Dr Scott Winterstein for statistical advice, and the U.S. Navy and the U.S. Fish and Wildlife Service for allowing our research to proceed on Midway Atoll.

REFERENCES

Bourne, W. R. P. and Imber, M. J., 1982. Plastic pellets collected by a prion on Gough Island, Central South Atlantic Ocean. *Mar. Pollut. Bull.* **13:** 20–21.

Carpenter, E. J., Anderson, S. J., Harvey, G. R., Miklas, H. P. and Peck, B. B., 1972. Polystyrene particles in coastal waters. *Sci.* **178:** 749–50.

Colton, J. B., Knapp, F. D. and Burns, B. R., 1974. Plastic particles in surface waters of the northwestern Atlantic. *Sci.* **185:** 491–97.

Connors, P. G. and Smith, K. G., 1982. Oceanic plastic particle pollution: Suspected effect on fat deposition in red phalaropes. *Mar. Pollut. Bull.* **13:** 18–20.

Croxall, J. P., 1997. Research and conservation: a future for albatrossses? Pp. 267-88 *in* Albatross Biology and Conservation ed by G. Robertson and R. Gales. Surrey Beatty & Sons: Chipping Norton.

Day, R. H., 1980. The occurrence and characteristics of plastic pollution in Alaska's marine birds. Master's Thesis, University of Alaska, Fairbanks. 111 pp.

Dickerman, R. W. and Goelet, R. G., 1987. Northern gannet starvation after swallowing styrofoam. *Mar. Pollut. Bull.* **18:** 293.

Fry, M. D., Fefer, S. I. and Sileo, L., 1987. Ingestion of plastic debris by Laysan albatrosses and wedge-tailed shearwaters in the Hawaiian Islands. *Mar. Pollut. Bull.* **18:** 339–43.

Furness, R. W., 1985. Ingestion of plastic by seabirds at Gough Island, South Atlantic Ocean. *Environ. Pollut.* **38:** 261–72. (Series A).

Hardy, J. T., 1982. The sea surface microlayer: biology, chemistry and anthropogenic enrichment. *Prog. Oceanog.* **11:** 307–28.

Harrison, C. S., 1990. Seabirds of Hawaii: Natural History and Conservation. Cornell University Press: New York. 249 pp.

Iwata, H., Tanabe, S. and Tatsukawa, R., 1995. Global contamination by persistent organochlorines in the ocean. Pp. 15 *in* Abstracts of the Second SETAC World Congress, Abstract 80, Vancouver, B.C. Canada. November 5–9, 1995.

Jones, P. D., Hannah, D. J., Buckland, S. J., Day, P. J., Leatham, S. V., Porter, L. J., Auman, H. J., Summer, C. L., Ludwig, J. P. Colborn, T. L. and Giesy, J. P., 1996. Persistent synthetic chlorinated hydrocarbons in albatross tissue samples from Midway Atoll. *Environ. Toxicol. Chem.* **15:** 1793–800.

Kenyon, K. W. and Kridler, E., 1969. Laysan albatrosses swallow indigestible matter. *Auk* **86:** 339–43.

Napolitano, G. E. and Richmond, J. E., 1995. Enrichment of biogenic lipids, hydrocarbons and PCBs from stream-surface foams. *Environ. Toxicol. Chemist.* **14:** 197–201.

Pettit, T. N., Grant, G. S. and Whittow, G. C., 1981. Ingestion of plastics by Laysan albatross. *Auk* **98:** 839–41.

Ryan, P. G., 1987. The effects of ingested plastic on seabirds: Correlations between plastic loads and body condition. *Environ. Pollut.* **46:** 119–25.

Ryan, P. G., 1988. Effects of ingested plastic on seabird feeding: Evidence from chickens. *Mar. Pollut. Bull.* **19:** 125–28.

Ryan, P. G., Connell, A. D. and Gardner, B. D., 1988. Plastic ingestion and PCBs in seabirds: Is there a relationship? *Mar. Pollut. Bull.* **19:** 174–76.

Ryan, P. G. and Jackson, S., 1987. The lifespan of ingested plastic particles in seabirds and their effect on digestive efficiency. *Mar. Pollut. Bull.* **18:** 217–19.

Sievert, P. R. and Sileo, L., 1993. The effects of ingested plastic on growth and survival of albatross chicks. Pp. 212–17 *in* The status, ecology, and conservation of marine birds of the North Pacific ed by K. Vermeer, K. T. Briggs, K. H. Morgan and D. Siegal-Causey. Canadian Wildlife Service Special Publication: Ottawa.

Sileo, L. and Fefer, S. I., 1987. Paint chip poisoning of Laysan albatrosses at Midway Atoll. *J. Wildl. Disease* **23:** 432–37.

Sileo, L., Sievert, P. R. and Samuel, M. D., 1989. Prevalence and characteristics of plastic ingested by Hawaiian seabirds. Pp. 665–81 *in* Proceedings of the Second International Conference on Marine Debris ed by R. S. Shomura and M. L. Godfrey. 2–7 April 1989, Honolulu, Hawaii. U.S. Dept. of Commerce, NOAA Technical Memorandum.

Sileo, L., Sievert, P. R. and Samuel, M. D., 1990. Causes of mortality of albatross chicks at Midway Atoll. *J. Wildl. Disease* **26:** 329–38.

van Franeker, J. A., 1985. Plastic Ingestion in the North Atlantic fulmar. *Mar. Pollut. Bull.* **16:** 367–69.

Perspectives

Most of the preceding chapters in this book have been drawn from the biological sciences. While conservation is underpinned by scientific knowledge the practitioners of conservation may require different skills to those essential to the proper execution of science. Non-government organizations have an important role to play in the resolution of issues involving human impacts on the earth's natural systems, including the oceans and dependent species such as albatrosses (Chapter 21). Challenging existing norms and developing alternative philosophical approaches to conservation issues are one facet of that role. Likewise, the resolution of seabird bycatch issues in fisheries is not wholly science based, and requires a multi-dimensional approach, an essential part of which is the existence of mechanisms (e.g., treaties, conventions and agreements) that facilitate dialogue and aid in the attainment of desired conservation outcomes (Chapter 22).

Waved Albatrosses sharing closeness during courtship.
TUI DE ROY/HEDGEHOG HOUSE

CHAPTER TWENTY-ONE

Wither the albatross: NGO perspectives on albatross conservation

MICHAEL EARLE[1] and BRUCE McKAY[2]

Due to their low reproductive rates and clumped concentrations in breeding and feeding areas albatrosses are particularly vulnerable to anthropogenic disturbances. The threats posed by human activities are now of a scale at which the impacts are global. "Global change" is occurring in two ways: as a systemic change to the atmosphere and the oceans, and as a cumulative process by which the sheer intensity and pervasiveness of human activity has become a globally significant phenomenon. Examples of the first include climate change and ozone layer depletion, and examples of the second include fisheries, land-use change, introduced species and pollution. Most seabird species are likely to be affected by at least some of these phenomena.

Key words: Albatrosses, Longline Fishing, Bycatch, Fisheries Management, Pollution, Climate Change.

INTRODUCTION

In this chapter we give a non-governmental organizational (NGO) perspective on three components of global change with relevance to seabirds: fisheries, pollution and climate change. We briefly review the threats to albatrosses and discuss some of the problems associated with their resolution, including the limitations of science and international agreements in environmental and species protection. We highlight some requirements for addressing these issues.

FISHERIES BYCATCH

By the late 1980s scientists became increasingly worried about the status of various populations of albatrosses throughout the Southern Ocean. Results from two independent long-term studies demonstrated decreases among Wandering Albatross *Diomedea exulans* populations on Iles Crozet (Weimerkirsch and Jouventin 1987) and South Georgia (Croxall *et al.* 1990). Incidental catches of albatrosses in longline fisheries were implicated in the decreases. The first documentation of albatrosses caught by Southern Bluefin Tuna *Thunnus maccoyii* longline vessels was provided by Brothers (1991) in the waters around Tasmania. Subsequently, Murray *et al.* (1992) described seabird captures in New Zealand waters and Dalziell and DePoorter (1993) observed albatrosses being hooked by Soviet longliners targeting Patagonian Toothfish *Dissostichus eleginoides* around South Georgia. These authors focused attention on the scale of albatross mortality in longline fisheries, and observations quickly accumulated of bycatch of these and other seabirds in fisheries throughout the Southern Hemisphere. It is likely that all albatross species are affected to some extent by longline fisheries (Gales 1997).

Most hooking of birds occurs while the longlines are being set, when the birds attack the baited hooks. Removal of a bait means one less hook to catch a fish, and thus a reduced profitability for the fishermen. The interests of conservation (the prevention of the deaths of seabirds) and the objectives of the fishermen (the maximization of profits), coincide in this instance, thus creating a potentially favourable situation for the resolution of a widely recognized and serious problem. Techniques have been devised to reduce seabird mortality (e.g., Alexander *et al.* 1997) and handbooks have been produced to educate fishermen on methods to minimize seabird bycatch in longline fisheries (e.g., Brothers 1996; CCAMLR 1996). However, the existence of mitigation measures does not always mean their adoption. For example, the Convention for the Conservation of Antarctic Marine Living Resources (CCAMLR) currently requires longline vessels fishing in waters under its jurisdiction to use streamer lines, thawed baits and the discharge of offal outside the times

[1]Green Group in the European Parliament, Rue Belliard, 97–113, 1047, Brussels, Belgium.
Present address: Route d'Amonines 15, 6987 Magoster, Belgium.
[2]SeaWeb, 1731 Connecticut Avenue NW, Washington DC 20009 USA.
Present address: 4058 rue Dorion, Montreal QC, Canada H2K 4B9.

Pp. 247–54 *in* ALBATROSS BIOLOGY AND CONSERVATION ed by G. Robertson and R. Gales. Surrey Beatty & Sons, Chipping Norton. 1997.

of line setting and the setting of lines during the day. However, scientific observers placed on longline vessels fishing for Patagonian Toothfish have reported vessels contravening these regulations (CCAMLR 1995). Thus there appears to be considerable resistance on the part of some fishermen to the use of even relatively simple measures to reduce seabird mortality, despite the fact that these same measures would lead to reduced bait loss and improved fishing efficiency. One reason may be that while the absolute number of seabirds caught in longline fisheries is high, catch rates per vessel are usually low, and the scale of the problem is not immediately evident to the fishermen. Another has to do with the economic pressures on fishing vessels due to the decline in the world fish stocks.

Trends in fish stocks and fisheries wastage

Many of the world's fisheries are in a perilous state. In 1950 the total catch of marine fish and shellfish was 18.5 million tonnes. Catches increased steadily in the following years, except for a brief dip in the early 1970s when the Peruvian Anchovy *Engraulis ringens* stock collapsed. By 1989, 86.4 million tonnes of seafood were harvested globally, an almost five-fold increase in 40 years. Catches have fluctuated around that level ever since, according to the UN's Food and Agriculture Organization (FAO). The FAO data do not include fish and other species discarded at sea (due to either their low economic value or to landing restrictions) which have been estimated at 27.0 million tonnes annually (range: 17.9–39.5 million; Alverson *et al.* 1994). Thus the global catch of marine fish and shellfish is about 110 million tonnes, about 25% of which is discarded, much of it dead or injured.

The impact of such intensive fishing has been very destructive. Of the world's fish stocks for which assessment data are available, 25% are classified by FAO as being over-exploited, depleted or recovering from depletion. A further 44% are fully or heavily exploited, meaning that increases in fishing effort will not lead to similar increases in catch. Fewer than one stock in three can be described as being lightly to moderately exploited (FAO 1995). Nonetheless, fishing fleets continue to grow in size and fishing efficiency. Between 1970 and 1989 the tonnage of the world's fishing fleet of vessels greater than 100 GRT (Gross Registered Tons) grew at 4.6%/year, from 13.5–25.5 million GRT. By comparison, the total catches grew at about half that rate, or 2.4%/year (Garcia and Newton 1994). As has often been said, there are too many boats chasing too few fish, and albatrosses and other seabirds find themselves among the many victims.

Problems in estimating seabird mortality

Estimating the extent of seabird mortality is complicated by several factors. One is the lengthy time lag, in this case measured in decades, between the onset of a problem and the development and implementation of solutions. Another is that the wide-ranging activity of the world's fishing fleets, the continual development of new fisheries and the paucity of detailed information on their activities suggests that any estimate of seabird mortality will probably be an under-estimate. Fisheries can develop rapidly, far faster than institutional mechanisms to monitor and control their activities. Also, there is no way of knowing how many other potentially serious environmental problems are posed by intensive, industrial-scale fisheries.

The first step towards the reduction of seabird mortality is the collection of detailed data on the fishing operations and resulting bycatch. These data are not easy to obtain. Bycatches in some fisheries, such as those for Bluefin Tuna and Patagonian Toothfish, have been documented by observer programmes. However, there are fisheries that currently catch seabirds, but for which no data on incidental catch are available, such as the tuna longline fleets of Taiwan, China and Korea (Alexander *et al.* 1997). In the 1980s there was also a large fleet of unmonitored driftnetters operating in the South Pacific Ocean. As albatross bycatch was observed in the driftnet fisheries in the North Pacific Ocean (Gould *et al.* 1997), it would be expected that seabirds were also caught in the driftnet fisheries in the South Pacific Ocean. Since published documentation of this fishery was limited to one observer report (Sharples *et al.* 1990) and independent observations by Greenpeace (Earle *et al.* 1994), the extent of any seabird mortality in this fishery will probably never be known.

Fishermen are notoriously shy about describing their activities, especially those involving incidental catches that could cause problems for them (IMALF 1994), so deploying dedicated observers on board fishing vessels is the most reliable way to properly document seabird catch rates. It is reasonable to assume tuna longlining vessels had been catching albatrosses for decades before being observed by Brothers (1991), though this was never recorded, due to the low catch rates of birds and observers being more interested in monitoring fish. This highlights the importance of observers and their mandate: it is impossible for them to record everything that happens, so unless data on incidental catch are explicitly required very little such information will be collected.

Further, seemingly small changes to fishing gear or practices may have a dramatic impact on the magnitude and species composition of seabird bycatch. For instance, increasing numbers of Japanese longliners operating in the Australian Fishing Zone are using monofilament lines, resulting in seabird catch rates which are three-times greater than those of conventional gear (Klaer and Polacheck 1995). Vigilance is necessary to ensure that progress made in reducing bycatch is not reversed by subsequent modification of fishing gears which aim to increase catch rates of the targeted fish species.

Albatrosses are not the only bycatch

Much of the discussion about seabird mortality in longline fisheries focuses on albatrosses. To an extent this is understandable, because they are a highly visible species and their life history characteristics render them particularly vulnerable to even low levels of mortality. It is important, though, that the solutions which are promoted to reduce albatross deaths do not increase the mortality of other species. One widely recommended measure is the setting of lines at night, when albatrosses do not forage. However, night sets have higher catches of White-chinned Petrels *Procellaria aequinoctialis* (CCAMLR 1995), so switching to night sets must be accompanied by adequate measures to eliminate the mortality of these and other night-feeding seabird species.

It is easy to get the impression that seabirds are the only bycatch in some of these longline fisheries, especially since longlining is reputed to be a relatively selective method of fishing (however, see Alverson *et al.* 1994). The mandate of the CCAMLR ad-hoc Working Group on Incidental Mortality Arising from Longline Fishing (IMALF) is to review data relating to "marine animals" within the Convention Area. The report of the 1994 meeting (IMALF 1994) made no mention of any taxon other than seabirds, except for very occasional references to marine mammals. Yet several working papers submitted to the meeting note that other species are caught, including sea turtles, sharks, skates, sunfish and other finfish (Imber 1994; Barea *et al.* 1994; Cherel *et al.* 1994), some of which may be seriously affected by longline fishing. The incidental catches of sharks is considered a serious conservation problem on a global scale (Bonfil 1994). CCAMLR observers, as well as those involved in the Bluefin Tuna fisheries, collect data on bycatch species, but fisheries managers and scientists seem to have largely ignored them. Analysis of these valuable data is urgently required as a first step to reducing the incidental mortality of other bycatch species.

The responsibility of fisheries management authorities

Traditionally, fisheries management has focused on attempts to prevent fishing fleets from over-exploiting target fish species. Rarely do management bodies concern themselves with incidental catches of species of no commercial value such as seabirds, and yet the management of a fishery determines its impact on non-target species. A case in point is the fishery for Southern Bluefin Tuna which began in the early 1950s. In 1994 Japan, Australia and New Zealand signed the Convention for the Conservation of Southern Bluefin Tuna (CCSBT), which has as its objective "the conservation and optimum utilization of Southern Bluefin Tuna". However, although management restrictions have been in place for over 10 years and there exists a legally binding agreement, the stock is severely depleted, with the parental biomass <10% of the 1960 level (Hayes 1997). At the 1996 meeting of the CCSBT Scientific Committee Australian and Japanese scientists presented differing interpretations on trends in Southern Bluefin Tuna stocks, with the former suggesting continued low parental biomass and the latter suggesting a high probability of recovery (Hayes 1997 and references therein). The implications of these management discussions for albatrosses and other seabirds are obvious. As CCSBT imposes no gear regulations to mitigate bycatch a higher catch limit for Southern Bluefin Tuna would lead to more seabirds being caught. From the seabird conservation perspective, such quota increases are undesirable and should not take place until sufficient progress has been made on mitigation measures to compensate for any increase in fishing effort. Indeed, some environmental NGOs have called for a closure of the fishery until the problem of albatross bycatch is resolved. The CCSBT is very much aware of the albatross bycatch situation, having created an Ecologically Related Species Group to examine the issue. As the mandate of the CCSBT concerns tuna, rather than seabirds, it remains to be seen whether seabirds will be a decisive factor in decisions on catch limits; thus far, they have not been.

CCAMLR is under different constraints, being charged with the "maintenance of the ecological relationships between harvested, dependent and related populations of Antarctic marine living resources and the restoration of depleted populations to levels [which ensure the greatest net annual recruitment]". This "ecological" aspect of CCAMLR sets it apart from other fisheries management bodies, and means that management must take into consideration the impact of fishing on incidentally caught species. This explains the speedy institution of measures to reduce seabird bycatch and the deployment of scientific observers, in addition to the development of a management regime for the Antarctic Krill *Euphasia superba* fishery which considers its impact on predator species.

Despite the publicity that seabird bycatch has received in recent years, and the supposed interest of the fishing industry in reducing bycatch rates, mitigative measures are still not consistently applied. This is at least partly due to the lack of enforcement of fisheries management bodies to adopted regulations. Among the most jealously guarded privileges under international law is the responsibility of the state whose flag the fishing vessel flies to prosecute alleged violations of regulations. Countries vary enormously in their willingness and/or ability to enforce the law, so successful prosecutions are few compared to actual violations.

The extensive illegal fishing of Patagonian Toothfish (e.g., CCAMLR 1995) has serious implications for the conservation of albatrosses. Many NGOs are convinced that the solution to illegal fishing does not lie with unilateral declarations of jurisdiction, such as was done by the United Kingdom in 1993 for fisheries around South Georgia and the South Sandwich Islands. Unless all states co-operate to exert the same degree of control throughout the entire range of the fishery, in both national and international waters, vessels will be able to avoid restrictions by simply moving to areas where enforcement is less strict, or even non-existent. The United Kingdom declaration led to bitter debate within CCAML and it would be irresponsible and extremely unfortunate if these heightened tensions, which are of a largely political nature, hindered the ability of CCAMLR to carry out its mandate of marine conservation.

The mobility factor

Article 64 of the 1982 United Nations Convention on the Law of the Sea classified the large tunas as "highly migratory species", a reflection of the fact that they pass through the jurisdiction of many coastal states, as well as the high seas. Albatrosses are also highly migratory species, as has been recognized by the listing in 1997 of the Amsterdam Albatross *D. amsterdamensis* (joins the Short-tailed Albatross *D. [Phoebastria] albatrus*) in Appendix 1 (migratory species which are endangered) and 12 albatross species in Appendix 11 (migratory species of unfavourable conservation status) of the Convention on the Conservation of Migratory Species. The high mobility of albatrosses means that they are vulnerable to capture by longline vessels throughout an extensive geographical area (Gales 1993).

The wide oceanic distribution and high mobility of tunas is reflected in the highly migratory nature of the fleets that fish on the high seas and within national economic zones, as well as the capital that drives them. As the fates of fish stocks wax and wane, these distant water fleets range the globe in search of stocks that will allow profitable fishing. A fishery develops and vessels quickly move in from other fisheries. In many instances, the flag these vessels fly or the company name of their registry belie their ownership, which frequently rests with large multinational companies. For instance, many of the vessels fishing for Patagonian Toothfish in the South Atlantic Ocean and flying the flags of Argentina and Chile are at least partially owned by Spanish, American or Norwegian interests.

This mobility of fish, birds and fishing fleets complicates the resolution of the albatross bycatch issue. Simply collecting data on the catch rates and colony of origin of the birds has required a high degree of co-operation and sharing of information among these bodies (IMALF 1994; CCAMLR 1995). That they have responded as they have is doubtless due to interest the general public has in seabirds and albatrosses in particular. Other incidentally caught species have not yet benefited to the same extent from this popularity.

The elimination of the threat to seabirds and other species posed by longline fishing requires a far deeper integration of the activities of the many agencies involved. Local and isolated initiatives will not be sufficient. For instance, attempts by one regulatory body to impose a limit on the number of albatrosses that can be caught by the fleets under its jurisdiction will come to naught if vessels simply shift their activities to less restricted waters. Only if all the relevant institutions act in a co-ordinated fashion, and on a geographical scale that is commensurate with the high mobility of the fish, birds and fleets, is there any hope for a comprehensive and lasting solution. Some imaginative and innovative approaches to the management of these fisheries are clearly necessary.

The United Nations fisheries agreement

An important development in attempts to reduce seabird bycatch was the negotiation of the United Nations Agreement on the Conservation and Management of Straddling Fish Stocks and Highly Migratory Fish Stocks, concluded in August 1995 (Doulman 1995). It made progress towards a global approach to fisheries management by establishing minimum standards for data collection, surveillance and enforcement, flag state control and other matters to be implemented by regional bodies.

Among the provisions which are directly relevant to the problem of albatross bycatch are requirements to:

- minimize the catch of non-target species
- collect and share data on catch of non-target species
- assess the impact of fishing on non-target and associated species and adopt plans to ensure the conservation of such species

This far-reaching Agreement promises to radically alter the way in which fisheries on straddling and highly migratory stocks are conducted. It remains to be seen how widely and readily this Agreement will actually be applied: there are already signs that some states wish to reduce its impact. The ink was barely dry on the document when, three months after the negotiations ended, a number of states stated that it was not applicable to the CCAMLR region. They argued that the UN Agreement dealt with stocks that straddled the high seas and the waters of coastal states, and that this situation does not pertain to CCAMLR. Other states were of the opinion that it was very relevant. Such arguments can keep lawyers occupied for years but, as many NGOs have noted, the UN Agreement is a very progressive document that must be widely and rapidly implemented if the situation for both fish and other marine species is to improve.

The precautionary approach to fisheries management

The UN moratorium on the use of large driftnets is often cited as an NGO success against destructive fishing practices. Yet many NGOs waged a single-minded campaign against driftnets, paying no attention to what the vessels would do if they stopped using driftnets, such as converting to the longline method of fishing. Perhaps the most noteworthy aspect of the UN moratorium was its wording, for driftnets could only be used if it could be scientifically demonstrated that they caused no "unacceptable impact".

The moratorium on driftnetting was an early manifestation of a precautionary approach to fishing, which advocates the prevention of environmental damage in the first place, rather than remedial action. It requires action before scientific proof of deleterious effects is available, and the reversal of the burden of proof onto the potential industry or regulatory body to show that the activity will not have a negative environmental or social impact. In effect the precautionary approach concedes the limitations of, and uncertainties inherent in, scientific knowledge and forces the re-assessment of the role of science in decisions that impinge on society, human health, and on the natural environment. The precautionary approach is an underlying principle that should enlighten the way in which fishing is conducted throughout the world's oceans. Important elements of a precautionary approach to fisheries would include (see also Cooke and Earle 1993 and the UN Agreement):

- fishing pressure should be constrained so as to maintain target and non-target species at high levels of abundance;
- new fishing gears and practices should be evaluated before being introduced to avoid bycatch and physical damage to the environment;
- closed areas should be established to protect the marine habitat.

Without a fundamental shift in thinking, from a reactive to a proactive view, the destructive impact of industrialized fisheries on the marine environment will only increase. For every high profile issue such as albatross bycatch several others are likely to appear in future. It is essential not to view these issues as problems to be resolved individually; rather, they are manifestations of wider problems regarding human attitudes to marine life that must be addressed in order to improve the overall management of fisheries.

POLLUTION

Plastics

Though longlining clearly poses the greatest known threat to many albatross populations plastic debris (Auman *et al.* 1997), oil (e.g., Stagi *et al.* 1997) and chemicals (Ludwig *et al.* 1997) also affect seabirds and other marine and coastal biota. Plastics have been found in most, if not all, species of albatrosses (Ryan 1987; McDermond and Morgan 1993). Fry *et al.* (1987) uncovered higher plastic loads in dead than in live Laysan Albatross *D. (Phoebastria) immutabilis* chicks and found that 8% of 24 chicks examined also had impacted proventriculi. Sievert and Sileo (1993) found that Laysan Albatross chicks with large volumes of proventricular plastic had significantly lower fledging masses than those with lower volumes. Laysan Albatross chicks with large volumes of plastic also had significantly lower survival rates than those with low volumes during an increased mortality event on Midway Atoll in 1986 (Sievert and Sileo 1993). While not an important direct cause of albatross chick death, ingested plastic may cause physiological stress, affect survival during the nestling period and perhaps after fledging (Sievert and Sileo 1993; Auman *et al.* 1997). Plastics may also transport chemical contaminants around the ocean and to albatrosses when ingested at sea (Auman *et al.* 1997).

Chemicals

The use and discharge of organochlorines has provoked a strong NGO focus due to these chemicals' toxicity, environmental persistence, and predisposition to bioaccumulate in biota. However, organochlorines continue to be extensively produced (Anderson 1990; Iwata *et al.* 1994; Marquenie and Reijnders 1989; Voldner and Li 1995) despite their highly publicized negative effects on humans and wildlife (Colborn and Clement 1992; Colburn *et al.* 1993), including seabirds (Bosveld and Van den Berg 1994). Toxic equivalence factors of PCBs, PCDD and PCDF in the eggs of Black-footed *D. (Phoebastria) nigripes* and Laysan Albatrosses from Midway Atoll in the central Pacific Ocean are now at levels where sensitive avian species show adverse reproductive effects (Jones *et al.* 1994). About 10% of productivity decreases since the 1990s in Black-footed Albatrosses at Midway Atoll have been attributed to the effects of organochlorine contaminants (Ludwig *et al.* 1997).

Overall, there are more than 65 000 chemical compounds in use in industry worldwide of which approximately 10 000 have regular application in one or more industrial processes (NRC 1984); and approximately 1 000 new chemicals are introduced into commerce annually (US Congress 1987). The sheer number of chemicals involved, the wide range of potential exposure routes and the individual and species-specific responses to exposure, the additive and synergistic dynamics of complex mixtures, and numerous other complexities inherent in the toxicological/biological interface has meant that relatively little can be said about the ultimate effects of chemical pollution. Indeed, information on toxicity for risk assessments is available for <2% of chemicals in use in the USA.

That the effects of contaminants and marine debris on albatross populations in particular are largely unquantified, and considered by biologists to be minor relative to the impact of fisheries (e.g., Ludwig *et al.* 1997), should not be reason for complacency. Rather, the problems associated with environmental contaminants are, in part, a reflection of the ongoing failure of society to fully appreciate the scale of human activity and the limitations of science in documenting its impacts. For various seabird populations, and those of other marine species, this has relevance not only in terms of the problems caused by pollution, fisheries, habitat destruction and introduced species, but is particularly urgent in this era of a warming climate.

CLIMATE CHANGE

Various authors (e.g., Bildstein 1991; Burger 1990; Duffy 1993; Nisbet 1994) have expressed concern about the potential threat posed to seabirds by climate change and the submergence of coastal areas, changes in marine productivity, and increases in ambient temperatures, storms (see Robertson 1997) and weather associated with the El Nino Southern Oscillation. A 1995 summary report by the Intergovernmental Panel on Climate Change (IPCC) stated that "the balance of evidence suggests a discernible human influence on global climate" and that global mean temperatures have increased by between 0.3°C and 0.6°C since the late 1800s. The IPCC statement also warned that "an immediate departure from business as usual is required" and that a changing climate increases the possibility of "surprises and unanticipated rapid changes" (Masood 1995; Pearce 1995). Others are more optimistic. Wigley *et al.* (1996) suggested that a concentration of CO_2 in the atmosphere about double the pre-industrial level is acceptable and that this level will not be reached for some 30 years. They assume that cleaner energy sources will have been developed by then, and that large cuts in CO_2 emissions will, in 30 years time, have less economic impact than at present (though see Pearce 1996; Romm and Curtis 1996).

Regardless of perceptions of urgencies, clear documentation of the effects of climate change will only come about after it has occurred; by then retreat may be impossible (Woodwell 1995). Here, it is hard to escape the conclusion that the situation at hand has a strong ethical component. However, the reduction of ethical questions to purely scientific ones, whether it involves fisheries and albatross by-catch, environmental contamination, or climate change, is to sabotage public debate by restricting the terms of reference to a language that few can follow. This is compounded by the blind faith of much of society, including policy makers and scientists, that scientific knowledge is not plagued by uncertainty and ignorance.

CONCLUSION

Numerous theories covering varying political, historical, economic, ethical, philosophical and social viewpoints have been developed that attempt to explain the human-related causes of global environmental change, of the difficulties in addressing it, and of the possible solutions. The complexity of this human-related component, and of the bureaucratic structures that have been erected to attempt human and environmental management or control, now appear to rival the complexity of the ecological processes of the biosphere itself. Faced with the enormous scope of the human-environment interaction, the understandable tendency for those involved with the fate of seabirds and other wildlife species is to restrict the context of the issues facing them in order to inform or derive some sort of management response. However, as Ehrenfeld (1978) and others have pointed out, information or solutions based on an arbitrarily restricted context may be worthless, or even destructive. For example, the belief that "informed policy" on climate change can only come about when it is defined by science assumes that science is not limited by its own ignorance nor bound by ethics.

The long-term impact of global threats to albatrosses, as with many seabird species, is difficult to quantify. While much continues to be done in some areas with regard to colony protection and local threats, such as from introduced species, habitat degradation, directed takes, and disturbance, climate change, fisheries and environmental contamination may swamp conservation initiatives. Any conservation approach requires acknowledgment of scientific uncertainty and ignorance and the limitations of international agreements, is precautionary in scope, and is fearless in asking questions concerning our use of and relationship with the natural environment.

REFERENCES

Alexander, K, Robertson, G. and Gales, R., 1997. The incidental mortality of albatrosses in longline fisheries. Australian Antarctic Division. 44pp.

Alverson, D. L., Freeberg, M. H., Murawski, S. A. and Pope, J. G., 1994. A Global Assessment of Fisheries Bycatch and Discards. FAO Fisheries Technical Paper 339. Food and Agriculture Organization of the United Nations, Rome. xxi+233 pp.

Anderson, E. V., 1990. Chemical Engineering News, May 21, 1990.

Auman, H. J., Ludwig, J. P., Giesy, J. P. and Colborn, T., 1997. Plastic ingestion by Laysan Albatross chicks on Sand Island, Midway Atoll, in 1994 and 1995. Pp. 239–44 *in* Albatross Biology and Conservation ed by G. Robertson and R. Gales. Surrey Beatty & Sons: Chipping Norton.

Barea, L., Loinaz, I., Marin, Y., Rios, C., Saralegui, A., Stagi, A., Vaz-Ferreira, R. and Wilson, N., 1994. Mortality of albatrosses and other seabirds produced by tuna longline fisheries in Uruguay. Paper presented to the CCAMLR Ad-Hoc Working Group on Incidental Mortality Arising from Longline Fishing, Hobart, Australia, 21–22 October 1994.

Bildstein, K. L., Bancroft, G. T., Dugan, P. J., Gordon, D. H., Erwin, R. M., Nol, E., Payne, L. X. and Senner, S. E., 1991. Approaches to the conservation of coastal wetlands in the western hemisphere. *Wilson Bulletin* **103**(2): 218–54.

Bonfil, R., 1994. Overview of World Elasmobranch Fisheries. FAO Fisheries Technical Paper 341. Food and Agriculture Organization of the United Nations, Rome. viii+119 pp.

Bosveld, A. T. C. and Van den Berg, M., 1994. Effects of polychlorinated biphenyls, dibenzo-p-dioxins, and dibenzofurans on fish-eating birds. *Environ. Reviews* **2:** 147–66.

Brothers, N., 1991. Albatross mortality and associated bait loss in the Japanese longline fishery in the Southern Ocean. *Biol. Conserv.* **55:** 255–68.

Brothers, N., 1996. Longline fishing dollars and sense. Tasmanian National Parks and Wildlife Service.

Burger, J., 1990. Seabirds, tropical biology and global warming: Are we missing the ark? *Colonial Waterbirds* **13**(2): 81–84.

CCAMLR, 1995. Report of the Fourteenth Meeting of the Commission, Hobart Australia, 24 October to 3 November 1995.

CCAMLR, 1996. Fish the Sea, not the Sky. CCAMLR, Hobart, Australia.

Cherel, Y., Weimerskirch, H. and Duhamel, G., 1994. Interactions between longline vessels and seabirds in Kerguelen waters and a method to reduce seabird entanglement. Paper presented to the CCAMLR Ad-Hoc Working Group on Incidental Mortality Arising from Longline Fishing, Hobart, Australia, 21–22 October 1994.

Colborn, T. and Clement, C. (eds.), 1992. Chemically induced alterations in sexual and functional development: the wildlife/human connection. Princeton Scientific Publishing, Princeton, NJ.

Colborn, T., vom Saal, F. and Soto, A., 1993. Developmental effects of endocrine-disrupting chemicals in wildlife and humans. *Environ. Health Perspect.* **101**(5): 378–84.

Cooke, J. G. and Earle, M., 1993. Towards a precautionary approach to fisheries management. *Rev. Euro. Comm. Internat. Environ. Law* **2:** 252–59.

Croxall, J. P., Rothery, P., Pickering, S. P. C. and Prince, P. A., 1990. Reproductive performance, recruitment and survival of wandering albatrosses *Diomedea exulans* at Bird Island, South Georgia. *J. Anim. Ecol.* **59:** 775–96.

Dalziell, J. and DePoorter, M., 1993. Seabird mortality in longline fisheries around South Georgia. *Polar Rec.* **29:** 143–45.

Doulman, D. J., 1995. Structure and Process of the 1993–1995 United Nations Conference on Straddling Fish Stocks and Highly Migratory Fish Stocks. FAO Fisheries Circular No. 898. Food and Agricultural Organization of the United Nations, Rome. iii+81 pp.

Duffy, D. C., 1993. Stalking the Southern Oscillation: environmental uncertainty, climate change, and North Pacific seabirds. Pp. 61–67 *in* The Status, Ecology and Conservation of Marine Birds in the North Pacific ed by K. Vermeer, K. T. Briggs, K. H. Morgan and D. Siegel-Causey. Canadian Wildlife Service Special Publication: Ottawa.

Earle, M., Woodley, T. H. and Hagler, M., 1994. An investigation of incidental catches of large-mesh driftnet fisheries from the South Pacific and North Atlantic. Report of the International Whaling Commission (Special Issue No. 15) pp. 529–32.

Ehrenfeld, D. W., 1978. The Arrogance of Humanism. Oxford University Press: New York.

FAO, 1995. Review of the State of World Fishery Resources: Marine Fisheries. FAO Fisheries Circular No. 884. Food and Agricultural Organization of the United Nations, Rome. 105 pp.

Fry, D. M., Fefer, S. I. and Sileo, L., 1987. Ingestion of plastic debris by Laysan albatrosses and wedge-tailed sheerwaters in the Hawaiian Islands. *Mar. Pollut. Bull.* **18**(6B): 339–43.

Gales, R., 1993. Co-operative Mechanisms for the Conservation of Albatross. Australian Nature Conservation Agency. 132 pp.

Gales, R., 1997. Albatross populations: threats and status. Pp. 20–45 *in* Albatross Biology and Conservation ed by G. Robertson and R. Gales. Surrey Beatty & Sons: Chipping Norton.

Garcia, S. M. and Newton, C., 1994. Current situation, trends and prospects in world capture fisheries. Paper presented at the Conference on Fisheries Management: Global Trends. Seattle, Washington, 14–16 June 1994.

Gould, P., Ostrom, P., Walker, W. and Pilichowski, K., 1997. Laysan and Black-footed Albatrosses: trophic relationships and driftnet fisheries associations of non-breeding birds. Pp. 199–207 *in* Albatross Biology and Conservation ed by G. Robertson and R. Gales. Surrey Beatty & Sons: Chipping Norton.

Hayes, E. A., 1997. A review of the Southern Bluefin Tuna fishery: Implications for ecologically sustainable management. Traffic Oceania, Sydney. 34 pp.

IMALF, 1994. Report of the CCAMLR Ad-Hoc Working Group on Incidental Mortality Arising from Longline Fishing, Hobart, Australia, 21–22 October 1994. SC-CAMLR-XIII/7. 36 pp.

Imber, M. J., 1994. Report on a tuna long-lining voyage aboard Southern Venture to observe seabird by-catch problems. New Zealand Department of Conservation Science and Research Series No. 65. 12 pp.

Iwata, H., Tanabe, S., Sakai, N., Nishimura, A. and Tatsukawa, R., 1994. Geographical distribution of persistent organochlorines in air, water and sediments from Asia and Oceania, and their implications for global redistribution from lower latitudes. *Environ. Pollut.* **85:** 15–33.

Jones, P. D. and nine authors, 1994. PCBs and 2,3,7,8-substituted polychlorinated dibenzo-p-dioxins and dibenzofurans in albatross tissues from Midway Island in the central Pacific Ocean. Extended abstract from Dioxin '94, Kobe, Japan. *Organohalogen Compounds* **20:** 137–40.

Klaer, N. and Polachek, T., 1995. Japanese longline seabird bycatch in the Australian Fishing Zone, April 1991 — March 1993. Report of Division of Fisheries, CSIRO. 95 pp.

Ludwig, J. P. Summer, C. L., Auman, H. J., Gauger, V., Bromley D., Giesy J. P., Rolland, R. and Colborn, T., 1997. The roles of organochlorine contaminants and fisheries bycatch in recent population changes of Black-footed and Laysan Albatrosses in the North Pacific Ocean. Pp. 225–38 *in* Albatross Biology and Conservation ed by G. Robertson and R. Gales. Surrey Beatty & Sons: Chipping Norton.

Marquenie, J. M. and Reijnders, P. J. H., 1989. PCBS, an increasing concern for the marine environment. C.M. 1989/N:12. ICES, Copenhagen. 12 pp.

Masood, E., 1995. Climate panel confirms human role in warming, fights off oil states. *Nature* **378:** 524.

McDermond, D. K. and Morgan, K. H., 1993. Status and conservation of North Pacific albatrosses. Pp. 70–81 *in* The Status, Ecology, and Conservation of Marine Birds of the North Pacific ed by K. Vermeer, K. T. Briggs, K. H. Morgan and D. Siegel-Causey. Canadian Wildlife Service Special Publication: Ottawa.

Murray, T. E., Taylor, P. R., Greaves, J., Bartle, J. A. and Molloy, J., 1992. Seabird bycatch by Southern Ocean fishery longline vessels in New Zealand waters. New Zealand Fisheries Assessment Research Document 92/22.

Nisbet, I. C. T., 1994. Effects of pollution on marine birds. Pp. 8–25 *in* Seabirds on Islands: Threats, Case Studies and Action plans ed by D. N. Nettleship, J. Burger and M. Gochfield. Birdlife Conservation Series No. 1. Birdlife International: Cambridge.

NRC, 1984. National Research Council, Commission on Life Sciences. Toxicity testing: strategies to determine needs and priorities. National Academy of Science: Washington, DC.

Pearce, F., 1995. Global warming "jury" delivers guilty verdict. *New Scientist* (9 December)**:** 6.

Pearce, F., 1996. Sit tight for 30 years, argues climate guru. *New Scientist* (20 January)**:** 7.

Robertson, C. J. R., 1997. Factors influencing the breeding performance of the Northern Royal Albatross. Pp. 99–104 *in* Albatross Biology and Conservation ed by G. Robertson and R. Gales. Surrey Beatty & Sons: Chipping Norton.

Romm, J. J. and Curtis, C. B., 1996. Mideast oil forever? *The Atlantic Monthly* **277**(4)**:** 57–74.

Ryan, P. G., 1987. The incidence and characteristics of plastic particles ingested by seabirds. *Mar. Environ. Res.* **23:** 175–206.

Sharples, P., Bailey, K., Williams, P. and Allan, A., 1990. Report of observer activity on board JAMARC driftnet vessel R.V. Shin-Hoyo Maru fishing for albacore in the South Pacific Ocean. 22 November — 23 December 1989 and 10 February — 3 March 1990. Working Paper No. 3 presented to the Third South Pacific Albacore Research Workshop, Noumea, 9–12 October 1990. 28 pp.

Sievert, P. R. and Sileo, L., 1993. The effects of ingested plastic on growth and survival of albatross chicks. Pp. 212–17 *in* The Status, Ecology, and Conservation of Marine Birds of the North Pacific ed by K. Vermeer, K. T. Briggs, K. H. Morgan and D. Siegel-Causey. Canadian Wildlife Service Special Publication: Ottawa.

Stagi, A., Vaz-Ferreira, R., Marín, Y. and Joseph, L., 1997. The conservation of albatrosses in Uruguayan waters. Pp. 220–24 *in* Albatross Biology and Conservation ed by G. Robertson and R. Gales. Surrey Beatty & Sons: Chipping Norton.

US Congress, Office of Technology Assessment., 1987. Wastes in Marine Environments. OTA-O-334. U.S. Government Printing Office: Washington, DC.

Voldner, E. C. and Li, Y-F., 1995. Global usage of selected persistent organochlorines. *The Sci. Total Environ.* **160/161:** 201–10.

Weimerskirch, H. and Jouventin, P., 1987. Population dynamics of the wandering albatross, *Diomedea exulans*, of the Crozet Islands: causes and consequences of the population decline. *Oikos* **49:** 315–22.

Wigley, T. M. L., Richels, R. and Edmonds, J. A., 1996. Economic and environmental choices in the stabilization of atmospheric CO2 concentrations. *Nature* **379**(6562)**:** 240–43.

Woodwell, G. M., 1995. Biotic feedbacks from the warming of the earth. Pp. 3–21 *in* Biotic Feedbacks in the Global Climatic System ed by G. M. Woodwell and F. T. Mackenzie. Oxford University Press: New York.

CHAPTER TWENTY-TWO

International legal and political bases to the management of the incidental catch of seabirds

MARCUS HAWARD,[1] ANTHONY BERGIN[2] and H. ROBERT HALL[1]

We examine the development of international legal and political responses to the problem of the incidental catch of seabirds, particularly albatrosses. Recent international legal and policy developments on the bycatch issue in general are surveyed. Although many of these developments may be categorized as "soft law" this is not to say that progress has not been made in dealing with this important issue in international fisheries management. The second part of the chapter focuses, in particular, upon the specific measures introduced by the Commission for the Conservation of Antarctic Marine Living Resources (CCAMLR) to reduce the incidental catch of albatrosses. While the CCAMLR measures concerning baited hooks, night-time line-setting, trash and offal dumping, streamer lines and caught live bird release efforts may have minimal direct effects in reducing the incidental catch of albatrosses — for the reason that most birds are caught in waters outside CCAMLR's geographic ambit of operation — their importance is significant in another sense. The CCAMLR measures offer a mitigation model for use by international fisheries management organizations employing longline techniques and having geographic ambits of operation adjacent to that of CCAMLR. The adoption of the CCAMLR mitigation model by other international fishery management organizations, together with the development of interorganizational collaborative efforts, may lead to the formation of a seabird bycatch regime which is patchwork in nature with its component norms, rules and decision making procedures articulated in separate agreements and managed by diverse international organizations. This prospect is discussed and a number of practical, process-oriented political lessons concerning regime formation are suggested to facilitate the effective management of seabird bycatch. These include the necessity of entrepreneurial leadership, the existence of salient solutions, the availability of equitable options and clear-cut and effective compliance mechanisms and the general perception of a sense of urgency triggered by a shock or crisis. Albatross conservation has a significant political dimension and is not solely a scientific matter. This means that attention must also be focused on identifying and engaging the political and economic factors driving ocean use.

Key words: International Law, International Regimes, Incidental Catch, Bycatch, CCAMLR, International Organizations, Mitigation Devices, Conservation Measures.

INTRODUCTION

In recent years a great deal of concern has been expressed by fisheries managers, scientists and conservationists about the effects of the incidental or bycatch of non-fish species, particularly marine mammals and seabirds. Expressions of this concern have been articulated in a number of international conferences and organizations. This chapter examines the development of international legal and political responses to the problem of bycatch focusing, in particular, upon the specific measures introduced by the Commission for the Conservation of Antarctic Marine Living Resources (CCAMLR) to reduce the incidental catch of albatrosses. CCAMLR's initiatives, which provide leadership on the bycatch issue, may facilitate the establishment of a broader international management regime. The chapter is divided into three sections. The first examines recent international legal and policy developments on the bycatch issue. This provides a contextual setting for the second part which outlines the initiatives undertaken by CCAMLR. The third section suggests some practical, process-oriented considerations concerning the formation of an international regime to manage incidental catch of seabirds.

INTERNATIONAL LEGAL AND POLICY DEVELOPMENTS ON BYCATCH

Bycatch has emerged as a major issue in international fisheries diplomacy and law. The tuna/dolphin, shrimp/turtle cases and the use of high seas pelagic driftnets have received widespread

[1]Department of Government, University of Tasmania, Hobart, Australia 7001.
[2]School of Politics, Australian Defence Force Academy, Canberra, Australia 2601.

Pp. 255–66 *in* ALBATROSS BIOLOGY AND CONSERVATION ed by G. Robertson and R. Gales. Surrey Beatty & Sons, Chipping Norton. 1997.

publicity in the international media and will not be dealt with here. Currently under increasing international scrutiny is pelagic and demersal longlining for its impact on seabird populations. The seabirds considered to be at greatest risk are albatrosses. Because albatrosses are long-lived and wide-ranging with low fecundity and breeding success they are extremely vulnerable to factors which threaten their survival.

In relation to the management of albatross bycatch the most significant developments has been the work undertaken by CCAMLR (see below), the inclusion of albatrosses as an "ecologically related species" (ERS) for study by the ERS working group of the Commission for the Conservation of Southern Bluefin Tuna CCSBT), and the listing of albatrosses on the Bonn Convention (Convention on the Conservation of Migratory Species of Wild Animals). These and other organizations and forums that have considered the issue of bycatch but with less emphasis on albatrosses are as follows (see also Alverson *et al.* 1994).

The Convention for the Conservation of Southern Bluefin Tuna (CCSBT)

The CCSBT, which entered into force in May 1994, has specific provision for consideration of conservation and management of "ecologically related species" (ERS). At the first meeting of the Commission established under the Convention in Wellington in May 1994, it was confirmed that an ERS working group would be formed to consider as a first priority seabird incidental catches. The Commission also decided to hold a special workshop on seabird incidental catches. The purpose of the workshop will be to consider standardizing data collection and analysis, assessment of the impact of incidental take, and programmes of mitigation devices and measures. The CCSBT parties have agreed that as a matter of priority other states fishing Southern Bluefin Tuna, *Thunnus maccoyii*, such as Taiwan, Korea and Indonesia should be drawn into the management effort so ensuring greater co-operation in adopting measures to reduce the catch of seabirds. A major task for the ERS group, once it is established, will be to assess the impact of tuna longlining on seabird populations throughout the range of the Southern Bluefin Tuna fishery.

The Bonn Convention

The issue of albatross bycatch has been a major impetus for ongoing developments within the Bonn Convention. Australia ratified the Convention in September 1991. In that year Australia raised concern regarding the level of albatross mortality associated with longline fishing. At the fifth meeting of the Conference of Parties to the Bonn Convention, held in April 1997, Australia nominated 11 southern hemisphere species and the Netherlands two northern hemisphere species for listing under the convention. The Short-tailed Albatross *Diomedea (Phoebastria) albatrus* had earlier been listed on Appendix I. The Amsterdam Albatross *D. amsterdamensis* joined the Short-tailed Albatross on Appendix I, with the remaining species listed on Appendix II of the Bonn Convention.

Appendix I of the Bonn Convention lists migratory species which are endangered. Range states are required to prohibit the taking of these animals, with few exceptions, so as to give them full protection. Appendix II lists migratory species which have an unfavourable conservation status and which require international agreements for their conservation and management, as well as those which would significantly benefit from the international co-operation that could be achieved by an international agreement. The Convention provides for two types of agreements for species listed in Appendix II. First, there are AGREEMENTs (the capitalization is intentional) intended to benefit migratory species — especially those with an unfavourable conservation status — over their entire range. Parties are also encouraged to conclude a second type of agreement (spelt in lower case, to differentiate them from AGREEMENTs) for populations of species that periodically cross national jurisdictional boundaries.

The Cancun Conference on responsible fishing and the Cancun Declaration

The Cancun Conference was an intergovernmental conference held at Cancun, Mexico, in 1992, organized by the Mexican Government in co-operation with the Food and Agriculture Organization (FAO). Its Declaration requested FAO to elaborate a Code of Conduct for Responsible Fishing. The Cancun Declaration pointed to the "vital need for fishing to continue and develop within a comprehensive and balanced system under the concept of responsible fishing". The Conference outlined two goals for coastal states in respect to incidental catch: (1) In the design and subsequent introduction of new fishing gear and practices, states should take into account qualified assessments of impacts on the sustainability of fisheries, giving due consideration to the specific biodiversity of different fishing areas; (2) States should promote systematic use of selective fishing gears and practices that minimize waste of catch of target species and bycatch of non-target species. The Conference also addressed non-tariff barriers to trade that can be tied to bycatch issues. This was in response to Mexico and other states' concerns over US efforts to impose its "conservation measures" on other nations. The Conference conveyed a message to UNCED that environmental concerns should not be used as justification for limiting exports of fishery products from developing countries.

The United Nations Conference on Environment and Development (UNCED) Agenda 21

The report of the UNCED conference in 1992 also addressed the issue of bycatch. Under Agenda 21 the report promotes the development and use of selective fishing gears and practices that minimize bycatch of non-target species (a repeat of the Cancun Declaration). The report also notes that countries should ensure that fishing vessels flying their flags on the high seas undertake fishing activities in a manner minimizing incidental catch and take measures to increase the availability of marine living resources as human food by reducing wastage, post-harvest losses, and discards.

The Organization for Economic Co-operation and Development (OECD)

The OECD have been organizing and continues to organize meetings relating to environment and trade and one of the sectors studied is fisheries. There is also a separate *ad hoc* expert group on fisheries studying the economics of fisheries management. From the *ad hoc* group a sub-group of four countries (Iceland, Canada, Australia and the US) was formed in 1994 to examine bycatch issues. This group produced a consensus report, delivered to the OECD in October 1995.

Elaboration of a Code of Conduct for Responsible Fisheries

The Code of Conduct for Responsible Fisheries developed by FAO and first envisaged by the Cancun Declaration endorses the precautionary approach to fisheries management (see below) and specifies that fishers should ensure that their gear, methods and practices are sufficiently selective as to minimize waste and discards and threat to endangered species, and that the intent of related regulations is not circumvented by technical devices. While non-binding, the code might become the future reference point for fishery managers and administrators throughout the world. The code covers all fishery operations, not simply high seas fisheries, and draws on the provisions of the 1994 Agreement to Promote Compliance with International Conservation and Management Measures by Vessels Fishing the High Seas. It is designed to be consistent with the 1982 Law of the Sea Convention, but also to take into account the 1992 Cancun Declaration, Agenda 21 as well as the conclusions and recommendations of the 1992 FAO technical consultation on high seas fishing and other relevant instruments.

The United Nations (UN) bycatch policies

The development of United Nations bycatch policies has been closely linked to the high seas driftnet issue. In late 1990 the UN was drawn into the driftnet conflict primarily based on the levels of discard (see Burke *et al.* 1994). Less than two years after its first resolution on the issue, the UN framed its 1991 resolution (44/225) recommending the termination of driftnetting on the basis that "some members of the international community have reviewed the best available scientific data on the impacts of large-scale pelagic driftnet fishing and have failed to conclude that this practice has no adverse impacts which threaten the conservation and sustainable management of living marine resources." The resolution called for high seas driftnets to be phased out beginning in January 1992, and a global moratorium was implemented effective 1 January 1993. A resolution on bycatch was also passed by the UN General Assembly in November 1994 calling on FAO, the UN high seas fish conference (see below) and other regional and sub-regional fisheries organizations to elaborate fisheries bycatch provisions and mitigation practices.

Research by the Food and Agriculture Organization

Apart from its work in elaborating the Code of Conduct for Responsible Fisheries and a Reflagging Agreement (see below), FAO has produced a major report on the bycatch issue, "A Global Assessment of Fisheries Bycatch and Discards" (Alverson *et al.* 1994). This seminal study considered 800 scientific papers on the topic and estimated that between 17.9 and 39.5 million tons (average 27.0 million tons) of fish are discarded each year in commercial fisheries. The highest quantities of discards are from the north west Pacific Ocean while tropical shrimp trawl fisheries generate a higher proportion of discards than any other fisheries type, accounting for one-third of the global total. Of four major gear groups, shrimp trawls stand alone at the top of the list; bottom trawls, longlines and pot fisheries come next. The third group consists of Japanese high seas driftnet fisheries, Danish seines and purse seines for Capelin *Mallotus villosus*. Relatively low levels result from pelagic trawls, small pelagic purse seiners and some of high seas driftnets. Not surprisingly the study concludes that quick solutions to the problem are unlikely and much more information is required.

The International Conference on the Sustainable Contribution of Fisheries to Food Security

The International Conference on the Sustainable Contribution of Fisheries to Food Security was organized by Japan and FAO and held in

Kyoto, Japan, in 1995, following the example of the Cancun Conference. The Conference discussed the sustainability of resources in the context of marine ecosystems, species utilization, international trade and shed light on resource management methods. It has served to focus greater attention on the bycatch issue. Japan is especially concerned about what is sees as the "over-protection" of marine living resources by environmentalists. It points to the campaigns on whaling and driftnetting in particular as showing that politics can run ahead of science in international fisheries diplomacy (see Bergin and Haward 1996).

The UN Conference on Straddling Fish Stocks and Highly Migratory Fish Stocks

The UN Conference on Straddling Fish Stocks and Highly Migratory Fish Stocks had its first substantive session in 1993 and concluded its final round of deliberations in 1995. The Conference produced a legally binding instrument for the management of such stocks (the "Agreement for the Implementation of the Provisions of the United Nations Convention on the Law of the Sea of 10 December 1982 Relating to the Conservation and Management of Straddling Fish Stocks and Highly Migratory Fish Stocks"). The Agreement adopts conservation and management measures for non-target species within the same ecosystem, promotes the development and use of environmentally safe fishing gear to minimize pollution, wastes and fisheries discards. For the first time in any fisheries instrument it spells out suggested guidelines for application of precautionary reference points in conservation and management of straddling fish stocks and highly migratory fish stocks. A number of articles in the Agreement involve measures to conserve "associated or dependent species" including seabirds such as albatrosses. The Agreement provides a strong framework for international action to reduce incidental catch of seabirds. As of August 1997, 59 countries have signed the Agreement, with 15 ratifications. The Agreement will enter into force with 30 ratifications.

The Law of the Sea Convention (LOSC)

The 1982 Law of the Sea Convention entered into force in November 1994. As of July 1997, 119 states have ratified the convention. The treaty has a separate provision that deals with bycatch within national jurisdiction. It is that in taking conservation measures for target species, coastal states

> shall take into consideration the effects on species associated with or dependent upon targeted species with a view to maintaining or restoring populations of such associated or dependent species above levels at which their reproduction may become seriously threatened.

It could be argued that the bycatch situation is also addressed in a special article on marine mammals. Article 65 declares that nothing in the various articles on national economic zones

> restricts the right of a coastal state or the competence of an international organization, as appropriate, to prohibit, limit or regulate the exploitation of marine mammals more strictly than provided for in this Part.

This means that the coastal state need not limit its regulations to protecting against over-exploitation to the point of endangering or threatening a population of marine mammals. Article 120 also emphasizes the obligations to conserve and co-operate specifically in respect of marine mammals in high seas areas. The LOSC recognizes freedom of fishing on the high seas but significantly adds "subject to the conditions laid down in section 2." These include the conservatory provisions in Articles 117 and 119. Moreover, Article 118 reinforces this concept of a conditional right, providing that "states shall co-operate with each other in the conservation and management of the living resources of the high seas", and that states exploiting the living resources of the same area must "enter into negotiations with a view to taking the measures necessary for the conservation of the living resources concerned" and that states must co-operate also in establishing appropriate regional or subregional fisheries organizations to this end.

The Biodiversity Convention

The Biodiversity Convention came into force in December 1993. Biodiversity is defined to include marine ecosystems and parties have a responsibility for the conservation and sustainable use of its own biological diversity. Parties are to co-operate in implementing the convention in areas beyond national jurisdiction such as areas of the high seas. Parties are also required to give emphasis to *in situ* conservation (that is, conservation of ecosystems, habitats and species in their natural environment). Enforcement is fairly weak; there is a requirement for periodic reports to a conference of parties for their review and comment.

The Convention on International Trade in Endangered Species of Wild Fauna and Flora (CITES)

The process for revising the criteria for listing species in the Convention on Trade in Endangered Species of Wild Fauna and Flora (CITES) was completed in Fort Lauderdale, USA, November 1994. The revised criteria (to be known as the *Everglades Criteria*) make it easier to follow in deciding whether a species warrants listing (Bergin and Haward 1995). CITES provides for the inclusion on Appendix I of species threatened with extinction that are or may be affected by trade and listing on Appendix I prohibits international trade in wild specimens. Appendix II, on the other hand, provides for the inclusion of species that are not necessarily threatened but may become so unless trade is strictly controlled and listing on Appendix II allows regulated trade to continue under a permit system which allows for trade monitoring.

The new listing criteria, the *Everglades Criteria*, include numerical examples as a guideline not a threshold, and make several references to the precautionary principle that scientific uncertainty should not be used as a reason for failing to act in the interests of conservation. Concern was expressed at the meeting that the level of shark bycatches in world fisheries may be unsustainable. The Animals Committee of CITES is to prepare a discussion paper on this issue so the next CITES meeting may result in species proposals on sharks. While at present no commercial marine fish are covered by CITES, marine wildlife species which are covered include cetaceans (listed on both Appendix I and Appendix II depending on the species), Dugong *Dugong dugon* (apart from the Australian population, this species is listed on Appendix I, while the Australian population is on Appendix II), and Short-Tailed Albatross *Diomedea (Phoebastria) albatrus* and Leatherback Turtle *Dermochelys coriacea* (both on Appendix I).

The precautionary principle in fisheries management

The precautionary principle — in case of doubt about what is happening, apply regulations in order to avoid a risk of a particular outcome — is gaining widespread acceptance internationally as having relevance in all fields of natural resource management, including fisheries (Bergin and Haward 1995; Dunlap 1995). As noted above, the new treaty regime on high seas fishing contains guidelines on how to operationalize this principle. In managing stocks states will be required to consider the impacts of fishing on associated ecosystems. A precautionary approach to fisheries management is likely to become adopted as guidelines under international management of high seas, straddling stocks, as well as by governments managing resources within their EEZs.

The Reflagging Agreement governing the flagging of fishing vessels

The Agreement to Promote Compliance with International Conservation and Management Measures by Fishing Vessels on the High Seas was concluded at the FAO in summer 1994 in pursuance of UNCED Agenda 21 to reinforce the concept of flag state responsibility in an attempt to "freeze out" flags of convenience (see Birnie 1993). Under this important new treaty regime each party is to take such measures as may be necessary to ensure that fishing vessels flying its flag do not engage in any activity that undermines the effectiveness of international conservation and management measures. Parties agree to establish proper authorization procedures for fishing boats flying their flags, to ensure that their boats are properly marked and to supply full details of their operations to state authorities to enable the Party to comply with obligations under the agreement. Parties also agree not to authorize vessels for fishing unless sufficient links exist to ensure that they can exercise its responsibilities effectively and to link the grant of authorization to fish with right to fly the flag. Parties agree not to grant fishing authorizations to vessels which have previously been registered in another state and which have fallen foul of fisheries management regulations, and also undertake to enforce those provisions against flag vessels with criminal sanctions of "sufficient gravity" including refusal, suspension or withdrawal of authorization to fish. The Agreement also introduces a limited form of port state control in that a port state shall notify flag states of vessels within its ports which are believed to have been used to violate international conservation and management measures. The Agreement requires acceptance by 25 states to come into force. As of July 1997, 10 states have ratified the Agreement.

It is evident from the foregoing overview that significant attention has been paid to the bycatch problem in recent years. It must be acknowledged, however, that much of the legal and policy developments described above fall within the "soft law" category of international law. That is, they are more in the nature of statements of intent or guidelines that states should rather than shall follow. While they are not formally binding, states that give a clear undertaking to act in accordance with these principles and general practices risk a loss of credibility and face high level political costs if they do not meet undertakings. And, of course, some of the soft law may harden into formal and precise treaty provisions at a later stage. What is clear now, though, is that there is considerable evidence suggesting that international law and diplomacy is increasingly making progress in dealing with one of the most important issues in international fisheries management in the 1990s. Exemplifying this progress has been the initiative undertaken by CCAMLR to reduce the incidental catch of albatrosses within its geographic ambit of operation and it is to this development that attention is now turned.

INCIDENTAL BYCATCH OF ALBATROSSES: THE DEVELOPMENT OF MANAGEMENT MEASURES BY CCAMLR

CCAMLR is an international organization established in 1982 through the Convention on the Conservation of Antarctic Marine Living Resources. This Convention, and the work of CCAMLR, applies to Antarctic marine living resources in an area bounded to the north by a line which approximates the position of the Antarctic Convergence.

Seabirds (particularly species of albatrosses and petrels) form the major non-fish bycatch from longline operations in the southern oceans. As intimated above, research on incidental catch of seabirds, and mitigation of this catch is ongoing (Brothers 1991; Murray *et al.* 1993). Indeed, the problem of incidental catch of albatross and petrel species has concerned scientists for several years and the rate of bycatch of these particular non-fish species has emerged as a major environmental issue.

CCAMLR has been considering the issue for over a decade and has established formal measures to mitigate the bycatch of seabirds. The 1994 meeting of CCAMLR (CCAMLR XIII) endorsed Conservation Measure 29/XIII governing longline fishing operations within the Convention's area of application (CCAMLR 1994). This measure endorses operational techniques which sink baited hooks as soon as possible after their placement in water; mandates setting lines at night; recommends that offal and trash should not be dumped while lines are set or hauled; requires that every effort be made to release live birds from lines; and mandates the use of streamer lines as bird scarers. Specifications for streamer lines are appended to the notice of this measure (CCAMLR 1994). An important adjunct to these operational guidelines is that CCAMLR (through its Secretariat) is to circulate the detail of the measure to relevant international organizations dealing with longline fisheries (mainly tuna fisheries) and having geographic ambits of operation adjacent to the CCAMLR Convention area.

CCAMLR (and its Scientific Committee, SC–CAMLR) is charged with using an "ecosystem" approach to the conservation of Antarctic marine living resources which has meant that populations of seabirds and their interactions with other species in the ecosystem have been the focus of considerable study. CCAMLR operates on a consensus model of decision making which has been criticized by some non-governmental organizations which believe this approach leads to weaker outcomes. In addition, some criticism has been levelled in the past at CCAMLR's concern with managing single species finfish rather than utilizing an ecosystem approach. These remarks may well have originated from the real difficulties CCAMLR experienced in its early years — difficulties associated with developing an innovative approach to resource management and difficulties in obtaining data on resource stocks and use. This led the first Executive Secretary of the Commission to comment that while CCAMLR's "first, faltering steps" had led to criticism, the "positive result was to focus more attention on multi-species and ecosystem aspects of the Scientific Committee's work" (Powell 1990).

The issue of incidental mortality associated with fishing operations was raised at CCAMLR's Third Meeting in 1984 when the United States delegation submitted a paper "Assessment and Avoidance of Incidental Mortality of Antarctic Marine Living Resources" (CCAMLR 1984). The Commission noted that while

> there does not seem to be any problem with entanglement of animals in lost or discarded fishing gear and other debris in the Convention Area at the present time, there is growing evidence that in other areas significant numbers of non-target fish, birds, marine mammals, turtles and other marine organisms are being caught and killed in lost and discarded fishing gear and other debris, *as well as caught and killed incidentally during certain fishing operations*." (CCAMLR 1984, emphasis added).

At the same time, SC-CAMLR had established working relationships with the BIOMASS Working Party on Bird Ecology and SCAR (the Scientific Committee on Antarctic Research) over the status and trends in Antarctic seabird populations. This provided further data and enhanced scientific linkages.

The need to collect data on incidental catch was discussed at both the Fourth and Fifth Meetings of CCAMLR in 1985 and 1986, respectively. In 1985, CCAMLR agreed that steps should be taken to record and report non-target species caught during fishing operations, and that the Executive Secretary should distribute material on measures developed by members to assess, avoid and mitigate accidental and incidental mortality of Antarctic marine living resources (CCAMLR 1985). At the 1986 Meeting, CCAMLR agreed that

> Members would take such steps as necessary to ensure operators of vessels engaged in fishing and related operations in the Convention Area maintain a record and report the number, species, and where appropriate, the age and sex and reproductive status, of any birds and marine mammals taken incidentally during fishing operations.

During the following two meetings (CCAMLR VI and VII, held in 1987 and 1988, respectively), members' reports on the assessment and avoidance of incidental mortality of Antarctic marine living resources were mainly concerned with marine debris and entanglement-related problems. In 1989, however, the issue of "incidental catch" emerged as a separate agenda item at the SC-CAMLR and CAMLR meetings, with SC-CAMLR agreeing to discuss incidental mortality on a regular basis. The development of a separate agenda item in SC-CAMLR meetings from this date indicates the increasing attention to the issue, the results of early data gathering and, significantly, concern with the introduction of longline fishing in the Convention area.

Indeed, longline fishing began in the Convention area in the 1988–89 fishing season and it was noted by CCAMLR then that such fisheries may experience substantial incidental mortality of seabirds (CCAMLR 1989). CCAMLR VIII requested the Scientific Committee to evaluate and provide advice on the ways to assess and minimize this mortality (CCAMLR 1989). Moreover, in adopting Resolution 5/VIII, "the Commission reiterated its concern at the commencement of an unregulated fishery of a type known elsewhere to cause substantial incidental mortality of seabirds" (CCAMLR 1989).

The Ninth Meeting of the Commission (CCAMLR IX) in 1990 saw the issue of incidental catches of albatrosses receive considerable attention. Australia presented a paper to the SC-CAMLR which described albatross mortality associated with longline tuna fisheries outside the Convention area; a estimate of 44 000 albatrosses killed annually. Australian delegates also presented information on Australian-Japanese efforts to reduce incidental catch of albatrosses through use of "bird scaring streamer lines". The significant decline in "catch rate" of 85% using these devices was presented to the Committee along with the economic benefits of such devices.

Recommendations from SC-CAMLR in relation to the issue of incidental catch of albatrosses were then adopted by the Commission, including a series of mitigation measures. A major emphasis was to ensure that information on incidental catch was collected and that members provide the data required to determine the best method of reducing the incidental mortality of seabirds. It was agreed that steps should be taken to place observers on longline vessels to ensure verifiable data. Although some concern was expressed over the lack of data on the problem or the effectiveness of mitigation devices, CCAMLR agreed that prior to evaluation of data, modifications of longline fishing techniques within the Convention area be implemented. These modifications reflected the experiences of Australia, New Zealand and Japan in tackling the problem within the Southern Bluefin Tuna fishery, and the development of the streamer line concept by the Japanese and the bait caster by Nigel Brothers, a Tasmanian scientist working on albatross bycatch. In short, the modifications included the deployment of "tori poles" and streamers; the requirement that fishing operations be conducted in such a way that baits sink immediately they are in the water; the setting of longlines only at night; and the prohibition of dumping trash or offal while longline operations are in progress. These recommendations were prepared as a draft Conservation Measure, although this proposed measure was not approved at this meeting of the Commission. Some delegations wanted more time to consider technical details, others wanted additional measures to be implemented as soon as possible (CCAMLR 1990). The delegates did agree, however, that formal adoption of a Conservation Measure would be discussed again at the next meeting of the Commission.

When this occurred, at the Tenth Meeting in 1991, SC-CAMLR noted that, despite a request for detailed information from members on incidental catch of seabirds from longlining, this data had not been provided. In addition, the Scientific Committee recorded that incidental mortality had been significantly under-reported (by 33%) in 1990–91, and that recommendations agreed in the previous year had not been followed by those undertaking fishing activities in the Convention area. A formal conservation measure was prepared by the Commission which noted extensive supporting documentation; for example Resolution 5/VIII, the draft conservation measure in Annex 6, CCAMLR IX; papers presented to CCAMLR X and advice of the Scientific Committee (SC-CAMLR 1991).

This formal measure, known as Conservation Measure 29/X, required specific action in the conduct of longline fisheries aimed at minimizing such incidental mortality. This action included mitigation methods concerning baited hooks, ship lighting, trash and offal dumping and the deployment of streamer lines that were precursors to the techniques later listed in Conservation Measure 29/XIII. The Chilean delegation, however, drew attention to the fact that effectiveness of these methods had yet to be established for vessels fishing within the Conservation area. France also gave notice that it would reserve application of Measure 29/X in the Economic Exclusion Zones of Iles Kerguelen and Iles Crozet "if they prove less strict or less appropriate that the measures adopted or considered by French authorities in these two areas". As a result of this reservation, Conservation Measure 29/X came into force in May 1992, except for waters around Iles Kerguelen and Iles Crozet.

An updated conservation measure was introduced at the Eleventh Meeting of the Commission in 1992. At this meeting too, members were encouraged to use the intercessional period to assess the effectiveness of methods to prevent incidental mortality and to report the results of these studies to the Scientific Committee meeting the following year. Conservation Measure 29/XI was subsequently adopted after an amendment removed ambiguities concerning some aspects of the use of streamer lines as a mitigation device.

At the Twelfth Meeting in 1993, the Commission noted the large amount of work on the issue of incidental catch of seabirds considered by the Scientific Committee during the intercessional period. The United Kingdom delegation recorded the important contributions which Australia and New Zealand had made to understanding interactions between albatrosses and fishing activities. Problems in the implementation of the Conservation Measure were, however, recognized — in particular, the lack of compliance with reporting requirements and a possible lack of compliance with Conservation Measure 29/XI (CCAMLR 1993). The Commission also endorsed the advice of the SC–CAMLR that the acquisition of statistically valid data may require the presence of scientific observers on board fishing vessels. Work on the incidental mortality of seabirds was enhanced by the establishment at this meeting of the Working Group on Incidental Mortality Arising From Longline Fishing (WG-IMALF) by the Scientific Committee. This group met for the first time in October 1994, before CCAMLR XIII and provided an important forum for the further development of Conservation Measure 29 in 1994.

It must be recognized, of course, that the measures adopted by CCAMLR may have minimal direct effects on reducing either the scale or extent of albatross bycatch — after all, most birds are "caught" in southern ocean waters outside the Convention area. This is not to decry, however, the significance of CCAMLR's initiative. Its Secretariat is contacting other international fishery management organizations to encourage the adoption of mitigation measures to reduce the bycatch of seabirds whose range extends beyond the Convention's area of application. Whether such proselytizing is, or shall be, effective remains to be seen. Moreover, there are few, if any, efficacious policy precedents to follow that may serve to minimize the problem. For example, international efforts to protect and manage the large populations of birds in the Bering Sea region have been relatively ineffective (Young 1994). Young contends that all of the region's bird populations continue to be at risk from such human activities as fishing and other resource developments despite the existence of protection measures contained in an international regime articulated in five bilateral agreements between Canada and the United States (1916), Mexico and the United States (1936), Japan and the United States (1972), Japan and Russia (1973) and Russia and the United States (1976); (see Lyster 1985). One factor that Young identifies as accounting for the deleterious outcome in the case of highly migratory species is the difficulty in achieving co-ordinated practices that cover the entire range of the birds. This is clearly a requirement in the case of albatross conservation. In the following section we outline several process-oriented considerations concerning interorganizational co-operation and international regime formation which may facilitate the effective management of seabird bycatch.

BEYOND CCAMLR: INTERORGANIZATIONAL CO-OPERATION AND PROCESS REQUIREMENTS FOR REGIME FORMATION

The need for co-operation among national and international fisheries management organizations to minimize the incidental catch of albatrosses and other seabirds is obvious given the overlap between the operational areas of longline fishing vessels and the foraging range of albatrosses. Collaboration between organizations can occur at both formal and informal levels and concern a number of activities. Collection and distribution of data on incidental catch from such tuna organizations as the International Commission for the Conservation of Atlantic Tunas (ICCAT), the Inter-American Tropical Tuna Commission (IATTC) and the Commission for the Conservation of Southern Bluefin Tuna (CCSBT) is an important starting point for such collaboration. Different organizations structure their scientific and data collection and dispersal processes differently and hence the need for co-ordination of such effort becomes important. It is clear that "good" science is a necessary but not a sufficient condition for "good" management, but equally that poor science or lack of information may have negative consequences (see Fløistad 1990; Bergin and Haward 1996). This heightens the importance of a co-ordinated approach to data collection; coordination which could be extended to other areas to reduce the negative effects of overlap between organizations.

Several mechanisms are available to facilitate interorganizational co-ordination. The first involves formal processes where management bodies are mandated to collaborate with other organizations with related objectives. The

Commission for the Conservation of Southern Bluefin Tuna, for example, is mandated (under Article 12 of the Convention for the Conservation of Southern Bluefin Tuna) to "collaborate with other inter-governmental organizations which have related objectives". This collaboration may extend to ensuring such related organizations have observer status at meetings of respective bodies. The collection of data and statistical summaries are also facilitated by the work of the Food and Agriculture Organization (FAO). In addition, scientific collaboration through participation in epistemic communities is another avenue to facilitate interorganizational co-ordination. An epistemic community is a group of individuals who share common values and views regarding causal mechanisms and appropriate responses to issues in general. The membership of an epistemic community usually transcends national boundaries and generally includes scientists or experts and government officials (for further discussion on this concept, see Haas 1989).

Interorganizational collaboration along such lines, together with CCAMLR's initiatives, may well turn out to be initial steps in the formation of an international regime to mitigate the incidental catch of seabirds. The concept of international regime refers to explicit and implicit norms, rules and decision-making procedures developed within an issue-area of world affairs which serve to govern the behaviour of participating actors (both state and non-state) and around which their expectations converge (Krasner 1982). The primary function of international regimes is to facilitate co-operative action as a means for achieving desirable goals or values that cannot be achieved by members of international society acting alone. Typically, they are responses to transboundary or common problems at the international level or to situations in which the pursuit of interests defined in individualistic terms regularly leads to socially undesirable outcomes (Young 1989).

Given the nature of the seabird bycatch problem, fishing industry interests, fisheries management regimes currently operating and developments concerning mitigation that have occurred to date, it seems less likely that a comprehensive, unified regime (established, for example, through the provisions of the Bonn Convention) will be designed than the evolution of a patchwork regime with its component norms, rules and decision-making procedures formally articulated in separate agreements and managed by diverse organizations. Such a development is a potential recipe for incoherence and ineffectiveness. How this kind of outcome is to be avoided is problematic. Moreover, political research on specific ideas than may be useful in regard to regime content is sparse. There has been research on regime formation, however, that we feel offers several process-oriented lessons warranting consideration by those with responsibilities in this area.

Young and Osherenko (1993) have recently developed a multivariate model of the process of regime formation which is based on the analysis of international co-operation on natural resource and environmental issues in the Arctic region concerning fur seals, polar bears, stratospheric ozone, the Svalbard Archipelago, Arctic haze and transboundary air pollution. At the core of this model is the political process called institutional bargaining defined as "efforts on the part of autonomous actors to reach agreement among themselves on the terms of constitutional contracts or interlocking sets of rights or rules [i.e., regimes] that are expected to govern their subsequent interations" (Young 1991).

Young and Osherenko identify three sets of major factors which may feed into this process of institutional bargaining: power factors, interest factors and knowledge factors. The first of these, power factors, refer to the actions of dominant actors (typically states) or consortia of dominant actors who may utilize their preponderant material resources to get others to conform to their regime requirements through some combination of coercion, co-optation or manipulation of incentives — as the United States did in the middle stages of negotiation of the 1987 Montreal Protocol to protect stratospheric ozone (Haas 1993).

The second set of factors, interest factors, refer to the actions of self-interested parties who may endeavour to co-ordinate their behaviour in order to reap joint gains through negotiation by securing explicit consent in regard to sets of rights and rules that constitute the emergent regime — as did the negotiators of the North Pacific Fur Seals regime (articulated in the North Pacific Sealing Convention of 1911), the Svalbard regime (articulated in the 1920 Treaty Relating to Spitzbergen) and the polar bear regime (formalized in the 1973 Agreement on the Conservation of Polar Bears).

Knowledge factors, the third set of factors, refer to cognitive considerations — shared ideas, beliefs, understandings and values held by epistemic communities generally consisting of scientists and government officials — that may also play a significant role in regime formation as in the creation of the polar bear regime, the transboundary air pollution regime (embodied in the 1979 Geneva Convention on Long-Range Transboundary Air Pollution) and the regime for the protection of stratospheric ozone.

Influencing the process of institutional bargaining through their capacity to shape how power factors are exercised, interest factors articulated and knowledge factors brought to bear on an issue are two cross-cutting factors: individual leadership and contextual considerations. In regard to leadership, Young and Osherenko propose that individuals may play critical roles in the process of institutional bargaining by (1) bringing a party's power to bear (i.e., the exercise of structural leadership); (2) exercising negotiating skill by agenda setting, devising options, popularising solutions, brokering interests (i.e., entrepreneurial leadership); and (3) generating and proposing ideas and systems of thought that influence the way in which issues at stake are understood (i.e., intellectual leadership).

But neither leadership nor the other factors described above operate in a vacuum — contextual factors to do with national or world circumstances and events which are seemingly unrelated to the issues under consideration may also play a major role in determining if institutional bargaining occurs and in shaping the content of any regime that forms. Such contextual factors include shifts in values and ideas (for example, growing environmental consciousness played an important role in the development of the Montreal Protocol) and political trends and changes (such as a period of thaw that occurred in the Cold War in the case of the polar bear regime).

The significance of understanding the process of regime formation in the way outlined above lies in the identification of key components and the linkages between them. While acknowledging that the theoretical model they have constructed is not highly developed, Young and Osherenko suggest that success in regime formation is likely to be achieved by participants who have a sophisticated, though often intuitive, grasp of the interplay of power, interests and knowledge and the way the cross-cutting influences of leadership and context shape the process of institutional bargaining (Young and Osherenko 1993). We concur: if the nascent seabird bycatch regime is to develop further, such components and linkages must be apprehended.

One critical consideration is the need for someone (or, perhaps, several individuals) to exercise entrepreneurial leadership. Young and Osherenko's research suggests that the presence of one or more individuals acting in such leadership roles is a necessary condition for regime formation to occur. Put another way, they argue that their research on the Arctic cases shows that institutional bargaining for the creation of international regimes cannot yield agreement in the absence of entrepreneurial leadership: someone must exercise negotiation skills by agenda setting, devising options, popularizing solutions and brokering interests; someone must help get the parties to the table and someone must keep them there until agreement is reached. Birdlife International's initiative in appointing a Global Seabird Conservation Co-ordinator from 1 December 1997 could well provide an example of such entrepreneurial leadership.

While entrepreneurial leadership is a critical consideration, intellectual leadership during the political process of institutional bargaining is not a precondition for regime formation. This is not to say that such leadership is unimportant. Young and Osherenko's research suggests that the role of intellectual leadership is more central in the earlier stages of regime formation, usually before explicit negotiations commence (Young and Osherenko 1993). This insight seems to be supported in the case of international efforts to date to mitigate the incidental catch of albatrosses. We regard the pivotal role Nigel Brothers (see earlier) has played in the development of measures to mitigate seabird bycatch as an example of such intellectual leadership.

The use of a state's material resources to exert bargaining leverage in regime formation (i.e., structural leadership) is a complex matter and fraught with difficulties. Young and Osherenko's research indicates that blatant and unsophisticated attempts to exercise power in institutional bargaining are likely to backfire, although they point out that participants should not overlook its utility but recognize that attempts to coerce, co-opt or manipulate incentives should be handled with care (Young and Osherenko 1993).

Several other considerations derived from the study of Arctic natural resource and environmental issues recommend themselves in regard to a seabird bycatch regime. The existence of a salient solution that offers a formula that is easy for policy makers to grasp and is appealing to the general public increases the probability of success in regime formation (Young and Osherenko 1993). Of course, a salient solution depends largely upon the problem addressed being formulated in an appropriate way. But this does not occur naturally. One of the important lessons for practitioners from Young and Osherenko's research is the need for parties involved to invest time and energy "to get the problem straight at the onset rather than assuming that the process of institutional bargaining will serve as a natural corrective to faulty initial formulations". (Young and Osherenko 1993). In the case of albatross bycatch, framing the problem in terms of conservation issues while failing to recognize the very real fishing industry interests involved is a recipe for failure.

The Arctic cases also suggest that regime formation can succeed only when all the major parties and interest groups involved feel that their primary concerns have been treated equitably or fairly, even at the expense of efficiency. Moreover, the chances of success rise when compliance mechanisms that the parties regard as clear-cut and likely to be effective are available (Young and Osherenko 1993).

Another important consideration is the notion that shocks or crises exogenous to the negotiating process trigger a sense of urgency thereby helping to promote agreement (Young and Osherenko 1993). For example, the reporting in mid-1985 of the discovery of the rapid thinning of the ozone layer over Antarctica during the Antarctic springtime revived stalled efforts to go beyond the general framework of the 1985 Vienna Convention for the Protection of the Ozone Layer. This revival culminated with the adoption of the Montreal Protocol. Furthermore, the publicization of "forest death syndrome" (or in German — *Waldsterben*) in Europe during the early 1980s was also a key development leading up to the 1985 protocol on sulphur emissions which augments the Convention on Long-Range Transboundary Air Pollution of 1979. Such considerations underscore the significance of publicity about the bycatch of albatrosses — the need to engender and maintain a sense of crisis, thereby spurring policy makers and negotiators to action to conclude international agreements concerning mitigation measures.

CONCLUDING REMARKS

This chapter began by surveying recent international legal and policy developments on the incidental or bycatch of non-fish species, particularly marine mammals and seabirds. Although it has been pointed out that many of these developments can be categorized as "soft law" this is not to say that progress has not been made in recent years in dealing with this important issue in international fisheries management. The initiative of CCAMLR of introducing measures to mitigate the bycatch of seabirds exemplifies this progress. While these CCAMLR measures may have minimal direct effects on reducing either the scale or extent of the albatross bycatch problem — for the simple reason that most birds are caught in waters outside the Convention's area of application — their importance is significant in another sense. The CCAMLR measures concerning baited hooks, night-time line-setting, trash and offal dumping, streamer lines and caught live bird release efforts offer a mitigation model for use by international fisheries management organizations employing longline techniques and having geographic ambits of operation adjacent to the CCAMLR Convention area.

How successful, or otherwise, CCAMLR Secretariat's encouragement of this model turns out to be lies, of course, in the future. We have suggested that such proselytization, together with interorganizational collaboration, may lead to the formation of an international regime which is patchwork in nature with its component norms, rules and decision-making procedures articulated in separate agreements and managed by diverse international organizations. As at the domestic level, patchwork provisions scattered among a number of agreements may well lead to an incoherent, ineffectual regime. But this is not a predetermined outcome — policy makers and negotiators need to be aware, however, of such a danger. Drawing on recent research about international regime formation we have suggested a number of practical, process-oriented, political lessons for those seeking to develop a seabird bycatch regime. These include the necessity of entrepreneurial leadership, the existence of salient solutions, the availability of equitable options and clear-cut and effective compliance mechanisms and the general perception of a sense of urgency triggered by a shock or crisis.

All these considerations point to an obvious conclusion — albatross conservation has a significant political dimension. It is not solely a scientific matter. This means that attention must also be focused on identifying and engaging the political and economic factors driving ocean use. And, echoing a strategy advocated by Bibby (1995), a wise approach to the task of albatross, and indeed, seabird conservation calls for far greater linkages between the social, economic and biological sciences than, perhaps, has been the experience in the past.

ACKNOWLEDGEMENTS

The authors wish to thank David Agnew, Bruce Davis, Eugene Sabourenkov, Donald Rothwell and Graham Robertson for their helpful suggestions.

REFERENCES

Alverson, D. L., Freeberg, M. H., Pope, J. G. and Murawski, S. A., 1994. A Global Assessment of Fisheries Bycatch and Discards. FAO Fisheries Technical Paper No. 339, Rome, FAO.

Bergin, A. and Haward, M., 1995. International environmental conventions and actions-implications for the fishing industry. Pp. 281–99 *in* Commodity Markets and Natural Resources. Proceedings of the National Agricultural and Resources Outlook Conference, Canberra, 7–9 February 1995. Australian Bureau of Agricultural and Resource Economics: Canberra, Australia.

Bergin, A. and Haward, M., 1996. Japan's Tuna Fishing Industry: Setting Sun or New Dawn? Nova Science Press: New York.

Bibby, C., 1995. A global view of priorities for bird conservation: a summary. *Ibis* **137:** 247–48.

Birnie, P., 1993. Reflagging of fishing vessels on the high seas. Review of European *Comm. Environ. Internat. Law* **2:** 270–76.

Brothers, N. P., 1991. Albatross mortality and associated bait loss in the Japanese longline fishery in the Southern Ocean. *Biol. Conserv.* **55:** 255–68.

Burke, W., Freeberg, M. and Miles, E., 1994. United Nations resolutions on driftnet fishing: an unsustainable precedent for high seas and coastal fisheries management. *Ocean Develop. Internat. Law* **25**(2)**:** 127–86.

CCAMLR, 1984. Report of the Third Meeting of the Commission. (Hobart, Australia, 3 September–14 September, 1984) CCAMLR: Hobart.

CCAMLR, 1985. Report of the Fourth Meeting of the Commission. (Hobart, Australia, 2–13 September, 1985) CCAMLR: Hobart.

CCAMLR, 1989. Report of the Eighth Meeting of the Commission. (Hobart, Australia, 16–17 November, 1989) CCAMLR: Hobart.

CCAMLR, 1990. Report of the Ninth Meeting of the Commission. (Hobart, Australia, 22 October–2 November, 1990) CCAMLR: Hobart.

CCAMLR, 1993. Report of the Twelfth Meeting of the Commission. (Hobart, Australia, 25 October–5 November, 1993) CCAMLR: Hobart.

CCAMLR, 1994. Report of the Thirteenth Meeting of the Commission. (Hobart, Australia, 26 October–4 November, 1994) CCAMLR: Hobart.

Dunlap, W. V., 1995. Bering Sea — The Donut Hole Agreement. *Internat. J. Mar. Coast. Law* **10**(1)**:** 114–26.

Fløistad, B., 1990. Communication between science and decision makers: The advisory function of the International Council for the Exploration of the Sea. *Internat. Chall.* **10:** 22–28.

Gales, R., 1994. Status and Conservation of Albatross and their Interactions with Fisheries: Final Report. 1994, Australian Nature Conservation Agency.

Haas, P., 1989. Do regimes matter? Epistemic communities and Mediterranean pollution control. *Internat. Organiz.* **43:** 377–403.

Haas, P., 1993. Stratospheric Ozone: Regime Formation in Stages. Pp. 152–85 *in* Polar Politics: Creating International Environmental Regimes ed by O. R. Young and G. Osherenko. Cornell University Press: Ithaca.

Krasner, S., 1982. Structural causes and regime consequences: regimes as intervening variables. *Internat. Organiz.* **36:** 185–205.

Lyster, S., 1985. International Wildlife Law. Grotius Publications Limited: Cambridge.

Murray, T. E., Bartle, J. A., Kalish, S. R. and Taylor, P. R., 1993. Incidental capture of seabirds by Japanese southern bluefin tuna vessels in New Zealand waters 1988–92. *Bird Conserv. Internat.* **3:** 181–210.

Powell, D., 1990. Antarctic marine living resources and CCAMLR. Pp. 61–70 *in* Antarctica's Future: Continuity or Change? ed by R. A. Herr, H. R. Hall and M. Haward. TGPO/AIIA: Hobart.

SC–CAMLR, 1991. Report of the Eighth Meeting of the Scientific Committee. (Hobart, Australia, 6–10 November, 1989) CCAMLR: Hobart.

Young, O., 1994. International Governance: Protecting the Environment in a Stateless Society. Cornell University Press: Ithaca.

Young, O. and Osherenko, G., 1993. International Regime Formation: Findings, Research Priorities, and Applications. Pp. 223–61 *in* Polar Politics: Creating International Environmental Regimes ed by O. R. Young and G. Osherenko. Cornell University Press: Ithaca.

Young, O. R., 1991. Political leadership and regime formation: on the development of institutions in international society. *Internat. Organiz.* **45:** 281–308.

Future Directions

This book provides accounts of albatross taxonomy and status, their population biology and ecology, and human-related effects on their survival. The list of topics covered is not exhaustive but does, to a large extent, reflect the current state of knowledge and the balance of interest in the albatross sciences. There are some glaring omissions, and with some species even the most basic information on their biology and ecology are lacking. Conservation, through public opinion, is shaped by science and if albatrosses are to maintain their current species richness, distributions and abundances into the next millennium a great deal must be done in the field of science to develop the knowledge upon which practical conservation decisions can be made. It is fitting, therefore, that this book ends with a sense of vision and guidance on the way forward in albatross research and conservation.

A Shy Albatross chick waits expectantly for the return of its parents to Albatross Island, Tasmania.
GRAHAM ROBERTSON

CHAPTER TWENTY-THREE

Research and conservation: a future for albatrosses?

J. P. CROXALL[1]

The major recent advances in studies of albatross biology, ecology and conservation are reviewed. In addition to highlights arising during the first Albatross Conference, pioneering research on topics little covered therein (e.g., behaviour, moult, diet, physiology) are summarized. Important topics and opportunities for future research are suggested. The major historical and current threats to albatrosses are reviewed and the priority initiatives needed to tackle the latest — and greatest — threat, that of incidental mortality associated with longline fisheries, summarized.

Key words: Research, Conservation, Systematics, Behaviour, Breeding Biology, Population Studies, Moult, Physiology, Diet, Foraging Ecology, Pollutants, Predators, Habitat Change, Longline Fishing.

INTRODUCTION

AT the conclusion of the first Conference and Workshop on the Biology and Conservation of Albatrosses in Hobart it was not difficult to recognize some of the outstanding successes of the meeting. First in illuminating and energising albatross research by bringing together people working — often in considerable isolation — on most of the world's species. Second, in harnessing the enthusiasm of the individuals engaged in albatross research and conservation in order to address goals of mutual concern.

In particular, it was inevitable and appropriate that attention should focus on conservation priorities, especially those relevant to interactions between albatrosses and fisheries. Here ornithologists, ecologists and conservationists were able to achieve a good measure of consensus (see Alexander *et al.* 1997) on priority issues for the present and future. The opportunity to discuss these ideas with fisheries scientists and managers was invaluable, because creating and sustaining the mutual need to resolve this problem is a pre-requisite for progress and essential for achieving any ultimate success.

However, much other material, possibly of less immediate relevance to these topics, was presented and discussed as oral, poster and informal contributions. Within the confines of a single volume it is impossible to do justice to much of this work. This chapter, which is a personal view, expands on the remarks in my concluding address to the Conference. It tries: (a) to identify some of the main achievements in the different areas of research on albatrosses, particularly those which received little attention at the Conference; and (b) to offer some suggestions on priorities for the future — in addition to the over-riding priority of ensuring the continued survival of all albatross taxa.

RESEARCH

Systematics

Until the Conference, consensus on the generic boundaries of albatrosses was essentially based on an evaluation published in 1866. Species limits reflected the conservative (lumping) treatment characteristic of most seabird taxa with mainly allopatric breeding distributions, few specimens in collections and little interest in, or opportunity for, critical study.

The recent first and major, step towards a new taxonomy for albatrosses (Nunn *et al.* 1996; Robertson and Nunn 1997) originated in the recognition that, at a time when conservation is focusing mainly on species as the unit of biological diversity, conservation at sub-Antarctic islands might depend critically on the correct ranking of albatross taxa. The outcome of the current research using molecular genetic

[1]British Antarctic Survey, Natural Environment Research Council, High Cross, Madingley Road, Cambridge, CB3 0ET, England, UK.

Pp. 269–90 *in* THE ALBATROSS BIOLOGY AND CONSERVATION ed by G. Robertson and R. Gales. Surrey Beatty & Sons, Chipping Norton. 1997.

techniques is a reinstatement of four genera within the family, the discovery that the sooty albatrosses *Phoebetria* are not primitive but related more closely to *Thalassarche* albatrosses than to the other groups and the suggested recognition of 10 new taxa at the species level. This re-emphasizes the importance of the New Zealand region, with no fewer than 10 endemic species. This new appraisal gives a much more realistic and appropriate view of the relationships of albatrosses, particularly for comparative studies of biology and ecology and for assessing threats and conservation priorities. The taxonomy and nomenclature adopted by the Workshop for this latter task is also used in this chapter.

This new taxonomy needs complementing with critical examination and correlation of the morphological, behavioural and ecological attributes of the constituent species and genera. Traditional and molecular techniques also need to be used to investigate the potential existence of other taxa, particularly within *Phoebetria* and Grey-headed Albatrosses *Thalassarche chrysostoma*, where genetic heterogeneity is greater than might be expected. Further work, utilizing more refined molecular techniques, should make it possible to characterize the genetic identity of the populations of albatrosses from different islands or island groups so that the origin of specimens found away from their breeding grounds can be determined. This would be of obvious relevance to beach-cast specimens and for determining the provenance of birds killed by fishing operations.

Unravelling the history of albatrosses from the fossil record is an even greater challenge, because of the fragmentary and disjunct material available and the difficulty of interpreting this in relation to change in marine systems at appropriate times and scales. While it seems certain that albatrosses were once much more widespread than today and suffered (e.g., the disappearance of the albatrosses of the North Atlantic Ocean) from environmental change in the Quaternary (Olson 1985), it may be a long time before we understand the circumstances which caused or contributed to this.

Behaviour

The suggested recognition of more species and the investigation of the potential for the existence of others may stimulate greater interest in the comparative behaviour of albatrosses. Only for eight species (Wandering *Diomedea exulans*, Northern Royal *D. sanfordi* Albatrosses, Black-browed *Thalassarche melanophrys*, Grey-headed *T. chrysostoma* and Buller's *T. bulleri* Albatrosses, Sooty *Phoebetria fusca* and Light-mantled *P. palpebrata* Albatrosses and Laysan Albatross *Phoebastria immutabilis*) are there reasonably comprehensive descriptions of the main breeding (display) behaviours and their contexts (Richdale 1949, 1950; Fisher 1971; Meseth 1975; Berruti 1981; Tickell 1984; Jouventin *et al.* 1985; Jouventin and Lequette 1990). Vocalizations have been even less studied and only for Wandering, Buller's, Laysan and Black-footed *Phoebastria nigripes* Albatrosses is even basic comparative material available (Sparling 1977; Warham and Fitzsimons 1987; Lequette and Jouventin 1991).

More surprising, perhaps, is the virtual absence of research on behavioural ecology. This may be partly because of the monogamous breeding system of albatrosses and partly because many important events take place at sea. Nevertheless the nature of courtship, particularly in the great albatrosses (*Diomedea* spp.), with numerous males attempting to display in relatively restricted areas to groups of females, deserves greater attention in terms of the process, signals for and consequences of mate selection. Although preliminary studies (see Croxall 1991) indicated that male attendance duration gave the best indication of likelihood of partner acquisition, further detailed studies are needed. Similarly, although monogamy may be the rule, extra-pair copulations are known to occur in some species (particularly in Wandering Albatrosses), but investigation of the cause and significance of these, including genetic studies to determine parentage, has yet to be undertaken. Because the role of the parent is usually perceived as decisive in determining the rate of delivery of energy to the offspring, the topic of parent-offspring conflict has received little study, except from a very recent experimental investigation of whether underfed chicks can induce parents to deliver more food (Weimerskirch *et al.*, in press). There is scope for further work in this research area.

Breeding biology

The potential recognition of a considerable additional number of albatross species should increase interest in documenting the basic features of the timing and duration of the breeding cycle. These data are not available for several of the species-level taxa previously recognized, so there is now considerable scope for undertaking novel comparative studies, including relating the timing of breeding to local environmental conditions (including food availability).

Population studies

Population studies take many forms, including censuses of breeding populations, demographic studies and investigation of age-related and individual-specific effects on reproductive performance and survival.

Determining the status and trends of albatross populations (see Fig. 1) is important for many research objectives and fundamental to detecting changes which may require more detailed investigation and/or management action. Some form of monitoring is needed for all the main populations of albatrosses. Sporadic counts at lengthy intervals and/or different times of year are unlikely to produce reliable results, because of the substantial within- and between-year variation in albatross breeding population size and success. A commitment to continuous studies (or at least three years of censuses repeated at intervals of five to 10 years) is therefore required; however, given the remote location of and difficult access to, many albatross breeding sites, this is a demanding requirement. Greater use of aerial photography and of remote-sensing techniques (e.g., Schwaller *et al.* 1989; Guinet *et al.* 1995), particularly as new satellite imagery promises resolution to ± 2 m, would seem appropriate. Such techniques will need careful calibration and ground truthing but could be a valuable adjunct to conventional studies and, ultimately, perhaps the most cost-effective means of monitoring at many of the remote sites.

Although a decrease in breeding population size may indicate that a problem exists, more detailed population data will often be required to define the nature and magnitude of the problem, to determine the element of the population at risk and to identify causal factors. However, the collection of detailed demographic data, especially on survival rates, is a major undertaking that often cannot be justified within a general monitoring programme, at least until a significant population decline has been detected. By this time, however, it may be too late, or very difficult, to determine the relevant proximate (i.e., demographic) and ultimate (i.e., environmental) factors responsible. Indeed, most, if not all, the long term studies of albatross populations started as investigations into aspects of age-related and/or lifetime reproductive success or other work for which a population of known-age birds was essential.

The current priorities for basic population studies are: a) for data holders to accord high priority to the analysis and publication of historical (and current) data on population change and demography; b) for species where no such data exist, to initiate work for a period adequate to provide a baseline against which future change can be measured.

The two most important variables to measure are adult and juvenile survival because, in broad terms, immigration/emigration effects and changes in age at first breeding (and even variation in chick-rearing success) have much less influence on the overall population. Of these, adult survival is much the easier to measure, requiring only a realistic sample size (sufficient to detect a 1–2% change in survival rate over the period until survival rate is re-estimated) and about three to five years of continuous effort (depending on the breeding frequency of the species). To estimate juvenile survival, however, needs several cohorts of fledged chicks to be banded and substantial recapture effort to be undertaken for several years, either around the modal age of first return to the breeding colony and/or to estimate the proportion surviving to enter the breeding population.

There are no short-cuts to producing reliable survival data capable of detecting the kind of changes that have biological significance (e.g., 1% and 5% for annual adult survival and recruitment rates, respectively). The need for more studies is demonstrated by Prince *et al.* (1994a, Table 14) finding only nine studies of seven species which had produced multi-year data on adult survival; only five of these had values for juvenile recruitment. The value of continuous records of survival also lies in the ability to relate these to changes in environmental effects, whether these are natural (e.g., climate, sea-surface temperature, etc.) or human-induced (location and intensity of fishing) (see Robertson 1993; Weimerskirch *et al.* 1997).

Many of the original objectives of current long-term population studies (e.g., to examine longevity, variation in lifetime reproductive success, senescence, etc.) have been compromised by the artificially high levels of mortality operating in recent years (see later). Nevertheless, studies that have investigated age/experience-specific patterns of reproduction and survival have broadly concluded that, at least in Wandering Albatrosses, effects are limited to the first year or so of reproduction (Lequette and Weimerskirch 1990; Weimerskirch 1992) and that this may reflect the lengthy experience acquired in pair-bonds prior to commencing reproduction (Croxall 1991; Croxall *et al.* 1992; Weimerskirch 1992). In studying egg size, which changes systematically, Croxall *et al.* (1992) showed that age, rather than experience, was the more important influence; however, between-individual variation was considerably more important than either. Although Fisher (1975) obtained broadly similar results from the limited data for Laysan Albatrosses, it should not be assumed that all albatrosses, particularly those with shorter duration of pair-bonds prior to laying and faster reproductive rates, will show the same patterns. However, the substantial difference in patterns of age-related reproductive success between albatrosses and shearwaters *Puffinus* (e.g., Wooller *et al.* 1990) suggests that events ashore during

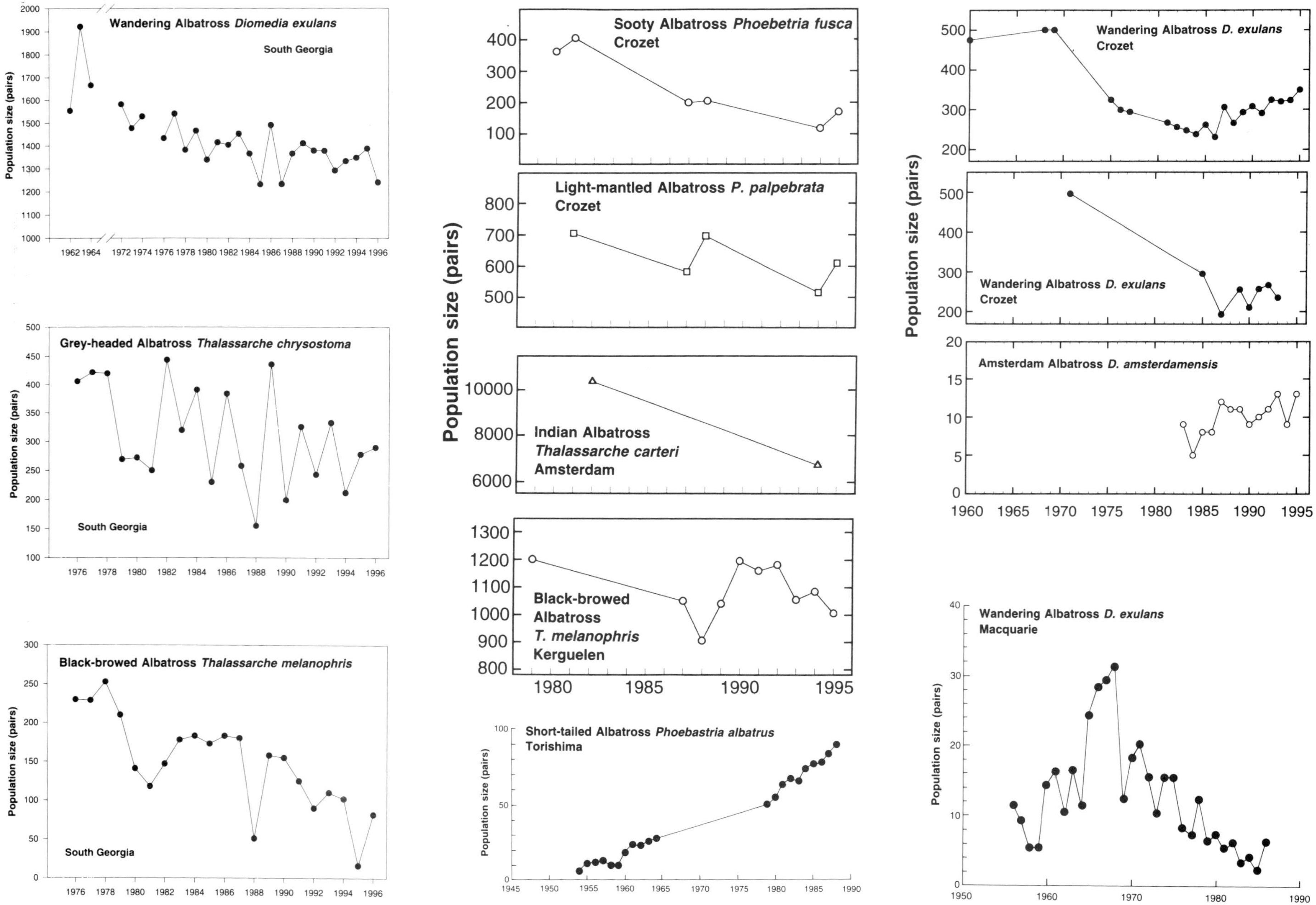

Fig. 1. Population trends of some albatross species. Each graph shows the site and species. Data from (a–c) Croxall *et al.* 1997; (d) de la Mare and Kerry (1994); (e) Hasegawa (1984 and unpubl.); (f–k) Weimerskirch and Jouventin (1997).

the pre-laying attendance periods may be of disproportionate importance in albatrosses — and still need modern detailed studies (see Fisher and Fisher 1969; Pickering 1989).

There are still many interesting questions to be addressed by long-term studies of albatross populations. These include assessments of: (a) the costs of divorce (in terms of trade-offs between the time taken to acquire a new partner and any subsequent improvement in reproductive performance); (b) the relationship between breeding frequency and survival, including the consequences (whether random or based on choice) of avoiding breeding in years of inclement conditions or low food availability; and (c) trade-offs between age of first breeding (and duration of prior pair-bonds) and subsequent reproduction and survival.

It is increasingly clear, however, that many features of so-called population reproductive performance actually relate to variation in attributes of individuals. Thus individual "quality" of Wandering Albatrosses was the main source of variation in egg size (Croxall *et al.* 1992) and also implicated in aspects of condition, reproduction and survival (Weimerskirch 1992). Furthermore Grey-headed Albatrosses in the top and bottom quartiles (based on past reproductive success) were not distinguishable in terms of experience or number of breeding attempts (Cobley *et al.*, in press). As albatross populations probably depend on relatively small numbers of individuals producing a high proportion of the surviving progeny, studies of performance, quality and condition at the individual level are highly relevant. The use of genetic markers to examine relatedness amongst birds in breeding populations and to identify progeny where banding data are unavailable could provide unique insights and foundations for future work.

A major handicap for many such studies is time needed to develop a population of known-age birds. Although various methods have been tried (Klomp *et al.* 1995) there is still no single reliable ageing criterion for albatrosses. The best approaches at present, accurate to ± 1 year, combine moult pattern with bill and plumage characteristics (Prince and Rodwell 1994; Prince *et al.* 1997).

Finally, it is important that those scientists with adequate population data ensure that the implications of their data are fully utilized in conservation contexts. This can range from conducting analyses to determine the relative sensitivity of populations to change in vital rates (e.g., Croxall and Rothery 1991; Weimerskirch *et al.* 1997), to using demographic data in formal population viability analyses, an increasingly popular mathematical modelling tool for guiding conservation management and research (e.g., Hamilton and Moller 1995).

Moult

Albatross do not undertake wing moult while breeding. Despite the likely importance of moult in species that have substantial plumage changes during lengthy adolescence and which presumably require to maintain high levels of feather quality for efficient gliding flight, the pattern and rate of moult in albatrosses remained unknown until very recently.

In their pioneer comprehensive study, Prince *et al.* (1993) identified the main moult patterns of Black-browed and Grey-headed Albatrosses as descendent moult of outer and ascendent moult of inner primary feathers in alternate years (also showing how these patterns change with increasing age of juveniles and in relation to breeding success of adults). A more complex but basically analogous pattern is now known for the Wandering Albatross (Prince *et al.* 1997). However a different pattern of primary feather replacement, proceeding in opposite directions from the wing midpoint is reported for Laysan and Black-footed Albatrosses (Langston and Rohwer 1995), with the additional specialization of primary moult recommencing each year at the primary next to the one moulted last in the previous year — termed a wrap-around moult.

Apart from the intrinsic interest of understanding the strategies by which albatrosses try to preserve the condition of their outer primaries (basically by accumulating old feathers in the the inner wing), the number of feathers moulted annually may provide a sensitive index to relate to breeding performance and individual condition. Thus Weimerskirch (1991) found that the extent of moult was inversely related to mass change between years in non-breeding Wandering Albatrosses and that in adult females (but not males) it was directly related to the duration of the inter-breeding period. Understanding the nature and significance of the trade-offs between condition, moult and breeding is clearly a priority for future research and a field in which results from albatrosses may exemplify general principles relevant to many other seabirds. Langston and Hillgarth (1995) showed that another aspect of body condition — parasite burden — also affected moult: Laysan Albatrosses with more nematodes replaced fewer primaries, grew fewer feathers simultaneously and started moult later.

Physiology and energetics

Presentations at the Conference on these topics were confined to Green and Brothers (1995) and there have been relatively few publications in recent years. This is surprising, because knowledge of individual and population activity-specific energetics is fundamental to topics including costs of flight, cost-benefit analysis of foraging performance, energy budgets (and thereby growth strategies) of chicks, interaction between performance, condition, reproduction and survival and estimating food consumption (and seasonal peaks in demand) of albatross populations. In addition, in view of their size and tameness, albatrosses are excellent experimental animals for a range of physiological investigations.

Over the years promising research has been initiated, but often not continued, in several fields. Thus an excellent introductory study to the thermoregulatory behaviour of Laysan and Black-footed Albatrosses (Howell and Bartholomew 1961) has been extended neither comparatively nor in depth, despite the substantial improvements in instrumentation for temperature studies of free-ranging animals, the range of thermal biotopes that albatrosses inhabit and the indications that thermal stress can have significant effects on breeding performance (Robertson 1993, 1995, 1997).

Water and ionic (especially salt) balance is obviously of general importance to marine birds and particularly so for species breeding in exposed habitats and undertaking long fasts. Early work (Frings *et al.* 1958; Frings and Frings 1959), suggesting different consequences for albatrosses on invertebrate (e.g., squid) and vertebrate (e.g., fish) diets — relating to different levels of calcium, magnesium and sodium in their tissues — has never been followed up.

The nature of hormonal control of reproduction and its role in influencing the unusual breeding cycles of albatrosses was studied (including pioneer laparoscopy of free-living seabirds (Hector 1984)) in both annual and biennial breeding species (Hector *et al.* 1986a,b, 1990; Hector 1988) and indicated that female hormone cycles were decisive in determining breeding frequency. There should be renewed interest in extending this work to other species, particularly given the discovery that some female Southern Royal Albatrosses *Diomedea epomophora* can breed in the year immediately after successfully rearing a chick (Waugh *et al.*, in press).

The role of hormones and secondary metabolites in influencing the termination of incubation fasts when body reserves are seriously depleted (Hector and Harvey 1986) deserves further critical study in conjunction with accurate information on the levels of body protein and fat and the rate of metabolic turnover. High levels of corticosteroid hormones may also reflect periods of particular stress for seabirds (e.g., Smith *et al.* 1994) and this may represent a complementary research topic for which albatrosses are particularly suitable.

The condition of birds is obviously a vital element of their existing and potential performance (whether in respect of moult, incubation or foraging). While average patterns of mass change through a season are of interest (e.g., Fisher 1967), details of individual status are essential for full understanding both of constraints and of the process as a whole (e.g., Weimerskirch 1992; Cobley and Prince, in press). However information on mass may need to be supplemented by data on the composition (in terms of water, protein and fat and whether structural or reserves) of material gained or lost. Techniques for obtaining and analysing appropriate data (e.g., biopsy, labelled isotope studies of water flux, etc.) are rapidly becoming available in ways allowing their application to free-living albatrosses.

The ability to recognize (through labelling) layers representing daily deposition of yolk (Grau 1976), demonstrated the lengthy period (c. 20 days) and early initiation (*c.* 30 days before laying) of yolk formation in albatrosses (Grau 1984; Astheimer *et al.* 1985). Further comparative studies would be of interest, as might investigations of variation between years and individuals in the amount of yolk deposited (see Hatchwell and Pellatt 1990).

In estimating energy expenditure of albatrosses at different stages of their annual cycle, early work was limited to direct measurements only of embryonic oxygen consumption (Pettit *et al.* 1982) and rough estimates of energy costs of incubation from rates of mass loss (e.g., Croxall 1982). It took time for the turnover rates of body water (Lifson and McClintock 1966) to be widely used in studies of ecological energetics of free-living vertebrates (and longer still to afford to use double-labelled water on animals as large as albatrosses!). Surprisingly, however, the relatively simple (and cheap) estimation of energy during fasting (e.g., incubation) has still only been made for Wandering and Laysan Albatrosses (Brown and Adams 1984; Pettit *et al.* 1988). Estimates of the costs of foraging trips to sea have been made for Wandering Albatrosses and Grey-headed and Shy *Thalassarche cauta* Albatrosses (Adams *et al.* 1986; Arnould *et al.* 1996; Costa and Prince 1987; Green and Brothers 1995). Studies of chick energetics using these techniques have only been carried out on Shy Albatrosses (Green and Brothers 1995); other studies of chick metabolism have been based on frequent weighings and estimates of food conversion efficiency (e.g., Ricketts and Prince 1984).

Even with improved accuracy from the use of new equations for determining CO_2 production (Speakman 1993; Speakman *et al.* 1993), the problem with estimating energy expenditure using water turnover is that it integrates the cost

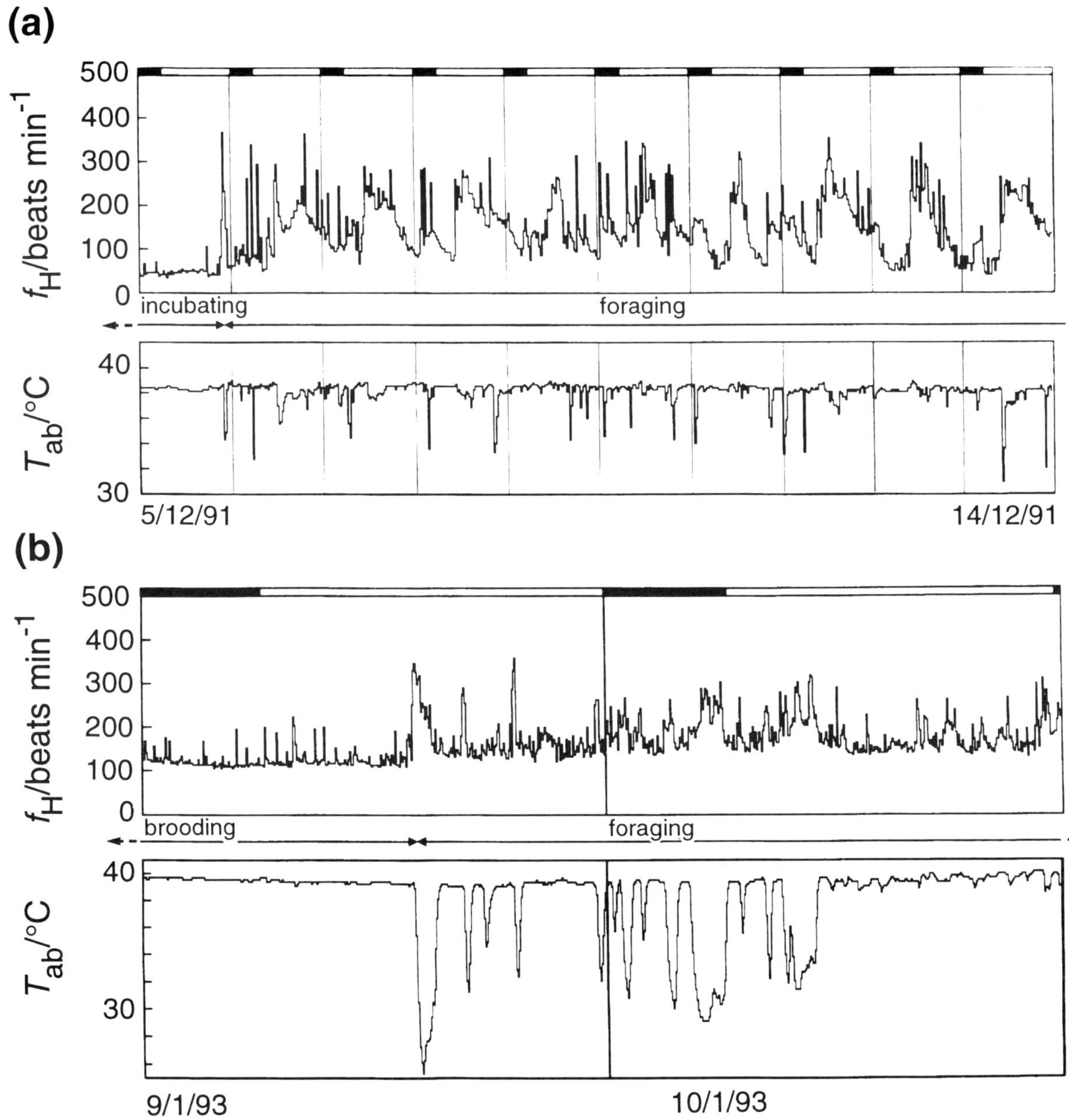

Fig. 2. Examples of simultaneously collected data on heart rate (beats min^{-1} sampled at 30 s intervals; top trace) and abdominal temperature (sampled every 60 s; bottom trace) obtained from free-ranging Black-browed Albatrosses acquired using an implanted recorder (from Bevan *et al.* 1995). The dark bars above the heart rate trace indicate periods of darkness. (a) Ten days of data obtained during incubation. (b) Two days of data obtained during brooding.

of all activities (e.g., resting, flying, diving) between the start and the finish of the experiment. It also has a restricted timespan over which measurements can be made before the levels of isotopically labelled water cannot be differentiated from baseline. Although overall water turnover values can be partitioned to some extent if obtained for birds with, for example, satellite tracking and activity recorder data (Arnould *et al.* 1996), these do not generate genuine activity-specific data.

Recent advances in recording of heart-rate (see Butler *et al.* 1995) and in the calibration of this against independent measures of energy expenditure, provide detailed and accurate records of energy expenditure which, if matched with simultaneous data on activity, could provide unique information on activity-specific energy budgets. Albatrosses have proved to be excellent subjects both for calibration studies (Bevan *et al.* 1994) and full-scale trials with free-flying birds (Bevan *et al.* 1995; see Fig. 2); there is immense scope for extending this work to other species and to birds carrying activity recorders of various kinds.

Studies of chick energetics, particularly involving relationships between parental provisioning rate and chick growth pattern (Prince and Ricketts

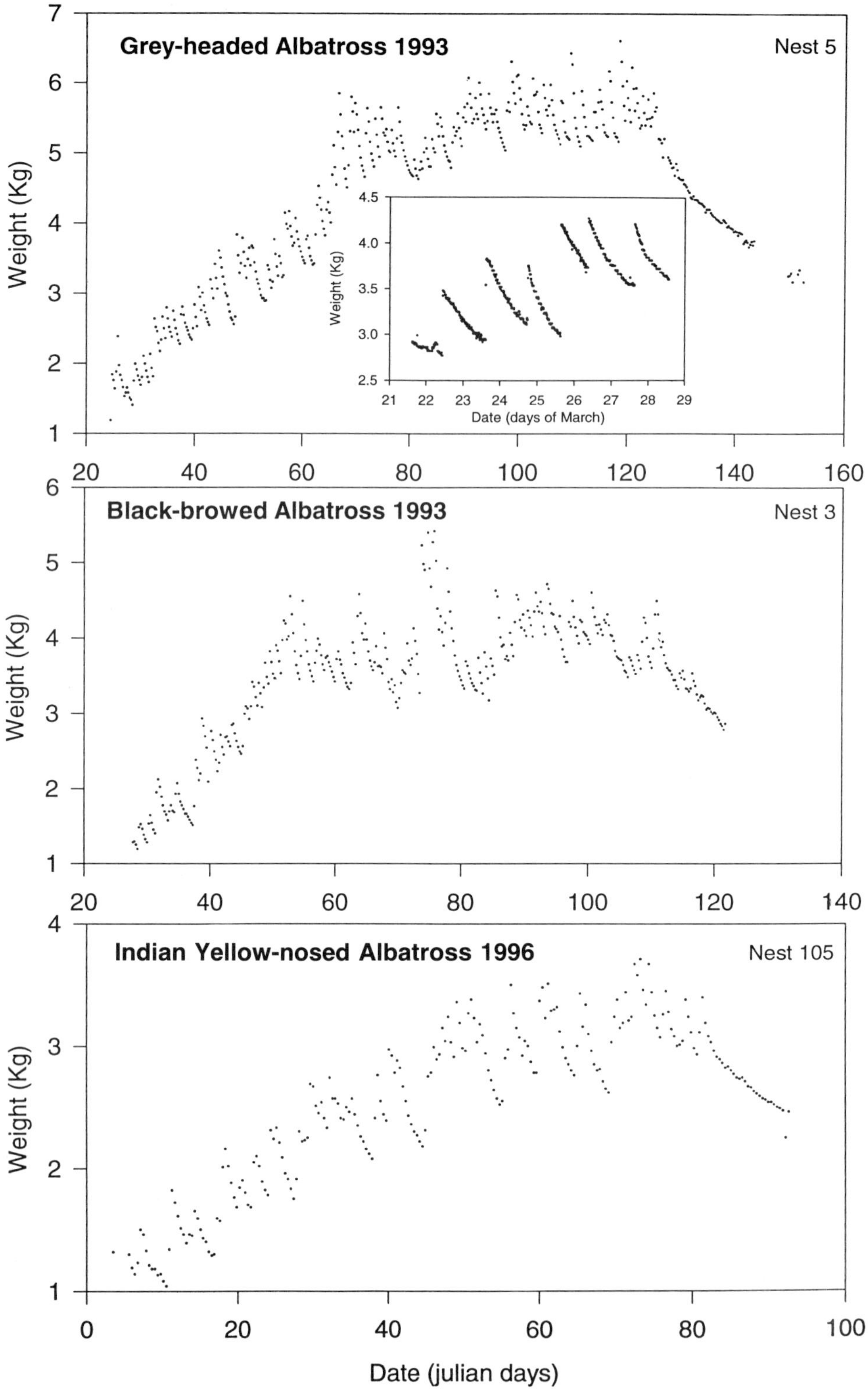

Fig. 3. Examples of data on growth patterns of albatross chicks, recorded using electronic weighing equipment.

(a) Grey-headed Albatross, South Georgia (Prince *et al.*, unpubl.)

(b) Black-browed Albatross, South Georgia (Prince *et al.*, unpubl.)

(c) Indian Yellow-nosed Albatross, Amsterdam Island (Weimerskirch and Prince, unpubl.)

Main illustrations show mass at 6 h intervals. The inset shows the raw data collected at 10 min. intervals. Dates are Julian days (i.e., 1 = 1 January), which roughly correspond to chick age in days.

1981) will be revitalized by the use of electronic balances deployed inside artificial nests. These have now been used on Black-browed, Grey-headed (Prince *et al.* 1994c), Indian (H. Weimerskirch and P. A. Prince, unpubl.) and Shy (R. Gales, A. Hedd *et al.*, unpubl.) Albatrosses (see Fig. 3).

Diet

Early work was largely based on anecdotal data from opportunistic sampling. The ability routinely to obtain regurgitated samples from adults about to feed chicks (Prince 1980) enabled the first quantitative descriptions of the composition of albatross diets. The development of techniques permitting reconstruction of the characteristics of prey from the beaks of cephalopods (Clarke 1986), otoliths of fish (Hecht 1987; Williams and McEldowney 1990; Reid 1996) and carapaces of crustaceans (Hill 1990) provided major additional opportunities for assessing the size (and sometimes reproductive status) of prey exploited by albatrosses.

Despite these developments, reviews of the fish (Reid *et al.* 1996), cephalopod (Croxall and Prince 1996) and overall (Cherel and Klages 1997) diet of albatrosses emphasize the generally inadequate basic quantitative knowledge of the diet of most albatross species. Indeed there are virtually no accurate quantitative data on the diet of Short-tailed *Phoebastria albatrus*, Amsterdam *Diomedea amsterdamensis* and Waved Albatrosses, nor for Atlantic *Thalassarche chlororhynchos*, Indian *T. carteri* or Shy Albatrosses, nor for most of the 10 species endemic to the New Zealand region. For the other species, few have had their diet studied in more than one year and within-year variation has rarely been described. To determine the diet of each species and to assess within-year and between-year variation, remains an important requirement. However, the new challenges are to investigate diet in the context of foraging ecology. Priorities are: (a) to use new techniques (e.g., stable isotopes (Hobson *et al.* 1994) and other biomarkers) to assess diet at times of year other than the chick-rearing period and in categories of birds other than breeding adults; (b) to relate information on diet to the time and locations where prey were obtained (e.g., by using a combination of satellite tracking, activity and dive recorders and stomach temperature recorders: see below); (c) to understand the extent to which diet reflects interactions between albatrosses and fishing vessels (see e.g.,Thompson 1992; Thompson and Riddy 1995; Gould *et al.* 1997); (d) to make realistic assessments not only of prey consumption by albatrosses but of the species and age of these prey and how this predation may relate to the demography and distribution of the prey; and (e) to quantify the extent to which albatrosses can be used as indicators of the status (e.g., abundance, distribution) of marine resources (Cherel and Weimerskirch 1995; Rodhouse *et al.* 1996).

At-sea distribution, behaviour and ecology

The ability to track albatrosses using satellites has already revolutionized our knowledge of their at-sea distribution and movements. However, the data painstakingly acquired by observers at sea and compiled into atlases of seabird distribution (e.g., Fisher and Fisher 1972; Weimerskirch *et al.* 1988) are still of value in providing a wealth of information on the global range of species. It will be a considerable time before satellite-tracking studies have followed sufficient birds of different species and age-classes (especially juveniles) and at different times of year to build up a picture which is both detailed and comprehensive.

Nevertheless, since the pioneer study of Jouventin and Weimerskirch (1990) at least 13 species of albatross have been tracked using satellites, three of these (Wandering, Black-browed and Light-mantled Albatrosses) from more than one site (see Table 1). The data on albatross distribution acquired via these studies (see Fig. 4) give substantial insights into the distribution of birds while breeding (particularly during chick-rearing but also during incubation) and, in a few cases, into the migration routes and location of "wintering" grounds (see Fig. 5). Much of this new research is described in detail elsewhere in this volume (e.g., Anderson *et al.* 1997; Prince *et al.* 1997; Weimerskirch and Jouventin 1997).

To extend the present picture of distribution, movements and foraging areas at different times of year will require studies of: (a) the remaining species and, where feasible, populations of all species which inhabit different island groups; (b) adults outside their breeding season; and (c) juveniles throughout as much of their immature life as possible. To carry out these studies will require not only intermittent duty cycles to extend battery life but also batteries that are rechargeable (e.g., via miniature solar panels) or with intrinsically longer life. Ultimately the use of satellite-linked transponders rather than transmitter systems will probably be necessary to acquire adequate information economically; better methods for long-term attachment of devices will also be needed.

Although characterizing the distribution and movements of albatrosses is of considerable interest, particularly in terms of assessing relationships with environmental (e.g., oceanographic, meteorological) features and events and with the timing and location of fisheries, more detailed interpretation requires data

Table 1. Satellite tracking studies of albatrosses.

Species	Site	Dates	Sample size: PTTs	Birds	Trips	Uplinks	Stage	Reference
Wandering Albatross *Diomedea exulans*	Iles Crozet	Jan.–Mar. 1989	6	6	7	1 561	Incubation	Jouventin and Weimerskirch 1990
	Iles Crozet	Jan.–Aug. 1990, Jan.–Feb. 1991	9	30	65	4 468	Incubation, chick-rearing	Weimerskirch *et al.* 1993
	Iles Crozet	Jan.–Mar. 1992	7	9	12	?	Incubation-brooding	Weimerskirch *et al.* 1994
	Iles Crozet	July 1992	6	7	15	?	Chick-rearing	Weimerskirch *et al.*, unpubl.
	Iles Crozet	Jan.–Mar. 1994	6	12	12	?	Incubation	Weimerskirch 1995
	South Georgia	Aug.–Sept. 1990	2	2	5	518	Chick-rearing	Prince *et al.* 1992
	South Georgia	Mar.–May 1991, 1992	11	11		1 063	Brooding	Croxall and Prince 1996
	South Georgia	Aug. 1990–Sept. 1994		56		10 581	Chick-rearing	Prince *et al.* 1997
	South Georgia	July–Dec. 1992	2	2	—	750	Post-breeding	Prince *et al.* 1997
	NSW, Australia	Aug.–Sept. 1993	5	5	—	1 056	Non-breeding	Nicholls 1995a
Amsterdam Albatross *D. amsterdamensis*	Amsterdam Island	Mar.–May 1996	5	7	8	?	Incubation	Weimerskirch *et al.*, unpubl.
Gibson's Albatross *D. gibsoni*	Auckland Islands	Jan.–May 1994	3	3	7	452	Incubation, brooding	Walker *et al.* 1995
	Auckland Islands	Jan.–Mar. 1995	2	2	—	?	Post-breeding	Nicholls *et al.*, unpubl.
	Auckland Islands	Jan.–Feb. 1996	11	15	15	2 740	Incubation	Walker *et al.*, unpubl.
	NSW, Australia	Sept.–Dec. 1994	1	1	—	?	Non-breeding	Nicholls *et al.*, unpubl.
Antipodean Albatross *D. antipodensis*	Antipodes Islands	Jan.–Feb. 1996	1	1	1	?	Post-breeding	Nicholls *et al.* 1996
	Antipodes Islands	Jan.–Mar. 1996	11	15	15	2 567	Incubation	Walker *et al.*, unpubl.
Northern Royal Albatross *D. sanfordi*	Taiaroa Head, NZ	Nov. 1993	2	2	5	?	Courtship/incubation	Nicholls *et al.* 1994
	Chatham Islands, NZ	Nov. 1994	2	2	?	?	Incubation	Nicholls *et al.*, unpubl.
	Chatham Islands, NZ	Nov. 1995 Feb.–June 1996	9	5	?	?	Incubation, chick-rearing	Nicholls *et al.*, unpubl.
Waved Albatross *Phoebastria irrorata*	Galapagos	Apr.–June 1995	5	5	7	229	Incubation	Anderson *et al.* 1997
Black-browed Albatross *Thalassarche melanophrys*	Iles Kerguelen	Feb. 1994	9	16	35	1 290	Chick-rearing	Cherel and Weimerskirch 1995
	South Georgia	Jan.–Mar. 1993	10	10	?	1 819	Chick-rearing	Prince *et al.* 1997
	South Georgia	Mar.–Apr. 1992, 1993	2	2	—	444	Post-breeding	Prince *et al.* 1997
Buller's Albatross *T. bulleri*	Snares Islands, NZ	Feb.–Mar. 1995	6	6	5	552	Incubation	Sagar and Weimerskirch 1996 Weimerskirch 1997
Indian Yellow-nosed Albatross *T. carteri*	Amsterdan Island	Dec. 1995–Jan. 1996	17	22	35	?	Incubation, chick-rearing	Weimerskirch *et al.*, unpubl.
Shy Albatross *T. cauta*	Albatross Island, Tas.	Sept.–Dec. 1993; Dec. 1994	14	30	48	694	Incubation, brooding	Gales *et al.*, in press
	Albatross Island, Tas.	Apr.–May 1994	5	5	5	604	Post-breeding	Gales *et al.*, in press
	Pedra Branca, Tas.	Dec. 1994	3	6	10	188	Incubation, brooding	Gales *et al.*, in press
	Pedra Branca, Tas.	Mar.–Apr. 1995	3	3	3	587	Post-breeding	Gales *et al.*, in press
Grey-headed Albatross *T. chrysostoma*	South Georgia	Jan.–Mar. 1993	10	10	?	1 919	Chick-rearing	Prince *et al.* 1997
	South Georgia	Jan.–Mar. 1994	5	5	17	435	Chick-rearing	Rodhouse *et al.* 1996
	South Georgia	Mar.–Apr. 1996	1	1	—	212	Post-breeding	Prince *et al.* 1997
Light-mantled Albatross *Phoebetria palpebrata*	Macquarie Island	Nov.–Dec. 1992	5	6	5	514	Incubation	Weimerskirch and Robertson 1994
	Iles Crozet	1993–1995	5	7	12	?	Incubation-brooding	Weimerskirch 1997
Sooty Albatross *P. fusca*	Iles Crozet	1992–1995	8	18	26	?	Incubation-brooding	Weimerskirch 1997

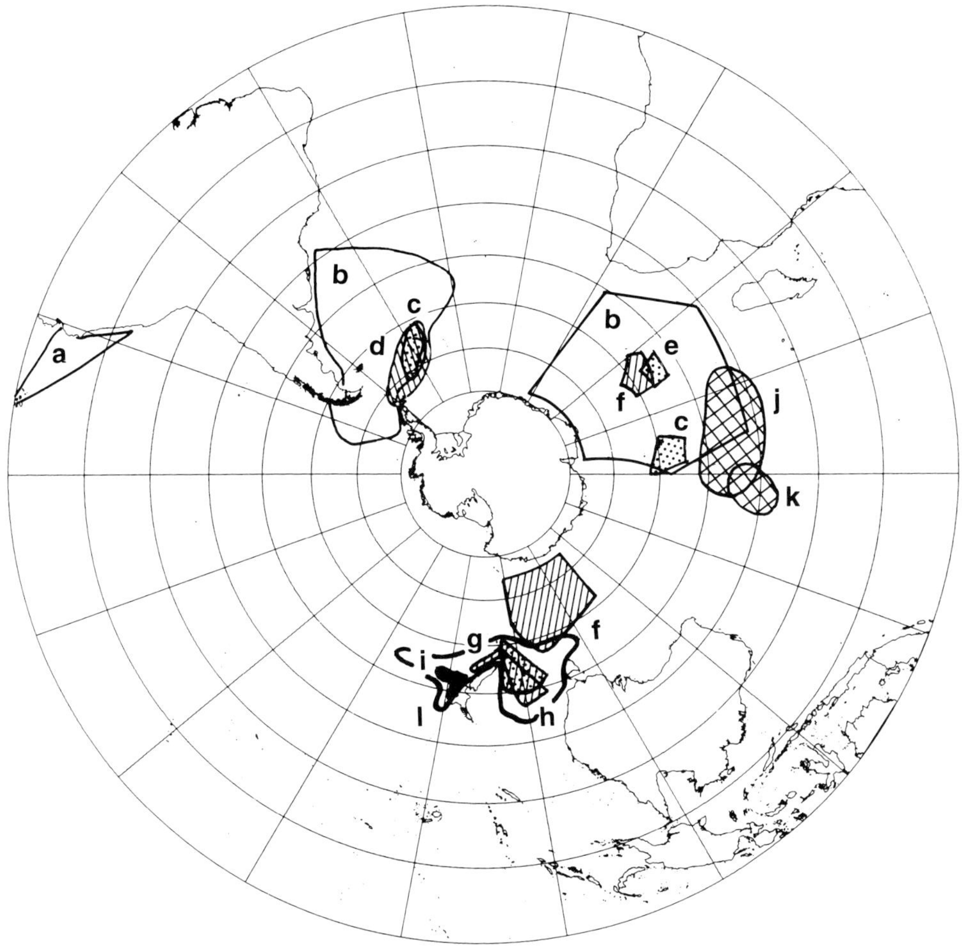

Fig. 4. Foraging ranges of breeding albatrosses as determined by satellite telemetry (data from sources listed in Table 1). a, Waved Albatross; b, Wandering Albatross; c, Black-browed Albatross; d, Grey-headed Albatross; e, Sooty Albatross; f, Light-mantled Albatross; g, Buller's Albatross; h, Antipodean Albatross; i, Northern Royal Albatross; j, Amsterdam Albatross; k, Indian Yellow-nosed Albatross; l, Gibson's Albatross.

complementary to those on location. The ideal, of course, would be simultaneously to acquire data on activity (flight, resting, diving, etc.), energy costs (e.g., via heart-rate), feeding rate (i.e., the time and location of each ingestion of prey — and its identity and size) and the nature of the oceanic and atmospheric environment (e.g., sea-surface temperature (see Fig. 6), salinity, barometric pressure, wind speed and direction). For studies of birds during chick-rearing, complementary data on the timing and size (and, ideally, energy density) of meals delivered to chicks and of their resulting growth performance (together with data on adult mass at the start and end of foraging trips) is also necessary. With such data investigation of many topics, hitherto largely unstudied in pelagic marine animals, becomes feasible. Examples include: (a) foraging efficiency (including differences between sexes, adults/juveniles and of the acquisition of feeding/foraging skills by the latter); (b) efficiency in provisioning offspring (and sharing of duties between sexes) and influence of provisioning rate on offspring growth and performance and on adult condition and subsequent performance/survival; and (c) influence of ocean and atmosphere features and events on distribution, range, foraging movements and performance.

Considerable progress has been made towards many of these objectives through the development of various devices to record: (a) activity at sea (e.g., time spent on water, orientation of flying birds and frequency and depth of diving (Prince and Francis 1984; Afanasyev and Prince 1993; Prince *et al.* 1994b; Hedd *et al.* 1997; Viswanathan *et al.* 1996; Huin and Prince 1997); (b) energy costs of foraging trips during chick-rearing (Bevan *et al.* 1995; Arnould *et al.* 1996); (c) prey ingestion (via changes in stomach temperature consequent on swallowing cold organisms) (e.g., Weimerskirch and Wilson 1992; Wilson *et al.* 1992, 1995a,b); (d) sea-surface temperature (Weimerskirch *et al.* 1995); (e) air temperature and barometric pressure (Nicholls *et al.* 1994, 1995b); (f) size and

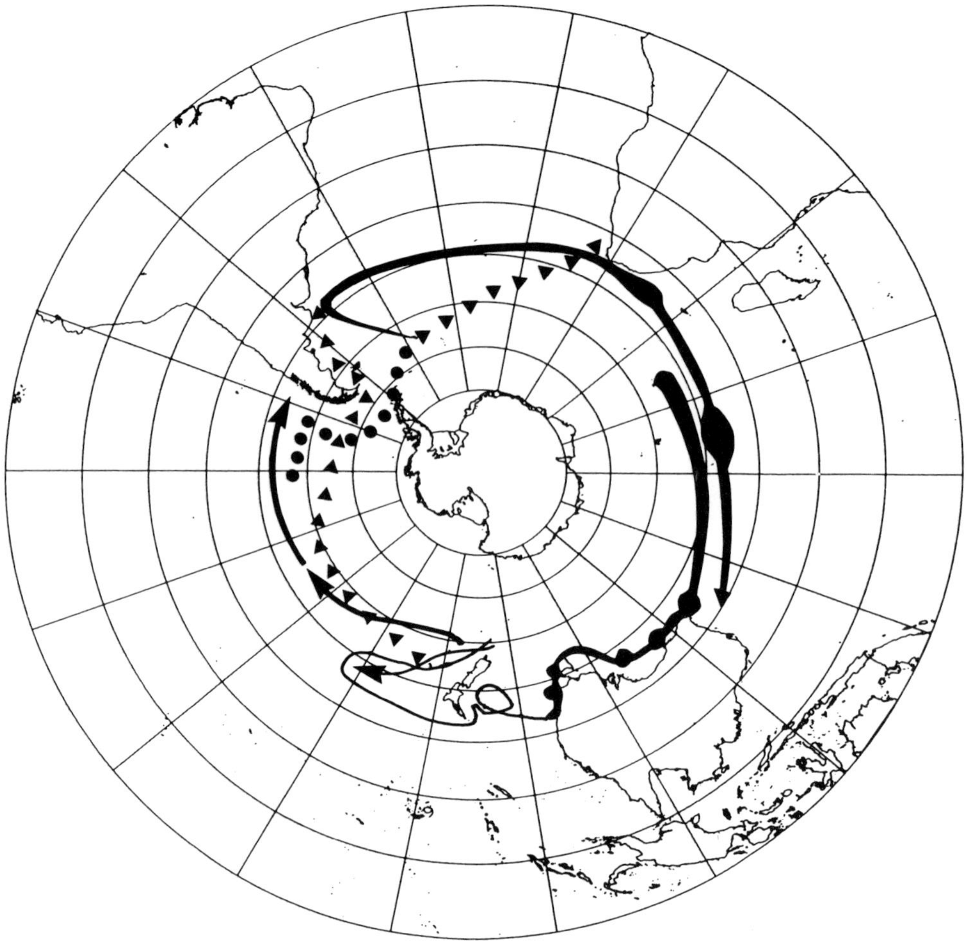

Fig. 5. Migration routes of albatrosses as determined by satellite telemetry (data from sources listed in Table 1). —— Wandering Albatross; ▶▶ Northern Royal Albatross (from New Zealand); ●● Grey-headed Albatross; —— Gibson's Albatross; ▶▶ Black-browed Albatross (from South Georgia); ➤ Antipodean Albatross.

timing of meals delivered to offspring and their resulting growth performance (Prince and Walton 1984; Prince *et al.* 1994c). These devices have produced much exciting data and new insights into the pelagic life of albatrosses. They have revealed unsuspected diving capabilities, indicated that feeding is mainly confined to daylight hours, confirmed that gliding flight is energetically cheap but that bouts of short duration activity, (associated with contact with water) are expensive, provided first estimates of the amount of prey ingested and the location of these events throughout complete foraging trips (see Fig. 7) and documented in detail the rate of provisioning of offspring.

Nevertheless, many of these studies are still relatively preliminary in nature and more development and calibration of instruments is required. Furthermore, only a few studies have used different instruments in combination and so quantification of the rates of energy expenditure and acquisition on foraging trips and of allocation of energy to various activities and between parents and offspring is only commencing. However, initial results examining foraging in relation to prey distribution (Weimerskirch *et al.* 1994; Cherel and Weimerskirch 1995; Rodhouse *et al.* 1996), estimating foraging effort at different stages of chick-rearing (Salamolard and Weimerskirch 1993) and relating it to mass balance (Weimerskirch 1995), demonstrate the considerable potential of these approaches.

Obviously we need to undertake more studies combining the use of different devices and to address specific hypotheses rather than produce descriptive results. However there will be continuing development of instrumentation in the search for smaller devices, with increased memories and including sensors to record a wider range of variables. This will pose a continuing requirement for: (a) validation studies of the performance of techniques and devices; (b) development of standardized or comparable procedures for the analysis and validation of data, especially for studies of heart-rate, prey ingestion and information derived from interpretation of the satellite-track; (c) critical comparison of similar data acquired by different devices and methods (e.g., positional/locational

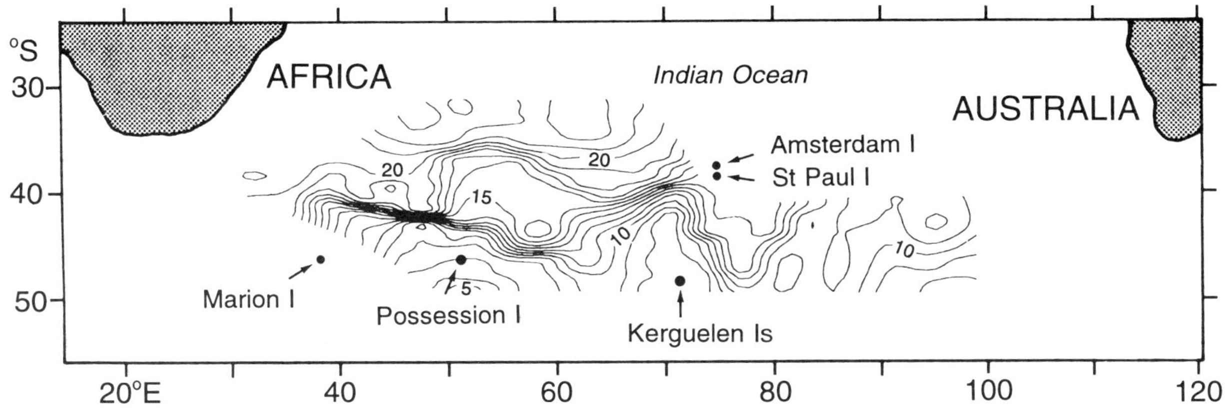

Fig. 6. Sea surface temperature (contours are 1°C isotherms) as measured by five Wandering Albatrosses breeding at Possession Island, Iles Crozet (from Weimerskirch *et al.* 1995).

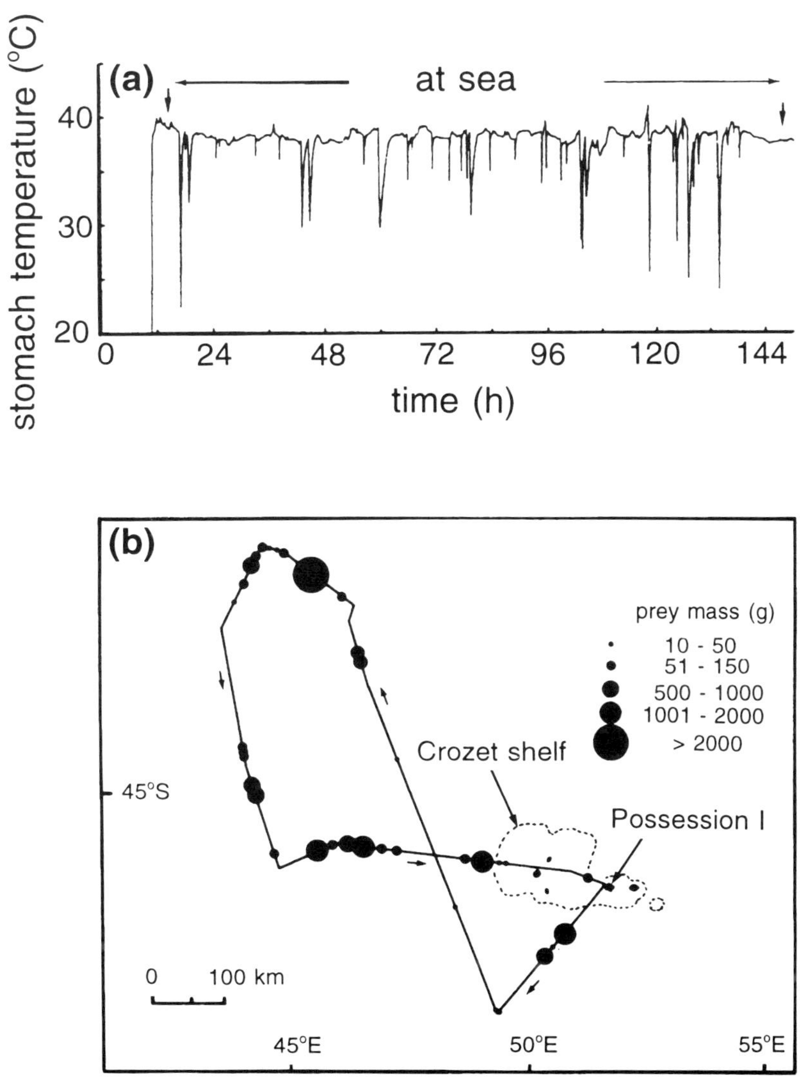

Fig. 7. Examples of the use of simultaneously collected data from stomach temperature recorder (a) and satellite transmitter in generating (b) a map of the location at which prey are ingested (together with an estimate of prey mass) by Wandering Albatrosses from Iles Crozet (from Weimerskirch *et al.* 1995).

data from geolocation, transponder, transmitter and dead-reckoning recorders, stomach temperature data from ingested pills, ingested cylinders, implanted electrodes, etc.).

One potential limitation could be the capacity of satellites to handle the reception and transmission of data from multi-functional recorders. The existing achievements of albatross researchers demonstrate the immense potential of these birds as paradigms for study of predator-prey interactions in marine systems and also as platforms for collecting a wide range of data on oceanography and prey distribution. We need to ensure that new generations of satellites for use with biological applications are designed with these abilities and requirements in mind. Above all the increasing challenges and complexities of albatross research will need greater co-operation and collaboration. Perhaps the time has come to develop some truly global projects (with levels of support similar to those for innovative ocean-atmosphere projects in the physical sciences) to integrate the abilities of and properly exploit the opportunities created by the many individuals and groups working on albatrosses.

CONSERVATION

Background

The foregoing sections testified to the excellence of albatrosses as experimental subjects and the fascinating insights into pelagic marine systems that they can provide. The research challenges for the future will keep albatross scientists fully occupied. However the greatest challenge, which is involving almost all albatross researchers (and many others besides) is that of ensuring that there are still viable populations of all albatross taxa to study through the 21st Century.

Much of the albatross conference, all of its workshop on albatross-fishery interactions and most chapters of this book address issues directly or indirectly relevant to albatross conservation, almost all of them resulting from problems created by mankind. The principal focus has been on the current major global problem — of interactions between albatrosses and longline fisheries. It is salutary to remember, however, that this is but the latest threat to the survival of albatrosses, that the variety and intensity of threats appear to be increasing and that we might do well to consider what the next substantial problem could be. It is depressing that nearly half the albatross populations with adequate documentation are decreasing and that most of those currently stable or increasing are recovering from past decreases (Gales 1997).

As Medway (1997) notes, albatrosses have been killed by humans since seafaring in their domains began (see Fig. 8). Many populations (e.g., Easter Island (Steadman 1995), Tristan da Cunha (Watkins 1987) were doubtless extirpated, or nearly so. These chronic (and sometimes acute) depredations probably reached their peak in the 19th Century when widespread low-level consumption of birds and eggs for food and sport (Medway 1997) coincided with the frequent and often long-term visits of sealers to many remote islands where albatrosses bred.

Just as the surviving albatross populations in the Southern Ocean and at staging posts to and from the Northern Hemisphere may have been starting to recover, those in the central North Pacific Ocean were subject to new threats. In the early 20th Century vast numbers of Laysan, Black-footed (Fisher 1949) and Short-tailed (Hasegawa 1984) Albatrosses were slaughtered by feather hunters. The populations of Laysan and Black-footed Albatrosses at Midway and Laysan Islands did partly recover by the 1950s. Then at Midway about 35 000 adults were killed in 1957 and 1958 (and possibly over 50 000 in total) to reduce air strikes (Robbins 1966), though without much success. Between the mid-1950s and 1967 Laysan Albatrosses at Midway were also being killed by collision with antenna arrays. In the only documented year (1964/65) at least 3 000 birds (82% adults) were killed (Fisher 1966). Ultimately the removal of the antennae (in 1967) and the final closure (in 1993) of the airbase, rather than concern for the birds, stopped the carnage.

As populations of albatrosses in the Pacific Ocean started to recover in the 1970s and 1980s they were subject to substantial mortality from new sources — gillnet and driftnet fishing. Pelagic gillnet fishing in the North Pacific started in 1952, peaked in the late 1950s and by the 1980s had decreased to about one-quarter of peak levels. Estimates of seabird mortality range from 214 000 to 760 000 birds per annum in the 1970s (see King 1984) and decreased to about 130 000 to 180 000 birds annually in 1978–1981. Probably no more than 1% of victims were Laysan Albatrosses; mortality would nevertheless have totalled at least 1 000–2 000 birds annually. Knowledge of albatross mortality in gillnet fisheries in other oceans of the world is virtually non-existent.

Large-scale driftnet fishing started in the late 1980s. In 1990 alone 4 426 Black-footed and 17 548 Laysan Albatrosses (2.5 and 1.5% of world populations) were killed (Northridge 1991; Johnston *et al.* 1993) and populations at Midway declined. The annual mortality of both species in the North Pacific Ocean from 1978 to 1990 was estimated as 2.2% of their populations (Gould and Hobbs 1993; McDermond and

Fig. 8. Exploitation of albatrosses. (a) Northern Royal Albatrosses collected for food at the Chatham Islands, New Zealand, in the 1930s (Photo by A. Wotherspoon, courtesy C. J. R. Robertson). (b) Laysan Albatross eggs collected at Laysan Island, Hawaiian Leeward Islands in the 1800s (Photo by W. A. Bryan courtesy of the University of Hawaii).

Morgan 1993); the magnitude of albatross mortality in the South Pacific Ocean (where extensive driftnet fisheries operated) is unknown. The global moratorium on driftnet fishing, effective from 1 January 1993, should have put an end to albatross deaths from this source.

Before dealing with the most recent global threat to albatrosses — longline fisheries — it may be appropriate briefly to review other recent and current threats.

Pollutants

Records of albatrosses ingesting plastics date back at least 30 years (Kenyon and Kridler 1969) and probably all species do so nowadays. Laysan Albatrosses (see Fig. 9) have the greatest incidence, widest variety and largest volume of ingested plastic of any seabird (Pettit *et al.* 1981; Fry *et al.* 1987; Sileo *et al.* 1989; Ludwig *et al.* 1997; Auman *et al.* 1997), much of which is egested and/or transferred to chicks. Even for Laysan Albatross chicks, mortality is relatively uncommon but the significance of physiological and mechanical distress is unknown (Sievert and Sileo 1993), except that dead chicks had more plastic and lighter body mass (Ludwig *et al.* 1997). Problems for other albatross species are probably rather less serious currently. There may, however, be a need to start collecting baseline data, e.g., by monitoring plastics (and other potentially harmful material) in the natural regurgitations which chicks of some albatross species provide before fledging (see Huin and Croxall, in press).

For over 20 years it has been likely, based on levels in Wandering Albatrosses (Risebrough 1977), that all albatross species would have detectable concentrations of DDT and PCBs. Few studies and no systematic monitoring of any species had been undertaken until very recently when Black-footed and Laysan Albatrosses were sampled at Midway (Ludwig *et al.* 1997). That Black-footed Albatrosses were sufficiently contaminated with PCBs and especially DDT-group chemicals to be at theoretical risk from egg-shell thinning and deleterious effects on embryos (Ludwig *et al.* 1997) is a cause for serious concern, especially given increasing empirical evidence consistent with these predictions.

Although albatrosses in the North Pacific Ocean may be at much greater risk from organochlorines and DDT (because of the extensive use of such compounds in south and south-east Asia), we should not be complacent elsewhere. It may be time to screen other species and to collect data on egg-shell thickness to provide baselines against which to measure subsequent change. Indeed, the potential of albatrosses to act as global indicators of the contamination of oceanic systems is exceptional. Thought should be given to the development of a worldwide sampling strategy using albatrosses.

Introduced predators

Albatrosses are sufficiently large to suffer less from introduced predators than many groups of seabirds. Nevertheless, both Black and Polynesian Rats *Rattus rattus* and *R. exulans* have been recorded eating eggs and killing chicks of Laysan Albatrosses (Moors and Atkinson 1984) and Norway Rats *R. norvegicus* killed chicks of Indian Albatrosses (Gales 1993). Rats presumably have the potential to do likewise to most other albatross species. Cats *Felis catus*, prior to their eradication at Marion Island, apparently contributed little to albatross chick mortality but could pose a threat to albatrosses elsewhere. Mustelids are currently under continual control at the Sanford's Albatross *Diomedea sanfordi* colony at Taiaroa Head, New Zealand (Robertson 1997) and would pose significant problems for other albatross species. Predation by pigs was believed to have prevented increase in the Wandering Albatross *Diomedea exulans* population at Inaccessible Island (Ryan *et al.* 1990) and pigs and cattle have caused disturbance, trampling and/or habitat degradation, giving rise to egg losses and/or restriction of breeding albatross populations at various sites. All these (and other) alien species need eradication from island groups where they occur. In addition to eradication programmes targeting alien species throughout the range of albatrosses, there is a need for persistent vigilance to ensure that potential predators are not introduced inadvertently to any island where albatrosses breed.

Habitat degradation

Change to habitats by humans (chiefly through stock or land management) where they have co-occurred with albatrosses may have restricted the area available to breeding populations. However, most such changes have probably been of much less significance than direct or indirect depredation by humans. There is some evidence (e.g., Robertson 1993, 1995, 1997) of climatically-induced changes in habitat that are detrimental, at least in the short term, to albatrosses. Better documentation of changes to the extent and quality of albatross breeding habitat (e.g., by using appropriate satellite and/or photographic imagery) might be useful for future studies of such effects.

The progressive degradation of albatross habitat at sea, however, is undoubtedly proceeding apace but at largely unknown rates and with uncertain consequences. Numerous manifestations of this were reviewed earlier — although not the potential consequences of worldwide over-fishing of many resources which are also important to food chains involving albatrosses. The chronic and pervasive deterioration in the quality of marine habitats may, however, have a more serious long-term effect on albatross populations than the more acute effects that we are aware of currently.

Longline fishing

Against this background the mortality of albatrosses caused by longline fishing (see Fig. 10) is but the latest — though potentially most serious — of the threats they have faced. The problem is, however, unique in several distinct dimensions. First, it is an actual or potential problem for nearly all species of albatross. Although the main focus has so far been in the Atlantic and Indian sectors of the Southern Ocean, the data provided at the Conference indicated substantial impact (probably equivalent to that sustained from driftnet fishing by Black-footed Albatrosses) in the North Pacific Ocean (Flint 1995) and potential impact throughout the Peru Current (Spear *et al.* 1995). Perhaps

Fig. 9. Plastic material regurgitated by Laysan Albatrosses at Midway Atoll, Hawaiian Leeward Islands.
FRANS LANTING

only for the Waved Albatross is there no current evidence of interactions with longline fisheries (Anderson *et al.* 1997; but see Merlen 1996).

Second, it has created enormous interest and publicity not just in the conservation press but also in the media generally. This publicity has helped the issue to rise towards the top of the agenda of marine conservation topics involving seabirds in countries as geographically remote from the main problem as the USA and UK.

Third, most of the albatross species known to be affected by longlines meet the new IUCN criteria for globally threatened status (Croxall and Gales 1997). Taxa in special danger from longlining, because of their tiny populations, include the Amsterdam Albatross (at least one killed on longlines (Gales 1993)) — one of only two albatross species with Critical status — and the Short-tailed Albatross (several individuals of which are known to have been killed on longlines (P. Gould, in litt.)). That several other globally threatened species continue to be killed in unacceptable numbers should create considerable concern and action by international conservation agencies and lead to more effective action at government and intergovernmental levels.

Fourth, there is interest in solving the problem on the part of some of those causing it. This is because the removal of bait from hooks by albatrosses has a significant effect on the catch rate of fish, reducing the profitability of the fishery. Recognition of this has accelerated attempts to reduce the severity of the problem. Thus innovative fishing masters were already aware of the problem and themselves devising ways to mitigate it, prior to the more systematic recent studies by scientists.

Fifth, a variety of ways of reducing albatross mortality is already widely available. Those critically evaluated to date include: (a) setting longlines only at night — extremely effective in reducing albatross bycatch but possibly increasing the bycatch of some petrel species; (b) streamer lines on (tori) poles to frighten birds away from danger areas during longline setting (generally very effective when rigged correctly); (c) automatic machines casting bait in ways that reduce the time for which baited hooks are available to albatrosses (applicable mainly to Japanese tuna vessels); and (d) appropriate weighting of lines and streamlining of setting process to ensure rapid sinking of line (chiefly applicable to Spanish system of longlining); (e) using bait that is thawed (Brothers *et al.* 1996) and has deflated swim bladders (enhancing sink rate).

Potentially the most effective solution may be to set (and haul) longlines underwater, entirely out of the reach of albatrosses. Such systems for

Fig. 10. A Wandering Albatross caught by a longline tuna vessel of eastern Australia. GRAHAM ROBERTSON

setting are now being built into the design of new longline vessels and adapted for use in older designs. Initial reports of their use are encouraging but require full-scale testing under a range of conditions with independent observers. Even if they prove to be 100% effective it may still take a considerable time to ensure that all longlining is conducted in this fashion.

Sixth, as a global problem it needs a global solution. For most albatrosses their migratory journeys take them into the coastal waters of a considerable number of states and into high-seas areas outside national jurisdiction. National and even regional, agreements to operate longline fisheries so as to minimize or eliminate mortality of albatrosses are likely to be insufficient on their own.

At present as many avenues as possible need exploring to assist national and regional initiatives. These should include: (a) educating fishermen to the benefit they can derive from no longer catching seabirds, especially albatrosses; (b) creating and financing schemes using specialist scientific observation and assistance to help fishermen use devices to reduce albatross mortality and to monitor their effectiveness; and (c) using the full range of international conventions and agreements to protect albatrosses, to regulate longline fisheries in ways designed to reduce (and ultimately eliminate) albatross mortality and to induce States not signatory to these conventions and agreements to abide by their provisions. To achieve all this will undoubtedly require considerable publicity, persistent advocacy and substantial patient negotiation; it is important that fisheries owners, managers and employees should be allies in this process. To give effective protection to albatrosses we need to be able simultaneously to set and work towards targets in seabird research, fishing technology, conservation action, public awareness and national and international legislation. We need to instill into everyone a regard for albatrosses as the symbol and the spirit of the status of the global marine system. The health of this system and the fate of albatrosses are inextricably linked.

ACKNOWLEDGEMENTS

I thank Peter Prince for sharing his knowledge of albatrosses with me over the last 20 years. I thank all those involved in organising the International Albatross Conference — a vital forum at a critical time for the future of albatrosses — and especially Rosie Gales and Graham Robertson. I am grateful for comments on this chapter, whose drafts were typed by Julie Leland, from John Cooper, Beth Flint, Rosemary Gales, Peter Prince, Chris Robertson, Graham Robertson, Kath Walker and Henri Weimerskirch. For provision of illustrations and/or permission to use unpublished material I am particularly grateful to Nigel Brothers, Beth Flint, Rosemary Gales, April Hedd, Frans Lanting, Durno Murray, David Nicholls, Peter Prince, Paul Sagar, Chris Robertson, Kath Walker, Sue Waugh and Henri Weimerskirch.

REFERENCES

Adams, N. J., Brown, C. R. and Nagy, K. A., 1986. Energy expenditure of free-ranging Wandering Albatrosses *Diomedea exulans*. *Physiol. Zool.* **59:** 583–91.

Afanasyev, V. and Prince, P. A., 1993. A miniature storing activity recorder for seabird species. *Ornis Scand.* **24:** 243–46.

Alexander, K., Robertson, G. and Gales, R., 1997. The incidental mortality of albatrosses in longline fisheries. Australian Antarctic Division, Tasmania. Pp. 44.

Anderson, D. J, Schwandt, A. J. and Douglas, H. D., 1997. Foraging ranges of Waved Albatrosses in the eastern tropical Pacific Ocean. Pp. 180–85 *in* Albatross Biology and Conservation ed by G. Robertson and R. Gales. Surrey Beatty & Sons: Chipping Norton.

Arnould, J. P. Y., Briggs, D. R., Croxall, J. P., Prince, P. A. and Wood, A. G., 1996. The foraging behaviour and energetics of Wandering Albatrosses brooding chicks. *Antarc. Sci.* **8:** 229–36.

Astheimer, L. B., Prince, P. A. and Grau C. R., 1985. Egg formation and the pre-laying period of Black-browed and Grey-headed Albatrosses at Bird Island, South Georgia. *Ibis* **127:** 523–29.

Auman, H. J., Ludwig, J. P., Giesy, J. P. and Colborn, T., 1997. Plastic ingestion by Laysan Albatross chicks on Sand Island, Midway Atoll in, 1994 and, 1995. Pp. 239–44 *in* Albatross Biology and Conservation ed by G. Robertson and R. Gales. Surrey Beatty & Sons: Chipping Norton.

Bartle, J. A., 1991. Incidental capture of seabirds in the New Zealand subantarctic squid trawl fishery, 1990. *Bird Conserv. Internat.* **1:** 351–59.

Berruti, A., 1981. Displays of the Sooty Albatrosses *Phoebetria fusca* and *P. palpebrata*. *Ostrich* **52:** 98–103.

Bevan, R. M., Woakes, A. J. and Butler, P. J., 1994. The use of heart rate to estimate oxygen consumption of free-ranging Black-browed Albatross *Diomedea melanophrys*. *J. Exp. Biol.* **193:** 119–37.

Bevan, R. M., Butler, P. J., Woakes, A. J. and Prince, P. A., 1995. The energy expenditure of free-ranging Black-browed Albatrosses. *Phil. Trans. Roy. Soc. Lond.* **350:** 119–31.

Brothers, N. P., 1991. Albatross mortality and associated bait loss in the Japanese longline fishery in the Southern Ocean. *Biol. Conserv.* **55:** 255–68.

Brothers, N. P., Foster, A. and Robertson, G., 1996. The influence of bait quality on the sink rate of bait used in the Japanese longline tuna fishing industry: an experimental approach. *CCAMLR Science* **2:** 123–30.

Brown, C. R. and Adams, N. J., 1984. Basal metabolic rate and energy expenditure during incubation in the Wandering Albatross *(Diomedea exulans)*. *Condor* **86:** 182–86.

Butler, P. J., Bevan, R. M., Woakes, A. J., Croxall, J. P. and Boyd, I. L. , 1995. The use of data loggers to determine the energetics and physiology of aquatic birds and mammals. *Brazilian J. Med. Biol. Res.* **28:** 1307–317.

Cherel, Y. and Klages, N., 1997. A review of the food of albatrosses (Diomedeidae). Pp. 113–36 *in* Albatross Biology and Conservation ed by G. Robertson and R. Gales. Surrey Beatty & Sons: Chipping Norton.

Cherel, Y. and Weimerskirch, H., 1995. Seabirds as indicators of marine resources: Black-browed Albatrosses feeding on ommastrephid squids in Kerguelen waters. *Mar. Ecol. Prog. Ser.* **129:** 295–300.

Clarke, M. R., 1986. A Handbook for the Identification of Cephalopod Beaks. Clarendon Press: Oxford.

Cobley, N. D. and Prince, P. A., in press. The effects of year, previous breeding performance and arrival weight on primary molt in the Grey-headed Albatross *Diomedea chrysostoma*. *Condor*.

Cobley, N. D., Prince, P. A. and Croxall, J. P., in press. Individual quality and reproductive performance in the Grey-headed Albatross *Diomedea chrysostoma*. *Ibis*.

Costa, D. P. and Prince, P. A., 1987. Foraging energetics of Grey-headed Albatross *Diomedea chrysostoma* at Bird Island, South Georgia. *Ibis* **129:** 149–58.

Croxall, J. P., 1982. Energy costs of incubation and moult in petrels and penguins. *J. Anim. Ecol.* **51:** 177–94.

Croxall, J. P., 1991. Reproductive constraints on albatrosses. *Proc. 20th Internat. Ornithol. Congr.* 281–302.

Croxall, J. P. and Prince, P. A., 1996. Potential interactions between Wandering Albatrosses and long-line fisheries for *Dissostichus* at South Georgia. *CCAMLR Science*. 101–10.

Croxall, J. P. and Prince, P. A., 1996. Cephalopods as prey: seabirds. *Phil. Trans. Roy. Soc. Series B* **351:** 1023–043.

Croxall, J. P. and Rothery, P., 1991. Population regulation of seabirds: implications of their demography for conservation. Pp. 272–96 *in* Bird Population Studies: Relevance to Conservation and Management ed by C. M. Perrins, J. D. LeBreton and G. M. Hirons. Oxford University Press: Oxford.

Croxall, J. P., Rothery, P. and Crisp, A., 1992. The effect of maternal age and experience on egg size and hatching success in Wandering Albatrosses *Diomedea exulans*. *Ibis* **134:** 219–28.

Croxall, J. P. and Gales, R., 1997. An assessment of the conservation status of albatrosses. Pp. 46–65 *in* Albatross Biology and Conservation ed by G. Robertson and R. Gales. Surrey Beatty & Sons: Chipping Norton.

de la Mare, W. K. and Kerry, K. R., 1994. Population dynamics of the Wandering Albatross *(Diomedea exulans)* on Macquarie Island and the effects of mortality from longline fishing. *Polar Biol.* **14:** 231–41.

Fisher, H. I., 1949. Populations of birds on Midway and the man-made factors affecting them. *Pac. Sci.* **3:** 103–10.

Fisher, H. I., 1966. Midway's deadly antennas. *Audubon Mag.* **68:** 220–23.

Fisher, H. I., 1967. Body weights in Laysan Albatrosses *Diomedea immutabilis*. *Ibis* **109:** 373–82.

Fisher, H. I., 1971. The Laysan Albatross: its incubation, hatching and associated behaviors. *Living Bird* **10:** 19–78.

Fisher, H. I., 1975. The relationship between deferred breeding and mortality in the Laysan Albatross. *Auk* **92:** 433–41.

Fisher, H. I. and Fisher, M. L., 1969. The visits of Laysan Albatrosses to the breeding colony. *Micronesica* **5:** 173–221.

Fisher, H. I. and Fisher, M. L., 1972. The oceanic distribution of the Laysan Albatross, *Diomedea immutabilis*. *Wilson Bull.* **84:** 7–27.

Flint, E. N., 1995. Albatross populations and conservation issues in the Northern Hemisphere. P. 9 *in* Abstracts of the First International Conference on the Biology and Conservation of Albatrosses, 28 August–1 September, 1995, Hobart, Australia.

Frings, H. and Frings, M., 1959. Observations on salt balance and behavior of Laysan and Black-footed Albatrosses in captivity. *Condor* **61:** 305–14.

Frings, H., Anthony, A. and Schein, M. W., 1958. Salt excretion by nasal gland of Laysan and Black-footed Albatrosses. *Science* **128:** 1572.

Fry, D. M., Fefer, S. I. and Sileo, L., 1987. Ingestion of plastic debris by Laysan Albatrosses and Wedge-tailed Shearwaters in the Hawaiian Islands. *Mar. Pollut. Bull.* **18:** 339–43.

Gales, R., 1993. Co-operative Mechanisms for the Conservation of Albatross. Australian Nature Conservation Agency and Australian Antarctic Foundation, Hobart.

Gales, R., Hedd, A. and Brothers, N., in press. Foraging movements of the Shy Albatross *Diomedea cauta* breeding in Australia: implications for interactions with longline fisheries. *Ibis* **140.**

Gales, R., 1997. Albatross populations: status and threats. Pp. 20–45 *in* Albatross Biology and Conservation ed by G. Robertson and R. Gales. Surrey Beatty & Sons: Chipping Norton.

Gould, P. and Hobbs, R., 1993. Population dynamics of the Laysan and other albatrosses in the North Pacific. *Internat. Nth Pac. Fish. Comm. Bull.* **53**(III): 485–97.

Gould, P., Ostrom, P., Walker, W. and Pilichowski, K., 1997. Laysan and Black-footed Albatrosses: trophic relationships and driftnet fisheries associations of non-breeding birds. Pp. 199–207 *in* Albatross Biology and Conservation ed by G. Robertson and R. Gales. Surrey Beatty & Sons: Chipping Norton.

Grau, C. R., 1976. Ring structure of avian egg yolk. *Poultry Science* **55:** 1418–422.

Grau, C. R., 1984. Egg formation. Pp. 33–57 *in* Seabird Energetics ed by G. C. Whittow and H. Rahn. Plenum Press: New York.

Green, B. and Brothers, N., 1995. Field energetics of White-capped Albatross, *Diomedea cauta cauta* and growth of chicks. P. 17 *in* Abstracts of the First International Conference on the Biology and Conservation of Albatrosses 28 August–1 September, 1995, Hobart, Australia.

Guinet, C., Jouventin, P. and Malacamp, J., 1995. Satellite remote sensing in monitoring change of seabirds: use of Spot Image in King Penguin population increase at Ile aux Cochons, Crozet Archipelago. *Polar Biol.* **15:** 511–15.

Hamilton, S. and Moller, H., 1995. Can PVA models using computer packages offer useful conservation advice? Sooty Shearwaters *Puffinus griseus* in New Zealand as a case study. *Biol. Conserv.* **73:** 107–17.

Hasegawa, H., 1984. Status and conservation of seabirds in Japan, with special attention to the Short-tailed Albatross. Pp. 487–500 *in* Status and Conservation of the World's Seabirds ed by J. P. Croxall, P. G. H. Evans and R. W. Schreiber. ICBP: Cambridge.

Hatchwell, B. J. and Pellatt, J., 1990. Intraspecific variation in egg composition and yolk formation in the common guillemot *Uria aalge*. *J. Zool.* **220:** 279–86.

Hector, J. A. L., 1984. Techniques for the serial collection of blood samples and inspection of gonads in free-living albatrosses. *Brit. Antarc. Surv. Bull.* **61:** 127–33.

Hector, J. A. L., 1988. Reproductive endocrinology of albatrosses. *Proc. XIX Internat. Ornithol. Cong.* 1702–709.

Hector, J. A. L. and Harvey, S., 1986. Corticosterone secretion through long incubation shifts in *Diomedea* albatrosses. *J. Gen. Comp. Endocrinol.* **62:** 349–52.

Hector, J. A. L., Croxall, J. P. and Follet, B. K., 1986a. Reproductive endocrinology of the Wandering Albatross *Diomedea exulans* in relation to biennial breeding and deferred sexual maturity. *Ibis* **128:** 9–22.

Hector, J. A. L., Follet, B. K. and Prince, P. A., 1986b. Reproductive endocrinology of the Black-browed Albatross *Diomedea melanophris* and the Grey-headed Albatross *D. chrysostoma*. *J. Zool.* **208:** 237–54.

Hector, J. A. L., Pickering, S. P. C., Croxall, J. P. and Follet, B. K., 1990. The endocrine basis of deferred sexual maturity in the Wandering Albatross *Diomedea exulans*. *Funct. Ecol.* **4:** 59–66.

Hecht, T., 1987. A guide to the otoliths of Southern Ocean fishes. *Sth Afr. J. Antarc. Res.* **17:** 1–87.

Hedd, A., Gales, R., Brothers, N. and Robertson, G., 1997. Diving behaviour of the Shy Albatrosses *Diomedea cauta* in Tasmania: initial findings and dive recorder assessment. *Ibis* **139:** 452–60.

Hill, H. J., 1990. A new method for the measurement of Antarctic Krill *Euphausia superba* Dana from predator food samples. *Polar Biol.* **10:** 317–20.

Hobson, K. A., Piatt, J. F. and Pitocchelli, J., 1994. Using stable isotopes to determine seabird trophic relationships. *J. Anim. Ecol.* **63:** 785–98.

Howell, T. R. and Bartholemew, G. A., 1961. Temperature regulation in Laysan and Black-footed Albatrosses. *Condor* **63:** 185–97.

Huin, N. and Croxall, J. P., in press. Fishing gear, oil and marine debris associated with seabirds at Bird Island, South Georgia, 1993/94. *Mar. Ornithol.*

Huin, N. and Prince, P. A., 1997. Diving behaviour of the Grey-headed Albatross *Diomedea chrysostoma* at Bird Island, South Georgia. *Antarct. Sci.* **9:** 243–49.

Johnston, D. H., Shaffer, T. L. and Gould, P. J., 1993. Incidental catch of marine birds in the North Pacific high seas drift net fisheries in, 1990. *Internat. Nth Pac. Fish. Comm. Bull.* **53**(III): 473–83.

Jouventin, P. and Lequette, B., 1990. The dance of the Wandering Albatross *Diomedea exulans*. *Emu* **90:** 122–31.

Jouventin, P. and Weimerskirch, H., 1990. Satellite tracking of Wandering Albatross. *Nature* **343:** 746–48.

Jouventin, P., de Monicault, G. and Blosseville, J. M., 1981. La danse de l'albatros, *Phoebetria fusca*. *Behaviour* **78:** 43–80.

Kenyon, K. W. and Kridler, E., 1969. Laysan Albatrosses swallow indigestible matter. *Auk* **86:** 339–43.

King, W. B., 1984. Incidental mortality of seabirds in gillnets in the North Pacific. Pp. 709–15 *in* Status and Conservation of the World's Seabirds ed by J. P. Croxall, P. G. H. Evans and R. W. Schreiber. ICBP: Cambridge.

Klomp, N., Gales, R. and Brothers, N., 1995. Determining the ages of adult albatrosses. P. 49 *in* Abstracts of the First International Conference on the Biology and Conservation of Albatrosses, 28 August–1 September, 1995, Hobart, Australia.

Langston, N. and Hillgarth, N., 1995. Moult varies with parasites in Laysan Albatrosses. *Proc. Roy. Soc. Lond. Series B* **261:** 239–43.

Langston, N. E. and Rohwer, S., 1995. Unusual patterns of incomplete primary molt in Laysan and Black-footed Albatrosses. *Condor* **97:** 1–19.

Lequette, B. and Jouventin, P., 1991. The dance of the Wandering Albatross II: Acoustic signals. *Emu* **91:** 172–78.

Lequette, B. and Weimerskirch, H., 1990. Influence of parental experience on the growth of Wandering Albatross chicks. *Condor* **92:** 726–31.

Lifson, N. and McClintock, R., 1966. Theory and use of the turnover rates of body water for measuring energy and material balance. *J. Theoret. Biol.* **12:** 46–74.

Ludwig, J. P., Summer, C. L., Auman, H., Rolland, R., Gilbertson, M. and Giesy, J. P., 1997. Recent studies of Laysan Albatross and Black-footed Albatross at Midway Islands, North Pacific Ocean, 1992–1995: Effects of organochlorine contaminants, plastics ingestion and fishing bycatch. Pp. 225–38 *in* Albatross Biology and Conservation ed by G. Robertson and R. Gales. Surrey Beatty & Sons: Chipping Norton.

McDermond, D. K. and Morgan, K. H., 1993. Status and conservation of North Pacific albatrosses. Pp. 70–81 *in* The Status, Ecology and Conservation of Marine Birds of the North Pacific ed by K. Vermeer, K. T. Briggs, K. H. Morgan and D. Siegel-Causey. Canadian Wildlife Service, Special Publication.

Medway, D. G., 1997. Human induced mortality of southern albatrosses at sea in the, 19th century — a brief historical review. Pp. 189–98 *in* The Albatross Biology and Conservation ed by G. Robertson and R. Gales. Surrey Beatty & Sons: Chipping Norton.

Merlen, G., 1996. Scavenging Waved Albatross around the Galapagos Archipélago: at risk from long-liners. *The Seabird Group Newsl.* **74:** 2–6.

Meseth, E. H., 1975. The dance of the Laysan Albatross, *Diomedea immutabilis*. *Behaviour* **54:** 217–57.

Moors, P. J. and Atkinson, I. A. E., 1984. Predation on seabirds by introduced animals and factors affecting its severity. Pp. 667–90 *in* Status and Conservation of the World's Seabirds ed by J. P. Croxall, P. G. H. Evans and R. W. Schreiber. ICBP: Cambridge.

Nicholls, D. G., Murray, M. D. and Robertson, C. J. R., 1994. Oceanic flights of the Northern Royal Albatross *Diomedea epomophora sanfordi* using satellite telemetry. *Corella* **18:** 50–52.

Nicholls, D. G., Murray, M. D., Battam, H., Robertson, G., Moors, P., Butcher, E. and Hildebrandt, M., 1995a. Satellite tracking of the Wandering Albatross *Diomedea exulans* around Australia and in the Indian Ocean. *Emu* **95:** 223–30.

Nicholls, D. G., Howey, P. and Warne, J., 1995b. A satellite transmitter with a miniature weather station on a seabird. P. 55 *in* Abstracts of the First International Conference on the Biology and Conservation of Albatrosses, 28 August–1 September, 1995, Hobart, Australia.

Nicholls, D. G., Murray, M. D., Elliott, G. P. and Walker, K. J., 1996. Satellite-tracking of a Wandering Albatross from the Antipodes Islands, New Zealand to South America. *Corella* **20:** 28.

Northridge, S. P., 1991. Driftnet fisheries and their impacts on non-target species: a world-wide review. FAO Fisheries Technical Paper, No. 320. Rome, FAO.

Nunn, G. B., Cooper, J., Jouventin, P., Robertson, C. J. R. and Robertson, G. G., 1996. Evolutionary relationships among extant albatrosses (Procellariiformes: Diomedeidae) established from complete cytochrome-*b* gene sequences. *Auk* **113:** 784–801.

Olson, S. L., 1985. The fossil record of birds. Pp. 79–238 *in* Avian Biology Vol. VIII ed by D. S. Farner, J. R. King and K. C. Parkes. Academic Press: Orlando.

Pettit, T. N., Grant, G. S. and Whittow, G. C., 1981. Ingestion of plastics by Laysan Albatross. *Auk* **98:** 839–41.

Pettit, T. N., Grant, G. S., Whittow, G. C., Rahn, H. and Paganelli, C. V., 1982. Embryonic oxygen consumption and growth of Laysan and Black-footed Albatross. *Amer. J. Physiol.* **242:** R121–R128.

Pettit, T. N., Nagy, K. A., Ellis, H. I. and Whittow, G. C., 1988. Incubation energetics of the Laysan Albatross. *Oecologia* **74:** 546–50.

Pickering, S. P. C., 1989. Attendance patterns and behaviour in relation to experience and pair-bond formation in Wandering Albatrosses *Diomedea exulans* at South Georgia. *Ibis* **131:** 183–95.

Prince, P. A., 1980. The food and feeding ecology of Grey-headed Albatross *Diomedea chrysostoma* and Black-browed Albatross *D. melanophris*. *Ibis* **122:** 476–88.

Prince, P. A. and Francis, M., 1984. Activity budgets of foraging Grey-headed Albatrosses. *Condor* **86:** 297–300.

Prince, P. A. and Ricketts, C., 1981. Relationships between food supply and growth in albatrosses: an interspecies chick fostering experiment. *Ornis Scand.* **12:** 207–10.

Prince, P. A. and Rodwell, S. P., 1994. Ageing immature Black-browed and Grey-headed Albatross using moult, bill and plumage characteristics. *Emu* **94:** 246–54.

Prince, P. A. and Walton, D. W. H., 1984. Automated measurement of feed size and feeding frequency in albatrosses. *J. Appl. Ecol.* **21:** 789–94.

Prince, P. A., Wood, A. G., Barton, T. and Croxall, J. P., 1992. Satellite tracking of Wandering Albatrosses *Diomedea exulans* in the South Atlantic. *Antarct. Sci.* **4:** 31–36.

Prince, P. A., Rodwell, S., Jones, M. and Rothery, P., 1993. Moult in Black-browed and Grey-headed Albatrosses *Diomedea melanophris* and *D. chrysostoma*. *Ibis* **135:** 121–31.

Prince, P. A., Rothery, P., Croxall, J. P. and Wood, A. G., 1994a. Population dynamics of Black-browed and Grey-headed Albatrosses *Diomedea melanophris* and *D. chrysostoma* at Bird Island, South Georgia. *Ibis* **136:** 50–71.

Prince, P. A., Huin, N. and Weimerskirch, H., 1994b. Diving depths of albatrosses. *Antarct. Sci.* **6:** 353–54.

Prince, P. A., Briggs, D. R., Socha, D. G., Croxall, J. P., Huin, N. and Rothery, P., 1994c. Provisioning rates and growth patterns in Black-browed and Grey-headed Albatrosses. P. 222 *in* Proceedings of the 6th SCAR Biology Symposium: Antarctic Communities: Species, Structure and Survival, Venice, Italy, 30 May–3 June, 1994.

Prince, P. A., Croxall, J. P., Trathan, P. N. and Wood, A. G., 1997. The pelagic distribution of South Georgia albatrosses and their relationships with fisheries. Pp. 137–67 *in* Albatross Biology and Conservation ed by G. Robertson and R. Gales. Surrey Beatty & Sons: Chipping Norton.

Prince, P. A., Weimerskirch, H., Huin, N. and Rodwell, S., 1997. Primary moult and maturation of plumage in the Wandering Albatross *Diomedea exulans chionoptera*. *Condor* **99:** 58–72.

Reid, K., 1996. A guide to the use of otoliths in the study of the diet of predators at Bird Island, South Georgia. British Antarctic Survey: Cambridge.

Reid, K., Croxall, J. P. and Prince, P. A., 1996. The fish diet of Black-browed and Grey-headed Albatrosses *Diomedea melanophris* and *D. chrysostoma* at South Georgia. *Polar Biol.* **16:** 469–77.

Richdale, L. E., 1949. The pre-egg stage in Buller's Mollymawk. Biological Monographs No. 2, Dunedin.

Richdale, L. E., 1950. The pre-egg stage in the albatross family. Biological Monographs No. 3, Dunedin.

Ricketts, C. and Prince, P. A., 1984. Estimation by use of field weighings of metabolic rate and food conversion efficiency in albatross chicks. *Auk* **101:** 790–95.

Risebrough, R. W., 1977. Transfer of organochlorine pollutants to Antarctica. Pp. 1203–210 *in* Adaptations within Antarctic Ecosystems ed by G. A. Llano. Smithsonian Institution: Washington.

Robbins, C. S., 1966. Birds and aircraft on Midway islands, 1959–63 investigations. *U.S. Fish. Wildl. Special Sci. Rpt* **85:** 63.

Robertson, C. J. R., 1993. Survival and longevity of the Northern Royal Albatross *Diomedea epomophora sanfordi* at Taiaroa Head,, 1937–1993. *Emu* **93:** 269–76.

Robertson, C. J. R., 1995. Effects of extreme climatic events on the population dynamics of the Northern Royal Albatross. P. 56 *in* Abstracts of the First International Conference on the Biology and Conservation of Albatrosses, 28 August–1 September, 1995, Hobart, Australia.

Robertson, C. J. R., 1997. Factors influencing breeding performance of the Northern Royal Albatross. Pp. 99–104 *in* The Albatross Biology and Conservation ed by G. Robertson and R. Gales. Surrey Beatty & Sons: Chipping Norton.

Robertson, C. J. R. and Nunn, G. B., 1997. Towards a new taxonomy for albatrosses. Pp. 13–19 *in* The Albatross Biology and Conservation ed by G. Robertson and R. Gales. Surrey Beatty & Sons: Chipping Norton.

Rodhouse, P. G., Prince, P. A., Trathan, P. N., Hatfield, E. M. C., Watkins, J. L., Bone, D. G., Murphy, E. J. and White, M. G., 1996. Cephalopods and mesoscale oceanography at the Antarctic Polar Front: satellite tagged predators locate pelagic trophic interactions. *Mar. Ecol. Progr. Ser.* **136:** 37–50.

Ryan, P. G., Dean, W. R. J., Moloney, C. L., Watkins, B. P. and Milton, S. J., 1990. New information on seabirds at Inaccessible Island and other islands in the Tristan da Cunha group. *Mar. Ornithol.* **18:** 43–54.

Sagar, P. M. and Weimerskirch, H., 1996. Satellite tracking of Southern Buller's Albatrosses from The Snares, New Zealand. *Condor* **98:** 649–52.

Salamolard, M. and Weimerskirch, H., 1993. Relationship between foraging effort and energy requirement throughout the breeding season in the wandering albatross. *Funct. Ecol.* **7:** 643–52.

Schwaller, M. R., Olson, C. E. Jr., Ma, Z., Zhu, Z. and Dahmer, P., 1989. Remote sensing analysis of Adelie Penguin rookeries. *Remote Sensing Environ.* **28:** 199–206.

Sievert, P. R. and Sileo, L., 1993. The effects of ingested plastic on growth and survival of albatross chicks. Pp. 212–17 *in* The status, ecology and conservation of marine birds of the north Pacific ed by K. Vermeer, K. T. Briggs, K. H. Morgan and D. Siegel-Causey. Special Publication of the Canadian Wildlife Service.

Sileo, L., Sievert, P. R. and Samuel, M. D., 1989. Prevalence and characteristics of plastic ingested by Hawaiian seabirds. Pp. 665–81 *in* Proceedings of the Second International Conference on Marine Debris, 2–7 April, 1989, Honolulu, Hawaii ed by R. S. Shomura and M. L. Godfrey. US Department of Commerce.

Smith, G. T., Wingfield, J. C. and Veit, R. R., 1994. Adrenocortical response to stress in the Common Diving Petrel, *Pelecanoides urinatrix. Physiol. Zool.* **67:** 526–37.

Sparling, D. W. Jr., 1977. Sounds of Laysan and Black-footed Albatrosses. *Auk* **94:** 256–69.

Speakman, J. R., 1993. How should we calculate CO_2 production in doubly labelled water studies of animals? *Funct. Ecol.* **7:** 746–50.

Speakman, J. R., Nair, K. S. and Goran, M. I., 1993. Revised equations for calculating CO2 production from doubly-labelled water in humans. *Amer. J. Physiol.* **264:** E912–E917.

Spear, L. B., Ainley, D. G. and Webb, S. W., 1995. Distribution abundance and behaviour of Buller's, Chatham Island and Salvin's Albatrosses off Chile and Peru: potential interaction with longliners. P. 25 *in* Abstracts of the First International Conference on the Biology and Conservation of Albatrosses, 28 August–1 September, 1995, Hobart, Australia.

Steadman, D. W., 1995. Prehistoric extinctions of Pacific Island birds: biodiversity meets zooarcheology. *Science* **267:** 1123–131.

Thompson, K. R., 1992. Quantitative analysis of the use of discards from squid trawlers by Black-browed Albatrosses *Diomedea melanophris* in the vicinity of the Falkland Islands. *Ibis* **134:** 11–12.

Thompson, K. R. and Riddy, M. D., 1995. Utilization of offal and discards from "finfish" trawlers around the Falkland Islands by the Black-browed Albatross *Diomedea melanophris. Ibis* **137:** 198–206.

Tickell, W. L. N., 1984. Behaviour of Black-browed and Grey-headed Albatrosses at Bird Island, South Georgia. *Ostrich* **55:** 64–85.

Viswanathan, G. M., Afanasyev, V., Buldyrev, S. V., Murphy, E. J., Prince, P. A. and Stanley, H. E., 1996. Lévy flight search patterns of wandering albatrosses. *Nature* **381:** 413–15.

Walker, K. A., Elliott, G., Nicholls, D. G., Murray, M. D. and Dilks, P., 1995. Satellite tracking of Wandering Albatross *Diomedea exulans* from the Auckland Islands: preliminary results. *Notornis* **42:** 127–35.

Warham, J. and Fitzsimons, C. H., 1987. The vocalizations of Buller's Mollymawk, *Diomedea bulleri* (Aves: Diomedeidae), with some comparative data on other albatrosses. *NZ J. Zool.* **14:** 65–79.

Watkins, B. P., 1987. Population sizes of King, Rockhopper and Macaroni Penguins and Wandering Albatrosses at the Prince Edward Islands and Gough Island, 1951–1986. *Sth Afr. J. Antarc. Res.* **17:** 166–62.

Waugh, S. M., Sagar, P. M. and Paull, D., in press. Laying dates, breeding success and annual breeding of Southern Royal Albatross *Diomedea epomophora epomophora* at Campbell Island during, 1964 to, 1969. *Emu.*

Weimerskirch, H., 1991. Sex-specific differences in moult strategy in relation to breeding in the Wandering Albatross. *Condor* **93:** 731–37.

Weimerskirch, H., 1992. Reproductive effort in long-lived birds: age-specific patterns of condition, reproduction and survival in the Wandering Albatross. *Oikos* **64:** 464–73.

Weimerskirch, H., 1995. Regulation of foraging trips and incubation routine in male and female Wandering Albatrosses. *Oecologia* **102:** 37–43.

Weimerskirch, H. and Jouventin, P., 1997. Changes in populations sizes and demographic parameters of six albatross species breeding on the French sub-Antractic islands. Pp. 84–91 *in* Albatross Biology and Conservation ed by G. Robertson and R. Gales. Surrey Beatty & Sons: Chipping Norton.

Weimerskirch, H. and Robertson, G., 1994. Satellite tracking of Light-mantled Sooty Albatrosses. *Polar Biol.* **14:** 123–26.

Weimerskirch, H. and Wilson, R. P., 1992. When do Wandering Albatross *Diomedea exulans* forage? *Mar. Ecol. Prog. Ser.* **86:** 297–300.

Weimerskirch, H., Bartle, J. A., Jouventin, P. and Stahl, J.–C., 1988. Foraging ranges and partitioning of feeding zones in three species of southern albatrosses. *Condor* **90:** 214–19.

Weimerskirch, H., Salamolard, M., Sarrazin, F. and Jouventin, P., 1993. Foraging strategy of Wandering Albatrosses through the breeding season: a study using satellite telemetry. *Auk* **110:** 325–42.

Weimerskirch, H., Doncaster, C. P. and Cuenot-Chaillet, F., 1994. Pelagic seabirds and the marine environment: foraging patterns of Wandering Albatrosses in relation to prey availability and distribution. *Proc. Roy. Soc. Lond., Series B* **255:** 91–97.

Weimerskirch, H., Wilson, R. P., Guinet, C. and Koudil, M., 1995. Use of seabirds to monitor sea-surface temperatures and to validate remote-sensing measurements in the Southern Ocean. *Mar. Ecol. Prog. Ser.* **126:** 299–303.

Weimerskirch, H., Brothers, N. and Jouventin, P., 1997. Population dynamics of Wandering Albatross *Diomedea exulans* and Amsterdam Albatross *D. amsterdamensis* in the Indian Ocean and their relationships with longline fisheries: conservation implications. *Biol. Conserv.* **79:** 257–70.

Weimerskirch, H., Mougey, T. and Hindermeyer, X., in press. Foraging ecology of Black-browed Albatrosses in relation to the requirements of the chick: natural variation and experimental study. *Behav. Ecol.*

Williams, R. and McEldowney, A., 1990. A guide to the fish otoliths from waters off the Australian Antarctic Territory, Heard and Macquarie Island. ANARE Research Notes No. 75.

Wilson, R. P., Cooper, J. and Plötz, J., 1992. Can we determine when marine endotherms feed? A case study with seabirds. *J. Exp. Biol.* **167:** 267–75.

Wilson, R. P., Putz, K., Gremillet, D., Culik, B. M., Kierspel, M., Regel, J., Bost, C. A., Lage, J. and Cooper, J., 1995a. Reliability of stomach temperature changes in determining feeding characteristics of seabirds. *J. Exp. Biol.* **198:** 1115–135.

Wilson, R. P., Weimerskirch, H. and Lys, P., 1995b. A device for measuring seabird activity at sea. *J. Avian Biol.* **26:** 172–76.

Wooller, R. D., Bradley, J. S., Skira, I. J. and Serventy, D. L., 1990. Reproductive success of Short-tailed Shearwater *(Puffinus tenuirostris)* in relation to their age and breeding experience. *J. Anim. Ecol.* **59:** 161–70.

Index

C

Q

R

S

T